Herbicide Resistance in Weeds and Crops

Herbicide Resistance in Weeds and Crops

J. C. Caseley, G. W. Cussans and R. K. Atkin

GRANT AIDED

Butterworth-Heinemann Ltd
Linacre House, Jordan Hill, Oxford OX2 8DP

 PART OF REED INTERNATIONAL BOOKS

OXFORD   LONDON   BOSTON
MUNICH   NEW DELHI   SINGAPORE   SYDNEY
TOKYO   TORONTO   WELLINGTON

First published 1991

© Butterworth-Heinemann Ltd 1991

British Library Cataloguing in Publication Data
Long Ashton International Symposium
(11th 1989 Bristol)
  Herbicide resistance in weeds and crops.
  I. Title   II. Caseley, John C.
  III. Cussons, G. W.   IV. Atkin, Roger K.
  632

ISBN 0 7506 1101 4

Library of Congress Cataloguing in Publication Data
A catalogue record for this book is available
from the Library of Congress

Printed and bound in Great Britain.

CONTENTS

PREFACE

This volume contains papers and poster abstracts presented at the Eleventh Long Ashton International Symposium in September 1989 which was convened to assess and review the importance of the increasing incidence of herbicide-resistant weeds and to consider progress in the development of herbicide-resistant crops. The importance of bringing together the agronomic, mechanistic and genetic aspects of herbicide resistance was recognised from the outset in planning the Symposium.

In his inaugural lecture Professor J. Gressel set the scene by outlining the development of herbicide resistance and identifying selection pressure, competitiveness, fitness and negative cross-resistance as key factors in this development. He stressed the importance of crop and land management in preventing or delaying the evolution of resistance.

In the second session the incidence and biology of herbicide-resistant weeds were considered. Both Dr H.M. LeBaron's world survey and the papers concerned with individual case histories clearly indicated the increasing seriousness of the problem. Of particular concern is the increasing incidence of cross-resistance and the appearance, within the short time of three years, of resistance to some of the newer highly active compounds, such as sulfonylureas.

The third session, on mechanisms of herbicide resistance, considered altered sites of action, degradation, conjugation and sequestration. It is well established that an altered site of action os responsible for most cases of weed resistance to triazines and there is only circumstantial evidence, based on identification of metabolites, that degradation and conjugation are responsible for cross-resistance in several weeds e.g. *Alopecurus myosuroides*. Generally, herbicide metabolism pathways are better defined in crops than in weeds and further research is required on resistance and especially cross-resistance mechanisms in weeds to provide a logical basis for remedial action in the field.

The final session embraced several aspects of herbicide resistance in crops including conventional plant breeding, *in vitro* techniques and genetic engineering. The commercial opportunities for genetically engineering herbicide-resistant crops, the field evaluation and legislation covering their release were considered.

All the sessions included thought-provoking discussions and were supplemented by 41 posters, abstracts of which appear in these Proceedings. The final paper by Dr B. Rubin gave a balanced overview of the Symposium and outlined some promising avenues for further research. The international perspective of the Symposium was reflected in the attendance of over 220 researchers from 23 countries.

Particular thanks are due to all contributors for providing excellent verbal presentations and posters. The timely production of most papers was appreciated.

We are grateful to many of our colleagues at Long Ashton for unstinting help at all stages of organising the Symposium especially Mrs S.E. Child and Mrs J. Knights who dealt with Symposium administration; we also acknowledge Drs M.S. Kemp and P.J. Lutman and Mr S.R. Moss for their help with programme organisation, and Mrs J.M. Llewellyn for her essential role in the production of the camera-ready copy. We specially appreciate the contribution of Mrs C.M. Bond for her assistance with graphics, detailed editing and indexing.

J.C. Caseley
G.W. Cussans
R.K. Atkin
Long Ashton Research Station
University of Bristol
March 1991

WHY GET RESISTANCE? IT CAN BE PREVENTED OR DELAYED

Jonathan Gressel

Dept. of Plant Genetics, The Weizmann Institute of Science, Rehovot, Israel

It is not hard to learn from pesticide use history how to prevent or delay resistance. Resistance has predominantly evolved where a single herbicide chemical or group was used annually with high selection pressure herbicides having high residual activities, or when ephemeral herbicides were repeatedly used. Resistances have not evolved where meaningful rotations or mixtures were used, despite multiple treatments. Analysis of this history allows modelling potential strategies to delay or prevent resistance. Three factors have powerful resistance delaying effects: selection pressure, fitness, and the use of herbicides exerting 'negative cross-resistance' (i.e. that control resistant biotypes at lower rates than the wild types). The broad cross- resistances to wheat selective herbicides that evolved in two grasses are disturbing as these metabolically mimic wheat in degrading the herbicides. The overlap of herbicide spectra precludes using negative cross-resistance for delaying resistance in monoculture wheat. Herbicide mixtures or rotations cannot be designed (without crop rotation) precluding metabolic cross-resistance. Using low selection pressure herbicides should help. It is imperative to engineer new modes of resistance into wheat to prevent major problems.

INTRODUCTION

The farming community has realized the value of herbicides over the last 20 years. Usage of herbicides now exceeds that of fungicides and insecticides combined. This is irrespective of whether "use" is measured in area, tons, or expenditures. In contrast, the means devoted to weed science are less than either entomology or plant pathology, whether this is measured as the numbers of weed scientists employed, or expenditures on basic or applied research.

Farmers have taken to herbicides because of 'cost-effectiveness'. The competition from weeds for space, water and fertilizers is fierce. The breeders developing newer higher yielding crops have inadvertently come

1

up with varieties that are less competitive with weeds. Herbicide usage is far cheaper than cultivation, uses far less fuel and usually does less damage to the crop roots, and to soil structure. This is 'ecologically' positive: less CO_2 from fossil fuel into the environment, less disruption of soil, and thus less wind and water erosion of soil.

Thus, the use of herbicides has generally been environmentally benign and their efficacy has allowed agriculture to expand to keep pace with developing populations.   However, there is a negative side.   Crop rotations have been reduced, alternative techniques have been abandoned and this has led to more intensive herbicide use.   This intensive use has led to resistance and to real or imagined concerns about contamination of ground water and toxicity.   We now face demands for decreased use from various sources.

Resistance management will allay many of the environmental problems associated with herbicide use.   This perspective should be used in considering the material in other chapters, to extrapolate beyond resistance.   The thesis to be presented below is that judiciously using smaller amounts of many herbicides, coupled with intelligent agronomy will prevent or delay resistance, without reducing crop yields beyond the savings in herbicide costs.

Intelligent management requires that herbicides and weeds must be better understood than at present.   Far more research effort from both the public and private sectors will be required.   Farmers will have to get down from their air-conditioned tractor cabins and look well at their fields and stop prophylactic dumping of herbicides.   The farmer is not fully to blame; we in the academic community were guilty of complacency and opted too often for the easy way out in advising the growers. Prophylactic applications worked, so why interfere?   Conversely, the farmer was subjected to massive commercial campaigns that "more weed control is better".   This brought about both the environmental problems and resistant weeds.   There can be some dire consequences to our food supply by some of the resistance problems.   The metabolic cross-resistance to almost all of wheat selective herbicides continues to evolve elsewhere.   Novel answers must be found and the farmers must be re-educated, to keep good herbicides around longer by using them more intelligently.

Definitions

Resistance must be defined, for uniform usage, although too much time is wasted on semantics.   Resistance is clearly "the inherited ability to survive treatment by a herbicide". The word resistance should always be followed or preceded by modifiers. An important modifier is the rate of application. Agricultural rates are assumed here, but they should be stated.   Thus, with a selective herbicide, the crop is naturally resistant.   In evolution, one must talk about naturally resistant individuals and resistant populations. Evolution of resistance is then the process whereby the rare resistant individual becomes the majority - i.e., a resistant-population.   We can talk about resistance factors, or "X" fold resistance.   This is usually described as the I_{50} of the resistance individuals divided by the I_{50} for the susceptibles.   The harder to measure factors of I_{90} to I_{99} for resistant divided by susceptible are more significant for field situations, and need not be the same as for I_{50}s. There may be full resistance at an agricultural use-rate – or partial resistance. The latter occurs when the plant is severely inhibited but still produces some seed. Tolerance was previously used for partial resistance (LeBaron and Gressel, 1982), but this meant other things to other researchers, and has been dropped (Gressel *et al.*, 1990).

MODELLING THE FACTORS CONTROLLING EVOLUTION OF RESISTANT POPULATIONS

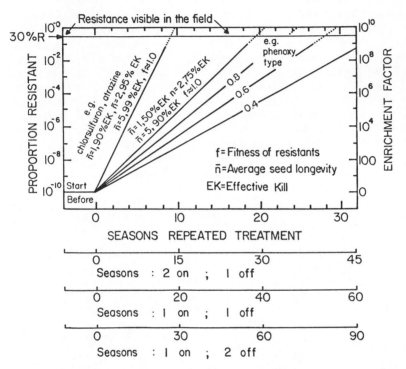

Fig. 1. <u>Presumed effects of herbicide rotations using the original model.</u> Overall average effect of scenarios with different selection pressures. (α = 0.1 = 90% effective kill (EK); α = 0.01 = 99% EK, α = 0.5 = 50% EK), seed bank dynamics (n = the average seed duration in the soil and f = differential fitness. The different scales give the different rotational scenarios from mono-herbicide to one treatment in three seasons. Source: Calculated from equations in Gressel and Segel (1978). From Gressel and Segel (1990a) by permission of the American Chemical Society.

If the importance of each of the factors controlling rates of evolution is understood, ways to modify cultural practices can be found. This is best done by modelling – the only problem is that most of the specific data that need to be inserted into the models are lacking. We started modelling about 12 years ago (Gressel and Segel, 1978). Most of the data available then were from Bradshaw's group in Wales (now Liverpool) that dealt with evolution of heavy metal resistance on mine tailings from Roman times on. Our basic model has remained (Fig. 1), although we understand the implications better, as more data on resistance appeared. A major advantage of models, besides their assistance in prediction, is that they show which types of data are missing and tell how important

3

they can be. This in turn can suggest priorities for research. Other groups have rightly started designing more sophisticated models (Maxwell *et al.*, 1989, and others). Surprisingly, there are no major differences in the predictions. The models in general refer to resistance inherited on one or a few major genes and not resistance that is polygenically inherited. Triazine, sulfonylurea, paraquat and dinitroaniline resistances are monogenically inherited. There is insufficient information about other types, but there is no reason to preclude polygenic resistance or resistance inherited due to gene amplifications. Models must explain why the vast majority of cases of resistance appeared in monoculture, monoherbicide situations, where monoherbicide is defined as using one or more herbicides with the same precise site of action, or one or more herbicides that are degraded in the same manner, or whose toxic products are degraded in the same manner. There are many cases where a large number of generations were treated with a given herbicide, but herbicides were rotated and resistance did not evolve. The same numbers of treatments in monoherbicide culture elicited resistance. The rate of evolution of resistance in a monoculture situation is mathematically described in a (simplified) equation.

$$N_n = N_o \left[1 + \frac{f(\alpha_r/\alpha_s)-1}{\bar{n}} \right]^n \qquad \text{Eq. 1}$$

The factors in the equation are:
N_n is the proportion (frequency) of resistant individuals (N) after (n) years of treatment.
N_o is the initial frequency of resistant individuals in the population.
f is the overall competitive fitness of the resistant individuals compared to the wild type, when the herbicide is absent. In the vast majority of atrazine-resistant cases this ranges from 0.1 to 0.5, with exceptions. Fitness may be much higher, with other herbicides with implications to be described later.
$\bar{n}$ is the average residence time (years) in the soil seed bank. In later models (for rotational situations) we used in a modified equation with σ, the fraction of seeds leaving the seedbank (Gressel and Segel, 1990a). Selection pressure is the single most important factor. It is defined by:

α_r – the proportion of resistants remaining divided by
α_s – the proportion of resistants remaining after herbicide treatment.

This is actually measured as control of seed output over a season. If there is a 99% effective control of susceptible seed output and no control of resistant seed output, $\alpha_r = 1$ and $\alpha_s = 0.01$ giving an overall selection pressure (α) of $\alpha = 1 \div 0.01 = 100$. If instead susceptible seed output is reduced only 50%, then $\alpha_s = 0.5$ and with $\alpha_r = 1$, the overall selection pressure (α) is only = 2. The model is plotted in Figure 1, inserting various selection pressure, fitness and seed bank characters. The way the various parameters interact with different herbicide use practices is discussed historically. History will then be used to predict the future – if only we can learn from history.

There is an exponential enrichment of resistant individuals with every treatment (Fig. 1). One cannot discern resistance by visual observation the season before there are 10–30% resistant individuals in a population in the field. Similarly, less than 0.1% resistance in a population cannot be easily discerned in laboratory experiments. One cannot follow this exponential increase while it is happening, as the small proportion of plants, probably derived from one founder are initially scattered and may go unnoticed. Resistance usually appears suddenly over a relatively large area in a field. The size of this area could be due to spread by crop harvesting equipment that spreads seed.[1]   Grain harvesting equipment is probably the major force in spreading (weed) seed throughout fields (Porterfield, 1989).

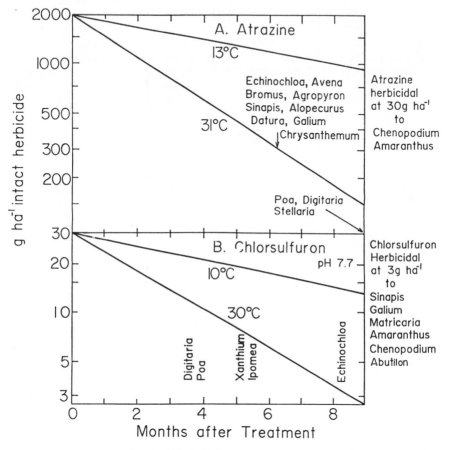

Fig. 2.   Dissipation of herbicides from the environment, effect of varying selection pressure on late germination of weed species.
A.   Atrazine, (plotted from data in Zimdahl *et al.*, 1970).
B.   Chlorsulfuron, (plotted from data in Thirunarayanan *et al.*, 1985). Weed germination at particular levels of pre-emergence herbicide treatment are taken from data kindly provided by P.F. Bocion (pers. comm.).

---

[1] The best way to guess if a patch of uncontrolled weeds in a field is due to resistance, or more commonly to missed spraying, is to observe it carefully. If the patch contains a few species, the sprayer probably missed. If it is comprised of one species, resistance may have evolved.

Selection pressure (α) can be measured, and it is clear that it is the major determinant of the steepness of the slope. The angle of slopes allow a comparison of what might happen with various herbicides and rates. Selection pressure will be highest when there is season long weed control by a pre-emergence herbicide. Resistance will evolve rapidly, as susceptible weeds never produce seed. It is the same high rate giving season long control, that gives ground water contamination in parts of the world with heavy rains at the time of pre-emergence applications. Seeds of many species germinate in flushes during the year. Seeds can germinate before or after a low residue post-emergence herbicide treatment, greatly lowering selection pressures. For example, *Amaranthus* and *Chenopodium* spp. produce very small, rapidly seeding plants under early spring and autumn photoperiods. A summer, post-emergence herbicide treatment allows a crop of susceptible seed to be produced before and/or after herbicide application, heavily diluting any seed from resistant plants. This severely lowers selection pressure. Thus, the length of residual activity and the timing of application can affect the rate of evolution of resistance.

The selection pressures of herbicides can be approximated by comparing the degradation curves of the herbicides with acute herbicide toxicity to various species. It is assumed that applications are made before weeds can germinate and produce seeds. Degradation rates of pre-emergence treatments of atrazine and chlorsulfuron are illustrated in Figure 2. This can be roughly compared with I_{80-100} values for various weeds. In a 5-month season, atrazine does not degrade sufficiently to allow germination of susceptible seed of *Amaranthus* spp. but has dissipated sufficiently to allow some susceptible weeds to germinate (Fig. 2A).

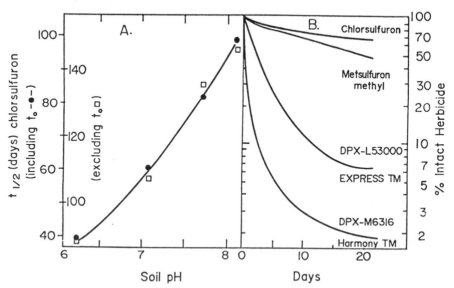

Fig. 3. <u>Degradation of sulfonylurea herbicides</u>. <u>A. Degradation of different sulfonylureas in the same soil type</u>. (Replotted from Beyer *et al.*, 1987). <u>B. Degradation of chlorsulfuron in soils of varying pH</u>. (Plotted from tabular data in Thirunarayanan *et al.*, 1985).

The selection pressure of atrazine is highest for *Amaranthus* spp. and least for the others. Triazine resistance evolved first among the broad leaves such as *Amaranthus* spp. and *Chenopodium* spp. and later to grasses, as expected. Grass resistance evolved and appeared first in orchards and roadsides, where either higher rates of atrazine were used and/or simazine, a slower degrading herbicide was used. Selection pressure is very high with chlorsulfuron (Fig. 2B) and should be greatest in high pH soils where degradation is slowest (Fig. 3A). Selection pressure should be almost negligible if some of the more rapidly dissipating analogs of chlorsulfuron are used (Fig. 3B). The choice is there whether farmers should be encouraged to get sulfonylurea resistance or not, by 'prescribing' the rapidly degraded analogs and limiting use of the slowly degrading sulfonylureas.

The following question is often posed: "it is easy to select for glyphosate resistance in the laboratory; can we get it in the field?" The answer from the models is "no", as glyphosate has a very short field persistence. How then did paraquat resistance evolve with a shorter biological persistence than glyphosate? Farmer persistence replaced chemical persistence; paraquat resistance only evolved where farmers sprayed paraquat 6–10 times a year for a few years. Presumably glyphosate resistance can also be obtained with multi-annual treatments. Resistant populations can theoretically be expected to evolve with any low persistence herbicide in weeds that have a single major productive flush of germination, if the timing of full germination and herbicide treatment coincide.

The initial frequencies of resistance were assumed to be the same in Figure 1 (left scale). This is clearly not the case in nature. There may be no genes for resistance to a pesticide. No species have evolved resistance to Bordeaux mixture in >100 years after monoculture use. Some weed species have genes for resistance to heavy metals and some seem to lack this trait in their genetic variability (Bradshaw and Hardwick, 1989). This had been true for *Gonococcus* and penicillin, but a plasmid from another bacterium with a gene for penicillinase transformed this species. There may be no genes in most weed species for 2,4-D resistance, a herbicide used for 45 years in monoculture wheat. It is more likely that 2,4-D has a low selection pressure, which may be getting lower. Soil microorganisms decay 2,4-D faster in a soil with many treatments than a pristine soil (Smith *et al.*, 1989). No species that were well controlled 45 years ago in vast areas of Canada have evolved resistance in wheat fields continually treated with 2,4-D (Hume, 1987).

There are clearly differences in the initial frequency of resistant genotypes of weeds towards different herbicides. Plastid genome (plastome) inherited triazine resistance is probably found at some minute frequency – probably between 10^{-10} and 10^{-20}. Note how poor the guesses are. A nuclear gene coding for a "plastome mutator" has been found in atrazine-resistant populations increasing the frequency (Arntzen and Duesing, 1983). If all else is equal for two herbicides (i.e., the slopes in Fig. 1 are the same), if the frequency of resistance to herbicide "A" is 10^{-18} and to herbicide "C" is only 10^{-6}, populations resistant to "C" will appear 3 times earlier than to "A". With acetolactate synthase inhibiting herbicides, the initial frequency is c. 10^{-6}, both in the laboratory (Chaleff and Ray, 1984) and in a field experiment (Stannard and Fay, 1987). This clearly explains why sulfonylurea-resistant populations appeared in 3 years while it took over 10 years and far larger areas for triazine resistance to appear.

It has been simple to select for resistance to other single target herbicides such as acetyl-CoA carboxylase inhibitors in the laboratory (Parker *et al.*, 1989). Resistance has evolved, probably by a similar target site modification in the field conferring cross-resistance to all aryloxyphenoxy proprionates (-fops) and cyclohexanediones (-dims) in *Avena fatua* (Powles and Howat, 1990) and to -fops alone in *Lolium multiflorum* (Gronwald *et al.*, 1989). This shows us how wary one must be about single target herbicides. Target site resistance also evolved to the tubulin binding and highly residual dinitroanilines in *Eleusine indica* (Vaughn, *et al.*, 1987; Vaughn and Vaughan, these Proceedings), and in *Setaria viridis* (Morrison and Beckie, these Proceedings).

The frequency of resistance may vary depending on the selection pressure as different alleles confer different levels of resistance. At low herbicide rates, more alleles will confer resistance than at high rates. All triazine resistant *psbA* genes from weeds that have been sequenced all have a transversion at the same amino acid - no. 264. This confers 500-2000 fold resistance to atrazine. It is clear from the data in Figure 2A and Figure 1 that *Amaranthus* spp. for example had to evolve a more than 90-fold resistance to atrazine. Algae and photosynthetic bacteria resistant to triazines have been selected for and isolated in the laboratory, i.e., they must be in populations at a frequency closer to 10^{-6} than 10^{-20}, as 10^{20} cells weigh thousands of tons. None have as high resistance factors as the weeds, and resistance codes to many different sites on the same *psbA* gene or its bacterial analogue.

It is valid to question whether lowering the selection pressure of triazines would actually delay resistance; there would possibly be 10 available resistant alleles (based on algal studies) of the *psbA* gene using lower selection pressure vs. seemingly one at present. With more alleles to select for, resistance should evolve faster. Probably not much in this case. It would shift the initial frequency tenfold, say from 10^{-15} to 10^{-14}. Compare that with the slopes for lowering selection pressure from 99% to 90% kill (Fig. 1), and note that even 10 multiple alleles with lower resistance negligibly reduce the time to resistance, compared to the delaying effect of lowering selection pressure.

Seed bank dynamics

The longer the life in the seedbank, the greater the buffering effect of susceptible seed from previous years, decreasing the rate of evolution of resistance. *Senecio vulgaris* has evolved triazine resistance in orchards, nurseries and roadsides where there was no mechanical cultivation, but not in cultivated maize fields. The *Senecio* seed is incorporated into the soil seed bank in such maize fields, where it is viable for many years (Watson *et al.*, 1987). All *Senecio* seed falling on undisturbed soil on roadsides or orchards either germinates, or dies during the following season (Putwain *et al.*, 1982). Resistance thus evolved where there was the lowest average seed bank life time (n), as predicted. Such information must be considered in formulating strategies for resistance management. Many other species do not have a seed-bank under specific agronomic situations, i.e., in minimum - till agriculture where (n=1). Indeed, the

8

first cases of clear MCPA resistance appeared in *Ranunculus acris* and *Carduus nutans* in New Zealand pastures, where there is no seed bank and there were repeated treatments for more than 15 years (Bourdot *et al.*, 1989).

The lack of competitive fitness (f) of resistant individuals can have a strong dampening effect on the evolution of resistance, but only when it can be expressed, i.e., when the herbicide is not present. Thus, the lack of fitness can have little influence with persistent herbicides, in mono-herbicide culture, but could be effective with the less persistent herbicides.   This is another reason to avoid persistent compounds, especially in monoculture.

The potentially complicated interactions between the parameters affecting the rates of evolution of resistance were described above.   The data demonstrate that even in the monoherbicide systems there can be vast differences that can be manipulated to delay resistance.   To use the atrazine-maize monoherbicide culture as an example, it is apparent that there are at least two ways to lower selection pressure to delay resistance: (a) use post-emergence treatments instead of pre-emergence treatments. Even with atrazine this would allow an early spring 'crop' of weed seeds; (b) use less atrazine or atrazine mixed with a more rapid degrading triazine such as cyanazine to allow a late flush of weeds, or use less atrazine with a grass killer. High rates of atrazine used are usually used so that atrazine will control both broad leaves and grasses. The data in Figure 3A clearly demonstrate that too much atrazine is applied.   The use of heterologous mixtures will be discussed in a later section.

DELAYING RESISTANCE BY AVOIDING MONOHERBICIDE CULTURE

The brief history of herbicide resistance has taught us that resistance has not appeared where herbicides were rotated or where mixtures were used. The models as initially interpreted predicted that resistance should have evolved in rotations, but later. Vast areas of the US corn belt are still devoid of resistance where rotations and/or mixtures were used. An analysis follows showing that resistant populations should not have evolved in these areas. This allows us to optimize rotations and mixtures to further elude resistance.

Rotations to avoid resistance

The model as shown in Figure 1 does not adequately account for events in the "off-years" during rotations when the competitive fitness of the resistant biotype is low. Resistance is shown in the model to evolve at a fixed rate as a function of the number of generations or seasons a weed was treated with a particular herbicide irrespective of intervening treatments (Fig. 1). This means that if it would take 8 years for resistance to occur in monoculture maize with atrazine as the sole herbicide, it would take 12 years with a maize/maize/wheat (or soybean) rotation where atrazine is used for control 2 of every 3 years; or 16 years in a maize/wheat (or soybean) rotation where atrazine is used every other year; or 24 years in a corn/wheat/soybean rotation where atrazine is used once in 3 years (see lower scales on Fig. 1). When the model was formulated 10 years ago, it appeared that the time was ripe for triazine resistance to break out in vast areas of the US corn belt where such rotations were used, as there had been 6-10 years of atrazine usage since it was introduced.   Yet, resistant populations only appeared in monoculture, monoherbicide maize.

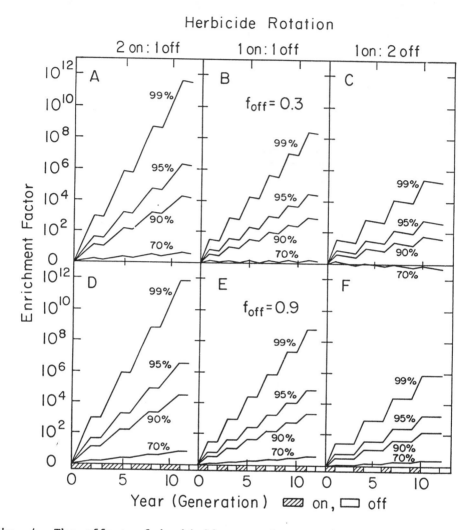

Fig. 4. <u>The effect of herbicide rotation on the rate of resistance enrichment</u>.   Three rotational scenarios are shown for herbicides with different selection pressures.   The two fitnesses represent triazine resistance (A-C; f=0.3) and sulfonylurea/ALS resistance, (D-F; f=0.9). The f_{on}= 1.   The weed seeds in the seed bank are presumed to have a 2 year residence time.   Source: Calculated from equations in Gressel and Segel, (1990a).   From Gressel and Segel, (1990b). (Reproduced by courtesy of the Weed Science Society of America.)

Better data incorporated into an updated model show how rotation has been a better strategy than previously predicted.   The newer data and model emphasize the highly reduced fitnesses of the triazine-resistant biotypes, which are of greater magnitude and importance than had initially been expected.

10

Lack of fitness of resistant weeds – a major consideration

The initial model used an average fitness differential for all the generations treated (Fig. 1). The fitness differential between resistant and susceptible individuals essentially can never become apparent with herbicides such as triazines that give season long weed control, as there is no time without herbicide for this differential to be expressed. Only resistant biotypes can survive when the herbicide is present. Thus, the fitness differential is unimportant with triazines in monoherbicide culture, but seems to be an important factor in delaying resistance to less persistent herbicides. The fitness differential is very important when herbicide usage is stopped for a season or more. Resistant biotypes are often more susceptible to some of the herbicides and cultivation procedures used in the rotational years (negative cross-resistance) as discussed later. The model was modified to consider what happens to resistant individuals in the "off" years when a herbicide is not used (Gressel and Segel, 1990a,b).

Only intraspecific competition during the evolution of resistant populations has been considered, except for one study (Warwick and Black, 1981). More data are needed from the agro-ecologists on the importance of interspecific competition, including that with crops and weeds. Still, wherever measured, the fitness of triazine-resistant individuals was about 10-50% of the wild type, when measured by competing the resistant with the wild type. There are a few exceptions with weeds (Yaacoby *et al.*, 1986) and photosynthetic bacteria (Brown *et al.*, 1988).

The effect of rotation, plotted using a modification of the earlier equation, describes the effects of fitness when the herbicide is used, and a different fitness for seasons when not used (Fig. 4). There is hardly any real delaying effect of fitness when the resistant individuals have near normal fitness (Fig. 4, D–F). Actually, there is no agronomically significant difference between Figure 1 and Figure 4 D–F, if the lines in Fig. 4 are smoothed. This high fitness may well be the situation with the weeds that evolved resistance at the level of acetolactate synthase. Thus, in such cases, rotation is of little assistance in truly delaying resistance. The only delay is for the number of generations the particular herbicide is not used. In such cases, only lowering the selection pressure delays resistance.

When there is a large fitness differential between resistant and susceptible individuals (as with most triazine-resistant weeds), there will be a greater delaying effect due to fitness (Fig. 4, A–C), and the effect is greatest when selection pressure is lowest. The plots describe the reduction of the proportion of resistant individuals in the "off" years. There are even some situations at low selection pressures where resistant individuals disappear in "off" years more rapidly than they are enriched for in "on" years. Thus rotation can clearly be advantageous. When there is negative cross-resistance, the fitness differential is even greater.

With a very slow rate of overall enrichment, the model suggests that it will take very many years for resistant populations to become a major problem. The (log) factors of enrichment at the end of 9 and 15 year periods are given in Table 1. When this factor is compared with the initial frequency of resistance (N_0), it can be estimated whether resistant populations should evolve in that time frame. If N_0 is 10^{-20}, (a guess for the N_0 of triazine resistance), and fitness is 0.3 or less, and selection pressure is low, i.e., the effective kill is less than 95%, we see that resistant populations will not evolve under any of these scenarios. Triazine resistance would only appear in 15 years, in monoherbicide culture, where the effective kill is 99% (with an enrichment of factor of 22.3). This enrichment factor makes up for the resistance frequency of 10^{-20}. With chlorsulfuron, N_0 should be 10^{-6} to 10^{-8}, explaining why resistance evolved so rapidly. It is not clear that one need actually consider whether mutations to resistance are dominant or recessive, as there may be only a small frequency difference between the two types in diploid organisms (Williams, 1976). This is because recombination (somatic-crossing over) increases homozygous recessive frequencies. Further ramifications of rotation are described in the papers describing the modified models from which Table 1 was derived (Gressel and Segel, 1990a,b). These include graphs showing in a continuous manner how the various factors can interact to give a rotational system, where there is no enrichment for resistance, or more feasible designs with yearly doublings of the frequency of resistance.

Table 1. <u>(Log) Enrichment of resistant individuals in weed populations over a 9- and 15-year period under different herbicide rotations</u>.

Rotation strategy	α	Effective kill (%)	9-year-period			15-year-period		
			Fitness in "off years"					
			0.9	0.5	0.3	0.9	0.5	0.3
			Log_{10} of enrichment factor					
No	2	50	1.0	1.0	1.0	1.7	1.7	1.7
rotation	10	90	5.1	5.1	5.1	8.5	8.5	8.5
	20	95	7.4	7.4	7.4	12.4	12.4	12.4
	100	99	13.4	13.4	13.4	22.3	22.3	22.3
2 on;	2	50	0.6	0.5	0.4	1.1	0.8	0.6
1 off	10	90	3.4	3.2	3.1	5.6	5.3	5.2
	20	95	4.9	4.7	4.6	8.2	7.9	7.8
	100	99	8.9	8.7	8.6	14.3	14.5	14.4
1 on;	2	50	0.5	0.3	0.2	0.8	0.4	0.2
1 off	10	90	2.8	2.6	2.4	4.4	4.0	3.8
	20	95	4.1	3.8	3.7	6.5	6.1	5.9
	100	99	7.4	7.1	7.0	11.8	11.4	11.2
1 on;	2	50	0.3	-0.1	-0.3	0.4	-0.1	-0.4
2 off	10	90	1.6	1.3	1.1	2.7	2.1	1.8
	20	95	2.4	2.0	1.9	4.0	3.4	3.1
	100	99	4.4	4.0	3.8	7.3	6.7	6.4

If the enrichment factor is larger than the log of the mutation freqency, resistance is to be expected. Effective kill of weeds is for a whole season, assuming no effect on resistant weed seeds. A seed bank half-life of 2 years is used. The fitness in "on years" is assumed. (Gressel and Segel, 1990b).

Negative cross-resistances in rotations

Table 2. <u>Negative cross-resistance of herbicide resistant biotypes.</u>

Primary Resistance species	Negative cross-resistance	Parameter measured	Rate giving 50% inhibition (R/S)	Ref.
<u>s-Triazines</u>	dinoseb	fresh weight	.27	b
Amaranthus	flumeturon	thylakoids	.22	c
retroflexus	DNOC	thylakoids	.5	c
Chenopodium album	dinoseb	"	.27	b
Brassica napus	DNOC	"	.66	b
Senecio vulgaris	dinoseb	"	.21	b
Conyza canadensis	DNOC	"	.1	b
Epilobium ciliatum	chlorpropham	fresh weight	.46	d
Brassica napus	dinoterb	thylakoids	.07	e
	dinoseb	"	.12	e
	medinoterb	"	.20	e
	pyridate[a]	"	.21	e
	ioxynil	"	.45	e
	bromoxynil	"	.71	e
Kochia scoparia	2,4-D	fresh weight	R inhibited	f
Epilobium ciliatum	oxyfluorfen	"	more than	g
	paraquat	"	S at a	g
	pyridate	"	single rate	g
<u>Dinitroaniline</u> *Eleusine indica*	chlorpropham		single rate	h
<u>Mecoprop</u> *Stellaria media*	benazolin	fresh weight	.53	i
<u>MSMA-DSMA</u> *Xanthium pensylvanicum*	paraquat		.50	j
	bentazon		.65	j
<u>Paraquat</u> *Conyza canadenis*	glufosinate	PS	.26	k

[a] Pyridate is metabolically activated to CL 9673, which was used directly with isolated thylakoids. [b] Fuerst *et al*, 1985; [c]Oettmeier *et al.*, 1982; [d]Bulcke *et al.*, 1987; [e]Durner *et al.*, 1986; [f]Salhoff and Martin, 1986; [g]Clay, 1987; [h]Vaughn *et al.*, 1987; [i]Lutman and Snow, 1987; [j]Haigler *et al.*, 1988; [k]Pölös *et al.*, 1987.

Many herbicides are more toxic to resistant individuals than to susceptible ones (Table 2). Table 2 only contains data for such "negative cross-resistances", but these are not the preponderant cases and are clearly not universal. Still, negative cross-resistance should be elucidated and incorporated into rotational strategies for preventing resistance, both before and after populations become preponderantly resistant. The delaying effect of negative cross-resistance must be added or compounded to the lack of fitness of resistant weeds, when considering rotations. The negative cross-resistances in atrazine-resistant weeds include herbicides that act at or near the same site in photosystem II (DNOC and dinoseb) as well as herbicides acting on other photosystems (paraquat) or at totally different sites. The phenolic PSII inhibitors listed in Table 2 have all been removed from the market leaving only pyridate. Possibly some of the newer alkylamino-alkyliden-pyrandiones (APs), that are thought to act on the *psbB* gene product of PSII (Asami *et al.*, 1988) will exert negative cross-resistance on triazine-resistant weeds. There was negative cross-resistance to other tubulin-binding herbicides in dinitroaniline-resistant *Eleusine indica* (Table 2), but not to six commercial herbicides on this weed (Vaughn *et al.*, 1987; Vaughn and Vaughan, these Proceedings). It should be carefully pointed out that most of the data in Table 2 on negative cross-resistance are from *in vitro* studies, the rest from small-scale pot studies, often without full dose response curves. We have a lack of field data on this potentially powerful tool to prevent resistance from evolving.

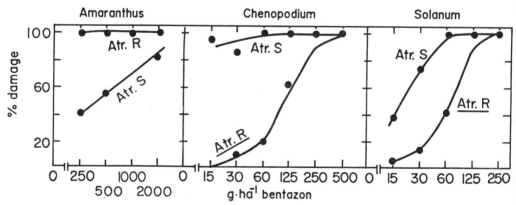

Fig. 5. <u>Negative and positive cross-resistance with bentazon of atrazine-resistant biotypes</u>. Bentazon was applied post-emergence in pot experiments in the glasshouse. *Amaranthus retroflexus* resistant biotypes show a very strong negative cross resistance. *Chenopodium album* and *Solanum nigrum* resistant biotypes all showed positive cross-resistance to bentazon at use rates that were lower than those normally applied. Source: Plotted from data kindly provided by Dr B. Würzer, BASF AG, Limburgerhof.

Negative cross-resistance may not be a simple tool to use. In one study with three weed species having triazine-resistant biotypes, there were differing cross-resistances to bentazon at the whole plant level. Triazine-resistant biotypes of *Chenopodium album* and *Solanum nigrum* were (positively) cross-resistant to bentazon (Fig. 5B, C), whereas only *Amaranthus retroflexus* exhibited the desired negative cross-resistance (Fig.

5A). Such differences have been previously reported at the thylakoid level; A triazine-resistant *Brassica* was five times more sensitive to BNT than the wild type, yet resistant and susceptible *Chenopodium album* were almost equally affected (Thiel and Böger, 1984). Negative cross-resistance is well documented in the literature on antibiotic-resistant bacteria in medicine, and in fungicide- and insecticide-resistant pathogens and arthropods, and is used as part of resistance management strategies. We clearly have to learn how to use it with herbicides. Suggestions for mixtures utilizing negative cross-resistance are described below.

Mixtures as a tool to delay/prevent resistance

Mixtures are divided (somewhat) arbitrarily into 3 types: (a) simple heterologous mixtures; (b) synergistic mixtures; (c) synergistic mixtures having a component exerting negative cross-resistance. These can be interrelated and/or overlapping.

Heterologous mixtures are more than one herbicide acting on the same weed at different sites of action. The best documented case is that of alachlor and atrazine. Alachlor is mainly used as a grass killer, but it is quite effective on *Amaranthus* spp. and *Chenopodium* spp., species often evolving triazine resistance.

The effect of heterologous mixtures is many-fold. One effect is to lower the initial frequency by compounding the frequencies of resistance to both herbicides. If the frequency to herbicide "A" is 10^{-8} and that of "B" is 10^{-6}, the compounded frequency is 10^{-14}, i.e., if all else is equal, it will take about twice as long for resistance to evolve. In the case of the alachlor/atrazine mixture, there is no information on the frequency of resistance to alachlor (or to atrazine). No alachlor resistance has evolved, despite its widespread use. Perhaps there can be no resistance? There have been no reports of *in vitro* attempts to isolate resistance to chloroacetamide herbicides.

A synergistic mixture in the general sense is one where the herbicidal effect of the mixture is greater than the effect of the sum of the components. This allows using less of each component, often giving an economic advantage to a mixture. The atrazine-alachlor mixture has been shown to be synergistic against some weeds (cf. Gressel, 1990). Assuming this to be so for *Amaranthus* and *Chenopodium* spp., then the rate of use of atrazine can be dropped. This can clearly be done in the field unless high levels of atrazine are required to control grasses not controlled by alachlor. The lowering of the rates reduces the selection pressure for resistance to each herbicide, slowing the rate of evolution for each. This effect is synergistic beyond the compounded frequencies.

Another type of synergistic mixture is one containing a herbicide and an otherwise inactive adjuvant. Synergistic adjuvants are used to control insecticide-resistant insects that evolved high levels of monooxygenases. Piperonylbutoxide inhibits monooxygenases, allowing both resistant and susceptible arthropods to be controlled by lower doses. The addition of monooxygenase inhibitors allows control of the *Alopecurus myosuroides* and *Lolium rigidum* that evolved cross-resistance to wheat selective herbicides

15

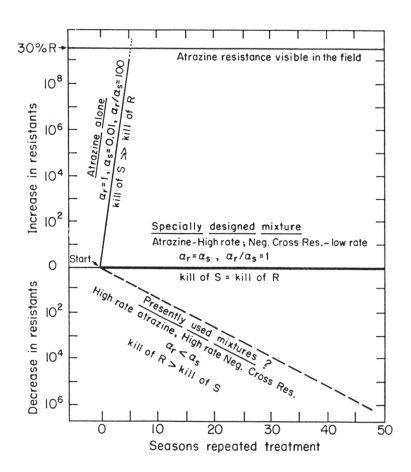

Fig. 6. **Modelling the effect of mixtures containing a herbicide exerting negative cross-resistance on the rate of evolution of a common herbicide, e.g. Atrazine.**

Thin line: The presumed rate of evolution of atrazine resistance when no other herbicide is used. It is assumed that atrazine has no effect on resistant biotype individuals and 99% effective kill (season long suppression of seed output) on susceptible biotype individuals.

Thick line: The effect of herbicides having strong negative cross-resistance when mixed with atrazine at normal use rates. These control resistant biotypes more effectively than atrazine controls the susceptible biotype.

Dashed line: The effect of specially balanced mixtures of atrazine (normal rate) and a herbicide having negative cross-resistance (used here at a low rate). The rates are balanced such that atrazine controls the susceptible biotype to the same extent that the mixed herbicide controls the resistant biotype. It is assumed (for the sake of simplicity) that the mixed herbicide is ineffective on the susceptible biotype at such a low rate.

(Kemp and Caseley, 1987; Powles *et al.*, 1989).   The problem with the synergists tested so far is that they abolish selectivity in wheat.

Chelators that remove copper and/or zinc from the enzymes that detoxify active oxygen species act as synergistic adjuvants, allowing control of paraquat resistant weeds at 10-fold lower concentrations of paraquat than were previously used (Gressel and Shaaltiel, 1988).

Synergistic mixtures exerting negative cross-resistance

When one herbicide exerts negative cross-resistance over individuals resistant to another herbicide in a mixture, it can exert quite a depression in the rate of enrichment of resistance.   There can even be a negative enrichment at normally used rates i.e., the resistant individuals are depleted to a lower than natural frequency.   This can be imagined by plotting the type of data shown in Table 2, using the model equation. For example, the mixtures presently used may well affect atrazine-resistant biotypes to a far greater extent than the sensitive biotypes.   The selection pressure is negative as α_r is less than α_s, as shown in Figure 6 (dashed line).   This is probably the case with pyridate, the main herbicide used (in mixture with atrazine) in Europe once resistance has evolved.   Other herbicides such as alachlor and metolachlor, which normally control *Amaranthus* and *Chenopodium* spp. may exert negative cross-resistance, but no evidence is available.   These chloroacetanilides are the herbicides mostly widely mixed with atrazine by virtue of their superior grass weed control.

It is not known whether the 2- to 10-fold lower rates of herbicide required to affect resistant vs. susceptible biotypes (Table 2) will carry over to the field situation. One could use lower rates of these herbicides to achieve the identical level of weed control of the resistant biotype as atrazine gives with susceptible biotype. When rates are thus balanced, there will be no selection pressure for atrazine resistance and no enrichment of resistance in the population (Fig. 6, thick line). This could be used prophylactically, i.e., before resistant populations predominate or just after resistance occurs. Lower rates will be much more "cost effective" than full rates. The low rates may also allow using herbicides that at normal rates are phytotoxic to maize.

The real field and even pot data needed to elucidate the magnitude of synergistic negative cross-resistance in mixtures are lacking.   The mathematical modelling shows how such data can be used to prevent resistant populations from evolving or strategies to manage the most widespread resistance that has already appeared.   Those involved with production of herbicides and those engaged in management of weeds are encouraged to try to obtain the necessary data and test the strategy in field situations. In places where resistant populations have evolved, the depletion of resistant populations should be followed, using various rates of the herbicide exerting negative cross-resistance.

CAN ROTATIONS OR MIXTURES BE INEFFECTIVE?

There are a few cases when rotations or mixtures can be classified as non-mixtures and non-rotations to the point of not delaying evolution by the number of seasons a herbicide is not used: (a) This can occur when the herbicides, even of different chemistry have precisely the same site of action. Resistant populations of *Chenopodium album* evolved in rotations of atrazine on maize and pyrazon (chloridazon) on sugar beets. Both herbicides bind to PSII at the same site, and cross-resistance is expected; (b) This can be ineffectual when the herbicides produce the same toxic moieties. Photosystem I, PSII herbicides and herbicides that cause protoporphyrin IX accumulation (nitrodiphenyl ethers and many others) all cause the generation of active oxygen species, the basic cause of phytotoxicity. All a weed has to do is to evolve higher levels of pre-existing enzymes of an active oxygen detoxification pathway that keep the plant alive until the herbicide has dissipated. Various species resistant to one active oxygen-generating xenobiotic were cross-resistant to others (Shaaltiel *et al.*, 1988), supporting this hypothesis; (c) Ineffectiveness can occur when the herbicides are degraded in the same manner. There is ample supportive evidence for a hypothesis that wheat, unlike other studied crops, degrades herbicides by a single type reaction; oxidation by monooxygenases. All weeds had to do was to evolve a 'biochemical mimicry', just as many weeds have evolved morphological and phenological mimicries, allowing them to follow wheat around the world (Barrett, 1983). This seems to be what has happened in Australian *Lolium rigidum* and English *Alopecurus myosuroides*. In these cases, there are few ways out; one is 'make' wheat to be like other crops; by genetically engineering it to be resistant to other herbicides. The ideal choice of genes is reviewed elsewhere (Gressel, 1988a). There may be a way out with a major wheat weed, *Avena fatua*. For reasons unknown, this weed is controlled by difenzoquat while for equally unknown reasons, wheat is spared (Shaner, 1984). This herbicide is not metabolized in wheat. Thus, there is good reason to use herbicides such as difenzoquat in a rotation to delay the evolution of resistance in monoculture wheat.

THE SPREAD OF RESISTANCE

It appears from the 'founder effect' studies of Gasquez and colleagues with triazine resistance in France, that each population evolved separately, not by transfer (Gasquez and Compoint, 1981; Darmency and Gasquez, 1983). The slight differences in the cross-resistance spectra to different herbicides in the Australian *Lolium* and the British *Alopecurus* indicate that there must have been co-evolution at many foci, and not spread of seed or pollen. Harvesting equipment does blow weed seeds within a field (Porterfield, 1989) and manure spreaders will spread undigested weed seeds from silage, for short distances, within fields.

Presumably vehicles will move resistant seed along rights of way. It would be most interesting if researchers would perform isozyme analyses *à la Gasquez* on the thousands of linear miles of triazine-resistant *Kochia scoparia* along railways to ascertain whether there was co-evolution, or the trains just spread the seeds throughout the midwest of the US. Interestingly, it took ten years before there were reports of this triazine-resistant *K. scoparia* appearing in neighboring maize fields (O. N. Burnside, pers. comm. 1989) and *K. scoparia* is a 'tumble weed'. Thus, usually 'spread' can be contained; it is the co-evolution that must be delayed or prevented.

The slowness of spread has other implications. There is considerable reluctance to release genetically engineered herbicide-resistant crops into the fields for fear that the weeds may 'catch' it by gene transfer. Elaborate tests (bordering the impossible) are called for to preclude gene transfer. No cases have been reported where weeds 'caught' herbicide resistance from crops that are naturally resistant despite considerable possibilities. The same possibilities are there. Even when the grasses seemingly evolved biochemical mimicry to wheat, the cross-resistance spectrum is greater in the grasses than the wheat, so it is highly unlikely that the genes came from wheat. The major forseeable problems are only where there are closely related crops and weeds, e.g. oats/*A. fatua*, rape/*B. campestris*. Such field experience should be considered as part of the risk assessment process. A separate problem might arise from crops becoming volunteer weeds, which can be handled as it now is, by alternative herbicides. Volunteer weeds could become a problem if multiple resistances are engineered into the same strains of one crop variety.

CONCLUDING REMARKS

Environmental consequences of resistance management

There is strong pressure to decrease herbicide use. We have seen from the above that decreasing usage is also desirable to successfully manage resistance. The only problem with demanding decreased herbicide use, is that demands are simplistically measured in tons. Tonnage was substantially decreased by switching from 2,4-D to chlorsulfuron; is that really better? Decreased usage should be measured by residual biological activity in the soil, or percent that gets to ground water, or into the crop, or on toxicological grounds, but clearly not in tons.

In the Introduction it was stated that ploys to prevent resistance would be good for the environment. In summary we see that this could be so. Our resistance management tools require using less total herbicides, and less of residual herbicides, and a greater variety of herbicides to lower selection pressure. These tactics will have less of any given herbicide leaching into ground water. Heterologous mixtures where less of each component is used, and synergistic mixtures using adjuvants or negative cross-resistance will also decrease amounts of any given herbicide. Rotations will also decrease the amount used of any given herbicide in a multiyear calculation.

The only issue where weed resistance managers and environmentalists may disagree is over which herbicides to use. If environmentalists persist that amount is 'bad', and amount not biological activity, and if they persist that all old herbicides are 'bad', there will be fewer compounds available. The 'oldies' often have multiple (unknown) sites of action along with low selection pressure; and resistances have not been found. The newer compounds have single sites of action, resistance is present at higher initial frequencies and they exert higher selection pressure. These contradictions must be reconciled.

Awareness, openness, gathering and sharing information

It is important to ascertain whether resistance can evolve to each herbicide. The newer herbicides seem only to have a single target site, which compounded with high selection pressure makes them vulnerable to evolution of resistance. How many laboratory or field experiments were performed to check vulnerability? Only recently was it shown that the nitrodiphenyl ethers (and compounds with dissimilar chemistry) may have a single site of action (Matringe *et al.*, 1989). Will resistance evolve?

Once resistant populations do evolve, they should not be shrugged off as 'only one location' at best.   At worst, the resistant material is destroyed instead of studied, and possibly put to use. Studies can indicate the mode of evolution, positive and negative cross-resistances, inheritance and fitness, all needed for resistance management to prevent further cases.   There are positive uses; triazine resistance was genetically transferred from *Brassica campestris* to rape (Souza-Machado and Hume, 1987) and from *Solanum nigrum* to potato by protoplast fusion (Gressel *et al.*, 1989).   The paraquat-resistant *Lolium perenne* was bred into pasture grass (Johnston and Faulkner, these Proceedings). This could have been done with paraquat-resistant *Poa annua* had not the population been obliterated.

The quicker resistance information gets to the weed science community at large, the faster it can react. Recent resistances that evolved were rumored to exist more than a year before they were released from secrecy; a year of lost management.

Management tactics

The best tactics to prevent or delay the appearance of resistant populations are:

(a) to ascertain vulnerability of each herbicide to resistance by elucidating its site(s) of action, mode(s) of degradation in crops and weeds, and the frequency of resistance.
(b) to employ herbicide treatments with the minimum selection pressure giving cost effective weed control. Such treatments will not give near total weed control, but will leave behind enough susceptible seeds each year to dilute out resistant seeds.   Modify harvesting equipment to prevent weed seed scatter over large distances.
(c) to use mixtures of compounds acting at different sites of action and having different modes of degradation, preferably with herbicides having large margins of negative cross-resistance, or synergy. Mix vulnerable herbicides with less vulnerable ones.
(d) to use rotations of herbicides having different sites of action and different modes of degradation; preferably where the weeds have negative cross resistance to the herbicides.   Rotate vulnerable with less vulnerable herbicides.
(e) to employ mechanical cultivations in the rotations, especially if they preferentially control unfit resistant biotypes.

It is hard to meet these criteria with monoculture wheat. As stated above, wheat probably has only a single mode of degradation (Gressel, 1988a,b). In such situations it is necessary either to (a) rotate crops to allow herbicide rotations; (b) rotate with herbicides having a placement selectivity (e.g. Glasgow *et al.* 1987) that is not related to herbicide metabolism in wheat; (c) find synergists that preferentially inhibit herbicide degrading enzymes in the weeds of wheat (Gressel, 1988a).

It is clear that there will be problems with those high selection pressure herbicides having resistant mutants that are fit. The only alternative in monoculture wheat is to replace these with less persistent herbicides of the same group, having less selection pressure.

ACKNOWLEDGEMENTS

The author is most obliged to Prof. L.E. Segel, of the Applied Mathematics Department, Weizmann Institute of Science, for collaborating in generating the models, and to Mr. P.F. Bocion, Dr. R. Maag Ltd. and Dr. B. Wurzer, BASF, for allowing the use of unpublished data. Useful editorial comments by G.W. Cussans are gratefully acknowledged. The author has the Gilbert de Botton Chair of Plant Sciences.

REFERENCES

Arntzen, C.J. and Duesing, J.H. (1983). Chloroplast-encoded herbicide resistance. In "Advances in Gene Technology" (F. Ahmand, K. Downey, J. Schultz, and R.W. Voellmy, eds) pp. 273-299. Academic Press, New York.

Asami, T., Koike, H., Inoue, Y., Takahashi, N., and Yoshida, S. (1988). Structure-activity relationship and physiological aspects of new photosynthetic electron transport inhibitors, 3-alkylaminoalkyliden- 2 -H-pyran-2,4-(3H)-diones (APs). Zeitschrift für Naturforschung **43c**, 857-861.

Barrett, S.C.H. (1983). Crop mimicry in weeds. Economic Botany **37**, 255-282.

Beyer, E.M., Brown, H.M., and Duffy, M.J.N. (1987). Sulfonylurea herbicide soil relations. British Crop Protection Conference - Weeds 531-546.

Bourdot, G.W., Harrington, K.C., and Popay, A.I. (1989). The appearance of phenoxy herbicide resistance in New Zealand pasture weeds. British Crop Protection Conference - Weeds 309-316.

Bradshaw, A.D. and Hardwick, K. (1989). Evolution and stress-genetic and phenotypic components. Biological Journal of the Linnean Society **37**, 137-155.

Brown, A.E., Truelove, B., Highfill, C.T., and Smith, S.G. (1988). Physiological competence of atrazine-resistant strains of the photosynthetic bacterium Rhodobacter sphaeroides (Rhodospirillaceae). Weed Science **36**, 703-706.

Bulcke, R., Verstraete, F., Van Himme, M., and Stryckers, J. (1987). Biology and control of Epilobium ciliatum Rafin. In "Weed Control on Vine and Soft Fruits" (R. Cavalloro and D.W. Robinson, eds) pp. 57-67. A.A. Balkema, Rotterdam.

Chaleff, R.S. and Ray, T.B. (1984). Herbicide-resistant mutants from tobacco cell cultures. Science **223**, 1148-1151.

Clay, D.V. (1987). The response of simazine-resistant and susceptible biotypes of *Chamomilla suaveolens*, *Epilobium ciliatum* and *Senecio vulgaris* to other herbicides. <u>British Crop Protection Conference - Weeds</u> 925-932.

Darmency, H. and Gasquez, J. (1983). Esterase polymorphism and growth form differentiation in *Poa annua* L. <u>New Phytologist</u> **95**, 289-297.

Durner, J., Thiel, A., and Boger, P. (1986). Phenolic herbicides: correlation between lipophilicity and increased inhibitor sensitivity in thylakoids from higher plant mutants. <u>Zeitschrift für Naturforschung</u> **41c**, 881-884.

Fuerst, E.P., Nakatani, H.Y., Dodge, A.D., Penner, D., and C.J. Arntzen. (1985). Paraquat resistance in *Conyza*. <u>Plant Physiology</u> **77**, 984-989.

Gasquez, J. and Compoint, J.P. (1981). Isoenzymatic variation in populations of *Chenopodium album* L. resistant and susceptible to triazines. <u>Agro-Ecosystems</u> **7**, 1-10.

Glasgow, J.L., Mojica, E., Baker, D.P., Tillis, H., Gore, N.R., and Kurtz, P.G. (1987). SC-0574 - a new selective herbicide for use in winter cereals. <u>British Crop Protection Conference - Weeds</u> 27-33.

Gressel, J. (1988a). "<u>Wheat Herbicides: The Challenge of Emerging Resistance</u>". Biotechnology Affiliates, Checkendon/Reading (UK), pp. 247.

Gressel, J. (1988b). Multiple resistances to wheat selective herbicides: new challenges to molecular biology. <u>Oxford Surveys of Plant Molecular and Cell Biology</u> **5**, 195-203.

Gressel, J. (1990). Synergizing herbicides. <u>Reviews in Weed Science</u> **5** (in press).

Gressel, J. and Segel, L.A. (1978). The paucity of genetic adaptive resistance of plants to herbicides: possible biological reasons and implications. <u>Journal of Theoretical Biology</u> **75**, 349-371.

Gressel, J. and Segel, L.A. (1990a). Herbicide rotations and mixtures; effective strategies to delay resistance. <u>In</u> "Managing Resistance to Agrochemicals: from Fundamental Research to Practical Strategies" (M.B. Green, H.M. LeBaron and W.K. Moberg, eds), pp. 430-458. American Chemical Society, Washington, D.C.

Gressel, J. and Segel, L.A. (1990b) Modelling the effectiveness of herbicide rotations and mixtures as strategies to delay or preclude resistance. <u>Weed Technology</u> **4**, 186-198.

Gressel, J. and Shaaltiel, Y. (1988) Biorational herbicide synergists. <u>In</u> "Biotechnology in Crop Protection" (P.A. Hedin, J.J. Menn and R.M. Hollingworth, eds), pp. 4-24. American Chemical Society, Washington D.C.

Gressel, J., Aviv, D., and Perl, A. (1989). Methods of producing herbicide resistant plant varieties and plants produced thereby. United States Patent No. 4,795,705, pp. 9.

Gressel, J., Georgopoulos, S.O., and Brattsten, L.B. (1990). "Pesticide Resistance". J. Wiley, New York (in manuscript).

Gronwald, J.W., Eberlein, C.V., Betts, K.J., Rosow, K.M., and Ehlke, N.J. (1989). Diclofop resistance in a biotype of Italian ryegrass. Plant Physiology 89S, 115.

Haigler, W.E., Gossett, B.J. Harris, J.R., and Toler, J.E. (1988). Resistance of common cocklebur (*Xanthium strumarium*) to the organic arsenical herbicides. Weed Science 36, 24-27.

Hume, L. (1987). Long-term effects of 2,4-D application on plants. I. Effects on the weed community in a wheat crop. Canadian Journal of Botany 65, 2530-2536.

Kemp, M.S. and Caseley, J.C. (1987). Synergistic effects of 1-aminobenzotriazole on the phytotoxicity of chlorotoluron and isoproturon in a resistant population of blackgrass. British Crop Protection Conference - Weeds 895-899.

LeBaron, H.M. and Gressel, J. (eds). (1982). "Herbicide Resistance in Plants". J. Wiley, New York. pp. 449.

Lutman, P.J.W. and Snow, H.S. (1987). Further investigations into resistance of chickweed (*Stellaria media*). British Crop Protection Conference - Weeds 901-908.

Matringe, M., Camadro, J.-M., Labbe, P. and Scalla, R. (1989). Protoporphyrinogen oxidase as a molecular target for diphenyl ether herbicides. Biochemical Journal 260, 231-235.

Maxwell, B.P., Roush, M.L., and Radosevich, S.R. (1989). A simulation model for predicting the development of herbicide resistance in weeds. Weed Science Society of America - Abstract No. 149.

Oettmeier, W., Masson, K., Fedtke, C., Konze, J., and Schmidt, R.R. (1982). Effect of different photosystem II inhibitors on chloroplasts isolated from species either susceptible or resistant toward S-triazine herbicides. Pesticide Biochemistry and Physiology 18, 357-367.

Parker, W., Sommers, D., Wyse, D., Gronwald, J. and Gengenbach, B. (1989). Weed Science Society of America - Conference Abstract 178.

Pölös, E., Mikulas, J., Szigeti, Z., Laskay, G., and Lehoczki, E. (1987). Cross-resistance to paraquat and atrazine in *Conyza canadensis*. British Crop Protection Conference - Weeds 909-916.

Porterfield, J.W. (1989). Harvest equipment should address weed seed problems. Agricultural Engineering (Jan/Feb), 11.

Powles, S.B. and Howat, P.D. (1990). Herbicide-resistant weeds in Australia. Weed Technology 4, 178-185.

Powles, S.B., Holtum, J.A.M., Matthews, J.M., and Liljegren, D.R. (1989). Multiple herbicide resistance in annual ryegrass (*Lolium rigidum*): the search for a mechanism. In "Managing Resistance to Agrochemicals: from Fundamental Research to Practical Strategies" (M.B. Green, H.M. LeBaron and W.K. Moberg, eds), pp. 394-406. American Chemical Society, Washington, D.C.

Putwain, P.D., Scott, K.R., and Holliday, R.J. (1982). The nature of resistance of triazine herbicides: case histories of phenology and population studies. In "Herbicide Resistance in Plants" (H.M. LeBaron and J. Gressel, eds) pp. 99-116. Wiley, New York.

Salhoff, C.R. and Martin, A.R. (1986). *Kochia scoparia* growth response to triazine herbicides. Weed Science **34**, 40-42.

Shaaltiel, Y., Glazer, A., Bocion, P.F., and Gressel, J. (1988). Cross tolerance to herbicidal and environmental oxidants of plant biotypes, tolerant to paraquat, sulfur dioxide and ozone. Pesticide Biochemistry and Physiology **31**, 13-23.

Shaner, D.L. (1984). Mode of action of difenzoquat. In "Wild Oats Symposium Proceedings" (A.E. Smith and A.I. Hsaio, eds) Vol. 2, pp. 49-57. Agriculture Canada Research Station, Regina

Smith, A.E., Hume, L., Biederbeck, V.O., and Lafond, G.P. (1989). Effects of long term 2,4-D applications on crop production, weed populations, soil biochemical activities and the rate of microbial degradation. 198th American Chemical Society Meetings, Miami Beach, Abstract No. 109.

Souza Machado, V. and Hume, D.J. (1987). Breeding herbicide-tolerant cultivars - A Canadian experience. British Crop Protection Conference - Weeds 473-477.

Stannard, M.E. and Fay, P.K. (1987). Selection of alfalfa seedlings for tolerance to chlorsulfuron. Weed Science Society of America Abstracts, St. Louis, p. 61.

Thiel, A. and Boger, P. (1984). Comparative herbicide binding by photosynthetic membranes from resistant mutants. Pesticide Biochemistry and Physiology **22**, 232-242.

Thirunarayanan, K., Zimdahl, R.L., and Smika, D.E. (1985). Chlorsulfuron adsorption and degradation in soil. Weed Science **33**, 558-563.

Vaughn, K.C., Marks, M.D. and Weeks, D.P. (1987). A dinitroaniline resistant mutant of *Eleusine indica* exhibits cross-resistance and supersensitivity to anti microtubule herbicides and drugs. Plant Physiology **83**, 956-964.

Warwick, S.I. and Black, L. (1981). The relative competitiveness of atrazine susceptible and resistant populations of *Chenopodium album* and *C. strictum*. Canadian Journal of Botany **59**, 689-693.

Watson, D., Mortimer, A.M., and Putwain, P.D. (1987). The seed bank dynamics of triazine resistant and susceptible biotypes of *Senecio vulgaris* implications for control strategies. British Crop Protection Conference - Weeds 917-924.

Williams, K.L. (1976). Mutation frequency at a recessive locus in haploid and diploid strains of a slime mould. Nature **260**, 785-786.

Yaacoby, T., Schonfeld, M., and Rubin, B. (1986). Characteristics of atrazine-resistant biotypes of three grass weeds. Weed Science **34**, 181-184.

Zimdahl, R.I., Freed, V.H., Montgomery, M.L., and Furtick, W.R. (1970). The degradation of triazine and uracil herbicides in soil. Weed Research **10**, 18-26.

DISTRIBUTION AND SERIOUSNESS OF HERBICIDE-RESISTANT WEED INFESTATIONS WORLDWIDE

Homer M. LeBaron

New Technology and Basic Research, Ciba-Geigy Corporation, Greensboro, NC 27419, USA

One hundred and seven herbicide-resistant weed biotypes have evolved in various locations around the world. This includes 57 species (40 broad-leaved weeds and 17 grasses) with biotypes resistant to triazine herbicides and 50 species (33 broad-leaved and 17 grasses) with biotypes resistant to 14 other classes of herbicides. However, because quite a number of these species have evolved biotypes that are resistant to more than one class of herbicides, the total number of species with resistance to one or more herbicides is 81 (56 broad-leaved and 25 grasses). Six weed species have evolved resistance to the ALS/AHAS inhibitors (e.g. sulfonylureas), mostly in Western US and Canada, and two other countries. There are now resistant weeds present in all but 10 of the 50 States of the USA, all but 2 provinces of Canada, 18 countries of Europe, and 10 other countries. Herbicide-resistant weed infestations are also increasing in seriousness in many areas. Of special concern is the occurrence of multiple- or cross-resistance to many herbicides within the same biotype, apparently by metabolic detoxification. It is essential that we learn quickly how to avoid or manage herbicide-resistant weeds. If we fail to respond with research and difficult decisions now, resistant weeds may have a greater economic impact on agricultural production and efficiency within 5 to 10 years than resistant insects and plant pathogens.

INTRODUCTION

The distribution and number of herbicide-resistant weeds have increased rapidly in recent years. Since LeBaron and Gressel (1982) published the first major review on the subject less than eight years ago, the numbers have more than tripled, and the area of land infested has probably increased more than 10 times. Other more recent reviews on the subject have been published by LeBaron (1989), LeBaron and McFarland (1990), Holt and LeBaron (1990) and van Oorschot (these Proceedings).

RESISTANCE TO TRIAZINE HERBICIDES

It has been 20 years since the first report of a triazine-resistant weed (*Senecio vulgaris*) was made in western Washington (Ryan, 1970). Within the first decade, about 30 resistant weed biotypes had evolved, mostly in North America (Bandeen *et al.*, 1982) and western Europe (Gressel *et al.*, 1982). The subsequent occurrences and distribution of triazine-resistant biotypes have continued steadily since that time (LeBaron, 1989; Holt and LeBaron, 1990; LeBaron and McFarland, 1990; van Oorschot, these Proceedings). According to present records and knowledge, at least 57 species of weeds (40 broad-leaved and 17 grasses) have evolved resistance to triazine herbicides somewhere in the world.

The land area covered by these resistant weeds has also continued to increase steadily, especially in recent years. An accurate number of hectares infested with triazine-resistant weeds is impossible to obtain because of the variability and difficulties of measuring and collecting these data. However, a very recent survey of each State in the US reveals that nearly one million hectares have some level of infestation of these weeds (see Table 1). This includes all crops and non-crop areas where these herbicides are used. Biotypes of *Chenopodium album*, *Amaranthus hybridus*, *A. retroflexus*, and *Kochia scoparia* are by far the most prevalent and serious triazine-resistant weeds in US.

In France, it is estimated that the area infested with triazine-resistant weeds has increased from about 200,000 hectares in 1981 to well over 1,000,000 hectares, mostly comprised of *C. album* (500,000), *Solanum nigrum* 400,000) and *A. retroflexus* (200,000). Nearly all French vineyards are now infested with resistant *S. vulgaris* or *Erigeron canadensis*. Switzerland reports most of their vineyards infested with resistant *A. retroflexus*, *S. vulgaris* and *A. lividus*; with *S. vulgaris* or other weeds through most of the orchards; and *C. album* or *A. retroflexus* invading about 50% of their corn. In the UK, resistant *S. vulgaris* or *Poa annua* are reported in virtually all fruit orchards and nursery stock. West Germany has at least six triazine-resistant weeds in corn and other crops, with infested areas of well over 100,000 hectares. Reports from most other areas are either incomplete or show relatively limited areas infested.

While it is obvious that the triazine-resistant weeds in North America and elsewhere are generally becoming more serious, this is not universally true. A closer analysis of the data from the US reveals some important observations (see Table 1). Of the 33 States where such weeds have occurred, 2 (confirmed to have resistant weeds in the mid-70s) report that no triazine-resistant weeds can now be found. In other States, some resistant biotypes previously reported cannot now be found. In 14 States, the contact person said that the triazine-resistant biotypes were growing or increasing in area within the State, while 16 said they were not expanding in area, and 3 said there was a slow increase. These observations were confirmed by the answers to a further question. Contacts in 9 States said triazine-resistant weeds were a serious weed problem that required special effort and extra work, while 18 said they were not serious, in that they were easily controlled with other herbicides and did not require special effort or expense. Six stated that they were a minor to moderate problem.

Table 1. Survey of the distribution of atrazine-resistant weeds in the US (June, 1989).

State	Species[1]	Confirmed	Hectares	Growing	Serious	Contact
California	2b 1g	1976	1,600	No	No	J.Holt, H.Agamalian
Colorado	3b	1977	40,000	Yes	No	P.Westra
Connecticut	3b	1980	4,000	Slowly	Moderate	J.Ahrens
Delaware	1b	1977	12,000	Yes	Yes	F.J.Webb
Hawai	2g	1988	40	Yes	Moderate	L.Santo
Idaho	1b	1976	200	No	No	R.Callihan
Illinois	2b	1982	12	No	No	E.Knake
Indiana	2b	1983	20,000	No	No	T.Bauman
Iowa	3b	1980	40	No	No	M. Owen
Kansas	1b 1g	1977	800	Yes	No	D.Marishita
Kentucky	1b	1985	4,000	Yes	No	M.Barrett
Maine	1b	1984	40	No	Yes	M.McCormick
Maryland	3b 3g	1972	80,000	Yes	Yes	R.Ritter
Massachusetts	2b	1978	20	No	No	P.C.Bhomick
Michigan	3b 1g	1975	40,000	Yes	Moderate	J.Kells
Minnesota	1b	1982	80	No	No	C.Kern
Montana	1b 1g	1977	0	No	No	P.Fay
Nebraska	2b 2g	1976	8,000	Slowly	Moderate	A.Martin
New Hampshire	1b	1984	800	Yes	Yes	J.Mitchell
New Jersey	1b	1985	40	No	No	J.A.Meade
New York	3b	1977	4,400	Yes	Yes	R.Hahn
N. Carolina	2b	1985	800	Yes	Yes	D.Worsham
Ohio	3b	1981	20,000	Yes	Minor	M.Loux
Oregon	3b 1g	1970	32,000	No	Moderate	A.Appleby
Pennsylvania	6b 1g	1978	80,000	Yes	Yes	N.L.Hartwig
Rhode Island	1b	1983	20	No	No	R.Wakefield
South Dakota	1b	1986	32	No	No	L.J.Wrage
Utah	1b	1976	0	No	No	S.Dewey
Virginia	2b	1976	100,000	Yes	Yes	S.Hagood
Washington	4b 1g	1968	16,000	No	Yes	W.Anliker C.Bucholtz
W. Virginia	2b 1g	1980	12,000	Slowly	No	C.B.Sperow
Wisconsin	3b	1978	400,000	Yes	Yes	R.E.Doersch
Wyoming	1b	1978	8	No	No	S.D.Miller

Total = 33 States reporting a total of 877,000 hectares (approximations).

[1] = number of species with resistant biotypes; b = broad-leaved species; g = grass species.

There is virtually no relationship between the length of time since the triazine-resistant weed was first reported in a State and the present area of infestation within the State. It appears that other factors such as tillage systems, climatic conditions, cropping practices and rotations, weed species and other herbicides used are more important in determining the extent of spread than the length of time since the first appearance of resistant weeds in the State.

In the US survey, every contact was asked if farmers had stopped using atrazine because of resistant weeds or for other reasons. Virtually all users are still applying atrazine because they consider there are no adequate or economical alternatives. There are always many other susceptible weeds that are controlled, and farmers handle the resistant biotypes as they do other weeds which are not controlled with atrazine (e.g. *Digitaria* spp., *Panicum dichotomiflorum*, *Sorghum halepense*, and *Sorghum bicolor*). They select a combination partner (e.g. metolachlor, alachlor, butylate, pendimethalin) or subsequent post-emergence herbicide (e.g. 2,4-D, dicamba, bromoxynil, MCPA) to control other weeds not controlled with atrazine (Parochetti *et al.*, 1982; Ritter, 1989; Stephenson *et al.*, 1990).

Extensive research has shown that most, if not all, of the weed biotypes which have evolved resistance to atrazine have been inferior in vigour, competitiveness and fitness compared to the wild-type or susceptible weeds of the same species (Jansen *et al.*, 1986; Stowe and Holt, 1988; Gressel, these Proceedings). This is apparently due to a less efficient photosynthetic system in the resistant weeds. The lack of fitness in most triazine-resistant weeds is a very important reason why they have been fairly easily controlled, and why there have been few problems of cross-resistance or multiple-resistance where both triazine and other types of herbicides have been used repeatedly together (Gressel and Segel, 1990).

Also, within the US, the close co-operation and communication between industry, State and university research, extension services, and farmers have been very important in avoiding, delaying and controlling atrazine-resistant weeds. With the first invasion of resistant weeds, prompt action is essential in order to avoid serious and more permanent problems. Preventive action to avoid herbicide-resistant weeds from developing in the first place is definitely the best strategy. It is virtually essential in all cases of herbicide resistance to have other classes or types of herbicides, with alternative sites and modes of action available.

Even in North America, triazine-resistant weeds have become a serious economic problem in a few areas. The corn growing areas of Wisconsin, Maryland, Pennsylvania, Virginia and Ontario (Canada) are of significant concern. The most serious problem is the control of *C. album* and *Amaranthus* spp. in continuous corn, under conservation tillage where cultivation is difficult or impossible, and where post-emergence herbicides are not used properly.

Triazine-resistant weeds have often required the farmers to increase their costs for weed control. This increase may range from very little, if they can obtain good control of all weeds with a herbicide combination applied at the same time, to $25 per hectare or more for separate treatments.

The current distribution of triazine-resistant broad-leaved weeds, from my surveys and records, is presented in Table 2. The distribution of triazine-resistant biotypes of grass species is presented in Table 3.

Table 2.   Distribution of triazine-resistant broad-leaved weeds (September, 1989).

	Species	Year found	Location
1.	*Abutilon theophrasti*	1984	Maryland
2.	*Amaranthus albus*	1988	Spain
3.	*Amaranthus arenicola*	1977	Colorado
4.	*Amaranthus blitoides*	1983	Israel
		1988	Spain
5.	*Amaranthus bouchonii*	1981	France
		1983	Switzerland
		1984	Hungary
		1984	Italy
6.	*Amaranthus cruentus*	1978	Italy
		1989	Spain
7.	*Amaranthus hybridus*	1972	Maryland
		1976	Virginia
		1977	Delaware
		1978	Massachusetts
		1978	New York
		1978	Pennsylvania
		1979	France
		1979	Switzerland
		1980	Connecticut
		1981	Hungary
		1981	Michigan
		1981	West Virginia
		1982	Illinois
		1984	Italy
		1984	Maine
		1985	Colorado
		1985	Kentucky
		1985	New Jersey
		1985	Wisconsin
		1986	Nebraska
		1987	Israel
		1988	Spain
8.	*Amaranthus lividus*	1978	Switzerland
		1981	France
		1988	Spain
9.	*Amaranthus powellii*	1968	Washington
		1970	Oregon
		1975	Ontario
		1982	Quebec
10.	*Amaranthus retroflexus*	1973	Austria
		1977	Hungary
		1978	Switzerland
		1979	France
		1979	Ontario
		1980	Connecticut

Table 2 (cont'd).

Species	Year found	Location
	1981	Ohio
	1982	New York
	1982	West Germany
	1983	Indiana
	1985	Bulgaria
	1986	North Carolina
	1986	South Dakota
	1987	British Columbia
	1987	Romania
	1987	Yugoslavia
	1988	Czechoslovakia
	1988	Spain
11. *Ambrosia artemisiifolia*	1975	Ontario
	1984	Pennsylvania
	1987	Yugoslavia
12. *Arenaria serpyllifolia*	1981	France
13. *Atriplex patula* (a)	1985	Switzerland
	1986	Austria
	1988	Poland
	1988	West Germany
14. *Bidens tripartita*	1976	Austria
15. *Brassica campestris* (or *B. rapa*)	1977	Quebec
16. *Chenopodium album*	1970	Ontario
	1973	Austria
	1973	Washington
	1975	Switzerland
	1977	France
	1977	New York
	1978	Nova Scotia
	1978	West Germany
	1978	Wisconsin
	1979	Belgium
	1979	Virginia
	1980	Michigan
	1980	Netherlands
	1980	Quebec
	1981	New Zealand
	1981	Ohio
	1982	Hungary
	1982	Maryland
	1982	Minnesota
	1983	Connecticut
	1983	Massachusetts
	1983	Pennsylvania
	1983	Rhode Island
	1983	West Virginia
	1984	Italy
	1984	Maine
	1984	New Hampshire

Table 2 (cont'd).

Species	Year found	Location
	1984	United Kingdom
	1985	Bulgaria
	1985	Illinois
	1985	Indiana
	1985	North Carolina
	1986	Iowa
	1987	East Germany
	1987	Spain
	1987	Yugoslavia
	1988	Czechoslovakia
	1988	Poland
17. *Chenopodium ficifolium*	1978	West Germany
	1983	Switzerland
	1985	Netherlands
18. *Chenopodium missouriense*	1978	Pennsylvania
19 *Chenopodium polyspermum*	1979	France
	1979	Switzerland
	1983	West Germany
	1987	Romania
	1988	Hungary
20. *Chenopodium strictum*	1978	Ontario
21. *Epilobium ciliatum*	1982	United Kingdom
(or *E. adenocaulon*)	1986	Belgium
	1986	Netherlands
	1987	Austria
	1988	Denmark
22. *Epilobium tetragonum*	1981	France
(or *E. adnatum*)	1987	West Germany
23. *Erigeron* (or *Conyza*)	1987	Spain
bonariensis		
24. *Erigeron* (or *Conyza*)	1978	Switzerland
canadensis	1981	France
	1983	Poland
	1984	Hungary
	1984	United Kingdom
	1987	Romania
	1987	Yugoslavia
	1988	Czechoslovakia
	1988	Denmark
25. *Galinsoga ciliata*	1980	West Germany
	1987	Netherlands
26. *Kochia scoparia*	1976	Idaho
	1976	Nebraska
	1976	Utah
	1977	Colorado
	1977	Kansas
	1977	Oregon
	1978	Wyoming
	1979	Montana

33

Table 2 (cont'd).

Species	Year found	Location
	1980	Iowa
	1980	Washington
	1984	California
	1985	Wisconsin
27. *Matricaria matricarioides* (or *Chamomilla suaveolens*)	1984	United Kingdom
28. *Myosoton aquaticum* (or *Stellaria aquatica*)	1983	West Germany
29. *Physalis longifolia* (a)	1984	Pennsylvania
30. *Polygonum aviculare*	1987	Netherlands
	1988	Poland
31. *Polygonum* (or *Fallopia*) *convolvulus*	1977	Austria
	1984	Pennsylvania
	1987	West Germany
	1988	Netherlands
32. *Polygonum lapathifolium*	1978	France
	1986	West Germany
	1988	Czechoslovakia
	1988	Netherlands
33. *Polygonum pensylvanicum*	1988	Iowa
34. *Polygonum persicaria*	1979	France
	1986	Netherlands
35. *Senecio vulgaris*	1968	Washington
	1973	Oregon
	1977	California
	1978	British Columbia
	1980	West Germany
	1981	France
	1981	Switzerland
	1981	United Kingdom
	1982	Netherlands
	1985	Bulgaria
	1986	Quebec
	1987	Belgium
	1988	Czechoslovakia
	1988	Denmark
	1989	Ontario
36. *Sicyos angulatus* (a)	1985	Ohio
37. *Sinapis arvensis*	1983	Ontario
38. *Solanum nigrum*	1978	France
	1978	Italy
	1978	West Germany
	1982	Netherlands
	1983	Switzerland
	1984	Michigan
	1984	United Kingdom
	1986	Austria
	1987	Belgium
	1987	Romania
	1988	Spain

Table 2 (cont'd).

	Species	Year found	Location
39.	*Sonchus asper*	1981	France
40.	*Stellaria media*	1974	West Germany
		1986	Austria

(a) These biotypes apparently do not have resistant chloroplasts, but are probably resistant due to metabolic detoxification or other mechanisms.

Table 3. Distribution of triazine-resistant grass weeds (Sept. 1989)

	Species	Year found	Location
1.	*Alopecurus myosuroides*	1982	Israel
2.	*Brachypodium distachyon*	1976	Israel
3.	*Bromus tectorum*	1977	Kansas
		1977	Montana
		1977	Nebraska
		1978	Oregon
		1978	Washington
		1981	France
4.	*Chloris barbata*	1988	Hawaii
5.	*Chloris radiata*	1988	Hawaii
6.	*Digitaria sanguinalis*	1983	France
		1985	Bulgaria
		1988	Poland
7.	*Echinochloa crus-galli* (a)	1978	Maryland
		1980	West Virginia
		1981	France
		1981	Ontario
		1982	Austria
		1984	Poland
		1985	Bulgaria
8.	*Lolium rigidum*	1982	Israel
		1988	Australia
9.	*Lophochloa phleoides*	1982	Israel
10.	*Panicum capillare*	1975	Michigan
		1980	Ontario
11.	*Phalaris paradoxa*	1982	Israel
12.	*Poa annua*	1976	California
		1977	France
		1978	Belgium
		1981	Netherlands
		1982	United Kingdom
		1983	Switzerland
		1983	West Germany

Table 3 (cont'd).

Species	Year found	Location
	1984	Japan
	1986	Austria
	1988	Czechoslovakia
13. *Polypogon monspeliensis*	1982	Israel
14. *Setaria faberi*	1984	Maryland
15. *Setaria glauca*	1981	France
	1981	Ontario
	1984	Maryland
	1984	Pennsylvania
	1987	Spain
16. *Setaria viridis*	1980	France
17. *Setaria viridis* var. *major*	1981	France

(a) Most of these and possible other resistant grass biotypes listed do not have resistant chloroplasts, but are probably resistant due to metabolic detoxification or other mechanisms.

RESISTANCE TO OTHER CLASSES OF HERBICIDES

Although reports of resistance or shifts towards more insensitive biotypes of some species to a few other classes of herbicides predated the triazine-resistant weeds (Bandeen *et al.*, 1982), it has only been within the last 10 years that most such resistant biotypes have evolved. There are now at least 50 weed biotypes which have been reported to be resistant to 14 other classes of herbicides somewhere in the world (see Table 4).

Table 4.   Occurrence and distribution of weed biotypes resistant to various other herbicides (non-triazines) (September, 1989).

Herbicide or Class	Species	Year found	Location
1. ALS/AHAS Inhibitors			
Sulfonylureas, Imidazolinones, & Triazolo-pyrimidine	*Ixophorus unisetus*	1988	Costa Rica
	Kochia scoparia	1987	Kansas
		1987	North Dakota
		1988	Colorado
		1988	Montana
		1988	New Mexico
		1988	South Dakota
		1988	Texas
		1988	Washington
		1989	Saskatchewan
	Lactuca serriola	1987	Idaho
	Lolium rigidum[d]	1986	Australia

36

Table 4 (cont'd).

Herbicide or Class	Species	Year found	Location
	Salsola iberica	1988	Kansas
		1988	North Dakota
		1989	California
		1989	Montana
		1989	Washington
	Stellaria media	1988	Alberta
2. Amides			
Propanil	*Echinochloa colonum*	1988	Colombia
	Echinochloa crus-galli	1986	Greece
3. Arsenicals			
MSMA & DSMA	*Xanthium strumarium*	1984	South Carolina
		1987	North Carolina
4. Carbamates	*Amaranthus hybridus*[a][b]	1988	Hungary
	Amaranthus retroflexus[a][b]	1988	Hungary
5. Dinitroanilines			
Trifluralin	*Amaranthus palmeri*	1989	South Carolina
	Eleusine indica	1973	South Carolina
		1982	Tennessee
		1987	Alabama
		1987	North Carolina
	Setaria viridis	1988	Manitoba
6. Bipyridiliums			
Paraquat and Diquat	*Arctotheca calendula*	1986	Australia
	Epilobium ciliatum	1983	United Kingdom
		1984	Belgium
	Erigeron bonariensis	1979	Egypt
		1984	Hungary
		1989	Kenya
	Erigeron canadensis	1980	Japan
		1988	Hungary
	Erigeron philadelphicus	1980	Japan
	Erigeron sumatrensis	1986	Japan
	Hordeum glaucum	1983	Australia
	Hordeum leporinum	1988	Australia
	Lolium perenne	1976	United Kingdom
	Parthenium hysterophorus	1989	Kenya
	Poa annua	1978	United Kingdom
	Solanum americanum	1988	Florida
	Youngia japonica	1986	Japan

Table 4 (cont'd).

Herbicide or Class	Species	Year found	Location
7. Nitriles			
Bromoxynil	*Chenopodium album*[a]	1988	West Germany
8. Phenoxyalkanoics			
2,4-D and MCPA	*Carduus nutans*	1989	New Zealand
	Cirsium arvense	1985	Hungary
	Daucus carota	1962	Ontario
	Ranunculus acris	1989	New Zealand
	Sphenoclea zeylanica	1982	Philippines
Mecoprop	*Stellaria media*	1985	United Kingdom
9. Picloram	*Centaurea soltstitialis*	1988	Idaho
10. Polycyclic Alkanoic Acids			
Diclofop-methyl	*Alopecurus myosuroides*	1988	United Kingdom
	Avena fatua	1985	South Africa
		1987	Australia
	Lolium multiflorum[c]	1987	Oregon
	Lolium rigidum[d]	1982	Australi
11. Chloridazon	*Chenopodium album*	1978	France[a]
		1984	East Germany
		1984	Hungary
12. Substituted Ureas			
Chlorbromuron, Chlorotoluron & Isoproturon	*Alopecurus myosuroides*[b]	1983	West Germany
		1984	United Kingdom
	Amaranthus hybridus[a b]	1988	Hungary
	Chenopodium album[a]	1988	Hungary
Diuron and Linuron	*Amaranthus bouchonii*[a]	1988	Hungary
	Amaranthus retroflexus[a b]	1988	Hungary
	Erigeron canadensis[a]	1988	Hungary
13. Triazoles			
Amitrole	*Lolium rigidum*	1988	Australia
	Poa annua	1986	Belgium
14. Uracils			
Bromacil	*Amaranthus hybridus*[b]	1988	Hungary
	Amaranthus retroflexus[a b]	1988	Hungary

A few of these have evolved resistance to some herbicide classes by cross-resistance, in which the weed first evolved resistance after frequent use of one herbicide, and it was then found to be resistant to one or more other types of herbicides that had not previously been used on the weed (Heap and Knight, 1986; Gressel, 1988; Solymosi and Lehoczki, 1989; Powles and Howat, 1990; Heap, these Proceedings; Moss and Cussans, these Proceedings; Dr Peter Solymosi, Hungary, personal communication).

When we consider all herbicide-resistant weeds one or more of these biotypes are present in 40 of the 50 US States, 8 provinces of Canada, 18 countries of Europe, and 10 other countries.

The list of 14 other classes of herbicides to which weed biotypes have evolved resistance includes many of our most important classes, including some multiple site of action herbicides and those predicted to be low risk for resistance (LeBaron and McFarland, 1990). Most of these, however, do not appear to be of serious potential or difficult to control. Almost without exception, the only serious resistant weeds are those for which few or no other effective or selective herbicides or other control measures are readily available.

The biotypes listed in Table 5 which are of greatest concern include those resistant to ALS/AHAS inhibitors (Mallory-Smith *et al.*, 1990; Thill *et al.*, these Proceedings), dinitroanilines (Gossett and Murdock, 1990; Morrison and Beckie, these Proceedings), polycyclic alkanoic acids (Heap, these Proceedings), and substituted ureas (Primiani *et al.*, 1990; Moss and Cussans, these Proceedings). Of utmost concern and potential danger to weed control technology are those biotypes possessing cross- and multiple-resistance to several classes of herbicides. The most noted examples are *Lolium rigidum* (annual ryegrass) in Australia (Heap and Knight, 1986; Powles and Howat, 1990; Heap, these Proceedings) and *Alopecurus myosuroides* (blackgrass) in the United Kingdom (Moss and Cussans, these Proceedings). It is a source of concern that multiple- and cross-resistance to herbicides can occur in weeds, apparently by mechanisms (metabolic detoxification, e.g. mixed function oxidases) similar to those by which insects rapidly evolve resistance to insecticides (Gressel, 1988). Such efficient oxidation of foreign organic chemicals may prevent almost any herbicide from reaching the target site intact. We must take much more aggressive and extreme precautions to prevent this type of resistance from occurring and to quarantine or eradicate such resistant weeds if and when they do appear. We need much more concentrated research efforts focused on herbicide resistance mechanisms, their genetics and biochemistry, especially on resistance to ALS/AHAS

Table 4 (cont'd).
[a] This biotype has developed cross-resistance from an atrazine-resistant biotype.
[b] This biotype usually shows varying degrees of cross-resistance to several classes of herbicides, including diclofop-methyl, pendimethalin, triazines and sulfonylureas.
[c] This biotype is cross-resistant to other polycyclic alkanoic acid herbicides.
[d] This biotype evolved resistance to diclofop-methyl after several years of use, and was then found to often be cross-resistant to most sulfonylurea herbicides.

inhibitors, other "low-rate" herbicides, and multiple- and cross-resistant weeds.

THE CHALLENGE AND URGENCY OF RESISTANT WEED MANAGEMENT

There are no clear relationships between families, genera or taxa in their tendency to evolve resistance. Predictability of gene frequencies is still very low. However, we can have a good degree of confidence that many weeds have sufficient genetic flexibility and biochemical adaptability to evolve resistance to most, if not all, single target site herbicides if the selection pressure is high enough for a long period of time.

We weed scientists must learn quickly what entomologists are still learning after 80 years and plant pathologists after 50 years of repeated experiences. Pests are here to stay. We will never find a potentially serious pest species on the endangered species list. It is a biological contradiction of terms. Their adaptability is one of the principal characteristics that make them pests. There is probably no chemical against which some pest cannot evolve resistance.

As weed scientists, we also face a major difference or requirement in contrast to applied entomologists and plant pathologists. While their pest control chemicals or tools are usually aimed at the control of one or few pests, we must provide the farmers or users of our technology effective means to control at least 6 to 10 weeds in almost all situations, and in some cases many more.

The evolution of weed biotypes resistant to sulfonylureas, diclofop-methyl and other herbicides, and especially biotypes having multiple- or cross-resistance to various classes of herbicides, have abruptly increased the level of concern, urgency and consequences of the phenomenon. This raised a special caution relative to the trend in the agricultural chemical industry towards new herbicides. They are mostly screening for and developing single target site herbicides. While this seems to be very desirable and successful from the perspective of aiming at very selective targets, reduced rates of application, and toxicological and environmental safety, such chemicals will be much more likely to evolve resistant weed biotypes.

It is obvious that we must apply all of the wisdom and understanding we have developed on pest resistance to pesticides, as well as exercising a greater degree of marketing control and self-restraint than has ever been demonstrated in our agriculture industry in order to protect, preserve, or prolong the use of our herbicides with a single site of action and high risk for resistance.

We should not forget or ignore that God put us all on notice in the beginning when he said to Adam: "Cursed is the ground for thy sake; -- thorns also and thistles shall it bring forth to thee; --in the sweat of thy face shalt thou eat bread;" (Genesis 3:17-19). Resistant weeds are another reminder that we will need to continue to sweat and must never become lazy or complacent. But remember that He did it for "our sake."

CONCLUSIONS

It is estimated that the land in the US now infested with triazine-resistant weeds approaches 1 million hectares. There are probably at least another 2 million hectares infested with triazine-resistant weeds in other countries, mostly in Canada and 17 countries of Europe. Weeds resistant to triazines have usually been managed or controlled with reasonable success. The evolution of weed biotypes resistant to sulfonylureas, diclofop-methyl, trifluralin, and other herbicides, and especially biotypes having multiple- or cross-resistance to various classes of herbicides, have abruptly increased the level of concern, urgency and consequences of the phenomenon. Not only are herbicide-resistant weeds appearing after fewer repeat annual applications of some of these newer herbicides, but there seem to be more species that have potential for resistance. In addition, the resistant biotypes are apparently equally fit and competitive once they evolve, unlike most biotypes resistant to triazine herbicides. It is doubtful whether we can be as successful in managing these resistant biotypes as we have been with triazine-resistant weeds in the past. The consequences of herbicide-resistant weeds are becoming more serious for everyone concerned. Farmers must learn to handle these resistant weeds as special problems which not only affect them by increasing the number and cost of herbicides, but which have special significance and importance to their neighbours and others.

Weed researchers and extension experts must become more sophisticated and careful about recommending combinations of herbicides that would minimize the opportunities for resistance to evolve, while still considering the lowest possible use rates and costs. Industry must consider resistant weed strategies even before developing or marketing the new herbicides with single target site modes of action. Regulatory agencies and government administrators must help to preserve alternative herbicides and other weed control options. Without acting promptly, agriculture, food production, and thereby, the world could be in serious jeopardy within 5 to 10 years.

ACKNOWLEDGEMENTS

The distribution data presented in this paper are mostly from surveys conducted by the author and supported by Ciba-Geigy Corporation.

REFERENCES

Bandeen, J.D., Stephenson, B.R. and Dowett, E.R. (1982). Discovery and distribution of herbicide-resistant weeds in North America. In "Herbicide Resistance in Plants" (H.M. LeBaron and J. Gressel, eds), pp. 9-30. John Wiley & Sons, Inc., New York.

Gossett, B.J. and Murdock, E.C. (1990). Palmer amaranth resistance to the dinitroaniline herbicides. Proceedings of the Southern Weed Science Society **43** (in press).

Gressel, J. (1988). Multiple resistance to wheat selective herbicides: new challenges to molecular biology. Oxford Survey of Plant Molecular Cell Biology 5, 195–203.

Gressel, J. and Segel, L.A. (1990). Modelling the effectiveness of herbicide rotations and mixtures as strategies to delay or preclude resistance. Weed Technology 4, 186–198.

Gressel, J., Ammon, H.U., Fogelfors, H., Gasquez, J., Kay, Q.O.N. and Kees, H. (1982). Discovery and distribution of herbicide-resistant weeds in North America. In "Herbicide Resistance in Plants" (H.M. LeBaron and J. Gressel, eds), pp. 31–55. John Wiley & Sons, Inc., New York.

Heap, I.M. and Knight, R. (1986). The occurrence of herbicide cross-resistance in a population of annual ryegrass, Lolium rigidum, resistant to diclofop-methyl. Australian Journal of Agricultural Research 37, 149–156.

Holt, J.S. and LeBaron, H.M. (1990). Significance and distribution of herbicide resistance. Weed Technology 4, 141–149.

Jansen, M.A.K., Hobe, J.H., Wesselius, J.C. and van Rensen, J.J.S. (1986). Comparison of photosynthetic activity and growth performance in triazine-resistant and susceptible biotypes of Chenopodium album. Physiologie Vegetale 24, 475–484.

LeBaron, H.M. (1989). Management of herbicides to avoid, delay and control resistant weeds: a concept whose time has come. Proceedings of the Western Society of Weed Science 42, 6–16.

LeBaron, H.M. and Gressel, J. (1982). "Herbicide Resistance in Plants". John Wiley & Sons Inc., New York.

LeBaron, H.M. and McFarland, J. (1990). Herbicide resistance in weeds and crops: an overview and prognosis. In "Managing Resistance to Agrochemicals: From Fundamental Research to Practical Strategies" (M.B. Green, H.M. LeBaron and W.K. Moberg, eds), pp. 336–352. American Chemical Society, Washington DC.

Mallory-Smith, C.A., Thill, D.C. and Dial, M.J. (1990). Identification of sulfonylurea herbicide-resistant prickly lettuce (Lactuca serriola) Weed Technology 4, 163–168.

Parochetti, J.V., Schnappinger, M.G., Ryan, G.F. and Collins, H.A. (1982). Practical significance and means of control of herbicide-resistant weeds. In "Herbicide Resistance in Plants" (H.M. LeBaron and J. Gressel, eds), pp. 309–323. John Wiley & Sons Inc., New York.

Powles, S.B. and Howat, P.D. (1990). Herbicide-resistant weeds in Australia. Weed Technology 4, 178–185.

Primiani, M.M., Cotterman, J.C. and Saari, L.L. (1990). Resistance of Kochia scoparia to sulfonylurea and imidazolinone herbicides. Weed Technology 4, 169–172.

Ritter, R.L. (1989). "Understanding herbicide resistance in weeds". Sandoz Crop Protection Corporation, Des Plaines, IL 60018.

Ryan, G.F. (1970). Resistance of common groundsel to simazine and atrazine. Weed Science **18**, 614–616.

Solymosi, P. and Lehoczki, E. (1989). Characterization of a triple (atrazine pyrazone-pyridate) resistant biotype of common lambsquarters (*Chenopodium album* L.). Plant Physiology **134**, 685–690.

Stephenson, G.R., Dykstra, M.D., McLaren, R.D. and Hamill, A.S. (1990). Agronomic practices influencing triazine-resistant weed distribution in Ontario. Weed Technology **4**, 199–207.

Stowe, A.E. and Holt, J.S. (1988). Comparison of triazine resistant and susceptible biotypes of *Senecio vulgaris* and their F_1 hybrids. Plant Physiology **87**, 183–189.

THE DEVELOPMENT OF HERBICIDE-RESISTANT POPULATIONS OF *ALOPECURUS MYOSUROIDES* (BLACK-GRASS) IN ENGLAND

R. Moss and G.W. Cussans

Department of Agricultural Sciences, University of Bristol,
AFRC Institute of Arable Crops Research, Long Ashton Research Station,
Long Ashton, Bristol, BS18 9AF, U.K.

Chlorotoluron-resistant *Alopecurus myosuroides* was first detected in England in 1982 and has now been found in 33 fields on 20 farms. There is no evidence that resistance has spread from a common source. The degree of resistance varies considerably between populations, the most resistant populations being found at Peldon, Essex. Most of the resistant populations are in fields where intensive winter cereal cropping and non-ploughing cultivation techniques have been practised. Herbicides to control *A. myosuroides* have been used frequently in these fields, on average 1.6 applications per year. Most chlorotoluron-resistant populations show cross-resistance to many other herbicides. Degree of resistance varies between herbicides, but is not related directly to chemical grouping or herbicide mode of action. The development of resistance in one field has been monitored and showed a gradual increase in resistance between 1976 and 1985. Tests on seed samples collected in 1978 and 1988 from two other fields, also showed that resistance had developed during this 10-year period. The genetic mechanism involved in herbicide resistance is not known, but it has been established that nuclear inheritance is involved. Integrated control strategies are suggested for the control of resistant *A. myosuroides*.

INTRODUCTION

Alopecurus myosuroides Huds. (black-grass) is an annual grass weed propagated solely by seeds. These are relatively non-dormant and most germination occurs in the autumn, from September to November. Consequently, *A. myosuroides* is mainly associated with autumn-sown crops, especially cereals (Moss, 1980). In the UK, cereals account for about 75% of the 5.2 million ha of arable cropped land, and there has been a trend towards growing more autumn-sown, and less spring-sown cereals. Consequently, *A. myosuroides* has become one of the most important weeds of cereals in England, although it is virtually absent from Wales, Scotland and Northern Ireland (Elliott *et al.*, 1979).

45

The herbicides, chlorotoluron and isoproturon, were introduced into t
UK in the early 1970s, mainly for the control of this weed. These t
herbicides were used on about 50% of the winter cereal area in Engla
and Wales in 1982 (Sly, 1984). Many fields have had successive, annu
applications for many years. Currently, isoproturon is the most wide
used herbicide for control of this weed in England.

Alopecurus myosuroides, as a weed of cereals, has many of the characteristi
listed by Harper (1956) which would favour the development of herbici
resistance: high reproductive capacity, absence of a large, dormant se
bank to buffer population changes, association with cereal monocultu
and intensive use of a single herbicide type.

OCCURRENCE OF RESISTANCE

In 1976, we collected seed samples from five fields where control of
myosuroides by chlorotoluron had been poor, in order to determine wheth
this was due to resistance. In pot experiments, no differences
response to chlorotoluron were detected between these populations, a
those from sites where control was normal. A further 19 samples we
collected in 1982/83 for testing for resistance to chlorotoluron. On
one of these showed any evidence of resistance. This was from Faringdo
near Oxford, a population which had shown no evidence of resistance
the tests using seeds collected in 1976. The degree of resistance w
modest, with the ratio of ED_{80} values (the dose required to redu
foliage fresh weight by 80% in pot tests) between the Faringdon and
standard susceptible stock averaging 2.6 (Moss and Cussans, 1985
Populations with a similar degree of resistance to chlorotoluron a
methabenzthiazuron had also been detected in Germany (Niemann a
Pestemer, 1984).

In 1984, a sample collected by the Agricultural Development and Adviso
Service (ADAS) from a farm at Peldon, Essex, where poor herbici
activity had been reported, was tested and showed a much greater degr
of resistance. In comparison with a susceptible stock, Peldon plan
required about 16 times the concentration of chlorotoluron to be damag
to the same degree (Moss and Cussans, 1985).

By 1988, seed samples from 296 fields had been tested for resistance
chlorotoluron by ADAS and Long Ashton Research Station (Moss, 1987; Mo
and Cussans, 1987; Orson and Livingston, 1987; Moss and Orson, 198
Clarke and Moss, 1989). Resistance was detected in 33 fields on 20 far
in England. These farms were located in seven counties: Buckinghamshi
(1 farm); Cambridgeshire (2); Essex (9); Lincolnshire (1); Oxfordshi
(3); Suffolk (3); Warwickshire (1) (Fig. 1). Resistance was not absolu
and varied considerably between populations. The most severe cases
resistance still occur on two farms at Peldon, Essex, but oth
populations in South Essex, Suffolk and Buckinghamshire, with
comparable degree of resistance,have been detected during the last t
years.

Most of the seed samples have not been collected at random but fr
fields where poor activity of herbicides has been reported. However,
1988 ADAS collected seeds from a randomised stratified sample of
fields, mainly within 80 km of Peldon (Clarke and Moss, 1989). Resista
A. myosuroides was found on 4 farms (5% of samples). This proportion w
similar to the 8% of samples showing resistance from the 188 separa

arms sampled on a non-random basis between 1982 and 1988. These results indicate that resistance now occurs widely in England, but that, at present a relatively small proportion of farms is affected.   It is probable that resistance is developing elsewhere. Movement of seeds of resistant populations in seed grain cannot be discounted entirely, but the wide geographical distribution of resistance suggests that each case is an independent development.

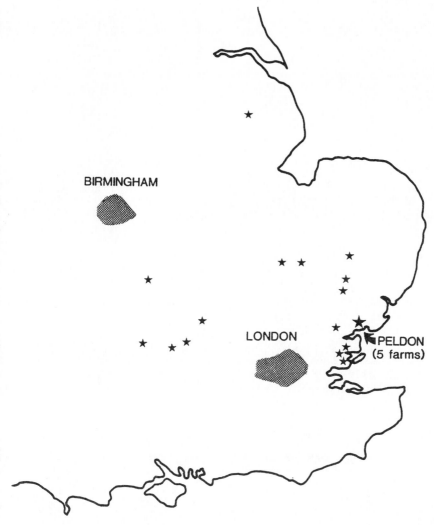

g. 1.   Location of farms with chlorotoluron-resistant *A. myosuroides* in gland (1988)

THOD OF TESTING FOR RESISTANCE

ere is a wide variation between populations in the level of resistance. is is coupled with the problem that tests in controlled environment oms are too expensive for routine screening work and, in glasshouse nditions, the absolute levels of herbicide activity vary considerably e to variations in temperature and lighting. It has, therefore, been cided to use a test for activity relative to standard populations.

The current testing procedure for resistance to chlorotoluron is based
sowing seeds in pots of sandy loam soil and spraying plants (six per po
at the two-leaf stage, with 2.5–3.5 kg ha^{-1} of chlorotoluron (Moss a
Orson, 1988). Foliage fresh weight is recorded as a measure of herbici
activity, 3–5 weeks after spraying. Three standard reference stocks a
included in all tests: Rothamsted 1987 (susceptible), Faringdon 19
(partially resistant), Peldon A1 1987 (resistant). Typically, the mean
reductions in fresh weight for the three standard stocks are: Rothamst
90%, Faringdon 70%, Peldon 20%.

The following arbitrary classification has been adopted to identi
different degrees of resistance, based on a comparison with t
percentage reduction in fresh weight values of the standard stocks:

Rothamsted value	Faringdon value					Peldon value
s | s	| * | ** | *** | **** | ****
susceptible	need further| partially ---increasingly resistant ---
|	| evaluation| resistant|

The mid value between Rothamsted and Faringdon is used to separate tho
stocks deemed to be susceptible and those requiring further evaluation
determine whether they exhibit a low level of resistance. Similarly, t
stocks in the Faringdon to Peldon range are separated into thr
categories. Populations are only termed 'resistant' (** or more) if th
show greater resistance than the Faringdon population, which h
consistently shown partial resistance to chlorotoluron in glasshous
controlled environment and simulated field conditions (Moss, 1987
Although results are analysed statistically, it is not thoug
appropriate to classify the resistance groupings using significance tes
(Clarke and Moss, 1989). In most tests, the resistance classes a
greater than the 95% confidence intervals.

Samples classed as * show small reductions in fresh weight, which a
usually statistically significantly lower (P ≤ 0.05) than the susceptib
standard stock (Rothamsted). These marginal levels of resistance mig
result in reduced field performance, especially under adverse conditio
for herbicide activity. Sites ranked * need monitoring to determi
whether resistance levels are increasing.

CULTIVATION, CROPPING AND HERBICIDE HISTORY OF FIELDS WITH RESISTA BIOTYPES

Most types of herbicide resistance are associated with the repeated u
of the same herbicide, often in situations of crop monoculture.
determine whether there was an association between resistance in *A. m
suroides* and cultural and herbicide history, information was obtained f
fields on 15 of the farms with resistant populations (Table 1).

All these fields were in a predominantly winter cereal rotation – wint
cereals were grown in 192 (91%) of the 211 crop years. There was a ve
low frequency of ploughing, in only 16 (8%) of the crop year
Non-ploughing tillage systems maximise the proportion of the we
population derived from seeds shed in the previous crop and, therefor
minimise the probability of back-crossing with earlier, unselect
generations. Herbicides to control *A. myosuroides*, mainly chlorotoluro
isoproturon and diclofop-methyl, were used frequently on all fields –

verage, 1.6 applications per year (range = 1.0-2.6). However, this is
ot an atypically high use of herbicides.

able 1. Cultivation, cropping and herbicide history for 15 fields with
erbicide-resistant _A. myosuroides_.

arm ode	Harvest years	Years ploughed	Years in winter cereal	No. of herbicide applications[1]		
				Chlorotoluron or isoproturon	Diclofop methyl	Other[2]
eldon A	1975-87	0	13	18	4	6
eldon B	1975-87	0	13	22	6	6
eldon C	1975-86	1	12	12	0	0
eldon D	1975-86	4	10	12	2	4
eldon E	1973-87	0	14	11	0	8
iptree A	1974-88	0	13	12	0	10
aringdon	1972-83	0	12	10	6	7
. Essex A	1972-88	0	15	14	7	7
. Essex B	1972-88	0	14	13	6	5
xford A	1975-88	2	9	14	3	4
xford I	1980-88	5	9	9	3	14
ucks. C	1978-88	1	11	11	0	0
arwick C	1975-88	1	12	5	4	8
incs. C	1970-88	2	17	18	7	10
ambridge A	1971-88	0	18	13	3	8
otals =		211	16	192	194	51	97

= only includes herbicides used for _A. myosuroides_ control.
= includes terbutryn and metoxuron in early/mid 1970s, tri-allate,
  chlorsulfuron/metsulfuron in mid-1980s, pendimethalin,
  flamprop-isopropyl, propyzamide, fluazifop-P-butyl.

he fields on two farms at Peldon where resistance is most severe, have
eceived very intensive herbicide treatment. It is difficult to know
nether this is a cause or a result of the presence of resistant _A. myo-_
uroides. In addition, the straw burning/minimum tillage system used can
esult in the development of an adsorptive surface layer which can reduce
ne activity of many soil-acting herbicides (Moss, 1984, 1985, 1987;
rson and Livingston, 1987). This would encourage the greater use of
erbicides even in the absence of resistance.

t is not possible to link directly the occurrence of resistance with
ntensity of herbicide use. Most of the herbicides used to control _A. myo-_
uroides are soil-acting and their performance is influenced greatly by
limatic and other environmental factors. Thus, the selection pressure
mposed by these herbicides may be poorly correlated with amount applied.

Table 2 Herbicides assessed for activity on chlorotoluron-resistant
A. myosuroides from Peldon (Moss, 1987; Moss and Cussans, 1987; Mos
unpublished; Kemp, unpublished).

Herbicide group	Herbicides for which resistance demonstrated	Herbicides with no evidence of resistance
Phenyl-ureas	chlorotoluron, diuron, isoproturon, linuron, metoxuron, methabenzthiazuron	-
Sulfonyl-ureas	chlorsulfuron	-
Triazines	cyanazine, simazine, terbutryn	-
Triazinones	SMY 1500	-
Aryloxyphenoxy -propanoates	diclofop-methyl fluazifop-P-butyl	quizalofop-ethyl
Cyclohexanediones	tralkoxydim	sethoxydim
Thiocarbamates	tri-allate	-
Carbamates	barban	carbetamide
Imidazolinones	imazamethabenz	-
Dinitroanilines	pendimethalin	trifluralin
Amides	-	propyzamide
Anilides	metazachlor	-
Aliphatics	-	glyphosate
Bipyridiliums	-	paraquat
Unclassified	-	ethofumesate

CROSS-RESISTANCE

Resistance to phenyl-ureas in _A. myosuroides_ from Peldon is associated wi
cross-resistance to other herbicides from different chemical groups wi
differing modes of action (Table 2).   A similar type of multi-herbici
resistance occurs in Australia with diclofop-methyl-resistant _Lolium rigid_
Gaud. (Heap, these Proceedings). The degree of resistance varies betwe
herbicides but is not related directly to chemical grouping or mode
action.   Although resistance is not absolute, experiments in simulat
field conditions show that substantial reductions in herbicide activi
can occur at recommended field rates (Moss, 1987; Moss and Cussan
1987).   Biochemical studies indicate that resistance is not due

ifferential uptake but to the enhanced ability of plants to detoxify
erbicides (Kemp and Caseley, 1987; Kemp *et al.*, 1990; Kemp and Caseley,
hese Proceedings).

reliminary tests indicate that most of the samples of chlorotoluron-
esistant *A. myosuroides* detected in England so far have a similar pattern of
ross-resistance to that of Peldon, showing cross-resistance to
erbicides such as pendimethalin and diclofop-methyl. However, more
xperimentation is needed to determine whether the degree of
ross-resistance is consistent between populations. One sample, from
iptree, Essex, shows a unique pattern of resistance. This population
hows partial resistance to chlorotoluron but no evidence of
ross-resistance to diclofop-methyl or pendimethalin (Moss and Orson,
988). The detection of this more specific form of resistance shows that
e cannot assume that a single detoxification mechanism occurs in all
esistant populations.

EVELOPMENT OF RESISTANCE

hen devising control measures to counteract resistance, it is useful to
ave information on the rate at which resistant populations develop. A
omparison has been made of the reponse to chlorotoluron of seven
opulations of *A. myosuroides* seeds, collected in 1978 and 1988 from the same
ields on seven farms in England. Plants grown from these seed samples
ere tested for sensitivity to chlorotoluron by growing seedlings in
iquid nutrient medium containing chlorotoluron at 250 µg l^{-1}. There was
o evidence of resistance to this concentration of herbicide in any of
he 1978 samples (Table 3). However, two of the corresponding 1988
amples, from Buckinghamshire (C) and Warwickshire (C), did show evidence
f resistance.

amples of seeds from one field (Faringdon), where partial resistance was
irst detected in 1982, were also tested for resistance to chlorotoluron.
he results provided evidence that resistance had increased in this field
etween 1976 and 1985 (Fig. 2).

hese observations show that herbicide resistance in *A. myosuroides* has
eveloped during the last 12 years, and current screening tests are not
imply detecting long-standing differences between populations.
evelopment of resistance in *A. myosuroides* has been much slower than the
nalogous evolution of resistance in *L. rigidum* (Heap, these Proceedings).

t present, we do not know why resistance occurs in some fields and not
n others which have received a similar number of herbicide applications.
t is possible that resistance is evolving slowly in some of the fields
ith susceptible populations, but has yet to reach detectable levels. At
resent, therefore, we cannot predict to what extent resistance will
ontinue to increase.

he extensive cross-resistance we have demonstrated has several important
mplications relevant to the development of resistance. It is probable
hat selection for resistance will occur despite using herbicides with
iffering modes of action. We need to determine whether herbicides with
iffering modes of action and efficacies are selecting for resistance in
he same way and to the same degree.

Table 3.   Response to chlorotoluron of seven populations of _A. myosuroi_
collected in 1978 and 1988.

Sampling site	% Reduction in foliage weight by chlorotoluron (250 μg l⁻¹)	
	1978	1988
Stockbed[1]	97	96
Bucks D	95	94
Oxford J	95	94
Oxford L	94	96
Bucks E	96	87
Warwick C	97	78
Bucks C	96	31

[1] = A seed production plot, established from seed
collected from a field in 1975. No herbicides
applied 1975–1988.

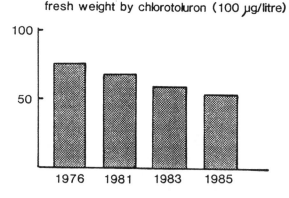

% reduction in foliage

fresh weight by chlorotoluron (100 μg/litre)

Date of seed collection

Fig. 2.   Change in sensitivity of _A. myosuroides_ to chlorotoluron betwe
1976 and 1985 (Faringdon population).

POPULATION STUDIES

Resistance of _A. myosuroides_ at the population level is due mainly to
increase in the level of resistance of all individuals within
population rather than to an increase in the proportion of very resista
types.   The variability in response to chlorotoluron between plan
within a susceptible (Rothamsted) and a resistant (Peldon) population h

een determined.   There was considerable plant-to-plant variation in
esponse to chlorotoluron but no overlap in sensitivity between the two
opulations – the most susceptible Peldon plants were more resistant than
ost resistant Rothamsted plants.

e have no evidence of any major fitness disadvantage linked with
esistance. In competition experiments between resistant and susceptible
opulations, no differences in competitive ability have been detected,
or were differences in growth detected in the absence of competition.

n uncontrolled crosses between a resistant and a susceptible population,
bout 25% of the progeny collected from the susceptible population showed
igher levels of resistance than progeny from crosses of susceptible
lants only (Fig. 3).

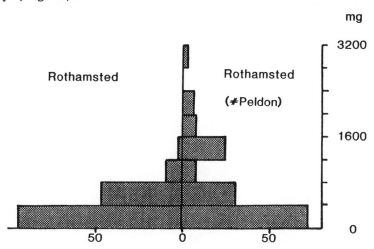

Numbers of plants in each weight category

(100 μg/litre chlorotoluron)

Fig. 3.   Response to chlorotoluron of progeny from uncontrolled crosses
within a susceptible population (Rothamsted) and between a susceptible
and a resistant (Peldon) population.

The mechanism of inheritance is not known at present, but these results
demonstrate that nuclear inheritance is involved. The introduction of
resistance into susceptible populations may occur via pollen, although
the importance of this will depend on the effective dispersal distance.
The range in plant response to herbicides suggests a polygenic mode of
inheritance. However, it is possible that a single gene is involved, if
phenotypic expression of resistance is modified by other factors.
Current crossing studies using clonal material should help clarify these
aspects.

CONTROL OF RESISTANT POPULATIONS

It is unlikely that the extent to which the field performance of
herbicides will be affected by resistance can be predicted accurately.
In conditions favourable for herbicide activity, resistance may not

always prevent a satisfactory level of control being achieved. I
contrast, in conditions unfavourable for herbicide activity, even partia
resistance may be sufficient to cause inadequate control.

If resistance develops slowly, despite high selection pressure, it may b
possible to reduce selection pressure enough to prevent furthe
resistance development. Field experiments conducted by ADAS at Peldon
where severe resistance occurs, show clearly the difficulty of achievin
an acceptable level of control from herbicides alone (Orson an
Livingston, 1987; Clarke and Moss, 1989). Ploughing was effective i
reducing *A. myosuroides* populations and resulted in increased activity o
herbicides, probably due to burial of the adsorptive surface laye
resulting from repeated straw burning. However, none of the 1
individual herbicides tested, nor sequences of three or four herbicides
has given consistently good weed control.

The most effective strategy to contain existing resistant populations i
likely to be a combination of ploughing and the inclusion in the arabl
rotation of crops on which effective herbicides can be used. I
addition, other cultural factors, such as delayed autumn drilling and th
growing of spring or other crops which discourage *A. myosuroides*, must b
considered. However, despite the use of an integrated system, control o
herbicide-resistant *A. myosuroides* is likely to be unreliable and expensiv
in a winter cereal dominated rotation. We need more information on th
factors which cause resistance to develop, so that sound guidelines ca
be produced to minimise the risk of resistance developing more widely.

ACKNOWLEDGEMENTS

We wish to thank Denise Bullock, Jill Hartnoll and Julia Balmford, fo
technical assistance, and the Home-Grown Cereals Authority for fundin
their placements at Long Ashton out of the Research and Development Levy

REFERENCES

Clarke, J.H. and Moss, S.R. (1989). The distribution and control o
herbicide resistant *Alopecurus myosuroides* (black-grass) in central an
eastern England. Proceedings 1989 Brighton Crop Protection Conference -
Weeds 301-308.

Elliott, J.G., Church, B.M., Harvey, J.J. Holroyd, J., Hulls, R.H. an
Waterson, H.A. (1979). Survey of the presence and methods of control o:
wild-oat, black-grass and couch grass in cereal crops in the Unite
Kingdom during 1977. Journal of Agricultural Science (Cambridge) 92
617-634.

Harper, J.L. (1956). The evolution of weeds in relation to thei
resistance to herbicides. Proceedings 3rd British Weed Contro:
Conference 179-188.

Kemp, M.S. and Caseley, J.C. (1987). Synergistic effects o
1-aminobenzotriazole (ABT) on the phytotoxicity of chlorotoluron an
isoproturon in a resistant population of black-grass. Proceedings 198
British Crop Protection Conference - Weeds 895-899.

Kemp, M.S., Moss, S.R. and Thomas, T.H. (1990). Herbicide resistance in *Alopecurus myosuroides*. In: "Managing Resistance to Agrochemicals: from Fundamental Research to Practical Strategies" (M.B. Green, H.M. LeBaron and W.K. Moberg, eds), pp. 376–393. American Chemical Society, Washington.

Moss, S.R. (1980). The agro-ecology and control of black-grass, *Alopecurus myosuroides* Huds., in modern cereal growing systems. ADAS Quarterly Review **38**, 170–191.

Moss, S.R. (1984). The influence of cultural practices on the activity of soil-acting herbicides. British Crop Protection Council Monograph No. 27, Symposium on Soils and Crop Protection Chemicals 77–86.

Moss, S.R. (1985). The effect of cultivation systems and soil factors on the performance of herbicides against *Alopecurus myosuroides*. Annals of Applied Biology **107**, 253–262.

Moss, S.R. (1987). Herbicide resistance in black-grass (*Alopecurus myosuroides*). Proceedings 1987 British Crop Protection Conference – Weeds 879–886.

Moss, S.R. and Cussans, G.W. (1985). Variability in the susceptibility of *Alopecurus myosuroides* (black-grass) to chlorotoluron and isoproturon. Aspects of Applied Biology 9, The Biology and Control of Weeds in Cereals 91–98.

Moss, S.R. and Cussans, G.W. (1987). Detection and practical significance of herbicide resistance with particular reference to the weed *Alopecurus myosuroides* (black-grass). In "Combating Resistance to Xenobiotics: Biological and Chemical Approaches" (M. Ford, D. Hollomon, B. Khambay and R. Sawicki, eds), pp. 200–213. Ellis Horwood, Chichester, England.

Moss, S.R. and Orson, J.H. (1988). The distribution of herbicide-resistant *Alopecurus myosuroides* (black-grass) in England. Aspects of Applied Biology 18, Weed Control in Cereals and the Impact of Legislation on Pesticide Application 177–185.

Niemann, P. and Pestemer, W. (1984). Resistenz verschiedener Herkunfte von Acker-Fuchsschwanz (*Alopecurus myosuroides*) gegenuber Herbizidbehandlungen. Nachrichtenblatt des Deutschen Pflanzenschutzdienstes **36** (8), 113–118.

Orson, J.H. and Livingston, D.B.F. (1987). Field trials on the efficacy of herbicides on resistant black-grass (*Alopecurus myosuroides*) in different cultivation regimes. Proceedings 1987 British Crop Protection Conference – Weeds 887–894.

Sly, J.M.A. (1984). Arable farm crops and grass 1982. Preliminary Report Pesticide Usage England and Wales No. 35. Ministry of Agriculture, Fisheries and Food, London, 30 pp.

RESISTANCE TO HERBICIDES IN ANNUAL RYEGRASS (*LOLIUM RIGIDUM*) IN AUSTRALIA

I.M. Heap

Department of Agronomy, Waite Agricultural Research Institute, Private Bag No. 1, Glen Osmond, S.A. 5064

Populations of annual ryegrass (*Lolium rigidum*) collected from southern Australia were tested for resistance to the herbicide diclofop-methyl. Over 40 resistant populations have been detected from widely separated areas extending from Western Australia to New South Wales. Populations differed quantitatively in their levels of resistance and there was a positive regression relating the level of resistance of the populations to their previous exposure to diclofop-methyl.
Eleven of the diclofop-methyl resistance populations and one known to be susceptible to the herbicide were evaluated for their cross-resistance to fluazifop-butyl, haloxyfop-methyl, sethoxydim, chlorsulfuron, glyphosate and propham. None of the populations was cross-resistant to glyphosate or propham. All the populations showed some level of cross-resistance to the other herbicides but there was considerable variation between populations in this resistance. The variation could not be related in any simple manner to the origins of the populations in Australia nor to their past histories of herbicide application. The results indicate that great complexity will be faced in resolving the biochemistry and genetics of the phenomenon and the formulation of advice to farmers.

INTRODUCTION

Annual ryegrass (*Lolium rigidum* Gaud.) is an annual, outcrossing, diploid grass weed of cereal crops in southern Australia. When uncontrolled, its high fecundity and competitiveness result in rapid population increases. Trends in the 1970s towards minimum tillage, stubble retention and earlier sowing of crops have reduced soil erosion and increased crop yields but have necessitated the use of selective herbicides to control ryegrass. The use of selective herbicides, such as trifluralin, diclofop-methyl, is now widespread practice. Diclofop-methyl was first registered and used in Australia for the control of annual ryegrass. A population resistant to this herbicide was found in 1981 on a farm where the herbicide had been applied over a four year period (Heap and Knight, 1982). It was shown subsequently that this population was cross-resistant to other herbicides to which the population had never been exposed (Heap and Knight, 1986).

This paper presents an overview of the most important results as greater detail will be published elsewhere.

MATERIALS AND METHODS

Bioassay

Representatives of chemical companies collected 179 populations of ryegras from fields in southern Australia. The collectors were asked to obtain see from at least 40 plants scattered throughout a field and to avoid areas wher spray patterns may not have been uniform, such as on headlands or aroun trees. The seed from a field was bulked to form a population. The histor of herbicide usage on these fields was obtained from farmer records.

Seedlings were tested for resistance to diclofop-methyl using a previousl described bioassay technique (Heap and Knight, 1986). To maximise th germination percentage of each population, seeds were placed on wet filte paper in an alternating light and temperature regime (8 h light, 20°C/16 dark, 15°C) for four days. Only those seeds which had germinated and had small root (2 mm) were tested for resistance in the bioassay.

The dosage (transformed to a $\log_e$ scale) of diclofop-methyl causing a 50 mortality (LD_{50}) in a population was estimated using a probit analysis. resistance index, RI, was calculated as the difference between the LD_{50} (log scale) of the population and a control susceptible population included i every bioassay. Thus the concentration required to achieve 50% mortality i a population having an RI of 3 is e^3 or 20 times that for the susceptibl control.

Cross-resistance

Eleven of the diclofop-methyl resistant populations and one known to b susceptible to the herbicide were evaluated for their cross-resistance t fluazifop-butyl, haloxyfop-methyl, sethoxydim, chlorsulfuron, glyphosate an propham.

Seeds were sown in potting soil in 1 l pots. The pots were placed on benche in the open during the normal growing season for annual ryegrass. Afte emergence, the seedlings were thinned to ten uniform plants per pot. All th herbicides are selective for grass weeds in crops, except for glyphosate an propham, and may be applied as foliar sprays. The lowest rate of applicatio of a herbicide was the one recommended by the manufacturers for weed control. except for sethoxydim, and the rates used therefore were 1, 2, 4 or 8 time the recommended rate. With sethoxydim they were 0.5, 1, 2 and 4 times th recommended rate. The herbicides were applied 24 days after sowing, when th plants were at the three-leaf stage, using a sprayer with an output of 125 ha^{-1}. The spreader, Agral 60, was added to all solutions at 2 ml l^{-1}. Th trial consisted therefore of 12 populations and seven herbicides at fiv rates, with three replicates of 10 plants each. Plant survival was recorde 49 days after sowing. The percentage survival was transferred to arcsines an subjected to analysis of variance (Table 1) (Snedecor and Cochran, 1980).

Similar techniques to those described above were used to evaluate the wid spectrum of cross-resistance of the first reported diclofop-methyl resistan population (Heap and Knight, 1982; Heap and Knight, 1986). A summary of thes experiments is reported in Table 2.

RESULTS

Bioassay

The variation in the resistance index (RI) values (Fig. 1) was found to be continuous with no evidence of grouping between populations, suggesting resistance is quantitative. The regression of RI on amount of diclofop-methyl previously applied was significant (r = 0.776, df = 177).

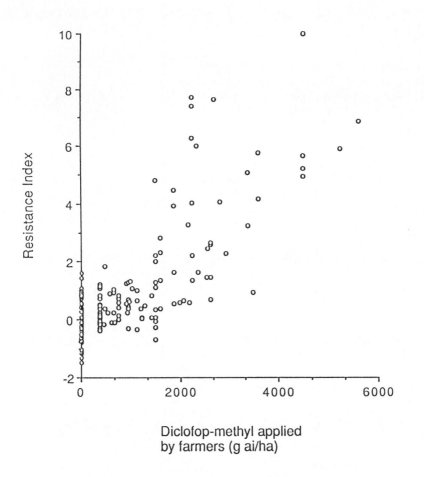

Fig. 1. <u>The relation between the resistance index and the sum of the previous exposure of the population to diclofop-methyl</u>.

In the total sample there were 34 populations with an RI greater than 1.8 and which may be considered as resistant to diclofop-methyl. They came from widely separated areas of southern Australia (Fig. 2). All populations with a RI above 2 had received at least 1,500 g a.i. ha^{-1} of diclofop-methyl (Fig. 1). This is 4 years of application if the farmer has applied the herbicide annually at the recommended rate of 375 g a.i. ha^{-1}.

Thirteen other populations which had received more than 1,500 g a.i. ha^{-1} had a low resistance index, between 0 and 1.8. These populations may be considered as still susceptible although none was more susceptible than the control used in the bioassay. These 13 populations are of special

significance because they indicate some farmers have applied a high tota[l]
level of herbicide but have avoided the problem of resistance. Mor[e]
populations have been detected since this survey and there are now over 4[0]
annual ryegrass populations with high levels of resistance to diclofop-methyl[.]

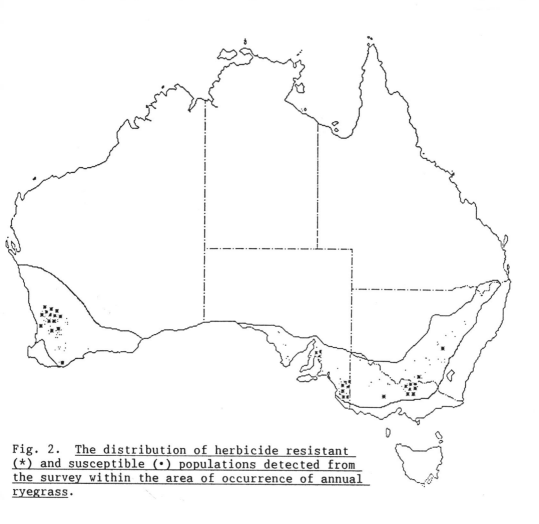

Fig. 2. The distribution of herbicide resistant
(*) and susceptible (•) populations detected from
the survey within the area of occurrence of annual
ryegrass.

Trifluralin had been used by many farmers for ryegrass control prior to th[e]
introduction of diclofop-methyl but there is no evidence in these result[s]
suggesting that trifluralin had influenced the rate of development o[f]
diclofop-methyl resistance.

Cross-resistance

All of the plants that were not treated with herbicides survived (i.e. th[e]
control plants of the 12 populations) and the mortality of the treated plant[s]
is attributed, therefore, to the herbicides.

All 12 populations were fully susceptible to glyphosate and propham and ther[e]
was no evidence of cross-resistance to these herbicides. For this reason n[o]
further results will be presented for glyphosate and propham.

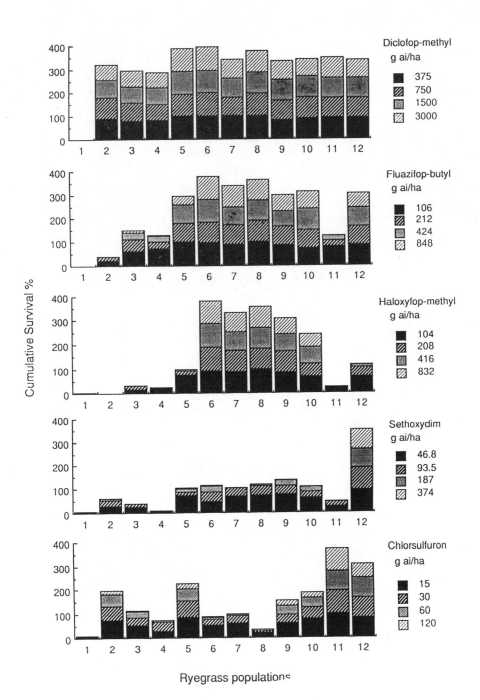

Fig. 3. The cumulative survival percentage for 12 populations of annual ryegrass when treated with herbicides. The survival percentage for each of the four levels of the herbicide is depicted as four parts of a column. The total height of the column indicates a cumulative percentage of survival for all four levels of the herbicide and it would have had a value of 400 if all plants survived.

For the other herbicides, survival percentages were obtained for eac
population at four rates of application and are presented in Figure 3.
statistical analysis of the survival results is presented in Table 1. Th
values for population 1 confirms that it was susceptible to diclofop-methyl
There was no survival at even the lowest rate of herbicide used (i.e. th
recommended rate of 375 g a.i. ha^{-1}). Less than 5% survival was found whe
this population was sprayed with the other herbicides at their recommende
rates, and no survival at higher rates (Fig. 3). The other 11 populations
known to be resistant to diclofop-methyl, showed only relatively smal
differences in survival following treatment with diclofop-methyl. The exten
of their resistance to this herbicide is evident from the fact that more tha
60% of the plants survived when sprayed with eight times the recommended rat
for the herbicide (3,000 g a.i. ha^{-1}).

Each population had a degree of cross-resistance to the other herbicides (Fig
3). The degree of this cross-resistance differed between the populations an
was not related to the small differences observed in diclofop-methy
resistance. For example, populations 2 and 9 were similar in their resistanc
to fluazifop-butyl and haloxyfop-methyl. Population 2 would be considered a
fully susceptible to haloxyfop-methyl whereas population 9 was resistant witl
a high survival at all rates of application. Nor was there a relation betwee
cross-resistances independent of diclofop-methyl resistance. Populations ?
and 11 showed opposite patterns in their resistances to haloxyfop-methyl an
chlorsulfuron although population 8 had been exposed only to diclofop-methy
and population 11 to diclofop-methyl and trifluralin. Trifluralin exposure
has now, in the other populations, led to resistance to chlorsulfuron.

Table 1.   Survival (as arcsine (%)) of 11 diclofop-methyl resistan
populations of annual ryegrass following treatment with 5 herbicides (mean
across all rates of application).

Herbicide	\multicolumn{11}{c}{Population}											LSD
	2	3	4	5	6	7	8	9	10	11	12	(0.05
Diclofop-methyl	64	29	59	81	87	71	79	67	68	70	68	10.0
Fluazifop-butyl	13	37	33	63	79	67	75	60	62	30	62	20.9
Haloxyfop-methyl	0	11	10	25	80	66	72	62	51	10	28	20.4
Sethoxydim	20	13	5	29	31	23	29	32	27	14	70	18.3
Chlorsulfuron	44	31	22	49	26	25	15	38	43	77	61	13.0

An attempt was made to divide the 12 populations into groups. Populations ?
to 5 inclusive showed a moderate degree of cross-resistance to fluazifop-
methyl, but had little resistance to haloxyfop-methyl and sethoxydim.
Populations 6 to 10 inclusive had a very high degree of resistance t
fluazifop-butyl and haloxyfop-methyl but a low resistance to sethoxydim an
chlorsulfuron. Population 11 had little cross-resistance to fluazifop-butyl,
haloxyfop-methyl and sethoxydim but the highest level of resistance t
chlorsulfuron. Population 12 was far more resistant to sethoxydim than an
other population but the population had received two applications of
sethoxydim. It needs to be emphasised, however, that nearly all of the

opulations that had resistance to diclofop-methyl had far more resistance to sethoxydim than population 1 (the diclofop-susceptible population). Cross-esistance to sethoxydim is therefore apparent.

able 2 presents the wide spectrum of cross-resistance exhibited by the first reported diclofop-methyl resistant population. The herbicides implicated in cross-resistance belong to different chemical groups and have different modes of action.

Table 2. The herbicides and chemical groups to which cross-resistance has or has not been found in diclofop-methyl resistant annual ryegrass.

Cross-Resistance	Moderate Cross-Resistance	No Cross-Resistance
Aryloxyphenoxypropionate	*Substituted Ureas*	*Carbamates*
Diclofop-methyl	Isoproturon	Carbetamide
Fluazifop-butyl		Asulam
Haloxyfop-methyl	*Miscellaneous*	Propham
Chlorazifop-propynil	Propyzamide	
Quizalofop-ethyl		*P-Nitro Diphenyl-ether*
Propaquizafop		Oxyfluorfen
Cyclohexanedione		
Alloxydim-sodium		*Triazine*
Sethoxydim		Simazine
Sulfonylureas		*Bypyridyls*
Chlorsulfuron		Paraquat
Metsulfuron-methyl		
Triasulfuron		*N-phospho-methyl glycine ester*
		Glyphosate
Dinitroaniline		
Trifluralin		
Triazinone		
Metribuzin		

DISCUSSION

Bioassay

As the resistant populations found in this study were obtained from herbicide sprayed areas and susceptible populations from unsprayed areas, often from the same properties, it is concluded that resistance is developing *in situ* as a result of the use of herbicides. This conclusion, together with the evidence that there appears to be no geographical limit to the occurrence of resistant populations (Fig. 2), suggests that genes for resistance may be present in all ryegrass populations.

It was evident from the results that some populations have received more than 1,500 g a.i. ha^{-1} of diclofop-methyl but remained susceptible, indicating that other factors in addition to the total amount of herbicide applied, are involved in the development of resistance. One factor which probably affects the rate of appearance of resistance in a weed is its density at the time of application of the herbicide. A large and dense population is more likely to contain the rare combinations of resistance genes that give rise to resistant populations. One strategy to delay the appearance of resistance is to reduce seed numbers of ryegrass by cultural techniques, such as by heavy grazing, spray-topping of pastures or cultivation before sowing the crop therefore exposing smaller populations to the selection pressure of the herbicide.

There was no evidence that application of selective herbicides other than diclofop-methyl, had led to the development of diclofop-methyl resistant populations. Most of these herbicides have been recently commercialised and insufficient time may have elapsed for resistance to be evident.

Cross-resistance

The results presented in Figure 3 indicate there is variation between populations in their cross-resistance to herbicides belonging to different chemical groups and with different modes of action. Although similarities were apparent between some populations in their cross-resistance patterns it is concluded that the populations cannot be allocated to discrete groups with similar responses within a group. In each instance the responses showed quantitative and qualitative differences.

In so far as populations 2 to 5 or 6 to 10 show some similarities it is evident that these bear no relation to the origin of these populations in Australia. In each instance the populations originate from farms in different States and there is no geographical association with cross-resistance. Nor is their any association with the spraying histories apart from resistance to sethoxydim in population 12. As population 12 had previously experienced two field applications of sethoxydim at high rates it may have responded to this selection pressure and its survival may be indicative of multiple resistance rather than cross-resistance.

The variation in cross-resistance implies that populations that have been selected by exposure to diclofop-methyl, and have evolved resistance, do not have a single common mechanism for metabolizing the other herbicides and several mechanisms must be involved. By extrapolation it might also suggest there may be more than one mechanism for metabolizing diclofop-methyl. Resolution of the issue of whether there are several mechanisms, and how resistance to diclofop-methyl results in cross-resistance to other herbicides to varying degrees, will depend on genetic and biochemical studies.

In regard to the biochemical nature of resistance, other studies have shown there is no detectable difference between resistant and susceptible annual ryegrass in the uptake and translocation of diclofop-methyl (Kocher, 1985, pers. comm.; Shimabukuro, 1986, pers. comm.; Holtum, 1987, pers. comm.). Recent experiments conducted by Shimabukuro indicate that diclofop-methyl disrupts the membrane integrity of both resistant and susceptible individuals in a similar manner, but resistant plants were more able to recover membrane integrity than susceptible plants. This observation does not explain the observed cross-resistance to other herbicides such as chlorsulfuron or trifluralin.

e results on variation in cross-resistance would be interpretable if
:sistance to diclofop-methyl were determined by several different mechanisms,
ιd therefore different genes, leading to a quantitative inheritance of the
ιenomenon. Cross-resistance to other herbicides could readily show variation
: a result of genetic linkage if several mechanisms were involved.

ιe first reported diclofop-methyl resistant population exhibited cross-
:sistance to 15 of the 22 herbicides tested. These 15 herbicides cover seven
ırbicide groups and five different modes of action (Table 2). The
·yloxyphenoxypropionate and cyclohexanedione herbicides both inhibit fatty
:id synthesis (Harwood *et al.*, 1987), the sulfonylureas block the essential
ιino acids valine and isoleucine by inhibiting the enzyme acetolactate
·nthase (Ray, 1985), trifluralin inhibits spindle microtubule formation
·arka and Soper, 1977), metribuzin and isoproturon are photosynthetic
ιhibitors (Pallet and Dodge, 1979) and propyzamide interferes with mitotic
ιvision (Bartels and Hilton, 1973). A similar situation of a grass weed,
lopecurus myosuroides, having a wide spectrum of herbicide cross-resistance has
ɛen reported by Moss and Cussans (1987).

ΟNCLUSION

ɛsistant populations appear to have originated independently, as a result of
ιclofop-methyl usage and differ quantitatively in their levels of resistance
ɔ the herbicide. Ryegrass populations resistant to diclofop-methyl may
xhibit cross-resistance to other herbicides, to which they have never been
xposed. Resistant populations vary in their levels of cross-resistance. If
ro012ss-resistance had been consistent across populations, farmers could have
ɛen informed where herbicides would be of value to apply and which would give
ɔ control. In view of the variation between populations, advice to farmers
ould only be given after testing their populations for cross-resistance. In
ddition, as cross-resistance has not been constant, conjectures cannot be
ade as to whether applications of the relatively newer herbicides - for
xample, chlorsulfuron - are likely to result in a reciprocal cross-resistance
ɔ diclofop-methyl in populations never treated with the herbicide.

ɛFERENCES

artels, P.G. and Hilton, J.L. (1973). Comparison of trifluralin, oryzalin,
ronamide, propham and colchicine treatments on microtubules. Pesticide
iochemistry and Physiology **3**, 462-472.

arwood, J.L., Walker, K.A. and Abulnaja, D. (1987). Herbicides affecting
ipid metabolism. British Crop Protection Conference - Weeds 159-169.

eap, J. and Knight, R. (1982). A population of ryegrass tolerant to the
ɛrbicide diclofop-methyl. The Journal of the Australian Institute of
gricultural Science **48**, 156-157.

eap, I. and Knight, R. (1986). The occurrence of herbicide cross-resistance
n a population of annual ryegrass, *Lolium rigidum*, resistant to diclofop-methyl.
ustralian Journal of Agricultural Research **37**, 149-156.

Moss, S.R. and Cussans, G.W. (1987). Detection and practical significance herbicide resistance with particular reference to the weed *Alopecurus myosuro* (Black-grass). In "Combating Resistance to Xenobiotics: Biological a Chemical Approaches" (M. Ford, D. Hollomon, D. Khambay and R. Sawicki, eds pp. 200-213. London: Society of Chemical Industry.

Pallett, K.E. and Dodge, A.D. (1979). Sites of action of photosynthet inhibitor herbicides: Experiments with trypsinated chloroplasts. Pestici Science 10, 216-220.

Parka, S.J. and Soper, O.E. (1977). The physiology and mode of action dinitroaniline herbicides. Weed Science 25, 79-87.

Ray, T.B. (1985). The site of action of the sulfonylurea herbicides. Briti Crop Protection Conference - Weeds 131-138,

Snedecor, G.W. and Cochran, W.G. (1980). Statistical Methods. The Iowa Sta University Press. Ames, Iowa, U.S.A.

THE OCCURRENCE OF TRIFLURALIN RESISTANT *SETARIA VIRIDIS* (GREEN FOXTAIL) IN WESTERN CANADA

I.N. Morrison, H. Beckie and K. Nawolsky

University of Manitoba, Department of Plant Science, Winnipeg, Manitoba, R3T 2N2, Canada

Since the early 1970s trifluralin has been used extensively to effectively control *Setaria viridis* (green foxtail, wild millet) in both oilseed and cereal crops in Western Canada. In early 1988, *S. viridis* from three fields in western and south-western Manitoba were confirmed to be resistant to the chemical. The seed was collected from fields where trifluralin had been applied in 1987 but failed to provide satisfactory control of the weed. From growth room experiments it was determined that the resistant populations were 4 to 5 times more resistant to trifluralin than susceptible populations, about twice as resistant as *Avena fatua* and approximately equal in sensitivity to wheat. The differential response between resistant and susceptible populations was confirmed under field conditions in 1989. An additional 26 of 70 samples collected from 'suspect' fields in the fall of 1988 were also confirmed to be resistant. The majority of these originated from fields in south-western Manitoba where trifluralin was applied frequently over the past 15 years both as a pre-plant incorporated treatment in rapeseed and flax and as a pre-emergence, shallowly incorporated treatment in wheat. The comparatively higher incidence of resistance in these areas undoubtedly is related both to the use pattern of the herbicide in the area, as well as to the fact that south-western Manitoba is normally drier than other areas of the Province. This, in turn, would result in greater carry-over of the herbicide from one year to the next and increased selection pressure on the weed.

INTRODUCTION

Since trifluralin was first introduced in western Canada in the early 1970s, it has consistently proven to be highly effective for controlling *A. fatua* and *S. viridis* (Rahman and Ashford, 1972; Chow, 1976). Currently the chemical is the active ingredient in three commercial products (Treflan, Triflurex and Rival) widely used as a pre-plant incorporated (PPI) treatment to control *A. fatua* and *S. viridis* in oilseed and pulse crops. Trifluralin alone or in combination with triallate is also used in some areas of the prairies to

67

control *S. viridis* in wheat and barley, primarily as a pre-emergenc
incorporated (PEI) treatment. As well, the chemical is the activ
ingredient in Heritage which is used mainly in southern Saskatchewan c
summer fallow.

Because of the persistent nature of trifluralin, sufficient amounts of th
chemical usually remain active in the soil to provide season-long contro
of *S. viridis* which characteristically emerges in several flushes during th
growing season. Indeed, many farmers have come to rely on trifluralin t
provide 'second year' control of the weed, recognizing that there i
significant carry-over of the chemical from one year to the next
Analytical results indicate that in most years between 10 and 30% of th
initial dosage persists into the next growing season (Smith, 1982).

In the spring of 1988, three samples of *S. viridis* seed from Lyleton, Killarne
and Oak River, Manitoba were determined to be resistant to triflurali
(Morrison *et al.*, 1989). The three resistant populations originated fro
fields in western and south-western Manitoba that had a history of repeate
trifluralin usage over the past 15 to 20 years. Dose response experiment.
conducted under growth room conditions established that the resistan
populations were approximately 5 times more resistant to trifluralin thai
susceptible populations, twice as resistant as *A. fatua* and about equal ii
sensitivity to wheat (Morrison *et al.*, 1989). Whereas dosages of between 0.'
and 0.3 kg ha^{-1} trifluralin controlled susceptible populations, resistan
plants were not completely killed at dosages up to 1.2 kg ha^{-1}.

The occurrence of trifluralin resistance in *S. viridis* marks the second instanc
of a major troublesome annual grass weed developing resistance to thi
herbicide. The first was *Eleusine indica* (goosegrass) in South Carolina wher
resistance was first suspected in the mid-1970s and positively confirmed ii
the early 1980s (Mudge *et al.*, 1984). In the case of the *E. indica*, the
appearance of resistant populations was associated with the repeated use o:
trifluralin in cotton over a period of 10 years.

FIELD SURVEY AND DISTRIBUTION

Following the initial confirmation of trifluralin resistance in *S. viridis* fron
the three sites, an additional 70 *S. viridis* seed samples were collected ir
August and September of 1988 from fields where trifluralin had not performe
well that year and resistance was suspected. 'Suspect' fields wer
identified as those where one or more of the following indicators applied:

1. Trifluralin had been applied either in the fall of 1987 or in the
   spring of 1988 and had controlled *A. fatua* and broadleaf weeds, but
   not *S. viridis*.

2. Irregular patches of *S. viridis* were present in a field where contro]
   was otherwise excellent.

3. Trifluralin had been applied for several years but *S. viridis* control
   seemed less and less satisfactory.

4.   *S. viridis* escapes did not appear to follow any pattern
      relating to application or incorporation of the herbicide.

From these and other samples collected directly from farmers or through the regional extension service, more than 40 samples have now been verified to be resistant to trifluralin, with the majority being from western and south-western Manitoba (Morrison *et al.*, 1989). Several of these were from fields that were inspected in June of 1988 by weed specialists from the University of Manitoba and Manitoba Agriculture where trifluralin had apparently failed to have any effect on *S. viridis* but had completely controlled other susceptible weed species including *A. fatua*. In none of these fields was it possible to attribute the lack of effective *Setaria* control to improper application or incorporation of the chemical or to other factors, e.g. excessively dry soil conditions that sometimes interfere with activity of the herbicide (Moyer, 1987).

From data compiled by Manitoba Agriculture using information supplied by the Manitoba Crop Insurance Corporation, it was evident that the distribution of resistant *Setaria* populations corresponded closely to the areas of greatest trifluralin usage. In the extreme south-west area of the Province in Crop Reporting District 1, approximately 30% of the area seeded to wheat and 85% of the area seeded to rapeseed was treated with trifluralin annually during the years 1982 to 1986 (Morrison *et al.*, 1989). Close to 36% of the four major crops grown in that district were sown on land treated with trifluralin during those years. This compares to less than 7% of the area in Crop Reporting District 12 in the Interlake Region north of Winnipeg, where no trifluralin-resistant *Setaria* populations have been identified to date.

In consideration of the results of the dose response experiments, the relatively large number of samples confirmed to be resistant from 'suspect' fields in 1988, and the importance of providing farmers with up to date information on the topic of resistance, Manitoba Agriculture (908 Norquay Bldg., Winnipeg, Manitoba, R3C 0P8) opted to include the following cautionary statement in the 1989 Guide to Chemical Weed Control:

'Caution: Populations of green foxtail tolerant to trifluralin have
             developed in a small number of fields, primarily in
             southwestern Manitoba. Rival, Treflan and Triflurex will
             not control trifluralin tolerant foxtail.'

In addition, several articles appeared in the farming press about the discovery of trifluralin-resistant *Setaria* in Manitoba and the topic was discussed at numerous extension meetings in areas of the Province considered to be most affected by the problem.

Reaction from farmers, as well as from both technical and sales staff working in the agricultural chemical industry varied from outright disbelief, through varying degrees of scepticism, to resigned acceptance. Much of the doubt surrounding the issue was expressed by individuals who had not actually seen the problem in farm fields in 1988, or who had little confidence that the results of the growth room procedures used to verify resistance could actually be extrapolated to field situations.

FIELD VERIFICATION

Any doubts about whether or not *S. viridis* populations had developed resistance to trifluralin were dispelled in the 1989 growing season from replicated field trials conducted at the University of Manitoba's field research station at Portage la Prairie and from additional experiments conducted in farm fields in south-western Manitoba.

At Portage, trifluralin was applied at nine dosages ranging from 0 to 3.0 kg ha^{-1} both as a PPI treatment in rapeseed (*Brassica napus* cv. Westar) and as a PEI treatment in wheat (*Triticum aestivum* cv. Katepwa). Prior to chemical application and incorporation, half of each plot was seeded with seed from a susceptible population of *S. viridis* and half from a resistant population. The plots were further subdivided such that half of each sub-plot was seeded to the crop and half was not.

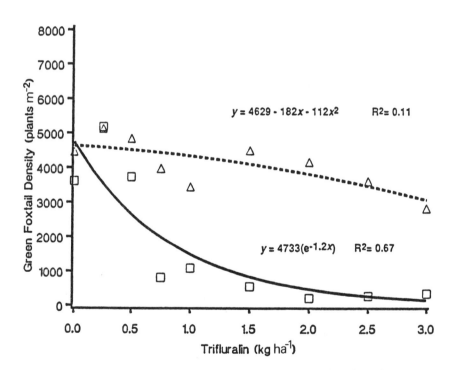

Fig. 1. Effect of PPI trifluralin on density of resistant (dotted line) and susceptible (solid line) *S. viridis* in a non-crop situation.

The recommended rates of trifluralin for rapeseed and wheat are respectively 0.8 to 1.4 kg ha^{-1} and 0.6 to 0.9 kg ha^{-1}. The differences in rates between the two crops relate to the differences in methods of incorporation. For each crop, the lower of the two rates is specified on light-textured soils with low organic matter (OM) contents (< 6%) and the higher rate on medium to heavy textured soils with OM contents from 6 to 15%. The soil where the trials were located is a Neuhorst clay loam (25% sand, 44% silt and 31% clay) with an OM content of 7.5%.

As evident from Figures 1 and 2 illustrating the effect of increasing dosages of trifluralin on the density of *S. viridis* seedlings under a non-crop situation, the susceptible *Setaria* was readily controlled by recommended dosages of the chemical, whereas the resistant population was not. Where the chemical was incorporated 5 to 8 cm deep using a tandem disc (Fig. 1), the LD_{50}'s calculated using regression equations fitted to the data were 0.6 kg ha^{-1} for the susceptible plants and 3.8 kg ha^{-1} for the resistant plants, indicating a 6-fold difference in tolerance. Where the herbicide was incorporated shallowly (2 to 3 cm) with harrows (Fig. 2), LD_{50}'s were 0.11 and 0.95 kg ha^{-1}, respectively, indicating close to a 9-fold difference in sensitivity. These differences closely parallel those reported in pot experiments under growth room conditions where susceptible *Setaria* proved to be 4 to 5 times more sensitive to trifluralin than resistant *Setaria* when the chemical was uniformly mixed throughout the soil, and up to 10 times more sensitive when the chemical was mixed only into the top 2 cm of soil (Morrison *et al.*, 1989).

Table 1. <u>Percent reduction in susceptible (S) and resistant (R) *S. viridis* seedling densities following preplant incorporated (PPI) and pre-emergence incorporated (PEI) treatments of trifluralin at recommended dosages.</u>

Application method and dosage (kg/ha)	% Reduction			
	No Crop		Crop[1]	
	S	R	S	R
PPI				
0.8	62	5	65	0
1.4	81	10	84	1
PEI				
0.6	100	34	100	19
0.9	100	48	100	28

[1] Where trifluralin was applied as PPI and PEI treatments, the cropped areas were seeded to rapeseed and wheat, respectively.

In the cropped areas of the plots, the results were comparable to those in the non-cropped areas although the weed densities were less in the former than in the latter. At the rate of trifluralin recommended as a PPI treatment in rapeseed, i.e. 1.4 kg ha^{-1}, the density of susceptible plants was reduced by 84%, whereas the density of resistant plants was reduced by 1% (Table 1). Similarly, in wheat where the chemical was applied at the recommended rate of 0.9 kg ha^{-1} as a PEI treatment, more than 99% of the susceptible plants were controlled compared to fewer than 28% of the resistant plants.

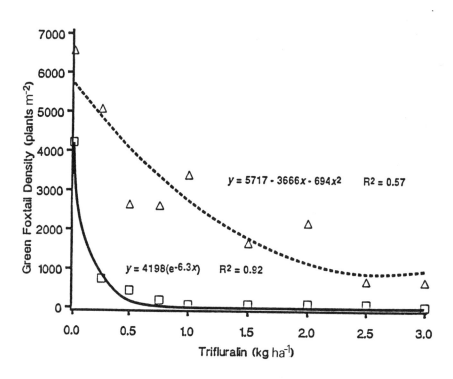

Fig. 2.   Effect of PEI trifluralin on density of resistant (dotted line) and susceptible (solid line) _S. viridis_ in a non-crop situation.

At the time the trials were established, seedbed conditions were ideal for seed germination and maximum effectiveness of trifluralin. It might well be expected that under more adverse conditions the efficacy of the herbicide would be reduced. Such would certainly appear to have been the case in 1988 when recommended rates of trifluralin had no noticeable effect on reducing the densities of resistant populations in farm fields. In that year seedbed conditions were inordinately dry during May and June during the period of crop and weed establishment.

While it is recognized that the resistant _Setaria_ populations identified in Manitoba do not exhibit the same extremely high degree of resistance to trifluralin comparable to certain crop plants and weeds that are unaffected by the herbicide even at very high dosages, it is appropriate within the context of this paper to refer to the resistant populations as being resistant as opposed to merely tolerant. It might be argued that this is incorrect if one adheres to the definition of resistance put forward by Gressel (1985) who proposes that truly resistant weeds are able to withstand

sages many times greater than the agricultural dosage with no observable
fect. However, using the definition that resistance marks a genetic
ange (within a population) in response to selection by toxicants that may
pair control in the field (Sawicki, 1987), it is entirely appropriate to
e the term resistant in reference to the *S. viridis* described in this paper.
rtainly from the practical and economic standpoint of the farmers who
countered the problem, the weed is nothing short of resistant!

HINDSIGHT

e appearance of trifluralin-resistant *S. viridis* might well have been
edicted based on characteristics of both the herbicide and the weed.
deed, in 1986, LeBaron[1] indicated that herbicides that are single site
xicants, that provide long soil residual and season-long control of
rminating weeds and that are applied frequently over several growing
asons without rotating or alternating with other types of herbicides pose
high risk for selection for resistance. Trifluralin was one of several
nitroaniline herbicides listed as falling into this high risk category.

milarly, *S. viridis* exhibits many characteristics that would predispose the
velopment of resistance within populations repeatedly exposed to
ifluralin. From both field and controlled environment studies the weed
s shown to be highly susceptible to trifluralin and effectively controlled
comparatively low dosages. Furthermore, where the weed is not controlled
can produce an abundance of seeds, most of which do not remain viable for
re than one or two years in cultivated soil (Douglas *et al.*, 1985). In
eoretical equations used to predict the rate of enrichment of herbicide
sistance within populations (Gressel and Segel, 1982), both the selection
essure and the average soil seed bank longevity figure prominently. Even
ere the initial frequency of resistant plants within the population is
w, a high degree of selection pressure and short seed longevity favour
pid appearance of resistance.

RACTICAL CONSEQUENCES

om a management standpoint, farmers who have been regular users of
ifluralin, particularly in the drier areas of the Province where
ifluralin residues carry over from one year to the next, are being advised
be on the lookout for new infestations of resistant *Setaria*. Where
sistance is suspected it is recommended that the affected areas be
ersprayed with a post-emergence product to control the weed. In wheat,
fective control of *S. viridis* can be obtained using propanil or fenoxaprop-
hyl which was introduced into the marketplace in 1989 and was

---

LeBaron, H.M. Resistance of Weeds to Herbicides. Presented to Illinois
ustom Spray Operators School, Jan. 8, 1986, Urbana, Illinois.

in extremely high demand in south-western Manitoba where the problem
trifluralin resistance is most common. In rapeseed, a number of broa
spectrum graminicides, including sethoxydim and fenoxaprop-ethyl a
recommended.

From field experience and limited greenhouse investigations, triflural
resistant *S. viridis* does not appear to be cross-tolerant to any of the common
used post-emergence herbicides registered for *S. viridis* control in weste
Canada. In contrast, the evidence to date indicates that triflural
resistant *Setaria* is also tolerant of ethalfluralin (Edge), but the differen
in sensitivity between resistant and susceptible populations is not as gre
as it is for trifluralin.

Considerable field experimentation under a variety of conditions will
required to determine whether or not it will be possible to obtain reliab
control with ethalfluralin which has recently been introduced to weste
Canadian farmers as a highly effective alternative to trifluralin for u
in rapeseed. From preliminary work conducted at the University of Manito
field station at Portage la Prairie, it seems improbable that ethalflural
will provide consistently acceptable control of trifluralin-resistant *Seta*
under a wide range of soil and climatic conditions.

REFERENCES

Chow, P.N.P. (1976). Dinitroaniline herbicides for grassy weed control
rapeseed. Canadian Journal of Plant Sciences **6**, 369–376.

Douglas, B.J., Thomas, A.G., Morrison, I.N., and Maw, M.G. (1985).   Tl
biology of Canadian weeds. 70. *Setaria viridis*.   Canadian Journal of Pla
Sciences **65**, 669–690.

Gressel, J. (1985) Herbicide tolerance and resistance: alterations of si
of activity.   In" Weed Physiology, Vol. II, Herbicide Physiology" (S.
Duke, ed) pp. 159–189. CRC Press, Baton Rouge.

Gressel, J. and Segel, L.A. (1982) Interrelating factors controlling tl
rate of appearance of resistance: The outlook for the future. In "Herbici
Resistance in Plants" (H.M. LeBaron and J. Gressel, eds) pp. 325–347. Wile
New York.

Morrison, I.N., Todd, B.G. and Nawolsky, K.M. (1989). Confirmation
trifluralin resistant green foxtail (*Setaria viridis*) in Manitoba.   We
Technology **3**, 544–552.

Moyer, J. (1987). Effect of soil moisture on the efficacy and selectivi
of soil-applied herbicides. Review of Weed Science **3**, 19–34.

Mudge, L.C., Gosset, B.J., and Murphy, T.R. (1984). Resistance
goosegrass (*Eleusine indica*) to dinitroaniline herbicides. Weed Science **3**
591–594.

hman, A., and Ashford, R. (1972). Control of green foxtail in wheat with ifluralin. Weed Science **20**, 754-759.

wicki, R.M. (1987).   Definition, detection and documentation of secticide resistance. In "Combating Resistance to Xenobiotics: Biological d Chemical Approaches" (M.G. Ford, D.W. Holloman, B.P.S Khambay and R.M. wicki, eds), pp. 105-117. Ellis Horwood Ltd., Chichester.

ith, A.E. (1982). Herbicides and the soil environment in Canada. Canadian urnal of Soil Science **62**, 433-460.

PARAQUAT RESISTANCE IN JAPAN

S. Matsunaka[1] and Kazuyuki Ito[2]

[1] Faculty of Agriculture, Kobe University, Nada-ku, Kobe 657, and
[2] Tropical Agriculture Research Center, MAFF, Tsukuba, Ibaraki 305, Japan

In Japan, from the beginning of the 1980s the control of *Erigeron philadelphicus* L. has become increasingly difficult in mulberry fields near to Tokyo. In 1982, paraquat resistant *E. canadensis* L. was also found in vineyards in Osaka Prefecture, and in 1987 *Conyza sumatrensis* Retz. and *Youngia japonica* (L.)DC. were found in mulberry fields. A survey in Spring 1989 showed that paraquat resistant *E. canadensis* was most widely distributed especially in the west half of Japan from Kanto area to Kyushu. The second paraquat resistant weed, *E. philadelphicus*, was found in the Kanto area and Yamanashi Prefecture. The other two species, *C. sumatrensis* and *Y. japonica* were restricted to only three prefectures in the Kanto area. Resistance has been observed in four biotypes of Compositae weeds in perennial crop farms where paraquat had been applied 2 or 3 times annually during the preceding 15 to 20 years. In *E. philadelphicus* and *E. canadensis*, the ecological and physiological fitness of the resistant biotype is lower than that of the normal types as is the case for atrazine-resistant weeds. Crossing experiments with *E. philadelphicus* provided evidence that a single dominant gene was responsible for paraquat resistance. Although there are reports showing that differences in the paraquat movement in weeds may be associated with paraquat resistance, we showed that the resistance to paraquat was observed at the protoplast level, by the measurements of chlorophyll bleaching, Mehler reaction and superoxide detoxification system.

INTRODUCTION

In this paper, the appearance of paraquat resistant weeds in Japan, their distribution and ecological fitness, genetical properties, and physiological aspects will be reviewed. Finally, possible future problems regarding paraquat and other herbicides will be discussed.

APPEARANCE OF PARAQUAT RESISTANT WEEDS IN JAPAN

Paraquat and diquat have been used in Japan to control weeds in perennial crops, such as fruit trees, tea and mulberry fields, or on levees of paddy fields, and in non-crop spaces.

From the beginning of the 1980s, the control of *Erigeron philadelphicus* L. (Philadelphia fleabane) in mulberry fields has been practised by sericultural farmers in Saitama Prefecture.

A survey of farmers by one of the authors (K.I.) revealed that paraquat had been applied 2 or 3 times annually during the preceding 8 to 11 years. It was deduced that new biotypes resistant to paraquat appeared approximately 5 or 6 years after paraquat was introduced to mulberry fields.

Using this information, field experiments were conducted at sites infested with paraquat resistant *E. philadelphicus*. The results are shown in Figure 1. At the recommended dosage, the weed showed high resistance to both formulations of paraquat, but it was very susceptible to bentazone, MCPA and glyphosate (Watanabe *et al.*, 1982).

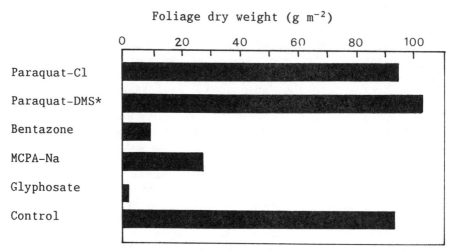

Fig. 1. <u>Control of paraquat-resistant</u> *Erigeron philadelphicus* <u>in a mulberry</u> <u>field with several herbicides applied at recommended dosages</u> (From Watanabe *et al.*, 1982).
*DMS: dimethylsulphate.

In a glasshouse experiment (Fig. 2), normal susceptible *E. philadelphicus* originating from a site where paraquat had not been applied, was killed at a dosage of 1 kg a.i.ha^{-1}. On the other hand, the resistant biotype showed no symptoms at the rates of 0.5–2.0 kg a.i.ha^{-1}, and retained a few green leaves even at a rate of 16 kg a.i.ha^{-1}. Such paraquat-resistant plants also developed resistance to diquat although the resistance was somewhat lower at the same dose.

A leaf disc test using a range of paraquat concentrations also showed that the level of resistance in the resistant biotype was 100 times higher than that of the susceptible one (Watanabe *et al.*, 1982).

Such paraquat resistance of *E. philadelphicus* was also observed in fruit and tea plantations.

Next, Kato *et al.* (1982) found a biotype of *Erigeron canadensis* L. (Horse weed) resistant to paraquat in vineyards in Osaka Prefecture. Even on our campus of Kobe University, we found a paraquat-resistant biotype of this weed.

Hanioka (1987) also reported biotypes of *Conyza sumatrensis* Retz., resistant to paraquat and diquat, and *Youngia japonica* (L.) DC (Asiatic hawkbeard) resistant to paraquat (Fig. 3).

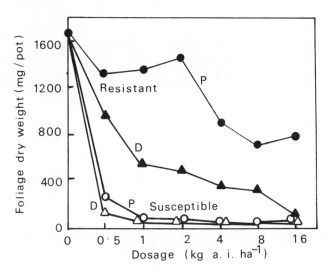

Fig. 2. <u>Relationship between applied dosages of paraquat (P) and diquat (D) and foliage dry weight of *E. philadelphicus* 10 days after application</u> (Watanabe *et al.*, 1982).

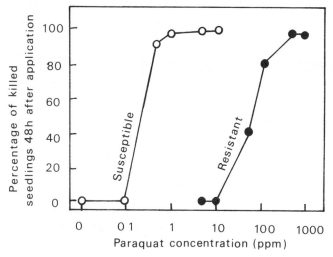

Fig. 3. <u>Mortality of *Youngia japonica* seedlings after foliar application of paraquat</u>(Hanioka, 1987).

Thus, we now have four Compositae weeds resistant to paraquat or diquat
in Japan. In April 1989 these four paraquat-resistant species were found
in Saitama Prefecture. At this site, paraquat had been applied 2 or ?
times annually during the preceding 15 years.

All over the world, paraquat-resistant biotypes have been detected in at
least ten weed species including 5 Compositae species and 3 grasses
belonging to 7 genera as shown in Table 1. Furthermore, a biotype
resistant to both paraquat and atrazine was reported in *E. canadensis* by
Pölös *et al*. (1986).

Table 1. <u>World distribution of paraquat-resistant weeds</u>.

	Species	Common name	Year found	Location
1.	*Arctotheca calendula*	Capeweed	1986	Australia
2.	*Conyza bonariensis* (*Conyza linifolius*) (*Erigeron bonariensis*)	Hairy fleabane	1979 1984	Egypt Hungary
3.	*Conyza sumatrensis* (*Erigeron sumatrensis*)		1987	Japan
4.	*Epilobium ciliatum*	American willowherb	1984	Belgium
5.	*Erigeron canadensis*	Canadian fleabane	1982	Japan
6.	*Erigeron philadelphicus*	Philadelphia fleabane	1980	Japan
7.	*Hordeum glaucum*	Barley grass	1983	Australia
8.	*Hordeum leporinum*		1989	Australia
9.	*Poa annua*	Annual meadow grass	1978	U.K.
10.	*Youngia japonica*	Asiatic hawkbeard	1987	Japan

ECOLOGICAL AND PHYSIOLOGICAL FITNESS

These Compositae species are dominant in the early stage of secondary
succession, and many species are plants that originated in North or South
America which became naturalized in Japan in modern times.

Figure 4 shows the distribution of a paraquat-resistant biotype of *E.
philadelphicus* in mulberry fields along the Arakawa River at Fukiage Town in
April, 1981. The ratio of resistant to susceptible was high in the
mulberry plots with a large population of this weed. On the other hand,
the ratio was low in the plots with a smaller weed population. This
Figure also shows that untreated areas have no paraquat-resistant *E.
philadelphicus* (Watanabe *et al*., 1982).

The survey of farmers by Ito (see above) showed that the paraquat-
resistant biotype was not detected in the areas where mulberry
cultivation had been discontinued only 2 or 3 years ago as shown in Table

2; areas treated every year showed a high ratio of resistant biotypes (Watanabe *et al.*, 1982).

We found that the growth of a paraquat-susceptible biotype of *E. canadensis* was more vigorous than that of the resistant biotype in the absence of paraquat (Matsunaka and Moriyama, 1987) (Fig. 5).

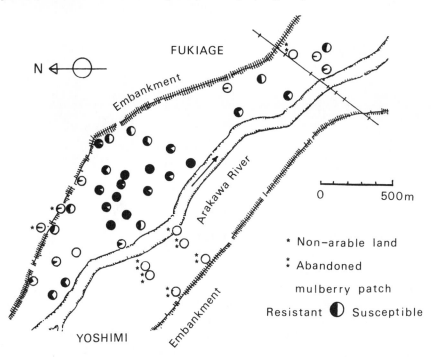

Fig. 4. <u>Distribution of paraquat-resistant biotypes of *Erigeron philadelphicus* in mulberry fields on the river land of the Arakawa River in April 1981</u> (revised from Watanabe *et al.*, 1982).

Table 2. <u>Percentage of paraquat-resistant biotypes in the total number of *Erigeron philadelphicus* plants examined under different frequencies of paraquat application, determined in April, 1982.</u>

Paraquat application	Land use	No. of samples	Resistant biotypes (%)
Never applied	Vacant*	10	2.1
2-3 years ago	Abandoned mulberry plots	2	0.0
Sometimes**	Vicinity of mulberry plots	14	47.5
2-3 times every year	Mulberry plots	24	80.5

* : Embankment, unused land etc.
** : sometimes receiving paraquat due to drift or boundary application in adjacent mulberry patches.

Recently Saka *et al.* (1989) reported that a susceptible biotype of *E. phila-delphicus* showed more vigorous photosynthetic activity than a resistant one at high light intensity and also high CO_2 concentrations.

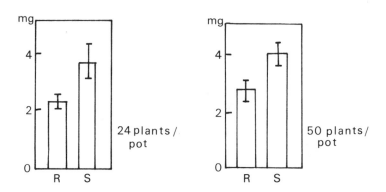

Fig. 5. <u>Competitive growth in growth chamber between paraquat resistant</u> <u>(R) and susceptible (S) biotypes of *Erigeron philadelphicus.*</u> Expressed in dry weight per plant (Matsunaka and Moriyama, 1987).

These facts seem to suggest that often the resistant biotype is weakly competitive with the normal susceptible one in the absence of paraquat application. It implies that the ecological fitness of the resistant biotype is lower than that of the normal susceptible one, as shown for the atrazine-resistant biotype of *Senecio vulgaris* and *Amaranthus retroflexus* by Conard and Radosevich (1979). The paraquat-resistant biotype was inferior before application of paraquat, but became dominant by selection pressure after long term application of this herbicide.

DISTRIBUTION OF PARAQUAT-RESISTANT WEEDS IN JAPAN

In late spring 1989, the scientists of the Ministry of Agriculture, Forestry and Fishery of Japan circulated a questionnaire on herbicide resistance covering weed species, herbicides, location and frequency of the treatment per year, to weed control researchers at agricultural experiment stations involved in sericulture, horticulture and tea cultivation in Japan. The full results will be published soon after their analysis. Some of their results (Satoh *et al.* pers. comm.) relating to paraquat revealed that paraquat-resistant *E. canadensis* is widely distributed in the south-west half of Japan from the Kanto area to Kyushu, except Okinawa. The second most widely distributed species is *E. philadelphicus*, which is located in the Kanto area and Yamanashi Prefecture. The distribution of the newly-discovered paraquat-resistant biotypes of *C. sumatrensis* and *Y. japonica* was more restricted than that of *E. philadelphicus*, being found only in Saitama, Gumma and Ibaraki Prefectures in the Kanto area. Thus the paraquat-resistant Compositae species are highly concentrated in the Kanto-Tosan area around Tokyo.

Analyses of herbicide use showed that in the Saitama Prefecture, where three Compositae biotypes resistant to paraquat were first found, the paraquat supply, total amount and amount per hectare, has increased considerably from the late 1970s; also, the total annual shipment into Japan of Gramoxone - the formulated product of paraquat - has been almost 5,000 kl. The Kanto-Tosan area is consuming about one third (32-38%), although the number of Prefectures in this area is only 19% of Japan. In this district sericulture is still active, so there is a relatively large area of mulberry tree cultivation.

The distribution of main Compositae in Japan is shown in Figure 6 (Kasahara, 1968). *E. canadensis* and *Y. japonica* occur all over Japan, but *E. philadelphicus* and *C. sumatrensis* show somewhat limited distribution. *Erigeron philadelphicus*, *C. sumatrensis* and *Y. japonica* have their paraquat-resistant biotypes only in E area, the Kanto-Tosan district described above. On the other hand. *E. canadensis* resistant to paraquat has been found in E, F, G, I and J areas from Kanto to Kyushu.

Species	Region									
	A	B	C	D	E	F	G	H	I	J
E. canadensis	5	5	5	5	4[a]	4[a]	5[a]	4[a]	3[a]	4[a]
E. philadelphicus		+			+[a]	+	+	+	+	
C. sumatrensis				+	+[a]	+	+	+	+	+
Y. japonica	3	3	3	3	3[a]	3	3	4	3	3
C. bonariensis	4	1	1	2	4	4	4	4	3	4
E. annuus	5	5	3	3	4	4	5	3	3	4

Population High (5)---Low (1)
+ : Density unknown.
[a] : Area where paraquat resistance found.

Fig. 6. <u>The flora of main Compositae weeds in Japan and areas where the paraquat-resistant weeds were found</u> (adapted from Kasahara, 1968).

GENETICAL PROPERTIES OF RESISTANCE

As to the genetical properties, Itoh and Miyahara (1984) reported that the segregation after many cross experiments provided evidence that a single dominant gene was responsible for the paraquat resistance in *E. philadelphicus*. Islama and Powles (1988) showed that the paraquat resistance in *Hordeum glaucum* was controlled by a single nuclear gene with incomplete dominance. Recently Shaaltiel *et al.* (1988) reported that the paraquat resistance in *C. bonariensis* which originated in Egypt is conferred by a single dominant gene.

PHYSIOLOGICAL ASPECTS OF THE RESISTANCE

As to the physiological aspects of the paraquat resistance, Tanaka *et al.* (1986) reported that difference in paraquat movement may be associated with the resistance in the two species, *E. philadelphicus* and *E. canadensis*.

However, in our experiments using *E. canadensis* in Kobe University, the resistance to paraquat measured by chlorophyll bleaching was observed at the protoplast level. Leaf disks and protoplasts showed a lower O_2 consumption as determined as the Mehler reaction in the presence of paraquat in the resistant biotype compared with the susceptible (Fig. 7), while in the intact chloroplasts there was no difference between biotypes (Kawaguchi, 1989).

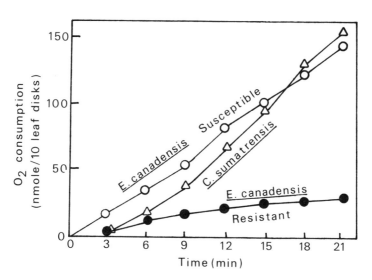

Fig. 7. <u>Oxygen consumption of leaf disks in the presence of paraquat</u> (Kawaguchi, 1989).

Table 3. <u>Activity of superoxide detoxifying enzyme system in extracts</u> <u>of intact chloroplasts of *E. canadensis*</u> (Kawaguchi, 1989).

Enzyme	Susceptible	Resistant (% of susceptible)
Superoxide dismutase	17.1*	28.9 (169)
Ascorbate peroxidase	5.2**	9.3 (179)
Glutathione reductase	1.6***	2.6 (161)
Monodehydroascorbate reductase	0.64***	1.2 (194)

*: unit mg protein^{-1}, **: μmol mg protein^{-1}, ***: mmol NADPH mg protein^{-1}

On the other hand, the activity of the enzymes involved in the superoxide detoxifying enzyme system, i.e. superoxide dismutase, ascorbate peroxidase, glutathione reductase and monodehydroascorbate reductase, was nearly two-fold greater in the resistant biotype compared with that of the susceptible biotype as shown in Table 3 (Kawaguchi, 1989). These data resemble those of Shaaltiel and Gressel (1986) and suggest that the resistance mechanism may be present at the protoplast level.

FUTURE PROBLEMS

From our knowledge of genetical properties, weed distribution, paraquat-diquat usage and the finding of resistance in Egypt and Hungary, we foresee that the next candidate for paraquat resistance in Japan may be *Conyza bonariensis* L. (Hairy fleabane).

Several years ago, paraquat formulations, which were widely distributed to farmer's stores, were sometimes abused or misused for suicide or murder. Then many counter-measures were taken, such as the addition of dark blue pigment and an emetic. Now the use of paraquat has been replaced by combined diquat plus paraquat formulation and at two-fold dilution compared with 1987. This is of significance since some paraquat resistant plants are susceptible to diquat.

Some of the farmers have begun to use other herbicides such as glyphosate, bialophos or glufosinate which can control perennial weeds by foliar application. However, the diquat plus paraquat combination will remain an important herbicide for farmers because of its effectiveness.

REFERENCES

Conard, S.G. and Radosevich, S.R. (1979). Ecological fitness of *Senecio vulgaris* and *Amaranthus retroflexus* biotypes susceptible or resistant to atrazine. Journal of Applied Ecology **16**, 171–177.

Hanioka, Y. (1987). Paraquat resistance in *Erigeron sumatrensis* Ritz. and *Youngia japonica* D.C. Weed Research, Japan **32** (Suppl.), 137–140 (in Japanese).

Islama, A.K.M.R. and Powles, S.B. (1988). Inheritance of resistance to paraquat in barley grass *Hordeum glaucum* Steud. Weed Research **28**, 393–397.

Itoh, K. and Miyahara, M. (1984). Inheritance of paraquat resistance in *Erigeron philadelphicus* L. Weed Research, Japan **29**, 301–307.

Kasahara, Y. (1968). In "Weeds of Japan Illustrated" pp. 518. Yokendo Ltd., Tokyo (in Japanese).

Kato, A., Okuda, Y., Juri, T., Dan, M. and Uejyo, Y. (1982). Resistance to paraquat and diquat in *Erigeron canadensis* L. Bulletin of Osaka Agricultural Research Center **19**, 59–64 (in Japanese with English summary).

Kawaguchi, S. (1989). Utilization of paraquat resistance in *Erigeron canadensis* for breeding of paraquat resistant crop and a study on the paraquat resistance mechanism. Masters' thesis, Kobe University, pp. 72 (in Japanese).

Matsunaka, S. and Moriyama, A. (1987). A property of paraquat resistant *Erigeron canadensis* L. Weed Research, Japan 32 (Suppl.), 141-142 (in Japanese).

Pölös, E., Mikuras, J., Szigeti, Z., Matkovics, B., Quyhai, D., Parducz, A. and Lehoczki, E. (1986). Paraquat and atrazine co-resistance in *Conyza canadensis* (L.) Conq. Pesticide Biochemistry and Physiology 30, 142-154.

Saka, H., Takanashi, J., Chisaka, H. and Uezono, T. (1989). Photosynthesis characteristics in paraquat-resistant biotype of fleabane (*Erigeron philadelphicus*). Abstracts of the 12th Asian-Pacific Weed Science Society Conference, 154.

Shaaltiel, Y. and Gressel, J. (1986), Multienzyme oxygen radical detoxifying system correlated with paraquat resistance in *Conyza bonariensis*. Pesticide Biochemistry and Physiology 26, 22-28.

Shaaltiel, Y., Chua, N-H., Gepstein, S. and Gressel, J. (1988). Dominant pleiotropy controls enzymes co-segregating with paraquat resistance in *Conyza bonariensis*. Theoretical and Applied Genetics 75, 850-856.

Tanaka, Y., Chisaka, H. and Saka, H. (1986). Movement of paraquat in resistant and susceptible biotypes of *Erigeron philadelphicus* and *E. canadensis*. Physiologia Plantarum 66, 605-608.

Watanabe, Y., Honma, T., Itoh, K. and Miyahara, M. (1982). Paraquat resistance in *Erigeron philadelphicus* L. Weed Research, Japan 27, 49-54.

CHLOROPLASTIC RESISTANCE OF WEEDS TO TRIAZINES IN EUROPE

J.L.P. van Oorschot

Centre for Agrobiological Research, P.O. Box 14, 6700 AA Wageningen, The Netherlands

This review is restricted to European weeds whose resistance to triazines is located in the chloroplast. The resistance factor is the ratio between the concentrations that produce the same effect in resistant and susceptible biotypes. It is very high for this form of resistance, especially when the inhibition of electron transport of isolated chloroplasts from both biotypes is compared. In intact plants such resistance factors are lower, but they are still considerably above those derivable from detoxification-induced resistance.
Not all methods used to determine herbicide resistance immediately discriminate between both forms of resistance. Field treatments, whole-plant studies, and also tests based on total photosynthesis such as the rate of CO_2 uptake or the sinking leaf disc technique will not lead to this distinction. The chloroplastic nature of the resistance can be demonstrated decisively in isolated chloroplasts treated with the herbicide, both by measuring electron transport and by inducing chlorophyll fluorescence after a period of darkness. A direct measurement of this fluorescence induction in detached leaves that have absorbed sufficient herbicide for electron transport to be completely inhibited, may also reveal whether the plants are resistant at the chloroplast level.
The number of species whose chloroplastic resistance is confirmed, and their distribution vary considerably over countries in Europe, largely because of variation in the extent of yearly treatments with triazines applied to continuously cropped maize in some countries. In other countries, perennial cultures such as orchards or vineyards are treated annually with triazines. How diligently this resistance is looked for also seems to be important. To date, 36 weed species with this resistance have been reported in 16 European countries. At present, farmers' growing familiarity with triazine resistance, and the use of alternative herbicides to overcome this resistance means that it is often no longer visible in the field. The incidence of cross-resistance to other herbicides complicates the picture.

INTRODUCTION

Resistance of plants to triazines has been known since these herbicides have been used for weed control. In fact, this phenomenon made it possible to use these compounds in resistant crops, but this means that certain weeds are also resistant. An important reason for this resistance is detoxification of these herbicides in some weed species. Furthermore, repeated use over several years led to a specific resistance appearing in the chloroplast. As the title of this paper indicates, resistance due to detoxification will be excluded from this review, but this is not because this form of resistance to triazines is unimportant. On the contrary, detoxification-induced resistance is also widespread, but published information on its distribution is very scattered and incomplete. It is obvious that the increase in maize cropping in Europe in recent years has led to a large increase in panicoid grasses such as *Digitaria* spp., *Echinochloa* spp. and *Setaria* spp., even in those countries where these genera used to be rare before the maize boom. The spread of these grasses to more northerly countries is probably not only connected with the soil management for this crop (which is also grown for forage), but also with the ability of these species to detoxify atrazine to some extent, in a similar way to maize (Jensen *et al.*, 1977).

This review will begin by discussing the various methods of determining resistance to triazines. Some of these methods do not discriminate between different mechanisms of resistance, but others unambiguously identify its chloroplastic nature. Then, the present occurrence of this chloroplastic resistance of weeds to triazines in various countries in Europe will be summarized. These data are based on published records, and on a recent inquiry among colleagues. In most of these studies the chloroplastic nature has been confirmed, but for some data the methods employed do not exclude other mechanisms. However, in these cases too, the assumption of chloroplastic resistance is obvious. Finally, cross-resistance to other herbicides is discussed, to complete the picture.

RESISTANCE FACTORS

The resistance of a weed to a herbicide can be quantified by the ratio between the concentrations required to produce the same effect in the resistant and susceptible biotype of a plant species. This factor has been introduced to compare the effect of triazines or other herbicides on electron transport in isolated chloroplasts of resistant and susceptible biotypes. Usually it represents the ratio between the concentrations that inhibit this transport by 50%. In this way, values between 100 and 1000 have been observed for the resistance factor in isolated chloroplasts of triazine-resistant biotypes of various weed species (Böger, 1981; Arntzen *et al.*, 1982).

For complete photosynthesis this factor appears to be much lower (van Oorschot and van Leeuwen, 1988). The reason for the discrepancy with that of electron transport in isolated chloroplasts is not clear. When the factor is determined from the growth of seedlings, there appears to be a large variation (from 38 to 1550) in various weed species (see Table 5). Here, the duration of the growth experiment will influence the results.

Similar factors for assessing detoxification-induced resistance to triazines are not directly available, but it is estimated that inactivation of atrazine in panicoid grasses (Jensen *et al.*, 1977), and of metamitron in sugar beet (van Oorschot and van Leeuwen, 1979), the most active plant species observed to date, is not more than 10 times that in plants with a weak ability to detoxify these herbicides. For intact plants, Beuret (1988) estimated a factor of 4 to 8 for a non-chloroplastic resistant biotype of *Amaranthus lividus* as related to the susceptible biotype. It is obvious that this type of resistance (also often named tolerance) is less effective than chloroplastic resistance.

88

METHODS USED FOR THE DETERMINATION OF RESISTANCE

Various methods (Table 1) have been used to determine whether plants are resistant to herbicides (Truelove and Hensley, 1982; Clay and Underwood, 1989), but not all of them enable the different mechanisms of resistance to be distinguished.

Table 1. <u>Methods for identification of resistant biotypes</u>.

Field treatments
Whole-plant studies
Floatation of leaf discs
Leaf photosynthesis
Electron transport of isolated chloroplasts
Chlorophyll fluorescence in isolated chloroplasts
Chlorophyll fluorescence in intact leaves

The first two methods (field treatments and whole-plant studies) will reveal neither the type of action of the herbicide, nor the mechanism of resistance that is involved here. In field treatments, failure of control is not always caused by resistance, even at increased dosage. For most whole-plant studies, seeds must be collected from plants suspected of being resistant, and then germinated and grown in comparison with a known susceptible biotype. A range of doses is either applied to the soil or dissolved in nutrient solution. Various herbicide concentrations have also been used in the Petri-dish method (Clay and Underwood, 1989), to compare the germination and early growth of seedlings of suspected biotypes with those of susceptible types.

The test with leaf discs floating on a phosphate buffer that contains the herbicide is more specific, because photosynthesis is involved in the buoyancy of these discs. Undisturbed photosynthesis of resistant biotypes produces sufficient oxygen in the intercellular air spaces to keep the discs floating. The method was first described by Truelove *et al.* (1974), and used by Gawronski *et al.* (1977) to reveal different tolerance of potato varieties to metribuzin. Obviously, this method does not discriminate between detoxification and chloroplastic resistance. The rapid and simple procedure can be carried out directly with leaves from suspected plants, but not all plant species can be used in this test. The direct measurement of leaf photosynthesis of plants treated with herbicide may reveal clear differences between resistant and susceptible biotypes (van Oorschot and van Leeuwen, 1988), but does not immediately indicate whether the resistance is of chloroplastic nature or should be ascribed to detoxification. However, additional recovery experiments after temporary inhibition of photosynthesis by root-applied herbicides may rule out the detoxification mechanism (van Oorschot, 1976).

Isolated chloroplasts can be used to ascertain whether the resistance to herbicides that inhibit electron transport is at the chloroplastic level. Measurement of electron transport in these chloroplasts will indicate herbicide activity at the specific Q_B protein, and inhibition of this transport (see Trebst, these Proceedings). Much higher concentrations will be needed for the same inhibition in isolated chloroplasts of resistant biotypes than in those of susceptible biotypes. Measurement of fluorescence induction of treated chloroplasts also clearly indicates whether or not electron transport is inhibited (e.g. Arntzen *et al.*, 1982). In the first seconds of illumination after a period of darkness, fluorescence rises rapidly to its maximum. It is followed by a gradual decrease to a lower level. Inhibition of electron transport in chloroplasts of treated, susceptible biotypes induces fluorescence to rise immediately to the final value in the first milliseconds, and to stay high because electron transport is blocked in the Q_B protein. Absence of this inhibition in chloroplasts of resistant biotypes results in the normal fluorescence transient curve (see also Fig. 1).

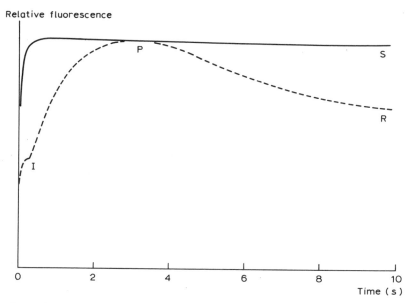

Fig. 1. <u>Induction curve of fluorescence of resistant (R) and susceptible</u> <u>(S) biotypes of *Solanum nigrum* treated with 0.1 mmol 1^{-1} of atraton for 16</u> <u>h, as measured with the SF-10 probe of Brancker</u>. (P = maximum, I = inflection point) (after van Oorschot and Straathof, 1988).

Special training is required, to be able to carry out the complex procedure to isolate chloroplasts, but fortunately, fluorescence induction can also be measured in intact leaves. If these leaves have been treated with enhanced herbicide concentrations prior to measurement, these induction curves will also indicate whether the plant is resistant at the chloroplast level (Ducruet and Gasquez, 1978). Here, complete inhibition of electron transport also leads to the immediate rise of fluorescence to its final and constant value, while absence of such inhibition results in a gradual rise and fall of fluorescence (Fig. 1).

Measurement of fluorescence of detached leaves is simple and can be carried out directly on leaves collected from suspected plants in the field. The only prerequisite is that pretreatment with herbicide should ensure sufficient absorption to inhibit electron transport. Control treatments of leaves from resistant, suspected plants with diuron will show the complete inhibition of this electron transport. The use of standardized conditions is essential. Resistance by detoxification will not be shown up by this method because electron transport in the chloroplasts of such plants is easily inhibited by high concentrations, as Ducruet and Gasquez (1978) observed for maize and atrazine.

OCCURRENCE OF CHLOROPLASTIC RESISTANCE IN EUROPE

The first published records on chloroplastic resistance of weeds to
triazines in Europe are from France (Ducruet and Gasquez, 1978), Germany
(Kees, 1978) and Switzerland (Ammon, 1977) where resistant *Chenopodium
album*, *Poa annua* and *Stellaria media* had been found in fields with continuous
maize cropping, and in cities where triazines had also been applied
yearly. These reports appeared some years after those in North America,
but the actual findings had been made earlier in the 1970s (for details,
see Gressel *et al.*, 1982). The spread of the maize crop across Europe, and
the increase in the continuous cropping system in recent decades caused
a rapid increase in the number of resistant weed species, and in the
number of countries and locations where they occur. Table 2 shows the
present situation in European countries. The data are derived from
published records and recent personal communications by colleagues.

Table 2 lists 36 weed species of which biotypes resistant to triazines
have been recorded in 16 European countries. In Belgium, Switzerland,
Czechoslovakia, Spain, France, Hungary, Italy, the Netherlands and
Yugoslavia the chloroplastic nature of this resistance has been confirmed
by measuring electron transport or fluorescence in isolated chloroplasts,
or fluorescence in intact leaves. In Austria and West Germany, the tests
were carried out by floating leaf discs on a solution of atrazine, while
whole plants were treated in Bulgaria, East Germany, Denmark and Poland.
In the United Kingdom, the resistant biotypes were identified with the
leaf disc flotation method, or in whole-plant studies including the
Petri-dish method.

Of this total, 29 species are dicotyledonous and 7 monocotyledonous,
divided over several families. In a worldwide survey, LeBaron and
McFarland (1988) listed in total 40 dicotyledonous and 15
monocotyledonous weed species with triazine resistance. In Europe some
genera have various species with triazine resistance such as *Amaranthus*
(7), *Chenopodium* (3), *Polygonum* (4) and *Setaria* (3). To date, the largest
numbers of resistant biotypes have been found in France (21 species),
Germany (14 species) and Switzerland (13 species). *C. album* has spread
almost everywhere (15 out of 16 countries). *Amaranthus retroflexus* has been
recorded in 10 countries, *Erigeron canadensis*, *Solanum nigrum* and *Senecio vulgaris* in
9, and *P. annua* in 8 countries. There are clear indications that the
intensity of searching for chloroplastic resistance is an important
factor in the known incidence of this phenomenon (van Oorschot, 1989).

To date, no direct evidence for resistance to triazines is available from
Ireland, Finland, Greece, Norway, Portugal, Romania, Sweden, the Soviet
Union and Turkey. The absence of extensive resistance in northern Europe
could be explained by the unsuitable climate for maize production, but
resistance of *Chamomilla suaveolens*, *Gnaphalium uliginosum* and *P. annua* is suspected
in nurseries in Norway (Lund-Høie, pers. comm.). Also subject to
suspicion are *Aphanes arvensis*, *Epilobium montanum*, *Galium aparine*, *S. vulgaris*, *Sherardia
arvensis* and *Viola arvensis* in fruit crops in Ireland (Mac Giolla Ri, 1989), and
A. retroflexus, *C. album*, *Echinochloa crus-galli* and *S.nigrum* in Portugal (Rosário,
pers. comm.). The absence of such resistant biotypes in Greek vineyards

Table 2. <u>Occurrence of chloroplastic resistance to triazines in various species in European countries</u>.

Weed species	A	B	BU	CH	CS	D	DD	DK	E	F	H	I	NL	PL	UK	Y
Amaranthus albus									+							
Amaranthus blitoides									+							
Amaranthus bouchoni				+						+	+	+				
Amaranthus cruentus									+		+					
Amaranthus hybridus				+					+	+	+	+				
Amaranthus lividus				+					+							
Amaranthus retroflexus	+		+	+	+	+	+		+	+	+					+
Ambrosia artemisiifolia																+
Arenaria serpyllifolia										+						
Atriplex patula	+					+										
Bidens tripartita	+															
Bromus tectorum										+						
Chamomilla suaveolens															+	
Chenopodium album	+	+	+	+	+	+	+			+	+	+	+	+	+	+
Chenopodium ficifolium			+		+					+	+		+			
Chenopodium polyspermum			+		+					+	+					
Digitaria sanguinalis										+			+			
Echinochloa crus-galli	+		+							+			+			
Epilobium ciliatum		+						+					+		+	
Epilobium tetragonum						+				+						
Erigeron bonariensis									+							
Erigeron canadensis				+	+	+	+			+	+		+		+	+
Galinsoga ciliata					+						+					
Myosoton aquaticum				+	+											
Poa annua	+	+		+	+	+				+			+		+	
Polygonum aviculare													+		+	
Polygonum convolvulus	+				+								+			
Polygonum lapathifolium				+	+					+			+			
Polygonum persicaria										+			+			
Senecio vulgaris		+		+	+	+	+	+		+			+		+	
Setaria glauca								+		+						
Setaria viridis										+						
Setaria viridis major										+						
Solanum nigrum	+	+		+		+			+	+	+	+			+	
Sonchus asper										+						
Stellaria media	+			+		+										

*A Austria (Neururer, 1986), B Belgium (Bulcke and van Himme, 1989), BU Bulgaria (Fetvadjieva *et al.*, 1985; Nikolova, pers. comm.), CH Switzerland (Ammon and Beuret, 1984), CS Czechoslovakia (Chodova, 1988 and pers. comm.), D West Germany (Kees, 1988), DD East Germany (Arlt and Jüttersonke, 1987; Arlt, pers. comm.), DK Denmark (Noyé, 1989; Rubow, 1989), E Spain (De Prado *et al.*, 1989; De Prado, pers. comm.), F France (Gasquez and Darmency, 1989), H Hungary (Solymosi and Lehoczki, 1989a; Solymosi *et al.*, 1986; Pölös *et al.*, 1988), I Italy (Zanin *et al.*, 1984; Zanin, pers. comm.), NL Netherlands (van Oorschot, 1989), PL Poland (Gawronski, 1986 and pers. comm.), UK United Kingdom (Clay and Underwood, 1989; Clay, 1989), Y Yugoslavia (Igrc, 1987; Janjic *et al.*, 1988; Sovljanski *et al.*, 1989; Arsenovic, pers. comm.).

(Damanakis and Giannopolitis, 1989) has been ascribed to the use of mixtures. In Romania *A. retroflexus*, *Chenopodium polyspermum*, *E. canadensis* and *S. nigrum* are suspected to be resistant (Ciorlaus, 1988), while in addition to the data for Poland given in Table 2 there is also suspicion for resistance of *A. retroflexus* (Gawronski, pers. comm.), *Capsella bursa-pastoris* (Lipecki, 1988a), *Polygonum aviculare* and *Atriplex patula* (Lipecki, 1988b).

The occurrence of chloroplastic resistance of weeds to triazines in a number of crops is given in Table 3. This shows that most of these resistant weeds occur in maize (28) and orchards (16), the crops that are ubiquitous and widespread in Europe, and that lead to the same areas being treated yearly with triazines because maize is cropped continuously, and trees are grown in the same place for many years. However, another reason could be that more studies have been done in these crops. Many weed species occur in several crops, especially *A. retroflexus*, *E. canadensis*, *P. annua*, *S. vulgaris* and *S. nigrum*.

Detailed surveys on resistance of weeds to triazines are scarce, and data on crop areas which are infested with resistant weed biotypes are fragmentary for many countries. An extensive survey made in France indicated that already in 1982 about one-quarter of the maize area was infested with resistant *C. album*, *S. nigrum* and *A. retroflexus* (Gasquez *et al.*, 1982). Now the infestation has more than doubled (Gasquez and Darmency, 1989). For Bavaria (West Germany), Kees (1988) estimated that in 1987 one-third of the maize area could have been infested, but this is now as much as two-thirds (Kees, 1989). In Belgium, Bulcke and van Himme (1989) estimated that more than 10% of the maize fields are infested with *C. album*, and about 50% with *S. nigrum*. In Austria, the most widespread resistant biotype of *A. retroflexus*, was estimated to occur on 40% of the maize area (Neururer, 1986). Zanin *et al.* (1984) estimated that 10% of the maize area in Italy was infested with resistant *C. album*, *S. nigrum* and *Amaranthus cruentes*, but this percentage is now probably lower, because of restrictions on the use of atrazine, and the replacement of maize by soybeans (Zanin, pers. comm.). The estimate that 4% of the maize area in the Netherlands is infested with resistant biotypes seems low. It was based on determinations from collected soil samples, and this could have underestimated the actual situation (van Oorschot and Straathof, 1988).

In orchards in Eastern Poland, Lipecki (pers. comm.) observed an increase of resistant *C. bursa-pastoris*. The infestation with resistant biotypes also seems to be increasing rapidly in Hungary (Solymosi and Kostyal, 1985) and Czechoslovakia (Chodova, pers. comm.). Herbicide resistance is now widespread in England in several weeds of fruit and ornamental crops (Clay and Underwood, 1989), while resistance in *S. nigrum* and *C. album* has been found recently in forage maize (Clay, 1989). The increasing familiarity of farmers with triazine resistance, and the possibility of using other pre-emergence herbicides to overcome this, means that resistance is less likely to show up in the field (Ammon, pers. comm.). This impedes further surveys of the phenomenon.

Table 3. <u>Occurrence of resistant weed species in various crops in European countries</u>[*].

Weed species	Maize	Orchards	Nurseries	Vineyards	Uncultivated
Amaranthus albus		+			
Amaranthus blitoides		+			
Amaranthus bouchonii	+				
Amaranthus cruentus	+				
Amaranthus hybridus	+				
Amaranthus lividus	+	+	+		
Amaranthus retroflexus	+	+	+	+	
Arenaria serpyllifolia				+	
Atriplex patula	+	+			
Bidens tripartita	+				
Bromus tectorum				+	
Chamomilla suaveolens		+			
Chenopodium album	+	+	+		
Chenopodium ficifolium	+				
Chenopodium polyspermum	+		+		
Digitaria sanguinalis	+	+			
Echinocloa crus-galli	+	+			
Epilobium ciliatum	+	+	+		
Epilobium tetragonum	+		+		
Erigeron bonariensis		+	+		
Erigeron canadensis	+	+	+	+	+
Galinsoga ciliata	+				
Myosoton aquaticum	+				
Poa annua	+	+	+	+	
Polygonum aviculare	+	+			
Polygonum convolvulus	+		+		
Polygonum lapathifolium	+			+	
Polygonum persicaria	+		+		
Senecia vulgaris	+	+	+	+	
Setaria glauca	+				
Setaria viridis	+				
Setaria viridis major	+				
Solanum nigrum	+	+	+	+	
Sonchus asper			+		
Stellaria media	+				

[*] Data derived from the same sources as given under Table 2.

CROSS-RESISTANCE

In practice, cross-resistance to other herbicides usually appears when the continuous maize cropping system is abandoned, and maize is alternated with other crops such as sugar beet. It may limit the efficacy of chloridazon and metamitron in this crop. It is also observed in orchards (Clay and Underwood, 1989). Cross-resistance to other herbicides which inhibit photosynthetic electron transport is not unexpected, because the target is the same, but that to herbicides with another action such as paraquat (Pölös *et al.*, 1988) and diclofop-methyl (Rubin *et al.*, 1985) is less obvious. To date, the phenomenon of cross-resistance has not always caused serious problems in practice (Solymosi and Lehoczki, 1987; Clay and Underwood, 1989), since alternative herbicides are available.

The resistance factor is used to quantify resistance and cross-resistance in isolated chloroplasts. A few examples for *Amaranthus hybridus* and *C. album* are given in Table 4. All data indicate a large factor for atrazine in both species. The factor is only somewhat lower for other herbicide groups such as triazinones (e.g. metribuzin). With one exception, a much lower factor is usually observed for uracils (e.g. lenacil and bromacil) and other herbicide groups, but there is still considerable cross-resistance. This is not so with diuron, whose effect on isolated chloroplasts of resistant and susceptible biotypes is very similar. With other urea herbicides the resistance factor is sometimes much higher; with monolinuron, for example, De Prado *et al.* (1989) found 103 for *A. hybridus* and 79 for *C. album*. Solymosi and Lehoczki (1989a) observed cross-resistance to diuron in isolated chloroplasts of *Amaranthus bouchonii*. Chloroplasts of resistant biotypes are not only more sensitive to bentazone (factor < 1, Table 4), but also to ioxynil, DNOC and pyridate. However, Solymosi and Lehoczki (1989b) found a biotype of *C. album* of which isolated chloroplasts also showed cross-resistance to pyridate.

The question of whether resistance and cross-resistance factors obtained from isolated chloroplasts can be extrapolated to field conditions remains open. In series of experiments where complete photosynthesis of intact leaves had been measured, van Oorschot and van Leeuwen (1988) found much lower resistance factors for 50% inhibition of photosynthesis in leaves of *A. retroflexus*, *S. nigrum* and *Brassica napus*. The resistance factor for atraton (comparable with atrazine) in these plant species varied between 26 and 30. Cross-resistance was found for metamitron (resistance factor 3-7), bromacil (factor 2-9), monolinuron (factor 3-5), but not for diuron (factor 0.8-1.1) and bentazone (factor 0.7-0.9). All values (except those for diuron and bentazone) are not more than one-tenth of those obtained from isolated chloroplasts. The reason for the discrepancy is not clear. Cross-resistance to some urea herbicides has also been found by measuring fluorescence induction of triazine-resistant *E. canadensis* (Pölös *et al.*, 1987), *C. album* and *Amaranthus* spp. (Solymosi and Lehoczki, 1989a), but this method does not provide quantitative comparisons.

Table 4. **Resistance factors for some herbicides in isolated chloroplasts of biotypes of** *Amaranthus hybridus* **and** *Chenopodium album* **(ratio between concentrations which cause 50% inhibition of electron transport, R/S).**

Plant species	Herbicides					References[*]
	atrazine	metri-buzin	lenacil	diuron	benta-zone	
Amaranthus						
hybridus	1000		22[+]	1.4	0.5	a
	830	260	20[+]	1.4	0.7	b
	883	190		1.6	0.7[x]	c
	1033		2040	1.3		d
Chenopodium						
album	215		26	1.6	0.3	a
	1300	33	88[+]		0.3	b
	542		167	1.2	0.8[x]	c
	1250	14		1.3		d
	430	100	56	1.2	0.4	e

[*]References: a Arntzen *et al.* (1982); b Fuerst *et al.*(1986); c De Prado *et al.* (1989); d Solymosi and Lehoczki (1989a); e Böger(1981).
[+]Bromacil and [x]ioxynil applied here.

Table 5. **Resistance factors for some herbicides in intact seedlings of biotypes of various weed species (ratio between concentrations which cause 50% inhibition of growth, R/S).**

Plant species	Herbicides					References[*]
	atrazine	metri-buzin	bromacil	diuron	dino-seb	
Amaranthus hybridus	485		24		0.3	a
Amaranthus retroflexus	125		24	1.0		b
Brassica napus	530		7		0.7	a
Chenopodium album	50		4		0.3	a
Chenopodium album	300	60	6[+]			c
Epilobium ciliatum	38					d
Senecio vulgaris	1550		10		0.2	a
Solanum nigrum	99			3[x]		e

[*]References: a Fuerst *et al.*(1986); b Arntzen *et al.* (1979); c van Dord (1982); d Bulcke *et al.* (1986); e Bulcke *et al.* (1985).
[+]Lenacil and [x]monolinuron applied here.

Resistance factors have also been determined in intact seedlings of resistant and susceptible biotypes of various plant species, but they should be interpreted with caution because the dose response curves of intact plants used to determine the 50% inhibition concentration for both biotypes are not independent of the duration of the growth experiments. There are not many data available on intact plants. Most of these are summarized in Table 5. The resistance factor for atrazine is in the same range as for isolated chloroplasts, and varies between 38 for *Epilobium ciliatum* and 1550 for *S. vulgaris*. There is a clear cross-resistance for metribuzin and bromacil (or lenacil), and less sensitivity for dinoseb. We need more data at the whole plant level. The so-called negative cross-resistance for bentazone has been used to make mixtures with atrazine to control also biotypes that are resistant in maize fields. However, this will probably not retard the further development of triazine resistance, because bentazone will not be active as long as atrazine.

Cross-resistance of triazine-resistant weeds to paraquat (not inhibiting electron transport) has been reported for *E. canadensis* (Clay and Underwood, 1989; Pölös *et al.*, 1988) and *E. ciliatum* (Clay and Underwood, 1989). It could be connected with the long-term use of both compounds at the same place. Rubin *et al.* (1985) observed an increased tolerance of triazine-resistant *Phalaris paradoxa* to diclofop-methyl.

REFERENCES

Ammon, H.U. (1977). Kombination chemisch-, mechanisch- und biologischer Methoden zur Unkrautbekämpfung in mehrjährigen Maisbau und erste Resultate über die Beeinflussung bodenphysikalischer Kenwerte. Proceedings EWRS Symposium Methods of Weed Control and their Integration, 243-254.

Ammon, H.U. and Beuret, E. (1984). Verbreitung Triazin-resistenter Unkräuter in der Schweiz und bisherige Bekämpfungserfahrungen. Zeitschrift für Pflanzenkrankheiten und Pfanzenschutz, Sonderheft X, 183-191.

Arlt, K. and Jüttersonke, B. (1987). Untersuchungen zur Resistenz der Sippen von *Chenopodium album* L. gegen Herbizide. Nachrichtenblatt für den Pflanzenschutz in der DDR 41 ,209-212.

Arntzen, C.J., Ditto, C. L. and Brewer, P.E. (1979). Chloroplast membrane alterations in triazine-resistant *Amaranthus retroflexus* biotypes. Proceedings National Academy of Science USA 76, 278-282.

Arntzen, C.J., Pfister, K. and Steinback, K.E. (1982). The mechanism of chloroplast triazine resistance: Alterations in the site of herbicide action. In "Herbicide Resistance in Plants" (H.M. LeBaron and J.Gressel, eds), pp. 185-214. Wiley-Interscience, New York.

Beuret, E. (1988). Cas particulier de résistance à l'atrazine et au linuron chez *Amaranthus lividus* L. et *Erigeron canadensis* L. Annales de l'Association Nationale pour la Protection des Plantes, 3, (2/1), 277-286.

Böger, P. (1981). Resistance against herbicides inhibiting photosynthesis. Plant Research and Development 13, 40-51.

Bulcke, R., Vleeschauwer, J. de, Vercruysse, J. and Stryckers, J. (1985). Comparison between triazine-resistant and -susceptible biotypes of *Chenopodium album* L. and *Solanum nigrum* L. Mededelingen Faculteit Landbouwwetenschappen Rijksuniversiteit Gent, **50/2a**, 211–220.

Bulcke, R. and Van Himme, M. (1989). Resistance to herbicides in weeds in Belgium. In "Importance and perspectives on herbicide-resistant weeds" (R. Cavalloro and G. Noyé, eds) pp. 31–39. CEC, Luxembourg.

Bulcke, R., Verstraete, F., Van Himme, M. and Stryckers, J. (1986). Variability of *Epilobium ciliatum* Rafin. (syn.: *E. adenocaulon* Hausskn.). Mededelingen Faculteit Landbouwwetenschappen Rijksuniversiteit Gent, **51/2a**, 345–353.

Chodova D. (1988). Researching for atrazine resistant weed populations in Czechoslovakia. Fragmenta Herbologica Jugoslavica **17**, 37–44.

Ciorlaus, A. (1988). Buruienile triazino-resistente, o problema de actualitate si pentru Romania. Al VI-lea Simpozion National de Herbologie, Bucuresti, 73–79.

Clay, D.V. (1989). New developments in triazine and paraquat resistance and co-resistance in weed species in England. Brighton Crop Protection Conference – Weeds 317–324.

Clay, D.V. and Underwood, C. (1989). The identification of triazine- and paraquat-resistant weed biotypes and their response to other herbicides. In "Importance and perspectives on herbicide-resistant weeds" (R. Cavalloro and G. Noyé, eds) pp. 47–55. CEC, Luxembourg.

Damanakis, M. and Giannopolitis, C.N. (1989). Resistant weed ocurrence in chemically weeded vineyards in Greece. In "Importance and perspectives on herbicide-resistant weeds" (R. Cavalloro and G. Noyé, eds) pp. 19–22. CEC, Luxembourg.

De Prado, R., Dominguez, C. and Tena, M. (1989). Triazine-resistant weeds found in Spain. In "Importance and perspectives on herbicide-resistant weeds" (R. Cavalloro and G. Noyé, eds) pp. 67–79. CEC, Luxembourg.

Ducruet, J.M. and Gasquez, J. (1978). Observation de la fluorescence sur feuille entière et mise en évidence de la résistance chloroplastique à l'atrazine chez *Chenopodium album* L. et *Poa annua* L. Chemosphere **8**, 691–696.

Fetvadjieva, N., Nikolova, G. and Konstantinov, K. (1985). Investigation of *Amaranthus retroflexus* L. and *Echinochloa crus-galli* L. resistance to herbicides. Proceedings Symposium of 12th Scientific Meeting of CMEA, Sofia, 37–50.

Fuerst, E.P., Arntzen, C.J., Pfister, K. and Penner, D. (1986). Herbicide cross-resistance in triazine-resistant biotypes of four species. Weed Science **34**, 344–353.

Gasquez, J., Barralis, G. and Aigle, N. (1982). Distribution et extension de la résistance chloroplastique aux triazines chez les adventices annuelles en France. Agronomie **2**, 119–123.

Gasquez, J. and Darmency, H. (1989). Herbicide resistance in France. In

"Importance and perspectives on herbicide-resistant weeds" (R. Cavalloro and G. Noyé, eds) pp. 81-94. CEC, Luxembourg.

Gawronski, S.W. (1986). Inheritance of resistance to triazine herbicides of *Echinochloa crus-galli*. Proceedings Fifth SABRAO Congress **1985**, 797-801.

Gawronski, S.W., Callihan, R.H. and Pavek, J.J. (1977). Sinking leaf-disc test for potato variety herbicide tolerance. Weed Science **25**, 122-127.

Gressel, J., Ammon, H.U., Fogelfors, H., Gasquez, J., Kay, Q.O.N. and Kees, H. (1982). Discovery and distribution of herbicide-resistant weeds outside North America. In "Herbicide Resistance in Plants" (H.M. LeBaron and J. Gressel, eds), pp. 31-55. Wiley-Interscience, New York.

Igrc, J. (1987). The importance of the species *Ambrosia artemisiifolia* in the world and in Yugoslavia. Fragmenta Herbologica Jugoslavica **16**, 47-55.

Janjic, V., Veljovic, S., Jovanovic, Lj., Plesnicar, M. and Arsenovic, M. (1988). Identification of *Amaranthus retroflexus* resistance to atrazine by the method of leaf fluorescence. Fragmenta Herbologica Jugoslavica **17**, 45-54.

Jensen, K.I.N., Stephenson, G.R. and Hunt, L.A. (1977). Detoxification of atrazine in three Gramineae subfamilies. Weed Science **25**, 212-220.

Kees, H. (1978). Beobachtungen über Resistenzerscheinungen bei der Vogelmiere (*Stellaria media*) gegen Atrazin in Mais. Gesunde Pflanzen **30**, 137-139.

Kees, H. (1988). Die Entwicklung triazinresistenter Samenunkräuter in Bayern und Erfahrungen mit der Bekämpfung. Gesunde Pflanzen **40**, 407-414.

Kees, H. (1989). Stand der Triazinresistenz bei der Unkrautbekämpfung im Mais. Bodenkultur und Pflanzenbau **1/89**, 23-26.

LeBaron, H.M. and McFarland, J. (1988). Overview and prognosis of herbicide resistance in weeds and crops. American Chemical Society Meeting, Los Angeles (in print).

Lipecki, J. (1988a). *Capsella bursa-pastoris* (L.) Med. – Another weed resistant to simazine? Acta Societatis Botanicorum Poloniae **57**, 187-189.

Lipecki, J. (1988b). Experiments on the resistance of some weed species occurring in orchards to simazine. Proceedings Xth Meeting of Herbicide Group, Committee of Horticultural Sciences, Polish Academy of Sciences 31-34.

Mac Giolla Ri, P. (1989). Herbicide programmes in soft fruit crops aimed at minimising the development of resistant weed populations. In "Importance and perspectives on herbicide-resistant weeds" (R. Cavalloro and G. Noyé, eds) pp. 111-119. CEC, Luxembourg.

Neururer, H.(1986). Zeitgemässe Unkrautbekämpfung in Mais. Pflanzenschutz 1986, 5-8.

Noyé, G. (1989). New herbicides for control of groundsel (*Senecio vulgaris* L.) in nursery cultures. In "Importance and perspectives on herbicide-resistant weeds" (R. Cavalloro and G. Noyé, eds) pp. 105-110. CEC, Luxembourg.

Pölös, E., Laskay, G., Szigeti, Z., Pataki, S. and Lehoczki, E. (1987). Photosynthetic properties an cross-resistance to some urea herbicides of triazine-resistant *Conyza canadensis* Cronq (L.). Zeitschrift für Naturforschung 42c, 783-793.

Pölös, E., Mikulas, J., Szigeti, Z., Matkovics, B., Do Quy Hai, Parducz, A. and Lehoczki, E. (1988). Paraquat and atrazine co-resistance in *Conyza canadensis* (L.) Cronq. Pesticide Biochemistry and Physiology 30, 142-154.

Rubin, B., Yaacoby, T. and Schonfeld, M. (1985). Triazine resistant grass weeds: cross-resistance with wheat herbicide, a possible threat to cereal crops. British Crop Protection Conference - Weeds 1171-1178.

Rubow, T. (1989). Triazine-resistant weeds, *Conyza canadensis* and *Epilobium ciliatum* : importance and control in young conifer plantations. In "Importance and perspectives on herbicide-resistant weeds" (R. Cavalloro and G. Noyé, eds) pp. 97-104. CEC, Luxembourg.

Solymosi, P. and Kostyal, S. (1985). Mapping of atrazine resistance for *Amaranthus retroflexus* L. in Hungary. Weed Research 25, 411-414.

Solymosi, P. and Lehoczki, E. (1987). New cases of co-resistance in weed stands resistant to atrazine. Novenyvedelem 23, 439-444.

Solymosi, P. and Lehoczki, E. (1989a). Co-resistance of atrazine-resistant *Chenopodium* and *Amaranthus* biotypes to other photosystem II inhibiting herbicides. Zeitschrift für Naturforschung 44c, 119-127.

Solymosi, P. and Lehoczki, E. (1989b). Characterization of a triple (atrazine-pyrazon-pyridate) resistant biotype of common lambsquarters (*Chenopodium album* L.). Journal of Plant Physiology 134, 685-690.

Solymosi, P., Kostyal, Z. and Lehoczki, E. (1986). Characterization of intermediate biotypes in atrazine-susceptible populations of *Chenopodium polyspermum* L. and *Amaranthus bouchonii* Thell. in Hungary. Plant Science 47, 173-179.

Sovljanski, R., Arsenovic, M., Klokocar Smit, Z., Janjic, V. and Ostojic, Z. (1989). Herbicide resistant weeds and search for biocontrol agent in Jugoslavia. In "Importance and perspectives on herbicide-resistant weeds" (R. Cavalloro and G. Noyé, eds) pp. 153-160. CEC, Luxembourg.

Truelove, B. and Hensley, J.R. (1982). Methods of testing for herbicide resistance. In "Herbicide Resistance in Plants" (H.M. LeBaron & J. Gressel, eds), pp. 117-131. Wiley-Interscience, New York.

Truelove, B., Davis, D.E. and Jones, L.R. (1974). A new method for detecting photosynthesis inhibitors. Weed Science 22, 15-17.

van Dord, D.C. (1982). Resistentie van *Chenopodium album* L. (melganzevoet) tegen atrazin en *Poa annua* (straatgras) tegen simazin in Nederland. <u>Mededelingen Faculteit Landbouwwetenschappen Rijksuniversiteit Gent</u> **47**, 37-44.

van Oorschot, J.L.P. (1976). Effects in relation to water and carbon dioxide exchange in plants. <u>In</u> "Herbicides, Physiology, Biochemistry, Ecology". Vol. 1. (L.J. Audus, ed.), pp. 305-333. Academic Press, London.

van Oorschot, J.L.P. (1989). Chloroplastic resistance of weeds to triazines in the Netherlands until 1988. <u>In</u> "Importance and perspectives on herbicide-resistant weeds" (R. Cavalloro and G. Noyé, eds) pp. 41-45. CEC, Luxembourg.

van Oorschot, J.L.P. and van Leeuwen, P.H (1979). Recovery from inhibition of photosynthesis by metamitron in various plant species. <u>Weed Research</u> **19**, 63-67.

van Oorschot, J.L.P. and van Leeuwen, P.H. (1988). Inhibition of photosynthesis in intact plants of biotypes resistant or susceptible to atrazine and cross-resistance to other herbicides. <u>Weed Research</u> **28**, 223-230.

van Oorschot, J.L.P. and Straathof, H.J.M. (1988). On the occurrence and distribution of chloroplastic resistance of weeds to triazines in the Netherlands. <u>Annales de l'Association Nationale pour la Protection des Plantes</u>, **3**, (2/1), 267-275.

Zanin, G., Molle, I. and Vazzana, C. (1984). Distribution et extension des dicotyledones adventices résistantes à l'atrazine en Italie. <u>7ème Colloque Internationale sur l'Ecologie, la Biologie et la Systématique des Mauvaises Herbes, COLUMA-EWRS</u>, 273-280.

MUTATION FOR TRIAZINE RESISTANCE WITHIN SUSCEPTIBLE POPULATIONS OF *CHENOPODIUM ALBUM* L.

Jacques Gasquez

INRA, Laboratoire de Malherbologie, BV 1540, 21034 Dijon Cédex, France

The first reported resistant species in France is *Chenopodium album* which is also the most widespread species all over the world. There is only one genotype within each resistant population and this genotype is the only resistant biotype (R) in the region. But the R genotype may be different according to different regions.
In order to understand how such resistant biotypes occur, many plants have been sampled within several susceptible populations which have never been treated with triazines. In almost all the populations, a high frequency ($\underline{c}$. 3 %) of mutated plants (I) appears among the offspring of some susceptible plants (Sp) grown without any selection pressure. Molecular analysis of ct DNA showed that the *psbA* gene of I plants carried the same mutation as typical R plants found in the field but when they are treated at seedling stage they are not so resistant as R plants. No heteroplasmy was detected within I plants. When I plants are treated with non-lethal doses at cotyledon stage their resultant progeny are resistant. This change is definitively inherited and is stable in further generations.
The appearance of resistance depends on the presence of the particular Sp genotype only. So a R biotype will spread only in a population having this Sp biotype. As it could have been killed by previous treatments used in other crops, some fields or regions may be prone to have resistant populations while others may remain free from this problem for many years.

INTRODUCTION

In many countries weeds have been controlled mainly by selective herbicides for more than 40 years. Actually weed densities have decreased during intensive agriculture ; but as certain species have disappeared other species have become dominant. Since the early 1970s numerous cases of herbicide resistance have been reported (LeBaron and Gressel, 1982). The resistant biotypes are due to genetic evolution of weed populations resulting from strong and constant selective pressure in monocultures treated with the same herbicide. Resistance has been reported for many herbicide families including recently released herbicides (LeBaron and McFarland, 1990).

The best documented case of resistance is the chloroplast triazine resistance now evident in more than 50 species in various countries (Le Baron, 1988) and in 21 species in 500,000 ha in France (Gasquez, 1988a).

This resistance, due to at least a point mutation on the chloroplast *psbA* gene, reduces the susceptibility of the photosynthetic electron transport to symmetrical triazines by roughly 1,000-fold (Hirshberg and McIntosh, 1983).

It has been assumed that the repeated use of the same chemical has favoured the selection of already existing resistant biotypes. Population genetic studies could be the best way to ascertain whether this hypothesis is correct. However very little data is available on the appearance and spread of such biotypes. Since these points are the very first steps of the explosion of any resistance, they need to be investigated in order to understand the reasons for herbicide failures in intensive agriculture.

This paper deals with data collected on population structures, genetic characteristics, mutation process and spread of resistance within the weed species, *Chenopodium album* (fat hen). This leads us to introduce a new hypothesis for resistance appearance which shows that the farmer is responsible for the spread of resistant biotypes as his weed management affects the genetic structure of the weed populations.

POPULATION STRUCTURE OF *CHENOPODIUM ALBUM*

The first report of triazine resistance in France was a *C. album* resistant biotype (Ducruet and Gasquez, 1978). Since this date, *C. album* has become the world's most widespread resistant (R) weed (LeBaron and McFarland, 1990); in at least 200,000 ha in maize fields and vineyards in France.

As some species have only one or few R populations we may assume that resistance frequency will depend on the genetic characteristics of the species population structure and mating system. Thus we have studied the polymorphism of *C. album* populations in various environments.

Ploidy levels

There are two ploidy levels in many populations of *C. album* (Al Mouemar and Gasquez, 1979) the most common karyotype is hexaploid (2n = 54) and corresponds to the species type. The other growing at low density (5 - 10 %) is a tetraploid (2n = 36) plus a variable number of supernumerary B chromosomes. Among the numerous accessions we have tested for resistance, none has been found to be tetraploid. It is surprising that no tetraploid plant has evolved towards resistance, although the presence of B chromosomes is generally correlated with a high evolutionary potential.

Although it is difficult to distinguish between hexaploid and tetraploid biotypes within a field, we can question if they belong to the same biological species. For instance, there is a difference in the flowering period of the two types. Artificial hybrids between the two are mostly sterile or produce seeds of poor viability (Al Mouemar, 1983). Whether or not there is only one species, the lack of R tetraploid biotypes could be the consequence of a genetic effect which associates resistance with some peculiar genotypes.

Isozyme polymorphism

We have studied isozyme polymorphism of various hexaploid populations of
C. album. The results show that four enzymes account for nearly 60
different phenotypes in a survey in Burgundy (Al Mouemar and Gasquez,
1983). However there are important differences according to the origin
of the populations. More than 20 phenotypes have been distinguished in
garden populations with very variable frequencies. In regularly treated
fields, an average of five phenotypes remain with only one at a 50 %
frequency indicating a high selective pressure (Al Mouemar and Gasquez,
1983).

Changes in the mating system could also explain these differences. In
pure stand *C. album* is an allogamous species, and the outcrossing rate
between closely adjacent plants is about 50 % (Gasquez, 1985), but this
value sharply decreases when plants are more distant : no hybrids could
be found when the mother plants grow five metres apart. Outcrossing rate
is reduced to only 5 % when two adjacent plants are growing in a maize
field (Gasquez, 1985). This discrepancy between private gardens and
fields is probably due to the increasing homogeniety of environmental
conditions caused by cultivation and weed management techniques. The
populations of gardens are obviously very close to "natural conditions".
As a farmer cannot tolerate a density of one *C. album* m^{-2}, field
population density is very reduced, leading to a very low heterozygosity
(Gasquez, 1985). Such a decrease of gene exchange reduces the
polymorphism and, according to the selective pressure, increases genetic
drift or heavy selection.

In R populations infesting maize fields polymorphism of *C. album* is
extremely reduced : only one phenotype was found within each population
(Gasquez and Compoint, 1981). Likewise, in Canada, R populations were
also shown to be monomorphic (Warwick and Marriage, 1982). This is the
consequence of a founder effect greatly favoured by repeated triazine
treatments.

Furthermore we have found in a given area in France, that there is always
the same phenotype throughout each population (Gasquez and Compoint,
1981). However R phenotypes differ from one region to another. This
indicates that the resistance in a country like France is not due to
founder effect but to the occurrence of many independent events in each
field. Early in the development of resistance, almost every new R
population represents a unique event. After many years of herbicide use
it is very difficult to distinguish R populations due to independent R
mutations and those resulting from accidental spread by man or animals.

So in fat hen, there are two ploidy levels in each population and
variable levels of polymorphism according to the environment. The high
density in garden populations preserves allogamy between many phenotypes.
In field populations only a few genotypes remain at low density,
reproduce by autogamy and are completely isolated in each field. In R
populations there is only one phenotype which is characteristic for each
region. So the correlation of resistance with a peculiar genotype is
certainly the consequence of genetic characteristics and the frequency
of such genotypes is higher in populations which have never been treated.

Presence of mutants

As susceptible populations and especially garden populations which have never been treated are highly polymorphic, we searched for mutant plants in such populations , in order to determine their density prior to any triazine treatment (i.e. in "natural conditions").

The first accession was a population from the Loire Valley where, out of 10,000 seedlings only 91 plants survived a 500 g ha^{-1} atrazine treatment (in the greenhouse, susceptible plants are killed with 150 g ha^{-1}) (Gasquez and Compoint, 1981). As this population was collected in a region where R populations were already known, such a high density of R plants could be due to external contamination.

We also studied populations in regions such as Burgundy and the Alps where no R populations were known. Seeds were collected plant by plant, or seedlings were transplanted to the greenhouse and allowed to set seeds. These plants were tested for resistance by a fluorescence test and were always found to be susceptible. Assuming that each seed set formed a family, samples of families were sown in the greenhouse and as previously, sprayed with 500 g ha^{-1} atrazine at the cotyledon stage. Few plants survived in some families (Table 1), with some rather high within-family frequencies (up to 12 %).

Table 1 : <u>Frequency of I mutant appearance in garden populations</u>.

	No. of families	No. of families having mutants (Sp)	Mean I frequency within Sp families
Burgundy	191	17	3.2
	180	8	5.3
	166	1	1.1
	76	5	0.1
	193	3	0.9
Alps	187	5	1.6
	145	2	1.7

In each population the R plants had the same phenotype, and this R phenotype differed from one region to another. These families giving mutant progenies have been called Sp. (Gasquez *et al.*, 1985). It is noteworthy that this matches very well with the founder effects previously observed in maize fields (Gasquez and Compoint, 1981).

Intermediary characteristics

Those surviving plants which are not affected by a 500 g ha^{-1} treatment, cannot survive as high rates of triazines as the typical R plants found in maize fields. The same is true for the Loire Valley population which was used as reference susceptible population because no plant could survive a 4.5 kg ha^{-1} atrazine treatment (Gasquez and Barralis, 1978).

Furthermore these plants show a fluorescence response different from the fluorescence curve of typical R plants : we called these plants "intermediate" (I). Crosses between I and susceptible (S) plants showed that the I characteristics were cytoplasmically inherited (Gasquez *et al.*, 1984). Molecular analyses of chloroplast DNA showed that the *psbA* gene of I plants carried the same mutation as typical R plants and no heteroplasmy was detected (Bettini *et al*, 1987).

What is surprising is that plants having the mutation correlated with triazine resistance (Mc Nally *et al*, 1987) are not highly resistant. This could be a genetic characteristic of the Sp genotype, or the genome of I plants could be still unbalanced. Preliminary data on crosses between I and R plants have shown some disjunctions in I and R phenotypes in hybrid progenies, indicating a nuclear control of the phenomenon.

Likewise, we found within Italian R populations of *Solanum nigrum* and *Amaranthus cruentus*, various levels of resistance at the seedling stage (Zanin *et al.*, 1981). Similarly, within *Poa annua*, we distinguished different R biotypes, by means of fluorescence curves and different slopes of triazine inhibition curves of isolated chloroplasts (Gasquez and Darmency, 1983). Furthermore intermediate *C. album* have been observed in Hungary (Solymosi *et al.*, 1986).

In garden conditions, I plants appear at frequencies ranging from 10^{-4} to 10^{-3}. When I plants are treated with sublethal doses of triazines (i.e. 500 g ha^{-1}) or with some systemic pesticides their progeny becomes completely resistant (Gasquez *et al.*, 1985) : they can survive doses as high as 40 kg ha^{-1} and they have typical resistant fluorescence curves. These characteristics are inherited unchanged.

Therefore, only one phenotype of *C. album* (but the same in a given region) per population produces spontaneously mutant plants. When these mutated plants are exposed to a chemical stress (i.e. pesticide treatment) a R biotype is produced. This occurs in garden populations, but nobody has ever found R plants in a field prior to triazine treatment.

So within *C. album* triazine resistance is due to a spontaneous mutation within a peculiar genotype. The Sp genotype might be a chimeric plant, i.e. some copies of *psbA* gene could be mutant, but the frequency of these mutant copies must be less than 10^{-3} (Bettini *et al.*, 1987). In some cases Sp plants produce no I plants whereas in other cases Sp plants produced a high frequency of I in their progenies. This could be due to a mutator system as previously suggested by Arntzen and Duesing (1983). If this is true, this system could probably act for other mutations at the chloroplast level but there are no reports of this.

All the *psbA* copies of I plants are mutated (P = 99.9 %) but these plants are not typical R plants and will produce R progenies only when treated. This could be due to nuclear control of chloroplast gene expression as there is some disjunction in hybrid progenies (Darmency and Gasquez, 1988). This seems to be one of the very few cases reported of environmentally induced genetic change (Durrant, 1962).

Relative fitnesses

Within several species, many authors have shown that R plants grow slower and produce less seeds than S plants, i.e. the R biotypes are less fitted than S ones (Conard and Radosevich, 1979). But in some cases, there is no difference or R plants may even have a higher fitness e.g. *Phalaris paradoxa* (Rubin *et al.*, 1985), and especially for *C. album* (Jansen *et al.*, 1986). As such discrepancies could be due to the comparison of very different genotypes and thus not only related to the resistance phenomena, it is necessary to compare either isogenic plants or very closely related plants such as F_1 hybrids. With reciprocal hybrids between R and S *S. nigrum*, we found that growth characteristics and seed production were subject to nuclear inheritance as there was no significant difference between R F_1 and S F_1 hybrids (Gasquez, 1988b). However in certain conditions, such F_1 hybrids of *Senecio vulgaris* may show some differences (Holt, 1989).

The three *C. album* biotypes Sp, I and R produced from treated I plants constitute a good model for this purpose : they are nearly isogenic (descendants of almost homozygous plants except for *psbA* mutation). Various experiments were set up in order to measure the fitness differences due to the *psbA* mutation or to complete resistance.

As we found differences in chloroplast membrane fatty acids of young leaves between R and S plants (Trémolières *et al.*, 1988), we studied growth characteristics of seedlings at 10 leaf stage. At two temperature conditions in a growth chamber, there was no significant difference except for the surface leaf area : I plants always have larger leaves than Sp plants, R plants have the smallest leaves and also the lowest dry weight.

In greenhouses, this difference only remains in older leaves up to flowering stage, but no difference seems to occur for other growth characteristics : all plants grow at the same speed except in the last stage when R plants grow faster and become taller but less branched. There is no significant difference in flowering date. When harvested at a given date, all plants have similar dry shoot weights, but Sp plants have heavier reproductive parts and a higher number of seeds (about a third) than I and R ones.

In controlled conditions, despite some small differences in growth characteristics, the three isogenic biotypes have a very similar rate of development. However, Sp plants seem to ripen their seeds faster since although they have similar flowering dates they produce more mature seeds. I plants produce less seeds because there is some sterility.

Outside in field conditions we have grown these three biotypes in a pure stand and in a maize field. The crop has no effect on growth rate and flowering date but in a pure stand all biotypes are shorter and more branched. R and I plants are generally very similar and are taller and less branched than Sp plants. These plants were left to flower and produce seeds until the maize harvest, at least three months, instead of the three weeks in greenhouse experiments. All the biotypes produced very high seed sets (more than 50,000 seeds/plant), but although R plants appear to produce more seeds there is no significant difference in seed records after the maize harvest. But probably because the seeds of Sp plants ripen faster, their germination is better than that of R and I ones (Sp = 92 % ; R = 75 % ; I = 65 %). In the following year we counted seedling emergence in all plots, and there was no significant difference between the three biotypes (about 150 seedlings m^{-2}).

Thus, in field conditions at low densities all the biotypes develop fully, produce large numbers of seeds and, in spite of differences in germination capacity, have similar emergence rates the next year. So if the fitness for annual species is the capacity to produce seedlings and fertile plants the next year, we must admit that there is very little difference in fitness between these isogenic biotypes. Significant differences could perhaps appear after several years with various climatic conditions or under competitive conditions (i.e. some R plants among Sp plants) as the differences between these biotypes are very variable according to the environmental conditions.

Dispersion of fat hen

In order to estimate the spread of *C. album* mutants in a maize field, we carried out a five year experiment by introducing one R plant in the centre of every 400 m^2 plot. Each resistant plant of the first year produced an average of 103,000 seedlings by the fifth year. But they were distributed in a very heterogeneous way : highest densities at the centre of the plots (about 1,000 m^{-2}) and very few plants at the periphery (less than 1 m^{-2}). Moreover the dispersion rate at right angles to the direction of cultivation was low (1.6 m year^{-1}), so the overall area covered following the introduction of a single R plant followed the equation $y = 25.5 \ x^2$ (with x, the number of years after the introduction of the resistant plant).

Thus the seed dispersal of the species is very limited and the increase of effective progeny is rather low (about 10 seedlings per mother plant).

HYPOTHESES FOR RESISTANCE APPEARANCE

Random appearance

The simplest hypothesis postulates that the mutation occurs at random at a given frequency (i.e. 10^{-6}) and that R biotypes remain at very low frequency before any triazine treatment. As resistance has appeared within 50 species and many times (thousands of populations all around the world) this implies that the R biotype must have a very low fitness in the absence of triazine treatment.

This has certainly not occurred everywhere for every species. There are some differences between this hypothesis and the spread of R populations since the first recorded occurrence.

- The fitness of R biotypes must be very poor in order to explain their constant very low frequency in the absence of triazine treatment. The selective value of R plants must be approximately similar to the mutation rate. Not all the present data corresponds to such a difference between the two biotypes ; especially within *C. album* whose isogenic plants Sp, I and R have almost identical fitness.

- In France, maize fields are invaded by at least 100 plants m^{-2} after five years treatment. If the initial R biotype frequency remains near the mutation frequency (i.e. 10^{-6}), not more than one R plant will appear per hectare in relation to *C. album* density in treated fields. As we have previously seen, the dispersal of one plant will not exceed 0.1 ha after five years. Such values would hardly account for the field distribution of R plants when the farmer first notices the failure of herbicide treatment.

- Moreover if R plants could appear in a random way and could invade the field in an average of five years it is very difficult to explain why two thirds of maize fields in France are free from any R weed in spite of nearly 20 years of continuous atrazine treatments every year.

- The mutation could have appeared only once so that its frequency in a region or a population would only depend on the past of the population and of the mating system of the species. But as we observed in garden populations, it is very easy to find mutant plants.

The lower relative fitness of R biotypes could be due to the "mutation cost" as the mutant may arise within the offspring of a rather rare susceptible plant (i.e. less fit). This could only be due to the fact that the mutation affects photosynthesis electron transport. This would be a characteristic of triazine treatment and not be related to a general genetic mechanism. Actually other natural resistance such as chlorotoluron resistance (Chauvel pers. comm.) or *in vitro* engineered resistance do not reduce the fitness of resistant biotypes.

The peculiar mutating genotype

As we have previously seen, the appearance of R plants of *C. album* is not correlated to the frequency of a natural mutation occuring by chance within all the species. On the contrary, a particular genotype produces high rates of mutated plants whose progeny may become resistant when treated.

So the appearance of R plants in a field depends on : 1) the presence of Sp plants which must pre-exist in the population before triazine treatment ; 2) a sufficient density of I plants from which some must survive, as they are generally killed by the usual atrazine treatment ; 3) treated plants must give enough seeds to produce R plants the following year.

It is only when all these conditions are fulfilled that R plants will spread, provided the farmer uses triazines for several years.

Few Sp plants give I mutants in garden populations, but we know that the polymorphism is reduced in field populations, so that some populations could remain free from R plants. Previous treatments in other crops may have killed Sp plants. In Burgundy each garden population we have studied has the same biotype Sp which produces I mutants, but up to now, triazine resistance has never been found in maize fields. This is certainly due to the lack of I plants in many fields. Fortunately this year we have found a field where maize has been grown only for three years on land which was previously a meadow. There were several large spots of R *C. album* which had the same isozyme phenotype as that of Sp plants growing in gardens some kilometers away. The polymorphism of this population was certainly very close to that of a garden population, so the probability of the appearance of I plants was very high. Thus this R population has certainly been produced from the many I plants which survived the first treatment and which produced a large amount of R progeny. Actually, in our studies, R populations usually developed in fields only recently cropped with maize.

Therefore triazine resistance due to a natural mutation has only appeared within weed populations where there was already a particular Sp genotype, producing high rates of mutant plants. Thus these populations, in fields only recently cropped with maize, still had high polymorphisms or had by chance kept the Sp genotype (by genetic drift or selection). Anyway, this fits well with all observations as the I frequency may be quite significant (10^{-3}) leading to a quick infestation (sometimes only three years). Furthermore many populations may still be free from any resistant weeds, either because the Sp genotype has disappeared or because its frequency is too low within a weed population due to good weed control.

The spread of resistance as a natural phenomenon is mostly the consequence of the weed management of a given field before and during maize cultivation. It is also noteworthy that it is impossible to estimate the rate of the area accidentally invaded by R weeds because the farmers take little care of the seed contamination during harvest or liquid manure spreading especially if they are stock breeders. Many pastures turned over to maize have been invaded from the very first year of cultivation.

CONCLUDING REMARKS

These data are related to triazine resistance of *C. album*. Even though there are other species having different R biotypes, each species may have a different way of mutation appearance according to its genetic characteristics. Could this hypothesis account for the appearance of any other resistance ? Is a mutation for resistance so particular that only one genotype could carry it ? Is the frequency of the mutation correlated to its consequences on the overall fitness?

We think that each resistance is a special phenomenon, creating a new genetic situation. Though the selected genotype could have a poor fitness, the high selective value conferred by the R gene in treated areas will lead to severe infestions as long as the field is treated.

Therefore the appearance and the spread of any herbicide resistance (as any other pesticide resistance) are only due to the farmer's practices. The longer he uses the same chemical in a given field, the higher is the probability of a heavy weed infestation. The appearance of resistance is correlated with the genetic characteristics and population structure of the weed. This may also happen if transgenic R crops are used repeatedly. Good weed management with herbicide and crop rotations will reduce weed density, reduce polymorphism and lead to important genetic drift. In these conditions weed density is generally too low and the population structure is destroyed. In such populations with much reduced variability the chance of resistance occuring is much reduced.

REFERENCES

Al Mouemar, A. (1983). Étude de l'évolution de populations de *Chenopodium album* L. en fonction de facteurs phytotechniques : Structure génétique, résistance aux herbicides. Thèse Doctorat Etat Besançon.

Al Mouemar, A. and Gasquez, J. (1979). Variations caryologiques et isoenzymatiques chez *Chenopodium album* L.. C.R. Acad. Sc. Paris 288 D, 677-680.

Al Mouemar, A. and Gasquez, J. (1983). Environmental conditions and isozyme polymorphism in *Chenopodium album* L. Weed Research **23**, 141-149.

Arntzen, C.J. and Duesing, J.H. (1983). Chloroplast encoded herbicide resistance. In "Advances in Gene technology" (F. Ahmand, K. Downey, J. Schultz and R.W. Voellmy, eds) pp. 273-299. Adacemic Press, New York.

Bettini, P., McNally, S., Sevignac, M., Darmency, H., Gasquez, J. and Dron, M. (1987). Atrazine resistance in *Chenopodium album*: Low and high levels of resistance to the herbicide are related to the same chloroplast *psbA* gene mutation. Plant Physiology **84**, 1442-1446.

Conard, S.G. and Radosevich, S.R. (1979). Ecological fitness of *Senecio vulgaris* and *Amaranthus retroflexus* biotypes susceptible or resistant to atrazine. Journal of Applied Ecology **16**, 171-177.

Darmency, H. and Gasquez, J. (1990). The fate of herbicide resistant genes in weeds. In "Fundamental and Practical Approaches to Combating Resistance", (M.B. Green, W.K. Moberg and H.M. LeBaron, eds), (in press). American Chemical Society, Washington, DC.

Ducruet, J.M. and Gasquez, J. (1978). Observation de la fluorescence sur feuille entière et mise en évidence de la résistance chloroplastique à l'atrazine chez *Chenopodium album* L. et *Poa annua* L.. Chemosphere **7**, 691-696.

Durrant, A. (1962). The environmental induction of heritable changes in *Linum*. Heredity **17**, 27-61.

Gasquez, J. (1985). Breeding system and genetic structure of a *Chenopodium album* population according to crop and herbicide rotation. In "Genetic Differentiation and Dispersal in Plants", (P. Jacquard, G. Heim and J. Antonovics, eds) pp. 57-66. Springer Verlag.

Gasquez, J. (1988a). Herbicide resistance in France. In "Herbicide-Resistant Weeds and Alternative Control Methods". Commission of the European Communities. Proceedings of a Meeting of the EC Experts' Group. Tølløse, Denmark 15-17 November 1988

Gasquez, J. (1988b). La résistance aux herbicides chez les angiospermes. In "Le Mode d'Action des Herbicides"(R. Scalla, ed.), (in press). INRA Paris.

Gasquez, J., Al Mouemar, A. and Darmency, H. (1984). Quels gènes pour la résistance chloroplastique aux triazines chez *Chenopodium album* ? 7ème Coll. Intern. Ecol. Biol. Syst. des Mauvaises Herbes, Paris, 281-286.

Gasquez, J., Al Mouemar, A. and Darmency, H. (1985). Triazine herbicide resistance in *Chenopodium album* L. Occurence and characteristics of an intermediate biotype. Pesticide Science **16**, 390-395.

Gasquez, J. and Barralis, G. (1978). Observation et sélection chez *Chenopodium album* L. d'individus résistant aux triazines. Chemosphere **11**, 911-916.

Gasquez, J. and Compoint, J.P. (1981). Enzymatic variations in population of *Chenopodium album* L. resistant and susceptible to triazine. Agroecosystem **7**, 1-10.

Gasquez, J. and Darmency, H. (1983). Variation for chloroplast properties between two triazine resistant biotypes of *Poa annua* L. Plant Science Letters **30**, 99-106.

Hirshberg, J. and McIntosh, L. (1983). Molecular basis of herbicide resistance in *Amaranthus hybridus*. Science **22**, 1346-1349.

Holt, J.S. (1990). Fitness and ecological adaptability of herbicide resistant biotypes. In "Managing Resistance to Agrochemicals: From Fundamental Research to Practical Strategies", (M.B. Green, H.M.LeBaron and W.K. Moberg, eds), pp. 419-429. American Chemical Society, Washington, D.C.

Jansen, M.A.K., Hobe, J.H., Wesselius, J.C. and van Rensen, J.J.S. (1986). Comparison of photosynthetic activity and growth performance in triazine-resistant and susceptible biotypes of *Chenopodium album*. Physiologie Végétale **24**, 475-484.

LeBaron, H.M. and Gressel, J. (1982). "Herbicide Resistance in Plants". Wiley & Sons, New York. pp. 401.

LeBaron, H.M. and McFarland, J. (1990). Herbicide resistance in weeds and crops: an overview and prognosis. In "Managing Resistance to Agrochemicals: From Fundamental Research to Practical Strategies", (M.B. Green, H.M.LeBaron and W.K. Moberg, eds), pp. 336-352. American Chemical Society, Washington, D.C._

McNally, S., Bettini, I P., Sevignac, M., Darmency, H., Gasquez, J. and Dron, M. (1987). A rapid method to test for chloroplast DNA involvement in atrazine resistance. Plant Physiology **83**, 248-250.

Rubin, B., Yaacoby, T. and Schonfeld, M. (1985). Triazine resistant grass weeds : Cross resistance with wheat herbicide a possible threat to cereal crops. British Crop Protection Conference - Weeds 1171-1178.

Solymosi, P., Kostyal, Z. and Lehoczki, E. (1986). Characterisation of intermediate biotypes in atrazine susceptible populations of *Chenopodium album* and *Amaranthus bouchonii* in Hungary. Plant Science **47**, 173-179.

Trémolières, A., Darmency, H., Gasquez, J., Dron, M. and Connan, A. (1988). Variation of transhexadecenoïc acid content in two triazine resistant mutants of *Chenopodium album* and their wild progenitor. Plant Physiology **86**, 967-970.

Warwick, S.I. and Marriage, P.B. (1982). Geographical variation in populations of *Chenopodium album* resistant and susceptible to atrazine I. Between and within population variation in growth and response to atrazine. Canadian Journal of Botany **60**, 483-493.

Zanin, G., Vecchio, V. and Gasquez, J. (1981) Indagini sperimentali su popolazioni di dicotiledoni resistenti all'atrazina. Rivista di Agronomia **XV** (3,4), 196-207.

SULFONYLUREA HERBICIDE RESISTANT WEEDS: DISCOVERY, DISTRIBUTION, BIOLOGY, MECHANISM, AND MANAGEMENT

Donald C. Thill[1], Carol A. Mallory-Smith[1], Leonard L. Saari[2], Josephine C. Cotterman[2], Michael M. Primiani[2] and John L. Saladini[2]

[1]University of Idaho, Department of Plant, Soil and Entomological Sciences, Moscow, Idaho, 83843, USA and [2]E.I. du Pont de Nemours & Co., Inc., Agricultural Products Department, Wilmington, Delaware 19898, USA

Sulfonylurea herbicide resistant *Lactuca serriola* L. (prickly lettuce) plants were discovered near Lewiston, Idaho in April 1987. This was the first confirmed occurrence of herbicide resistance resulting from the use of sulfonylurea. *Kochia scoparia* (L.) Schrad (kochia) plants from Liberal, Kansas were confirmed as resistant to chlorsulfuron and metsulfuron-methyl in 1988. Since then, sulfonylurea resistant *K. scoparia* has been identified in eight other States and one Canadian Province. *Salsola iberica* Sennen & Pau (russian thistle) plants from Kansas, Montana, North Dakota and Washington and *Stellaria media* (L.) Vill. (common chickweed) plants from Alberta, Canada have also been found to be resistant to sulfonylurea herbicides. Most resistant plants were collected from fields where dryland winter wheat (*Triticum aestivum* L.) was grown either continuously or in rotation with summer fallow and where chlorsulfuron or chlorsulfuron plus metsulfuron-methyl had been applied at 7 to 14 month intervals for 3 to 5 years. Total sulfonylurea herbicide used ranged from 52 to 200 g a.i.ha^{-1}. Resistant *K. scoparia* plants have also been collected from noncrop areas where sulfometuron-methyl was applied annually for 3 to 4 years. Total sulfometuron-methyl used ranged from 96 to 425 g ai ha^{-1}. The mechanism of resistance is an altered site of action, acetohydroxyacid synthase enzyme, which is inhibited less in resistant than in susceptible biotypes by sulfonylurea and imidazolinone herbicides. Resistance is not due to differences in herbicide absorption, translocation, or metabolism. Chlorsulfuron-resistant *K. scoparia* and *L. serriola* are resistant to several other sulfonylurea and imidazolinone herbicides. However, resistant and susceptible biotypes are usually controlled equally by herbicides with different modes of action. Resistance in *L. serriola* is controlled by one nuclear gene with incomplete dominance.

[Published with the approval of the Agricultural Experiment Station, University of Idaho, as Article Number 89733]

115

INTRODUCTION

The sulfonylurea herbicides were commercialized in 1982 when chlorsulfuron was registered for use on several small grain cereals for selective control of many broadleaf weeds and some grasses and when sulfometuron-methyl was sold as a noncrop herbicide (Beyer *et al.*, 1988). Since 1982, several other sulfonylurea herbicides have been used to control weeds in soybeans (*Glycine max* (L.) Merr.), rice (*Oryza sativa* L.), corn (*Zea mays* L.), and noncrop areas. By 1985, 14 agricultural chemical companies had filed sulfonylurea herbicide patents with 230 sulfonylurea herbicide patents issued in the USA by June 1987 (Beyer *et al.*, 1988).

The sulfonylureas are highly active herbicides, effective at use rates as low as 2 g a.i. ha^{-1} (Beyer *et al.*, 1988). This contrasts with most other herbicides, e.g. bromoxynil and 2,4-D are used at 0.56 to over 1.0 kg a.i. ha^{-1}, respectively. The sulfonylurea herbicides inhibit acetohydroxyacid synthase (AHAS) activity (also known as acetolactate synthase, E.C.4.1.3.18) resulting in decreased biosynthesis of branched chain amino acids (Ray, 1984). The imidazolinone herbicides also inhibit AHAS and have the same mechanism of action (Shaner *et al.*, 1984). Structurally, the sulfonylurea herbicides are composed of three parts: an aryl group, a sulfonylurea bridge, and a nitrogen-containing heterocycle. For a thorough review of sulfonylurea herbicide degradation pathways and mode of action, see Beyer *et al.*, 1988.

This chapter reviews the discovery, distribution, biology, and mechanism of resistance in sulfonylurea herbicide resistant weeds and then discusses strategies for avoiding and managing sulfonylurea resistance in weeds.

DISCOVERY

In April 1987, Mr Richard Lloyd who farms about 15 km south of Lewiston, Idaho, observed that 1986 fall applied DPX-G8311 (5:1 chlorsulfuron:metsulfuron-methyl, 26 g a.i. ha^{-1}) failed to control *L. serriola* in winter wheat (Thill *et al.*, 1989). Poor control was noted in the previous winter wheat crop from 16 g a.i. ha^{-1} of DPX-G8311. The cropping system on this farm has been continuous no-till winter wheat since 1983. Chemical fallow was used in rotation with winter wheat on some fields, but fields with shallow soil were recropped each year. Glyphosate and chlorsulfuron were used to control weeds during the fallow period. Average rainfall on the farm is 36 to 46 cm year^{-1} and soil pH ranges from 5.1 to 6.7. From 1982 to fall 1986, the grower applied a total of 52 to 200 g a.i. ha^{-1} of sulfonylurea herbicides to each of nine fields on his farm. The applications were made at 7 to 14 month intervals. Additionally, weed scientists from the University of Idaho have experimented with herbicide efficacy on this farm for more than 12 years. Chlorsulfuron was first applied in 1980 at rates as high as 137 g a.i. ha^{-1}, which is much higher than 9 to 26 g a.i. ha^{-1} used commercially.

DPX-M6316, a sulfonylurea herbicide under development, plus either metsulfuron-methyl or chlorsulfuron controlled 100% of the *L. serriola* in treated plots on the Lloyd farm during the 1983-84 winter wheat growing season (Schaat *et al.*, 1985). In 1987 and 1988, at the same location, tests of further sulfonylurea development products, DPX-G8311, DPX-R9674

116

and CGA-131036, failed to control the *L. serriola* (Mallory-Smith *et al.*, 1990).

In October 1987, *K. scoparia* seeds were collected near Liberal, Kansas from plants which had survived chlorsulfuron treatments (Primiani *et al.*, 1990). The field had been used for monoculture winter wheat production and had been treated with a total of 105 g a.i. ha^{-1} (about 15 g a.i. ha^{-1} year^{-1}) chlorsulfuron from 1982 to 1987. Subsequent greenhouse and laboratory experiments (discussed in following sections) confirmed sulfonylurea resistance in this *K. scoparia* biotype.

DISTRIBUTION

Table 1. <u>Distribution of weeds resistant to sulfonylurea herbicides as of August 1989.</u>

Weed species	Location	No. of sites[*]
Lolium rigidum[#]	Australia	
Alopecurus[#] *myosuroides*	England[#]	
Stellaria media	Alberta, Canada	2
Kochia scoparia	Colorado, USA	5
	Kansas, USA	1,1
	Montana, USA	19
	New Mexico, USA	0,2
	North Dakota, USA	9
	South Dakota, USA	1
	Texas, USA	9,2
	Washington, USA	0,4
	Saskatchewan, Canada	5
Latuca serriola	Idaho, USA	1
Salsola iberica	Kansas, USA	1
	Montana, USA	1
	North Dakota, USA	1
	Washington, USA	5

[#] Weeds that initially developed resistance to other classes or families of herbicides and also are resistant to the sulfonylureas.

[*] Where two numbers are presented per location, the first is for cropland and the second is for non-cropland sites. A single number represents cropland site (Schumacher, pers. comm.).

Sulfonylurea herbicide resistant weeds are known to occur in Australi (Powles and Howat, 1989), Canada (Schumacher, pers.comm.), England (Moss 1987), and the USA (Mallory-Smith *et al.*, 1990; Primiani *et al.*, 1990) These cases have resulted from two different types of selection Herbicide-resistant biotypes of *L. serriola* (Thill *et al.*, 1989), *K. scopar* (Primiani *et al.*, 1990), *S. iberica* and *S. media* (Saari, unpublished data) i North America were selected with sulfonylurea herbicides. *L. rigidum* i Australia (Powles and Howat, 1989) and *A. myosuroides* Huds. (Moss, 1987) i England were selected with nonsulfonylurea herbicides and are resistan to several families of herbicides, including the sulfonylureas. The sit of action located on the AHAS enzyme, is altered in the biotypes selecte with sulfonylurea herbicides. Resistance to several different herbicid families in *L. rigidum* and *A. myosuroides* may be due to enhanced metaboli activity of mixed function oxidases (Powles, pers. comm.; Kemp an Caseley, 1987), which may detoxify the sulfonylurea herbicides in manner similar to *Streptomyces griseolus* (O'Keefe *et al.*, 1987). In man tolerant species chlorsulfuron is detoxified through metabolism (Sweetse *et al.*, 1982; Hutchison *et al.*, 1984).

Weeds resistant to sulfonylurea herbicides as of August 1989 and thei locations are listed on Table 1 (Schumacher, pers. comm.). It i interesting to note that few new locations and no new species wer reported and confirmed in 1989. This may reflect grower acceptance o the sulfonylurea-resistant management strategies adopted during 1988.

Sulfonylurea-resistant *K. scoparia* biotypes have been collected from bot noncrop and crop sites (Schumacher, pers. comm.). In noncrop sites, total of 96 to 425 g a.i. ha^{-1} of sulfometuron-methyl was applied for to 4 years to roadsides and railroad rights-of-way. Resistant *S. media,* *K. scoparia* and *L. serriola* usually were collected from cropland where dryland cereals were grown continuously or in rotation with summer fallow and where chlorsulfuron or DPX-G8311 had been applied at 7 to 14 month intervals for 3 to 5 years. Total sulfonylurea herbicide use during the period ranged from 52 to 200 g a.i. ha^{-1}. Some fields are infested with resistant biotypes of *K. scoparia* and *S. iberica* where little or no sulfonylurea herbicide had been used previously (Schumacher, pers. comm.). Apparently, resistant plants and seeds from adjacent fields blew into these fields.

BIOLOGY

According to Hill (1982), plant populations that show resistance to herbicides are species with the following characteristics: predominantly herbaceous annuals; wholly or partially self-fertile; associated with agricultural habitat; colonizers; high reproductive capacity; complex genetic variability expressed as polymorphic phenotypes; evolutionary strategies selecting for fitness; periodic mass death of individuals; and rapid development of plants from seedling to maturity. The biology of *S. media, K. scoparia, L. serriola* and *S. iberica* is reviewed briefly in the following section.

S. media biotypes are resistant to sulfonylurea (Saari, unpublished data) and triazine (Kees, 1978) herbicides and mecoprop (Lutman and Lovegrove, 1985). Hill (1982) refers frequently to *S. media* in his discussion of taxonomic and biological considerations of herbicide-resistant and herbicide-tolerant biotypes indicating that this species adapts quickly

118

to frequent use of certain herbicides. The biology of *S. media* was reviewed by Turkington *et al.* (1980). Some biological characteristics of *S. media* cited in the review include: annual or winter annual with flushes of germination in the early spring and late fall; life cycle of 5 to 7 weeks; one to three generations of seed per year; flowers self pollinated before opening (cleistogamous); the plant flowers and sets seed throughout the year; seeds dispersed easily; high seed viability with little initial seed dormancy, but long seed bank life; 500 to 2,500 seeds per plant with 5 to 15 million seeds ha^{-1}; diploid with 2n chromosome counts ranging from 20 to 44; high degree of phenotypic plasticity and genotypic flexibility; and a pioneering species that rapidly colonizes open areas.

K. scoparia biotypes exist that are resistant to sulfonylurea (Primiani *et al.*, 1990), triazine (Johnston and Wood, 1976) and both triazine and sulfonylurea herbicides (Primiani *et al.*, 1990). *K. scoparia*, an introduced herbaceous annual ornamental, has escaped from cultivation to croplands, roadsides and waste areas (Eberlein and Fore, 1984). The species is a colonizer of saline and dry areas that germinates in the spring over a wide temperature range. The plant has a highly variable form, flowers about 8 to 12 weeks after emergence, and typically produces over 14,000 seeds per plant. Seed longevity in soil ranges from less than 2 years (Burnside *et al.*, 1981) to over 3 years (Zorner *et al.*, 1984). *K. scoparia* is diploid with a 2n chromosome number of 18 (Uhrikova, 1974). The plant is partially self-fertile, but predominantly open pollinated (Nalewaja, pers. comm.). The plant is a tumbleweed, which is an effective means of seed dispersal.

L. serriola biotypes are not known to be resistant to any class of herbicides other than AHAS-inhibiting herbicides. This colonizer of disturbed areas (Arthur, 1894; Prince and Carter, 1985) is an annual or winter annual with various polymorphs (Gaines and Swan, 1972). Seedlings emerge throughout the year with germination peaks in early winter and late spring (Marks and Prince, 1981). *L. serriola* flowers in July to August (Arthur, 1894). The species is self-fertile and out crossing is rare (Lindquist, 1960; Parrish and Bazzaz, 1978). Medium size plants can produce over 8,250 seeds (Arthur, 1894), which are small in size and beak-tipped with a cluster of fine white hairs that carry the seed in the wind (Gaines and Swan, 1972). Seeds have no primary dormancy and germination is rapid in suitable conditions (Marks and Prince, 1981). Buried seeds can acquire dormancy and have a half-life of about 1.5 years in soil (Marks and Prince, 1982). The chromosome number for *L. serriola* is 2n equal 18 (Whitaker and McCollum, 1954).

As with *L. serriola*, the only known resistance in *S. iberica* is to AHAS inhibiting herbicides. Chlorsulfuron can control *S. iberica* for more than one growing season (Young and Whitesides, 1987), which may result in increased selection pressure for the resistant biotype. *S. iberica* is a primary invader of disturbed sites and is the best known tumbleweed rolling across the country by wind and scattering its seeds (Stevens, 1943). Seeds begin to germinate in early spring and continue germinating throughout the growing season (Stevens, 1943), which is indicative of their ability to germinate over a wide range of alternating temperatures (Evans and Young, 1972). Flowers of *S. iberica* are protergynous (stigmas mature before the anthers), which is characteristic of wind pollinated flowers (Stevens, 1943). Flowering begins in July to early September and continues until frost (Stevens, 1943; Beatley, 1973). Later flowering usually is associated with plants shaded by cereal crops (Stevens, 1943). A single *S. iberica* plant growing in a crop free environment can produce 152,000 seeds, while plants growing in competition with spring wheat produce about 17,400 seeds/plant (Young, 1986). Seeds are short lived in soil with over 99% germinating the first year (Chepil, 1946; Ogg and Dawson, 1984).

These four species are annuals that germinate over a broad temperature range and at more than one time during the year. All have various polymorphs, mature fairly rapidly, are completely or partially self-fertile, have relatively high seed output, and have highly efficient seed dispersal systems. All of these weeds can colonize similar types of habitat. Sulfonylurea herbicides effectively control the susceptible biotypes of these weeds. Three of the four weeds are known to be diploid.

Evolution of herbicide resistance is partially dependent on the frequency of the resistant gene(s) in the initial population. For example, one out of every million cells derived from soybean leaf protoplasts are sulfonylurea resistant (Mauvais, pers. comm.). If resistant plants occur at this frequency, then the starting point for selection of sulfonylurea-resistant biotypes is very high. High seed production, high sulfonylurea susceptibility, and extended periods of germination all serve to create a high potential for the increase of resistant phenotypes. Obviously, the population genetics of sulfonylurea herbicide resistant species needs to be examined in greater detail.

L. SERRIOLA AND *K. SCOPARIA* RESISTANCE TO HERBICIDES

In field experiments, less than 30% control of the sulfonylurea herbicide resistant biotype of *L. serriola* was obtained from DPX–G8311, DPX–R9674, and CGA–131036 (Mallory–Smith *et al.*, 1990). MCPA, 2,4–D, bromoxynil, diuron, picloram, and clopyralid plus 2,4–D gave at least 86% control of *L. serriola* DPX–G8311 tank mixed with the above non–AHAS inhibiting herbicides controlled at least 90% of the weed. The Lloyd farm infested with the sulfonylurea-resistant biotype was treated with 2,4–D in 1987, picloram plus 2,4–D in 1988, and clopyralid plus 2,4–D in 1989. Very few *L. serriola* plants have seeded on the farm since 1986. However, a 1988 survey showed that seven of nine fields on the Lloyd farm are still infested with the resistant biotype (Mallory–Smith *et al.*, 1990). The grower plans to continue using non–AHAS inhibiting herbicides, either alone or in combination with sulfonylureas, to control the resistant biotype. Adjacent areas within about 3 km of the initial infestation have been surveyed for the resistant biotype of *L. serriola*. Except for one nearby location, no resistant plants were found off the farm property (Thill, unpublished data).

Non–AHAS inhibiting herbicides will be required to control the resistant *L. serriola* and to prevent its spread throughout northern Idaho. Non-resistant *L. serriola* infestations should not be sprayed with AHAS inhibiting herbicides used alone. An effective herbicide with a different mode of action should be included in the control programme to prevent or delay resistance at other locations.

Many glasshouse herbicide experiments have been conducted with resistant and susceptible biotypes of *L. serriola* (Mallory-Smith *et al.*, 1990) and *K. scoparia* (Primiani *et al.*, 1990). These include dose-response experiments with several sulfonylureas, imidazolinones, and herbicides with different modes of action.

The herbicide dose required to reduce growth of the resistant biotype of *L. serriola* by 50% was 1000 ppb a.i. w/v of metsulfuron-methyl applied in 30 ml as a post-emergence drench (Mallory-Smith *et al.*, 1990). The GR_{50} for the resistant biotype was greater than 1000 ppb of CGA-131036, chlorsulfuron, DPX-L5300, DPX-M6316, sulfometuron-methyl, bensulfuron-methyl and chlorimuron- ethyl. The GR_{50} of the susceptible biotype ranged between 10 and 100 ppb for all the herbicides except DPX-M6316 and bensulfuron-methyl, which required 850 to 1000 ppb, respectively (Mallory-Smith *et al.*, 1990).

The sulfonylurea dose required for 90% growth reduction (GR_{90}) of the resistant *K. scoparia* biotype ranged from 12 to greater than 62 g a.i. ha^{-1} (applied as a spray) for post-emergence applications and from 32 to greater than 250 g a.i. ha^{-1} for pre-emergence applications of chlorsulfuron, sulfometuron-methyl, DPX-L5300, DPX-M6316, metsulfuron-methyl, and CGA-131036 (Primiani *et al.*, 1990). The GR_{90} for susceptible biotypes ranged from 2 to 4 g a.i. ha^{-1} for post-emergence application and from 1 to 29 g a.i. ha^{-1} for pre-emergence applications of the same herbicides. This represented a 3- to greater than 31-fold increase for the post-emergence treatments and 2- to greater than 250-fold increases for the pre-emergence treatments. Resistance of *K. scoparia* and *L. serriola* biotypes to one sulfonylurea herbicide (e.g., chlorsulfuron) resulted in resistance to all other sulfonylurea herbicides.

Imidazolinone herbicides also inhibit the AHAS enzyme (Shaner *et al.*, 1984). In glasshouse studies, the response of resistant and susceptible biotypes of *L. serriola* varied with imidazolinone herbicide (Mallory-Smith *et al.*, 1990). The resistant biotype required four times the concentration of a post-emergence drench application of imazapyr and imazethapyr than the susceptible biotype to produce a 50% reduction in growth. Similar results were observed when imazapyr and imazethapyr were applied as a post-emergence spray (Thill, unpublished data). However, both biotypes were controlled equally (the GR_{50} was about 3600 ppb) with a drench application of imazaquin (Mallory-Smith *et al.*, 1990). These whole plant responses are similar to results reported by Saxena and King (1988) for *Datura innoxia* cell culture. The amount of imazapyr required to reduce growth 90% was greater than 3-(pre-emergence) and 6-times (post-emergence) more for the resistant *K. scoparia* plants than for the susceptible plants (Primiani *et al.*, 1990).

These three independent studies show that biotypes of three species resistant to sulfonylurea herbicides are also resistant to imidazolinone herbicides. Other plants resistant to one family of AHAS inhibiting herbicides should be tested for resistance to other AHAS inhibiting herbicides. Management strategies designed to avoid or delay weed resistance to AHAS inhibiting herbicides should incorporate the fact that resistance does occur between different classes of AHAS inhibiting herbicides.

Resistance to herbicides with different modes of action also has been examined in the glasshouse. Resistant and susceptible biotypes of *L. serriola* were controlled equally with several rates of 2,4-D, bromoxynil, dicamba, diuron, metribuzin and atrazine (Thill, unpublished data). Likewise, resistant and susceptible biotypes of *K. scoparia* were controlled similarly with bromoxynil, MCPA, and diuron (Primiani *et al.*, 1990). However, about two and one half times more atrazine was required to reduce growth of the sulfonylurea resistant *K. scoparia* than the susceptible biotype (Primiani *et al.*, 1990). Possibly, the sulfonylurea-resistant biotype also was triazine-resistant, while the susceptible biotype was not resistant to triazines (Primiani *et al.*, 1990). Further glasshouse work has confirmed that *K. scoparia* biotypes are either resistant to triazines and sulfonylureas, sulfonylureas only, triazines only, or susceptible to both herbicide classes (Thill, unpublished data). Multiple herbicide resistance, in species such as *K. scoparia* and *S. media* must be considered when developing management practices to control sulfonylurea resistant weeds.

MECHANISM OF SULFONYLUREA RESISTANCE

Sulfonylurea and imidazolinone herbicides inhibit AHAS enzyme (Ray, 1984; Shaner *et al.*, 1984), the first common step in biosynthesis of the branched chain amino acids, isoleucine, valine, and leucine. AHAS from resistant *K. scoparia* was less sensitive to inhibition by chlorsulfuron and other sulfonylurea herbicides than AHAS from susceptible *K. scoparia* (Table 2). For chlorsulfuron, metsulfuron-methyl, sulfometuron-methyl, and CGA-

Table 2. AHAS inhibiting herbicide concentration required to inhibit AHAS activity 50% (I_{50}) isolated from sulfonylurea susceptible (S) and resistant (R) *K. scoparia*.

Compound	I_{50} (nM)		Ratio
	S	R	(R/S)
Chlorsulfuron	22	400	18
Metsulfuron-methyl	26	130	5
Sulfometuron-methyl	10	280	28
CGA-131036	40	460	12
Imazapyr	6000	38000	6

131036, the herbicide concentration required to reduce enzyme activity 50% (I_{50}) was 5- to 28-fold greater using AHAS isolated from the resistant biotype compared to the AHAS from the susceptible biotype. The imidazolinone herbicide, imazapyr, also was less inhibitory (6-fold) to resistant *K. scoparia* AHAS. Similar decreases in sensitivity to AHAS inhibitors has been observed for AHAS isolated from sulfonylurea resistant *S. iberica* and *S. media* (Saari, unpublished).

Other possible resistance mechanisms in *K. scoparia* are differences in rate of chlorsulfuron uptake, metabolism, and translocation. Neither of the two biotypes metabolized ^{14}C-chlorsulfuron significantly during a 21 hour period (Table 3). In contrast, wheat, which is chlorsulfuron tolerant (Sweetser *et al.*, 1982), metabolized 99% of the chlorsulfuron to inactive products during the same period (Table 3). The uptake and subsequent translocation of ^{14}C-chlorsulfuron by susceptible and resistant *K. scoparia* were similar over the time period studied (Table 4) indicating that these processes were not involved in the mechanism of resistance.

Table 3. Metabolism of ^{14}C chlorsulfuron expressed as % chlorsulfuron remaining, by wheat and by sulfonylurea-susceptible (S) and resistant (R) *K. scoparia*.

Time (h)	Wheat	S *Kochia*	R *Kochia*
0	65	98	99
4	19	97	96
21	1	99	94

Table 4. Uptake (%) and translocation (%) of ^{14}C chlorsulfuron by sulfonylurea susceptible (S) and resistant (R) *K. scoparia*.

Time (h)	Uptake		Translocation	
	S	R	S	R
1	30	36	3	2
3	43	30	7	8
6	40	34	15	11
24	46	70	41	32
48	65	71	29	43

Sulfonylurea resistance in *K. scoparia* is due to decreased sulfonylurea herbicide sensitivity of the AHAS enzyme. This conclusion is corroborated by the lack of difference seen in the rate of uptake, metabolism, and translocation of chlorsulfuron between the resistant and susceptible biotypes.

INHERITANCE OF SULFONYLUREA RESISTANCE

Mazur and Falco (1989) recently reviewed the development of herbicide resistance in crops, including a section on sulfonylurea and imidazolinone herbicides. In corn, imidazolinone resistance was inherited as a single dominant trait (Anderson and Hibbard, 1985; Anderson and Georgeson, 1986). To determine inheritance in *L. serriola* pollen from a sulfonylurea resistant male parent was transferred to the stigmas of unpollinated susceptible *L. serriola* and to domestic lettuce

123

(*Lactuca sativa* L. var. 'Bibb') flowers (Mallory *et al.*, 1989). The F_1 seedlings were treated with metsulfuron-methyl, allowed to self pollinate, and seeds collected. The F_2 seedlings were treated with metsulfuron-methyl and scored as susceptible, intermediate, or resistant in their response to the treatment. The best fit for Chi Square Analysis of the F_2 generation was a 1:2:1 ratio indicating the trait is controlled by a single nuclear gene with incomplete dominance (Mallory *et al.*, 1989).

RESISTANT WEED MANAGEMENT STRATEGIES

Several strategies exist for delaying herbicide resistance in weeds (Gressel and Segel, 1982). First, use tank mixtures of herbicides with different modes of action. Mixture partners available, efficacy overlap, the timing of application, and the relative residual activities of the herbicides involved may limit effectiveness of this strategy. For example, if one herbicide in a mixture controls a different weed spectrum and/or has a longer soil residual activity than the herbicide partner(s), the delay in the appearance of herbicide resistant weeds will be compromised. Post-emergence or short residual herbicide mixtures must be applied after weed emergence for both herbicides to exert their action. Alternatively, sequential treatments of herbicides with different modes of action can be applied. Rotation of crops often accompanies this practice but may not be applicable in monoculture areas such as the dryland cereal growing areas in parts of the USA. In general, both herbicide mixtures and rotations have been effective in delaying development of herbicide resistance in weeds (Gressel, 1987).

Reduced selection pressure for resistant biotypes also is important to slow the development of herbicide resistant weeds. This can be accomplished by using herbicides with short soil persistence and reducing the application frequency and dose of each herbicide. The potential for herbicide resistance in weeds can be reduced further by geographically limiting where the herbicide is used. Tillage also removes weeds in a nondiscriminatory manner and should be used whenever possible. As previously covered, when developing herbicide resistant management strategies it is important to remember that resistance does occur between different classes of AHAS inhibiting herbicides and between herbicides with different modes of action.

REFERENCES

Anderson, P.C. and Georgeson, M. (1986). Selection and characterization of imidazolinone tolerant mutants of maize. In "Biochemical Basis of Herbicide Action". 27th Harden Conference Progress Abstracts, Wye College, Ashford, UK.

Anderson, P.C. and Hibbard, K.A. (1985). Evidence for the interaction of an imidazolinone herbicide with leucine, valine, and isoleucine metabolism. Weed Science 33, 479-483.

Arthur, J.C. (1894). Wild or prickly lettuce. Bulletin Number 52, Volume V, Agricultural Experiment Station, Purdue University, Lafayette, Indiana.

Beatley, J. (1973). Russian thistle (*Salsola*) species in western United States. Journal of Range Management **26**, 225–226.

Beyer, E.M., Jr., Duffy, M.J., Hay J.V., and Schlueter, D.D. (1988). Sulfonylureas. In "Herbicides: Chemistry, Degradation and Mode of Action, Volume 3" (P.C. Kearney and D.D. Kaufman, eds), pp. 117–189. Marcel Dekker, Inc., New York.

Burnside, O.C., Fenster, C.R., Evetts, L.L., and Mumm, R.F. (1981). Germination of exhumed weed seeds in Nebraska. Weed Science **29**, 577–586.

Chepil, W.S. (1946). Germination of weed seeds. I. Longevity, periodicity of germination, and vitality of seeds in cultivated soil. Science of Agriculture **26**, 307–346.

Eberlein, C.V. and Fore, Z.Q. (1984). Kochia biology. Weeds Today **15**, 5–6.

Evans, R.A. and Young, J.A. (1972). Germination and establishment of *Salsola* in relation to seedbed environment. II. Seed distribution, germination, and seedling growth of *Salsola* and microenvironmental monitoring of the seedbed. Agronomy Journal **64**, 219–224.

Farkas, A. and Amen, J. (1940). The action of diphenyl on *Penicillium* and *Diplodia* moulds. Palestine Journal of Botany, Jerusalem Series **2**, 38–45.

Gaines, X.M. and Swan, D.G. (1972). "Weeds of Eastern Washington and Adjacent Areas". pp. 310–311. Camp-Na-Bor-Lee Association, Inc., Davenport, Washington.

Georghiou, G.P. (1986). Introduction: The magnitude of the resistance problem. In "Pesticide Resistance: Strategies and Tactics for Management", pp. 14–43. National Academy Press, Washington, D.C.

Gressel, J. (1987). Appearance of single and multi-group herbicides resistances and stategies for their prevention. British Crop Protection Conference – Weeds 479–488.

Gressel, J. and Segel, L.A. (1982). Interrelating factors controlling the rate of appearance of resistance: the outlook for the future. In "Herbicide Resistance in Plants" (H.M. LeBaron and J. Gressel, eds) pp. 325–347. John Wiley & Sons, New York.

Gressel, J., Ammon, H.U., Fogelfors, H., Gasquez, J., Kay, Q.O.N., and Kees, H. (1982). Discovery and distribution of herbicide-resistant weeds outside North America. In "Herbicide Resistance in Plants" (H.M. LeBaron and J. Gressel, eds) pp. 31–55. John Wiley & Sons, New York.

Hill, R.J. (1982). Taxonomy and biological considerations of herbicide-resistant and herbicide-tolerant biotypes. In "Herbicide Resistance in Plants" (H.M. LeBaron and J. Gressel, eds), pp. 81–98. John Wiley & Sons, New York.

Hutchison, J.M., Shapiro, R., and Sweetser, P.B. (1984). Metabolism of chlorsulfuron by tolerant broadleaves. Pesticide Biochemistry and Physiology **22**, 243–247.

Johnston, D.N. and Wood, W.N. (1976). *Kochia scoparia* control on non-cropland. Proceedings North Central Weed Control Conference **31**, 126.

Kees, H. (1978). Beobachtungen uber resistenzer-scheinungen bei der vogelmiere (*Stellaria media*) gegen atrazin im mais. Gesunde Pflanzen **30**, 137.

Kemp, M.S. and Caseley, J.C. (1987). Synergistic effects of 1-aminobenzotriazole on the phytotoxicity of chlorotoluron and isoproturon in a resistant population of black-grass (*Alopecurus myosuroides*). British Crop Protection Conference - Weeds, 895–899.

LeBaron, H.M. (1989). Management of herbicides to avoid, delay and control resistant weeds: A concept whose time has come. Proceedings Western Society of Weed Science **42**, 6–16.

Lindquist, K. (1960). On the origin of cultivated lettuce. Hereditas **46**, 319–350.

Lutman, P.J.W. and Lovegrove, A.W. (1985). Variation in the tolerance of *Galium aparine* (cleavers) and *Stellaria media* (chickweed) to mecoprop. British Crop Protection Conference - Weeds, 411–418.

Mallory, C.A., Dial, M.J., and Thill, D.C. (1989). Inheritance of sulfonylurea herbicide resistance in prickly lettuce (*Lactuca serriola* L.). Proceedings of the Western Society of Weed Science **42**, 39.

Mallory-Smith, C.A., Thill, D.C., and Dial, M.J. (1990). Identification of sulfonylurea herbicide resistant prickly lettuce (*Lactuca serriola*). Weed Technology **4**, 163–168.

Marks, M. and Prince, S. (1981). Influence of germination date on survival and fecundity in wild lettuce, *Lactuca serriola*. OIKOS **36**, 326–330.

Marks, M. and Prince, S.D. (1982). Seed physiology and seasonal emergence of wild lettuce, *Lactuca serriola*. OIKOS **38**, 242–249.

Mazur, B.J. and Falco, S.C. (1989). The development of herbicide resistant crops. Annual Review of Plant Physiology **40**, 441–470.

Moss, S.D. (1987). Herbicide resistance in black-grass (*Alopecurus myosuroides*) British Crop Protection Conference - Weeds 879–886.

O'Keefe, D.P., Romesser, J.A., and Leto, K.J. (1987). Plant and bacterial cytochrome P-450: Involvement in herbicide metabolism. In "Phytochemical Effects of Environmental Compounds" (J.A. Saunders and L. Kosak-Channing, eds), pp. 151–173. Plenum Publishing Corporation, New York.

Ogg, A.G., Jr. and Dawson, J.H. (1984). Time of emergence of eight weed species. Weed Science **32**, 327–335.

Parrish, J.A.D. and Bazzaz, F.A. (1978). Pollination niche separation in a winter annual community. <u>Oecologia (Berlin)</u> **35**, 133-140.

Powles, S.B. and Howat, P.D. (1989). A review of weeds in Australia resistant to herbicides. <u>Weed Science Society of America Abstracts</u> **29**, 113.

Primiani, M.M., Cotterman, J.C. and Saari, L.L. (1990). Resistance of Kochia (*Kochia scoparia*) to sulfonylurea and imidazolinone herbicides. <u>Weed Technology</u> **4**, 169-172.

Prince, S.D. and Carter, R.N. (1985). The geographical distribution of prickly lettuce (*Lactuca serriola*). III. Its performance in transplant sites beyond its distribution limit in Britain. <u>Journal of Ecology</u> **73**, 49-64.

Ray, T.B. (1984). Site of action of chlorsulfuron. <u>Plant Physiology</u> **75**, 827-831.

Ryan, G.F. (1970). Resistance of common groundsel to simazine and atrazine. <u>Weed Science</u> **18**, 614-616.

Saxena, P.K. and King, J. (1988). Herbicide resistance in *Datura innoxia*. Cross-resistance of sulfonylurea-resistant cell lines to imidazolinones. <u>Plant Physiology</u> **86**, 863-867.

Schaat, B.G., Thill, D.C., and Callihan. R.H. (1985). Broadleaf weed control in winter wheat. Western Society of Weed Science Research Progress Report 320-321.

Shaner, D.L., Anderson, P.C., and Stidham, M.A. (1984). Imidazolinones. Potent inhibitors of acetohydroxyacid synthase. <u>Plant Physiology</u> **76**, 545-546.

Stevens, O.A. (1943). Russian thistle life history and growth. Agricultural Experiment Station Bulletin 326, North Dakota Agricultural College, Fargo, North Dakota.

Sweetser, P.B., Schow, G.S. and Hutchison, J.M. (1982). Metabolism of chlorsulfuron by plants: Biological basis for a new herbicide for cereals. <u>Pesticide Biochemistry and Physiology</u> **17**, 18-23.

Thill, D.C., Mallory, C.A., Saari, L.L., Cotterman, J.C., and Primiani, M.M. (1989). Sulfonylurea resistance - mechanism of resistance and cross resistance. <u>Weed Science Society of America Abstracts</u> **29**, 132.

Turkington, R., Kenkel, N.C., and Franco, G.O. (1980). The biology of Canadian weeds. 42. *Stellaria media* (L.) Vill. <u>Canadian Journal of Plant Science</u> **60**, 981-992.

Uhrikova, A. (1974). Index to chromosome numbers of Slovakian flora. pt. 4. Acta Facultatis rerum naturalium Universitatis comenianae.

Whitaker, T.W. and McCollum, G.D. (1954). Shattering in lettuce - its inheritance and biological significance. <u>Bulletin of the Torrey Botanical Club</u> **81**, 104-110.

Young, F.L. (1986). Russian thistle (*Salsola iberica*) growth and development in wheat (*Triticum aestivum*). Weed Science **34**, 901–905.

Young, F.L. and Whitesides, R.E. (1987). Efficacy of postharvest herbicides on Russian thistle (*Salsola iberica*) control and seed germination. Weed Science **35**, 554–559.

Zorner, P.S., Zimdahl, R.L., and Schweizer, E.E. (1984). Effect of depth and duration of seed burial on Kochia (*Kochia scoparia*). Weed Science **32**, 602–607.

MANAGING HERBICIDE RESISTANCE THROUGH FITNESS AND GENE FLOW

Steven R. Radosevich, Bruce D. Maxwell and Mary Lynn Roush

Oregon State University, Oregon, USA.

Understanding the dynamics of herbicide resistance must involve and link the processes in three basic plant science disciplines: physiology, genetics and ecology. Models serve to organise and focus research at the process level, and to provide a tool for evaluating resistance management tactics. Sensitivity and elasticity analysis of a model by Maxwell *et al.* (1990) identified two sets of life-history processes important for understanding and managing herbicide resistance. These are processes that influence fitness of resistant biotypes relative to susceptible biotypes, and processes that contribute to gene flow through space and time. Survivorship and fecundity, both influenced by plant competition, are important determinants of relative fitness. Herbicides influence survivorship, and, thus, result in an overwhelming increase in the relative fitness of the resistant type. However, when herbicide selection pressure is removed, population dynamics are determined by differences in all processes that contribute to fitness. The relative fitness of R and S biotypes has important consequences for the management of herbicide resistance. Reduced fitness in R-types infers that resistant plants will be replaced by susceptible individuals over time after herbicide use is abandoned. Alternatively, if fitness of the R-type is not less than the susceptible, resistance should decline slowly, if at all. Competition components of fitness will also be influenced by the presence of other species in the system. These premises have not been examined experimentally, although the alternatives lead to very different management tactics for managing resistance. Gene flow influences two important management scenarios: spread of resistance and use of S-genotype sources to slow development of resistance. Current attempts to manage resistance are dominated by tactics that only employ other herbicides to remove R-plants from populations that have developed resistance. Manipulation of S-type gene flow could be an equally valuable alternative, and be more cost effective than control measures that reduce both R- and S-type plants.

INTRODUCTION

Herbicides are used extensively in agriculture because they are cost effective and time efficient tools to reduce the abundance of weeds. Reduction in weed abundance usually results in improved crop yields and presumably increased economic benefits to farmers (Cousens, 1986, 1987). Recent trends in herbicide development have produced extremely specific and selective chemicals that are used routinely in cropping systems. However, the intensive use of such herbicides has precipitated an alarming increase in the development of herbicide resistance (LeBaron and Gressel, 1982; Radosevich and Holt, 1984; Gressel, 1987). The development and dissemination of herbicide resistant plants represents a significant threat to modern agriculture. The seriousness of this problem is aggravated because we understand poorly the processes that underlie the population dynamics of herbicide resistance, and subsequently fail to provide options for the prevention, delay, and reduction of herbicide resistance. Clearly, agricultural scientists need to consider how they are addressing this problem.

The development of herbicide resistance results from the nearly exclusive use of one type of tool for weed control. It is a symptom of over-use and reliance. We need to change the way we think about the role of herbicides in farm systems in order to make further progress in the delay and prevention of herbicide resistance. A common approach to manage resistance is to replace the herbicide that selected resistance with another that controls both resistant and susceptible weed-biotypes. This approach may provide a short-term solution but may only aggravate the problem in the long term because the biological processes responsible for resistance are ignored. In order to prevent, delay, or reduce the incidence of herbicide resistance, tactics must be devised that recognize and utilize the biological, ecological and economic processes responsible for herbicide resistance. In this paper we pose such tactics, utilizing a biological model for herbicide resistance (Maxwell *et al.*, 1990). The simulations of our model assume single gene nuclear inheritance of resistance by an outcrossing weed species that possesses a relatively short-lived seed bank. However, other simulations of herbicide resistance are possible by modifying these assumptions in the model (Maxwell *et al.*, 1990).

MODELS AS RESEARCH AND MANAGEMENT TOOLS

Weed science is an integrative science that encompasses three plant science disciplines: physiology, genetics, and ecology. A strategic biological approach to understanding the mechanisms and dynamics of herbicide resistance must integrate and link these disciplines. Consider the phenomenon of triazine resistance (Radosevich and Appleby, 1973 a,b; LeBaron and Gressel, 1982). The physiological process that results in resistance is a mutation in the single-protein binding site associated with photosystem II (Radosevich *et al.*, 1979; Pfister *et al.*, 1979). The function of this protein was linked to differences in chloroplast and whole-plant performance (Sims *et al.*, 1981; Holt *et al.*, 1983), which also influenced the relative fitness of the two biotypes (Conard and Radosevich, 1979; Holt and Radosevich, 1983). Recent investigations indicate that plant fitness is also influenced by interactions between the maternally-inherited single-locus gene that codes for the protein, and nuclear inheritance of other plant performance traits (Stowe and Holt, 1988). Integration of these genetic, physiological, and ecological processes ultimately influences the dynamics of resistant and susceptible populations.

Because the processes that influence the dynamics of herbicide resistance interact in a complex manner, it is important to organize information and focus research activities towards these processes. Models can serve such a function, as well as provide a tool for evaluating management tactics. Gressel and Segel, (1978, 1982) developed a model that suggests important factors that influence occurrence and development of herbicide resistance. Maxwell *et al.* (1990) have developed a population model that

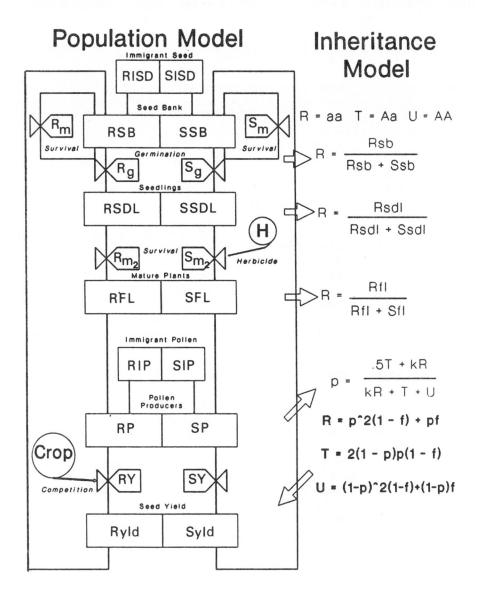

Figure 1. <u>A representation of the Maxwell *et al.* (1990) model that follows life-history stages of resistant and susceptible biotypes and incorporates the influences of fitness processes and gene flow.</u> Open arrows indicate the flow of information between the population model and the inheritance model. State variables are named above each box and processes are indicated in italics. Abbreviations are explained in Table 1.

simulates the development, spread, and dynamics of herbicide-resistant and -susceptible weed biotypes (Fig. 1 and Table 1). The Maxwell *et al.* (1990) model provides a basis to evaluate the relative importance of specific biological processes involved in the dynamics of herbicide resistance.

Table 1. <u>Definitions used in the development of the herbicide resistance model</u> (see Figures 1, 4 & 7, and Maxwell *et al.*, 1990).

Parameter	Definition and default values

Immigration parameters:

a_{sd}	The number of seed produced per unit area by the source population. a_{sd} = TOT = model input value.
a_p	The number of pollen-producing plants per unit area representing the total amount of pollen produced by the source population. $a_p = a_{sd}/10$.
x'	The distance from the center of the source population to the center of the treated population.
c_s	The radius of the total source area.
c_o	The radius of the interior source population that is equal to the treated population radius (0.5).
c	The square root of the scaling factor $[c = (c_s^2/c_0^2)^{1/2}]$.
bp	The steepness of the diffusion gradient for pollen (1.7)
bsd	The steepness of the diffusion gradient for seed (7.4).

Demographic parameters:

RSB	The number of R seed in the seed bank before germination.
SSB	The number of S seed in the seed bank before germination.
R_m	The proportion of R seed that die over 1 generation in the seed bank (0.7).
S_m	The proportion of S seed that die over 1 generation in the seed bank (0.7).
ISD	The total number of seed added to the seed bank through immigration over 1 generation.
RISD	The number of immigrant seed that is R-phenotype.
SISD	The number of immigrant seed that is S-phenotype.
R_g	The proportion of R seed that germinate over 1 generation (0.3).
S_g	The proportion of S seed that germinate over 1 generation(0.3).
RSDL	The number of R seedlings produced from seed germinating in the seed bank.
SSDL	The number of S seedlings produced from seed germinating in the seed bank.
R_{m2}	The proportion of R seedlings that die over 1 generation(0.75).
S_{m2}	The proportion of S seedlings that die over 1 generation(0.75).

Table 1 (cont'd).

Parameter	Definition and Default Values

RFL	The number of R mature (flowering) plants.
SFL	The number of S mature (flowering) plants.
h	Herbicide efficacy on seedlings (95%).
IP	The total number of pollen-producing plants represented by immigrant pollen.
RP	The number of R pollen-producing plants represented by immigrant pollen.
SP	The number of S pollen-producing plants represented by immigrant pollen.
RY	The number of R seed produced per R plant.
SY	The number of S seed produced per S plant.
Ryld	The R seed yield per unit area.
Syld	The S seed yield per unit area.
Tyld	The seed yield per unit area with genotype Aa.

Inheritance parameters:

Rsb	The proportion of the seed bank that is R-phenotype.
Tsb	The proportion of the seed bank that is Aa genotype.
Usb	The proportion of the seed bank that is AA genotype.
Ssb	The proportion of the seed bank that is S-phenotype.
Rsdl	The proportion of seedlings that are R-phenotype.
Tsdl	The proportion of seedlings that are Aa genotype.
Usdl	The proportion of seedlings that are AA genotype.
Ssdl	The proportion of seedlings that are S-phenotype.
Rfl	The proportion of mature (flowering) plants that are R-phenotype.
Tfl	The proportion of mature (flowering) plants that are Aa genotype.
Ufl	The proportion of mature (flowering) plants that are AA genotype.
Sfl	The proportion of mature (flowering) plants that are S-phenotype.
Rof	The proportion of R pollen-producing plants in the outside (source for immigration) population.
Sof	The proportion of S pollen-producing plants in the outside (source of immigration) population.
Tof	The proportion of pollen-producing plants with Aa genotype in the outside (source for immigration) population.
Rp	The proportion of R pollen-producing plants in the treated population.
Sp	The proportion of S pollen-producing plants in the treated population.
Tp	The proportion of pollen-producing plants with Aa genotype in the treated population.
Up	The proportion of pollen-producing plants with AA genotype in the treated population.
p	The probability of an individual's donating the a allele in a mating.

Table 1 (cont'd).

Parameter	Definition and Default Values

---

kR	The fitness of the R pollen relative to the S pollen (0.9).
f	The inbreeding coefficient.
m	The forward mutation rate to resistance (10^{-6}).
Ne	The effective size of the population (related to the number of reproducing adults).
R	The proportion of the total matings that will produce R seed.
T	The proportion of the total matings that will produce seed with Aa genotype.
U	The proportion of the total matings that will produce seed with AA genotype.
S	The proportion of the total matings that will produce S seed.

Competition parameters: see Fig. 4.

RY_{max}	The maximum yield per R plant (900 seeds/plant).
Sy_{max}	The maximum yield per S plant (1000 seeds/plant).
a_r	The area required to produce RY_{max} (1).$_i$
a_s	The area required to produce SY_{max} (1).$_i$
z_{sr}	The influence of S plant density on the seed yield of R plants (1).
z_{rs}	The influence of R plant density on the seed yield of S plants (1).
z_{cr} R	The influence of the crop plant density on the seed yield of plants (1.2).
z_{cs}	The influence of the crop plant density on the seed yield of S plants (1.1).
N_i	The density of the other weed or crop (200 plants/m²).
br	The coefficient which determines the form of the relationship between RY_t and the total density (0.8).
bs	The coefficient which determines the form of the relationship between SY_t and the total density (0.8).

Sensitivity and elasticity analyses (Maxwell *et al.*, 1988; Moloney, 1988) conducted by Maxwell *et al.* (1990) have identified two sets of life-history processes that are important for understanding and managing the dynamics of herbicide resistance: (1) processes that influence fitness of resistant biotypes relative to the susceptible biotypes and crop species, and (2) processes that contribute to gene flow in space and time. We propose that research focuses on these two sets of key biological processes to understand the dynamics of resistance, and to develop effective management strategies.

Fitness

Fitness describes the evolutionary advantage of a phenotype, which is based on its survival and reproductive success (Silvertown, 1982; Crow, 1986). Linkages between the disciplines of physiology and ecology are necessary to understand the processes that influence the relative fitness of resistant and susceptible genotypes.

The relative fitness of R and S genotypes has a powerful influence on the dynamics of mixed resistant and susceptible populations. Life-history processes that are important determinants of the relative fitness of biotypes are: survivorship processes (demography of seeds, seedlings, and mature plants), fecundity (pollen and seed production), and plant competition. Herbicide selection pressure influences one of these fitness processes (i.e. survivorship). When the herbicide is used, its selection pressure (reduced survivorship of susceptible seedlings) results in an overwhelming increase in the relative fitness of the resistant genotype (Gressel and Segel, 1978, 1982; Maxwell *et al.*, 1990). Thus, a potential management tactic to prevent resistance is to reduce the herbicide rate or number of applications to a field (Fig. 2). This tactic will require better information on economic weed thresholds. Better understanding about the influence of environment and specific weed species on herbicide effectiveness will also be necessary since herbicide rate and application frequency are now used to increase the reliability and spectrum of plants controlled by these tools.

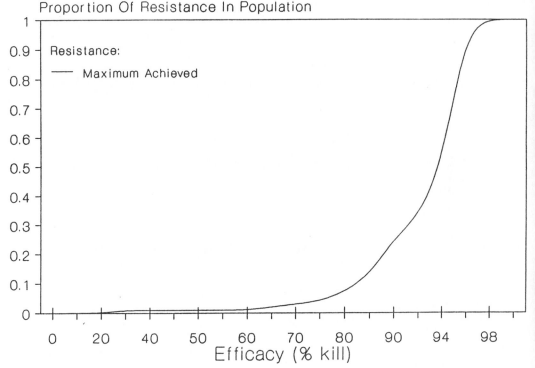

Figure 2. The simulated influence of herbicide efficacy on the maximum level of resistance achieved in a weed population that had 5 continuous years of herbicide applications (Maxwell *et al.*, 1990).

When herbicide selection pressure is removed, population dynamics are determined by differences in all processes that contribute to the fitness of each biotype (Fig. 3). Some instances have been reported in which herbicide resistance previously described can no longer be found in those fields (LeBaron, pers. comm.). This suggests that biological processes,

like fitness, may be used to remove resistance from fields after it has
developed.

Phenotype Proportions

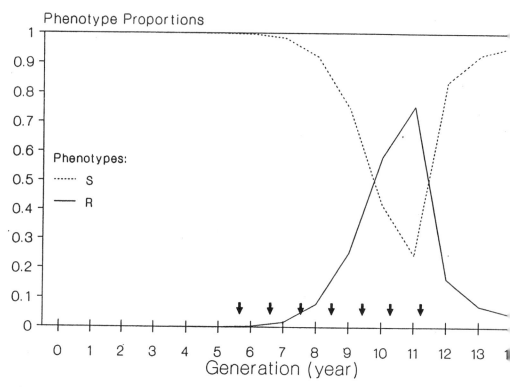

Figure 3. <u>Model simulations describing the evolution of resistance when
the herbicide is used in the system and the subsequent dynamics of
resistance in the weed population after the herbicide has been removed
(Maxwell *et al.*, 1990)</u>. The solid arrows indicate the years of continuous
herbicide use.

Many herbicide-resistant biotypes are less fit than their susceptible
counterparts (Conard and Radosevich, 1979; Holt and Radosevich, 1983).
Reduced fitness in the R-type infers that resistant plants will be
replaced by susceptible individuals over time after herbicide use is
abandoned. Alternatively, if the fitness of the R-type is not, or only
slightly, less than the susceptible (Rubin *et al.*, 1985; Valverde *et al.*,
1988), resistance should decline slowly, if at all. Competition
components of fitness will also be influenced by the presence of other
species (i.e. crop or other weeds) in the system, especially if they are
strong competitors. These premises have not been examined
experimentally, although the alternatives lead to very different tactics
for managing resistance (Gressel, 1987; DuPont, 1988). For example,
simulations by Maxwell *et al.* (1990) suggest that competition from S-
biotype or crop plants can significantly reduce the longevity of
resistance after it has developed (Fig. 4), if the mechanism of R
inheritance is a single recessive gene. Other simulations (Fig. 5)
question the merit of different herbicide treatments that control both
S- and R-types following resistance development because the suppressive
influence of the S-type upon the R-type can be eliminated by this
approach. Although weed abundance will be reduced by such a tactic, the

longevity of resistance to the original herbicide is increased (Fig. 5) because fitter S-types and a source of the S genotypes will also be removed from the weed population. The important question here is: are the yield losses caused by the escape of a few S-type plants sufficient to counter the direct expenses necessary to achieve near-perfect weed control and the potential for herbicide loss via resistan .e.

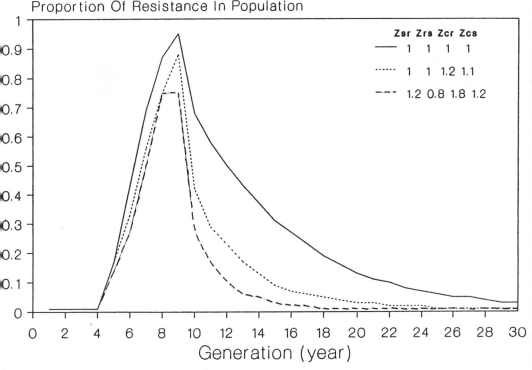

Proportion Of Resistance In Population

Figure 4. Model simulations demonstrating the influence of relative competitive abilities of resistant and susceptible phenotypes in the presence of the crop on evolution of and recovery from resistance (Maxwell *et al.*, 1990). Z_{rs}, Z_{sr}, Z_{cr} and Z_{cs} are relative competitive ability parameters in the competition equations.

Gene flow

In order to understand processes that influence the flow of genes through space and time, linkages must be established between the disciplines of ecology and genetics. Gene flow processes directly alter the frequencies of R and S alleles in populations (Levin and Kerster, 1974). Immigration of pollen and seed introduces genes into the population. Inbreeding and genetic drift result from limited flow of genes within the population. Seed dormancy slows the loss of genes from the population. Seed bank dynamics include gene flow processes, as well as seed survivorship (a fitness process).

Gene flow from outside populations (i.e. pollen and seed immigration) addresses two important management scenarios: (1) the spread of resistance over the landscape, and (2) the use of S-genotype sources (e.g. fence rows, untreated rows or fields, etc.) to slow the development of resistance. Current attempts to manage herbicide resistance are dominated by tactics that only employ other herbicides to remove R and S plants from populations that have developed resistance (Fig. 5). Our

model (Maxwell *et al.*, 1990) suggests that manipulation of S-type gene flow, as either seed or pollen, is an equally valuable alternative (Fig. 6). Such management tactics may also be more cost-effective than the addition of extra control measures to reduce R-type plant abundance.

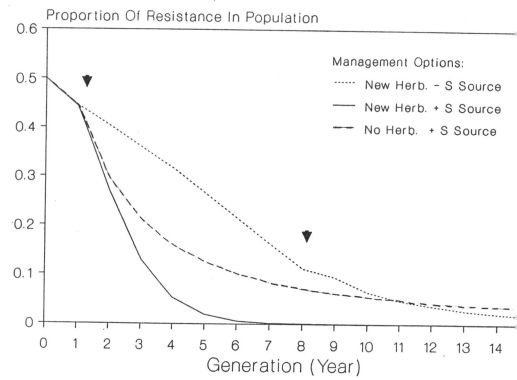

Figure 5. __Maxwell *et al.* (1990) model simulations demonstrating the use of a new herbicide (one that has 85% efficacy on both R and S individuals) and a S source population for managing recovery from herbicide resistance in a weed population__. Arrows indicate the beginning and end of continuous new herbicide applications.

VALUE OF THE S-TYPE

Current notions about weed control suggest that weeds have little or no value and, therefore, they can be prophylactically reduced in crops with no adverse consequences. However, weeds, i.e., the S-type plants, may have a value that is at least equivalent to the costs of controlling them or to the amount of crop that is lost if they are not controlled. The reduced number of S-type plants in treated fields, as a result of extreme herbicide selection pressure, may also account for the recent extensive development of herbicide resistance. Thus, herbicide resistance can be an additional cost of weed control that results from over-zealous application frequency or rate.

Herbicide resistance may be prevented, or at least delayed, if selection pressure from the herbicide is reduced (Fig. 2). A tactic suggested is to allow survival of some susceptible weeds in the treated fields to allow S-type weed gene flow and fitness processes to act against the R-type. However, such tactics will require a re-evaluation of our current

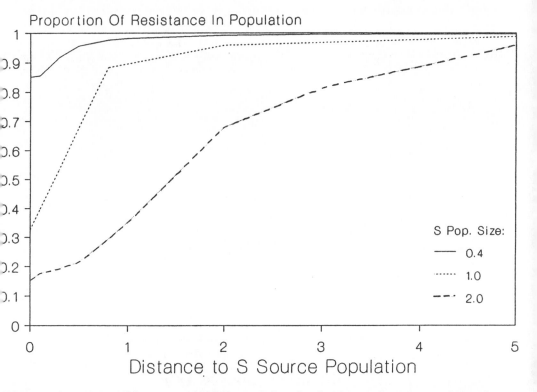

Figure 6.   Maxwell *et al.* (1990) model simulations demonstrating the influence of the size (relative to the treated population) and distance (in units of treated pop. diameter) to (edge of source to edge of treated) a 100% susceptible source population on the level of resistance achieved after five continuous years of herbicide applications.

perceptions about the impacts of weeds in crops. Critical to this thinking are the concepts of biological and economic thresholds (Auld and Tisdell, 1986; Auld *et al.*, 1987; Cousens, 1986, 1987; Cussans *et al.*, 1986). Most competition studies indicate that maximum crop production is achieved when all weeds are controlled (see Zimdahl, 1980 and Cousens *et al.*, 1987 for summaries). However, optimum levels of weed control result when both the costs and economic benefits of weed suppression are considered. Biological and economic thresholds for weed competition vary according to the density, spacing, and species of crop (Carlson and Hill, 1985; Auld *et al.*, 1987). The acceptance of economic weed thresholds, other than near zero, are likely to significantly reduce herbicide selection pressure and herbicide resistance (Maxwell *et al.*, 1990), and may not have a serious impact on crop productivity or economic returns (Auld *et al.*, 1987).

Management of herbicide resistance after it has developed must also consider threshold concepts because fitness and gene flow processes can depend on the distance and location of S-plants in relation to R-plants (Fig. 7). The value of the S-types may also be related to the breeding system and the ability of the weed species to disperse pollen and seed. However, little is known about any of these processes, or the density levels of weeds (S- and R-types) and crops that might influence them.

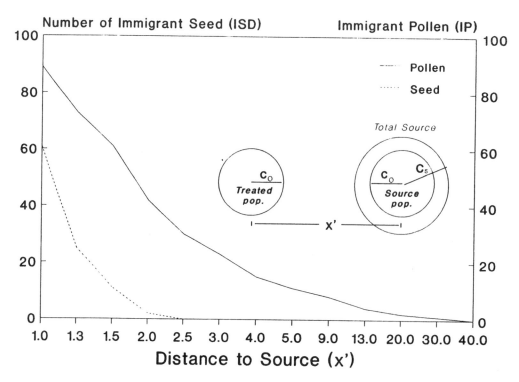

Figure 7. <u>Pollen and seed dispersal as a function of distance (in units of treated population diameter) to a source population (Maxwell *et al.*, 1990)</u>. See Table 1 for parameter.

THE RE-THINKING PROCESS

The understanding and acceptance of weed thresholds in crops may provide the means to effectively manage herbicide resistance and to reduce the costs of crop production. However, a greater effort to develop biological and ecological information about weed species will be necessary to develop rational weed management tactics involving thresholds. Biology and economics imply that some weeds will remain in fields. Therefore better information on the crop/weed interactions and the weed species' reproductive capacity, dispersal ability, and spatial distribution are necessary to devise management tactics involving weed thresholds. Such tactics, once devised, should prevent and reduce herbicide resistance.

Unfortunately, current attempts to manage resistance are dominated by approaches that employ other herbicides to remove plants from populations that have already developed resistance. These tactics are likely to only exacerbate the problem, as suggested by recent trends in the occurrence of multiple resistances and cross-resistance in the United Kingdom (Moss, 1987) and Australia (Heap, these Proceedings). The development of herbicide-resistant crop plants increases the probability that other "nonselective" chemicals can be used to combat resistance. Such tactics are double-edged, however, since they may select rapidly for herbicide-resistant weeds. If such crops are used in rotations to break the herbicide-monoculture pattern, their employment may delay resistance from occurring. On the other hand, such elaborate technological methods may not be necessary for many weed crop systems to prevent and manage resistance, because manipulation of S-type gene flow and the relative fitness among S, R, and crop plants may offer simple alternatives for the management of resistance.

REFERENCES

Auld, B.A. and Tisdell, C.A. (1986). Economic threshold/critical density models in weed control. In "Symposium on Economic Weed Control", pp. 261-268. European Weed Research Society Proceedings.

Auld, B.A., Mentz, K.M. and Tisdell, C.A. (1987). "Weed Control Economics". Academic Press, London. pp. 177.

Carlson, H.L. and Hill, J.E. (1985). Wildoat (*Avena fatua*) competition in spring wheat: Plant density effects. Weed Science 33, 176-181.

Conard, S.G., and Radosevich, S.R. (1979). Ecological fitness of *Senecio vulgaris* and *Amaranthus retroflexus* biotypes susceptible or resistant to triazine. Journal of Applied Ecology, 16, 171-177.

Cousens, R.D. (1986). The use of population models in the study of economics of weed control. In "Symposium on Economic Weed Control", pp. 269-276. European Weed Research Society Proceedings.

Cousens, R.D. (1987). The theory and reality of weed control thresholds. Plant Protection Quarterly 2(1), 13-20.

Cousens, R.D., Moss S.R., Cussans, G.W. and Wilson, B.J. (1987). Modelling weed populations in cereals. Reviews of Weed Science 3, 93-112.

Crow, J.F. (1986). "Basic concepts in population, quantitative, and evolutionary genetics". W.H. Freeman and Co., New York. pp. 273.

Cussans, G.W., Cousens, R.D. and Wilson, B.J. (1986). Thresholds for weed control-the concepts and their interpretations. In "Symposium on Economic Weed Control", pp. 253-260. European Weed Research Society Proceedings.

E.I. DuPont de Nemours and Company. (1988). DuPont weed resistance workshop. Denver, CO.

Gressel, J. (1987). Appearance of single and multi-group herbicide resistances and strategies for their prevention. British Crop Protection Conference - Weeds, 479-488.

Gressel, J. and Segel, L.A. (1978). The paucity of plants evolving genetic resistance to herbicides: possible reasons and implications. Journal of Theoretical Biology **75**, 349-371.

Gressel, J. and Segel, L.A. (1982). Interrelating factors controlling the rate of appearance of resistance: the outlook for the future. In "Herbicide Resistance in Plants", (H.M. LeBaron and J. Gressel, eds), pp. 325-348. John Wiley and Sons, New York.

Holt, J.S. and Radosevich, S.R. (1983). Differential growth of two common groundsel (*Senecio vulgaris*) biotypes. Weed Science **31**, 112-115.

Holt, J.S., Radosevich, S.R. and Stemler, A.J. (1983). Differential efficiency of photosynthetic oxygen evolution in flashing light intriazine-resistant and triazine-susceptible biotypes of *Senecio vulgaris* L. Biochimica Biophysica Acta **722**, 245-255.

LeBaron, H.M., and Gressel, J. (1982). "Herbicide resistance in plants". John Wiley and Sons, New York. pp. 449.

Levin, D.A., and Kerster, H.W. (1974). Gene flow in seed plants. Evolutionary Biology **7**, 139-220.

Maxwell, B.D., Roush, M.L. and Radosevich, S.R. (1990). Predicting the evolution and dynamics of herbicide resistance in weed populations. Weed Technology **4**, 2-13.

Maxwell, B.D., Wilson, M.V. and Radosevich, S.R. (1988). Population modeling approach for evaluating leafy spurge development and control. Weed Technology **2**, 132-138.

Moloney, K.A. (1988). Fine-scale spatial and temporal variation in demography of a perennial bunchgrass. Ecology **69**, 1588-1598.

Moss, S.R. (1987). Herbicide resistance in black-grass (*Alopecurus myosuroides*). British Crop Protection Conference - Weeds, 879-886.

Pfister, C., Radosevich S.R. and Arntzen, C.J. (1979). Modification of herbicide binding to photosystem II in two biotypes of *Senecio vulgaris*. Plant Physiology **64**, 995-999.

Radosevich, S.R. and Appleby, A.P. (1973a). Relative susceptibility of two common groundsel (*Senecio vulgaris* L.) biotypes to six S-triazines. Agronomy Journal **65**, 553-555.

Radosevich, S.R. and Appleby, A.P. (1973b). Studies on the mechanism of resistance to simazine in common groundsel (*Senecio vulgaris* L.). Weed Science **21**, 497-500.

Radosevich, S.R. and Holt, J.S. (1984). "Weed Ecology, implications for vegetation management". John Wiley and Sons, New York.

Radosevich, S.R., Steinback, D.W. and Arntzen, C.J. (1979). Effect of photosystem II inhibitors on thylakoid membranes of two *Senecio vulgaris* biotypes. Weed Science **27**, 216-218.

Rubin, B., Yaacoby, T. and Schonfeld, M. (1985). Triazine-resistant grass weeds: cross-resistance with wheat herbicide, a possible threat to cereal crops. British Crop Protection Conference - Weeds 1171-1178.

Silvertown, J.W. (1982). "Introduction to Plant Population Ecology". Longman, London. pp. 209.

Sims, J.D., Stemler, A.J. and Radosevich, S.R. (1981). Differential light response to photosynthesis by triazine resistant and susceptible *Senecio vulgaris* biotypes. Plant Physiology 67 (4), 744-748.

Stowe, A.E. and Holt, J.S. (1988). The relationship of triazine resistance to decreased productivity in F_1 hybrids of common groundsel (*Senecio vulgaris* L.). Weed Science Society of America abstracts pp. 66-67. Abstracts, Las Vegas, NV.

Valverde, B.E., Radosevich, S.R and Appleby, A.P. (1988). Growth and competitive ability of dinitroaniline-herbicide resistant and susceptible goosegrass (*Eleusine indica*)biotypes. Proceedings, Western Society of Weed Science, p. 81.

Zimdahl, R. L. (1980). "Weed/Crop Competition: A Review". International Plant Protection Center, Oregon State University, Corvallis. pp. 195.

THE MOLECULAR BASIS OF RESISTANCE OF PHOTOSYSTEM II HERBICIDES

Achim Trebst

Department of Biology, Ruhr-University, P.O.B. 10 21 48, D-4630 Bochum 1, Germany

Two groups of compounds with quite different essential chemical elements inhibit photosystem II: the urea/triazine and the phenol-type families. Both groups displace plastoquinone and bind to a site on the D-1 protein subunit of the reaction centre of photosystem II. The first examples of resistance to these herbicides – triazine in *Amaranthus hybridus* and diuron in *Chlamydomonas reinhardtii* – were found to be due to a change in the properties of the binding niche and eventually to a mutation in the *psbA* gene that is responsible for an amino acid substitution in the sequence of the D-1 protein. Further mutations in the *psbA* gene, both by site selected screening or site directed mutagenesis, have been described leading to amino acid substitutions in the D-1 protein at positions: phe 211, val 219, ala 251, phe 255, gly 256, ser 264, asn 266 and leu 275. Significantly, mutations in triazine/urea resistance lead to an increased sensitivity to phenol-type inhibitors. These amino acid changes in herbicide resistant mutations have been used to model the binding niche for the plastoquinone and herbicides on the D-1 protein. The modelling is based on the homology of photosystem II to the reaction centre of purple bacteria. The cross tolerance of various herbicides in the mutants can be used to identify interactions of side chains with specific amino acid residues. Thus, the molecular basis of herbicide tolerance in photosystem II can be well described.

THE D1-PROTEIN SUBUNIT OF PHOTOSYSTEM II AS THE TARGET FOR THE HERBICIDES

It is well known that PS II herbicides act by displacing the secondary plastoquinone Q_B from its binding site on photosystem II (Velthuys, 1981). The urea/triazine family and the phenol-type inhibitors (Trebst and Draber, 1979, 1986) may be differentiated on the basis of their binding and inhibitory pattern at photosystem II.

The protein target of the PS II herbicides has been identified as the D-1 polypeptide subunit of photosystem II. This is also called the herbicide or Q_B binding protein (Pfister *et al.*, 1981; for review see Kyle, 1985). The D-1 subunit of photosystem II is encoded by the chloroplast gene *psbA*. Its DNA and the protein sequence of the D-1 protein was first obtained by Zurawski *et al.* (1982). The *psbA* gene from a large number of species has been sequenced and found to be highly conserved from cyanobacteria to higher plants (see Wolfe, 1989; Morden and Golden, 1989). Whereas there usually appears to be one copy of the *psbA* gene in higher plants (Palmer, 1985), there are two in *Chlamydomonas* (Erickson *et al.*, 1984b) and *Prochlorothrix* (Morden and Golden, 1989) and three in *Synechococcus* (= *Anacystis*) (Brusslan and Haselkorn, 1988). The mature protein has 343 amino acids. The product of the *psbA* gene is processed on the N- and C-terminus in plants and algae (see Mattoo *et al.*, 1989), except for *Euglena* (Karabin *et al.*, 1984). Nine amino acids at the C-terminus (Reisfeld *et al.*, 1982) and sixteen in blue-green algae (Morden and Golden, 1989) resulting in the same terminus at ala 244 (Takahashi *et al.*, 1988) and the methionine on the N-terminus is cleaved off. The protein is modified by phosphorylation and acetylation at threonine 2 (now AS 1) at the N-terminus in the mature protein (Michel *et al.*, 1988). During assembly it is also intermittently palmitoylated (Mattoo and Edelman, 1987). The D-1 protein is found to be homologous to another chloroplast coded polypeptide, namely, the D-2 protein encoded by the psbD gene (Rochaix *et al.*, 1984; Alt *et al.*, 1984).

Photosystem II is homologous to the reaction centre of purple bacteria both in function (except for oxygen evolution) and structure. The homology of the sequence of the D-1 subunit of photosystem II to the L and M subunit of the photosynthetic reaction centre of purple bacteria (Williams *et al.*, 1984; Youvan *et al.*, 1984) suggests that the D-1 polypeptide is not only the Q_B binding protein of photosystem II, but together with the D-2 subunit forms the reaction centre core protein of photosystem II (Deisenhofer *et al.*, 1985; Trebst and Depka, 1985) . This changed dramatically concepts of photosystem II structure and function, including the mode of action of herbicides in photosystem II, and made possible the molecular modelling and genetic engineering of the herbicide-binding niche. Together with the D-2 polypeptide, the herbicide-binding protein D-1 binds the reaction centre chlorophyll P680 (a chlorophyll dimer), two monomeric chlorophylls, two pheophytins and Fe. Furthermore, two plastoquinones (Q_A at the D-2 and Q_B at the D-1 polypeptide), the primary electron donor to P680 = tyrosine 161 on the D-1 subunit (Debus *et al.*, 1988; Vermaas *et al.*, 1988), and also the Mn-binding sites are bound to or in these two subunits. In total about 100 amino acids each of the D-1/D-2 protein heterodimer carry eight redox centres. The homology of the amino acid sequence and folding symmetry of photosystem II to the reaction centre of the purple bacteria (Michel *et al.*, 1986; Michel and Deisenhofer, 1988), the X-ray structures of the crystallized *Rhodopseudomonas viridis* (Deisenhofer *et al.*, 1985) and of *Rhodobacter sphaeroides* (Allen *et al.*, 1987), allowed the identification of important conserved amino acids and the prediction of the folding and structure of the reaction centre of photosystem II (Trebst, 1986; Michel and Deisenhofer, 1988). The amino acid changes in herbicide-tolerant plants and algae provided further support for the folding model (Trebst and Draber, 1986; Trebst, 1986). In turn, the proposed structure of photosystem II and the folding of its D-1 protein subunit greatly aided understanding of the molecular mechanisms of herbicide resistance. The successful preparation of a purified PS II reaction centre that consisted just of the D-1 and the D-2 protein (+ cytochrome b-559) confirmed the hypothesis (Nanba and Satoh, 1987; Barber *et al.*, 1987). As these core preparations of photosystem II with only four subunits (the D-1, D-2 and cytochrome b-559 did not retain the structural conformation of the Q_B site, they did not bind herbicides with high affinity (Giardi *et al.*, 1988). Only a completely assembled photosystem II with all seven essential protein subunits (that bind the core reaction centre, both plastoquinones + 30°core antenna chlorophylls) have the structural and functional integrity required for high affinity herbicide binding (see Trebst and Depka, 1985).

HERBICIDE RESISTANT MUTANTS

The first triazine resistant mutant, recognized in 1979 in *Senecio vulgaris* L. to be due to a change in the primary structure of the target (rather than to a change in uptake or degradation), was essential for identifying the herbicide-binding protein (Radosevich and DeVilliers, 1976; Pfister *et al.*, 1979). Using this mutant and a radioactive azido-triazine Pfister and colleagues (Pfister and Arntzen, 1979; Pfister *et al.*, 1981) located the specific herbicide-binding protein as a 32 kDa molecular weight band among the many thylakoid proteins, separated on SDS gel electrophoresis. It was subsequently shown that this (highly trypsin sensitive) 32 kDa protein was identical to the "rapid turnover" product of a light-stimulated expression of a gene (later called psbA) in the chloroplast genome (Mattoo *et al.*, 1981). This important breakthrough in the identification of the herbicide-binding protein and the mapping and sequencing of the *psbA* gene led to the demonstration that the herbicide-binding protein D-1 is the reaction centre of photosystem II. The amino acid change in triazine-resistant *Amaranthus*, used by Pfister and Arntzen (1979), was determined by *psbA* gene sequencing by Hirschberg and McIntosh (1983) to be at serine 264. Although Hirschberg *et al.* (1984) had suggested that this serine might be involved in quinone binding, its significance was only appreciated after a folding prediction of the D-1 polypeptide according to the bacterial reaction centre (Trebst, 1986). Figure 1 shows the folding of part of the D-1 protein in two hydrophobic transmembrane helices and a parallel helix that is involved in Q_B and herbicide binding. Eight amino acid changes in the D-1 protein, brought about by single site mutations in the *psbA* gene in herbicide-tolerant plants and algae have been found in recent years. Table 1 lists these known mutations, including double mutations, sequenced so far. Many of these have been found in several organisms - as reviewed by van Oorschot in these Proceedings. In higher plants, only a substitution of serine 264 by glycine has been observed, while in cell cultures of *Nicotiana*, changes of serine 264 to threonine and asparagine have been noted (Sato *et al.*, 1988; Páy *et al.*, 1988). In triazine or triazinone resistant green and blue-green algae, the serine 264 change is the most prominent and also most easily obtained mutation. Due to the different codon usage for serine in higher plants (AGT) compared to algae (TCT), serine 264 is changed to alanine in green and blue-green algae and to glycine in higher plants. In the green algae *Chlamydomonas* or in blue-green algae (*Synechococcus* and *Synechocystis*) other mutants have been identified by both induced (site-selected) and in site-directed mutagenesis and by using a lower triazine or metribuzin concentration for selection pressure. With phe 211, val 219, ala 251, phe 255, gly 256, ser 264, asp 266 and leu 275, this yields eight amino acid substitutions. It is surprising that in spite of the high conservation of the D-1 protein throughout the plant kingdom, amino acids can be changed without a complete loss of function. All mutants grow photoautotrophically.

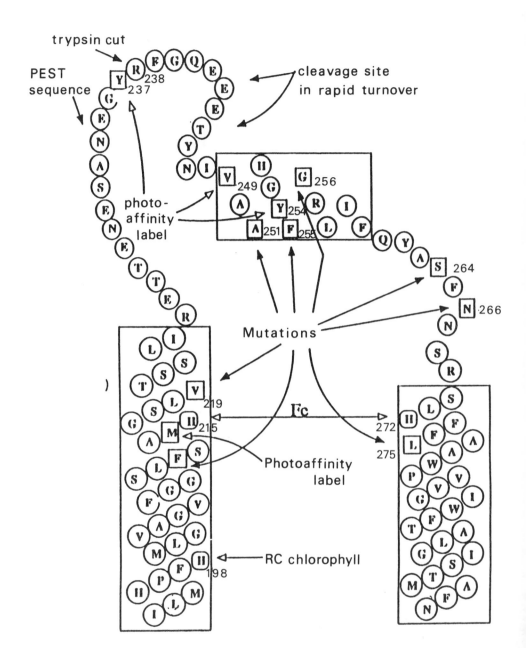

Fig. 1 <u>Model of the folding of the amino acid sequence in the binding niche of the quinone and herbicide binding site on the D-1 polypeptide subunit of photosystem II</u>. Indicated are two transmembrane helices (IV and V) and the parallel helix on the matrix side of the thylakoid membrane. Amino acids of importance for herbicide binding, as implied by mutations, photoaffinity labeling, effect on trypsin cut and on cleavage, and PEST-sequence in rapid turnover, and those in the binding of the central Fe and the reaction centre chlorophyll are marked.

Table 1. Amino acid changes in herbicide tolerant mutants in higher plants or in the L-subunit of the photosystem of purple bacteria.

Amino acid change	Tolerance to:	Organism and Reference
D-1 subunit of PS II:		
Phe 211 → Ser	atrazine/DCMU	*Synechococcus* (1)
Val 219 → Ile	metribuzin/DCMU/ioxynil	*Chlamydomonas, Synechococcus* (1,2,13)
Ala 251 → Val	metribuzin	*Chlamydomonas* (8)
Phe 255 → Tyr	atrazine/cyanoacrylate	*Chlamydomonas, Synechococcus* (2,4,13,22)
Gly 256 → Asp	atrazine/DCMU/bromacil	*Chlamydomonas* (2)
Ser 264 → Gly	atrazine	*Amaranthus* (6)5
Ser 264 → Ala	metribuzin/atrazine/	*Anacystis, Chlamydomonas,* (2,3,6,7)
Ser 264 → Thr	DCMU/bromacil	*Euglena, Synechocystis* (21,23)
Ser 264 → Asn	triazine	*Nicotiana, Euglena* (9,24)
	triazine	*Nicotiana* (10)
Ser 264 → Ala comparison of the	atrazine/DCMU	*Synechococcus* (11)
Ser 264 → Gly two changes	atrazine	
Asn 266 → Thr	ioxynil	*Synechocystis* (12)
Leu 275 → Phe	metribuzin/bromacil/DCMU	*Chlamydomonas* (2,13)
double mutants:		
Phe 255 → Tyr + Ser 264 → Ala	urea/triazine	*Synechococcus* (14)
Phe 255 → Leu + Ser 264 → Ala	DCMU + reversal of atrazine tolerance	*Synechocystis* (15)

149

Table 1 (cont'd)

Amino acid change	Tolerance to:	Organism and Reference
Phe 211 → Ser	atrazine	Synechocystis (15)
+ Ala 251 → Val		
L subunit of purple bacteria:		
Phe 216 → Ser	terbutryn	Rps. viridis (16)
Phe 216 → Thr or Leu, Val, Pro	triazine	Rb. capsulatus (20)
Tyr 222 → Phe	now DCMU sensitive!	Rps. viridis (16)
Tyr 222 → Gly	terbutryn	Rb. sphaeroides (17)
Ser 223 → Pro	"	Rb. sphaeroides (17)
Ser 223 → Ala	no change, incompetent	Rb. capsulatus (20)
Ser 223 → Asn	triazine	Rb. capsulatus (20)
Thr 226 → Ala or Met	"	Rb. capsulatus (20)
Gly 228 → Val or Arg	"	Rb. capsulatus (20)
Ile 229 → Met	"	Rb. sphaeroides (18)
Ile 229 → Met or Leu/Ala/Thr/Cys/Ser	atrazine	Rb. capsulatus (19)
double mutants:		
Phe 216 → Ser (L subunit)	terbutryn	Rps. viridis (16)
+ Val 263 → Phe (M subunit)	"	
Arg 217 → His	"	

150

Table 1 (cont'd)

1 = Gingrich *et al.*, 1988; 2 = Erickson *et al.*, 1989 and original literature quoted therein; 3 = Pucheu *et al.*, 1984; 4 = Ohad *et al.*, 1987; 5 = Hirschberg & McIntosh, 1983; 6 = Golden & Haselkorn, 1985; 7 = Astier *et al.*, 1986; 8 = Johanningmeier *et al.*, 1987; 9 = Sato *et al.*, 1988; 10 = Páy *et al.*, 1988; 11 = Ohad & Hirschberg, 1989; 12 = Ajlani *et al.*, 1989b; 13 = Wildner *et al.*, 1989a,b; 14 = Horovitz *et al.*, 1989; 15 = Ajlani *et al.*, 1989a; 16 = Sinning *et al.*, 1989a,b; 17 = Paddock *et al.*, 1989; 18 = Gilbert *et al.*, 1985; 19 = Bylina & Youvan, 1987; 20 = Bylina *et al.*, 1989.21 = Johanningmeier & Hallick, 1987; 22 = Hirschberg *et al.*, 1987; 23 = Osiewacz and McIntosh, 1987; 24 = Aiach *et al.*, 1989.

THE TOPOLOGY OF THE HERBICIDE BINDING SITE

As Figure 1 indicates, all of these changes fall into one area of the D-1 protein, the sequence of the amino acids from 211 to 275. This area is the Q_B site where a plastoquinone molecule is bound during photosynthetic electron flow. It consists of two transmembrane helices IV and V that are connected to each other by an Fe atom bridged via two histidines. The D-1 protein is also connected by the Fe to the D-2 subunit by two further histidine bridges. A parallel helix connects the two transmembrane helices in either protein.

Figure 2 gives the likely orientation of azido-triazine in the Q_B binding niche on the D-1 protein, using the model proposed already (Trebst, 1986). Whereas the alkyl side chains are oriented towards the stretched amino sequence which contains the serine 264, the azido-substituent is oriented towards met 214 on helix IV (Fig. 2). As discussed above, the model is based on the homology to the purple bacteria reaction centre. A comparison of Figures 2 and 3 shows the great homology of the model for PS II to the known terbutryn binding on the bacterial system (Michel *et al.*, 1986) and of the proposed triazine binding in photosystem II as deduced from the eight herbicide-tolerant plant mutants shown in Fig. 1. According to the X-ray structure of the reaction centre crystals of *Rps. viridis* soaked with terbutryn (Michel *et al.*, 1986), terbutryn has two or three hydrogen bridges to the L subunit. One hydrogen bridge to the serine 223 OH-group from the nitrogen in the ethyl side-chain of the triazine ring and a second from the backbone nitrogen from the peptide bond of ile 224 to the ring nitrogen (Fig. 3). The larger t-butyl substituent is oriented towards a cleft in the binding site.

Terbutryn is the only herbicide that blocks bacterial photosynthesis to any appreciable extent (Stein *et al.*, 1984). Terbutryn resistance in purple bacteria has been found and the amino acid change located by sequencing the gene for the L subunit. Several amino acids, including double mutants, have been found (see Table 1). Because of the great homology of the Q_B binding site of purple bacteria in photosystem II, these mutations help in describing the herbicide-binding niche on photosystem II. Indeed, quite homologous amino acids are changed: ser 223 in L vs ser 264 in D-1; phe 215 in L vs phe 255 in D-1. Other changed amino acids are at equivalent positions in the herbicide-binding niche in either system

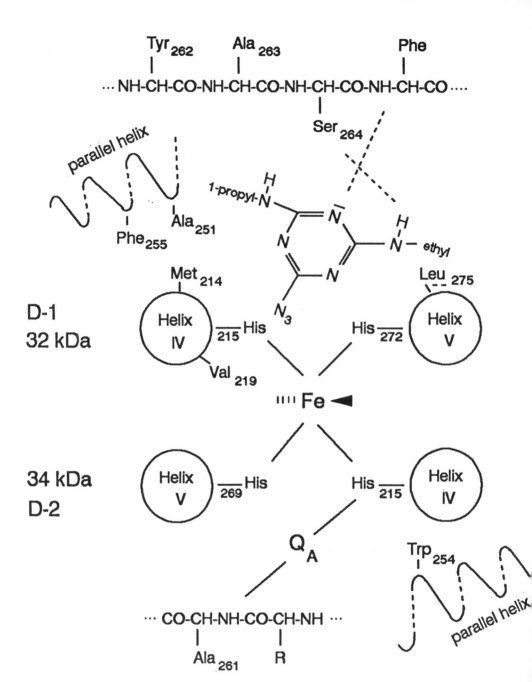

Fig. 2 <u>Model of azido–triazine in the binding niche of the $Q_B$site of the D-1 protein of photosystem II.</u> Possible hydrogen bridges towards the polypeptide chain backbone and the OH group of ser 264 as well as the orientation of the azido group towards met 214 are indicated. Four histidines bind Fe and this way connect the D-1 and D-2 polypeptide subunits. Parallel helices, perpendicular to the page plane place phe 255 and trp 254 below the triazine or Q_A binding site respectively. Some other amino acids in the binding niche in the D-1 protein changed in herbicide tolerant mutants are also oriented into the binding niche (val 219, ala 251, phe 255, ser 264, leu 275).

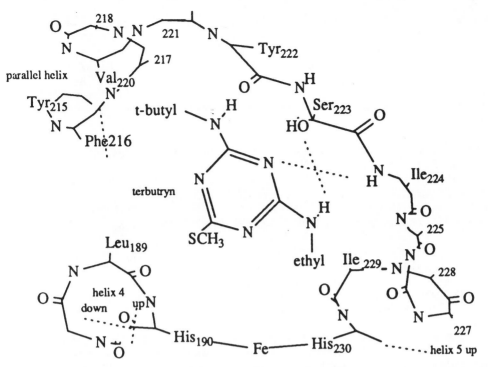

Fig. 3 <u>Terbutryn in the binding niche on the L subunit of <i>Rh.viridis</i> taken from the X-ray structure of terbutryn in the crystallized reaction centre of that purple bacterium.</u> Indicated are hydrogen bridges towards the nitrogen in the peptide bond of ile 224 and towards the OH-group of ser 223. A parallel helix towards the outside of the membrane and the ends of the two transmembrane helices D and E appear as viewed from the inside of the membrane.
Drawn from Fig. 4 of the paper by Michel <i>et al.</i> (1986).

(compare Figs 2 and 3). One mutant of <i>Rps. viridis</i> with a change at tyrosine 222 has become diuron sensitive (Sinning <i>et al.</i>, 1989b), and is so far the only case of urea sensitivity in purple bacteria. It seems that a greater change in the structure of the L subunit occurred than in the other mutants already described for the L or D-1 subunit. Among the <i>Rb. capsulatus</i> mutants, a ser 223 to ala change yielded a photosynthetically incompetent strain (Bylina <i>et al.</i>, 1989) (as against the equivalent ser 264 to ala change in the D-1 protein of algae). A ser 223 change to asn did not alter triazine sensitivity in <i>Rb. capsulatus</i>, nor ser to pro in <i>Rh. sphaeroides</i> or in the double mutant in <i>Rps. viridis</i> (see Table 1), and this is similar to the ser 264 changes in the D-1 protein, where inhibitor tolerance also depends whether gly, ala, thr or asn is formed.

Photoaffinity labeling with radioactive azido-derivatives has identified four more amino acids in the binding niche of photosystem II (see Fig. 1): met 214 in triazine (Wolber <i>et al.</i>, 1986), tyr 237 and 254 in urea (Dostatni <i>et al.</i>, 1988) and val 248 in ioxynil-binding (Oettmeier <i>et al.</i>, 1989). The ala 251, gly 256 and ser 264 change in the D-1 protein of photosystem II in higher plants and algae also reduces the photosynthetic potency of photosystem II (see Rochaix and Erickson, 1988). Nevertheless, the organisms still grow phototrophically (therefore the OH-group of ser 64 is not essential for photosynthetic electron flow). Other triazine and metribuzin resistant changes in *Chlamydomonas* show even less or no decrease in photosynthetic capacity (Erickson *et al.*, 1984,1985 and 1989; Rochaix and Erickson, 1988) like that of leu 275 (Wildner *et al.*, 1989a). This suggests that these amino acids are not of great importance in quinone binding and for photosynthesis, and can be replaced by another substituent. From these observations, amino acids are indicated in the herbicide target protein that can be exchanged to give herbicide tolerance without loss of catalytic property. Some mutations change specifically the tolerance to different compounds. This will help to identify the amino acids for genetically engineering crop plants tolerant to specific herbicides. Therefore, the cross-resistance data are important in this respect, but also for determining the orientation of the compounds in the Q_B site of the D-1 protein.

CROSS RESISTANCE IN HERBICIDE TOLERANT PLANTS AND ALGAE

The cross resistance of triazine- and triazinone-resistant organisms to other photosystem II herbicides has been widely studied. Although all herbicides bind to the same binding domain on the D-1 subunit of photosystem II (Fig. 1), each has its specific orientation in the binding niche. Substituents and chemically reactive groups of an inhibitor are involved in a chemical interaction with the protein with both the backbone polypeptide chain and the amino acid residues in the binding niche. As the substituent is different, so is the interaction. The compounds overlap, but each differently and specifically. Chemical structures cannot be simply superimposed, there are steric constraints: the binding niche towards ser 264 and the stretched short sequence between the parallel helix and transmembrane helix V is tight and does not allow large substituents. Those substituents following the isoprenyl side chain of plastoquinone in a cleft between the parallel helix and the coiled sequence towards helix IV, may, however, be large. These hydrophobic interactions of large lipophilic side chains contribute an appreciable part of the total binding energy, though not necessarily, in a specific way. The essential element, as postulated (for example by Hansch, 1969; Moreland, 1967; and Büchel, 1972) common to all herbicides of the urea/triazine family, as well as the essential chemical features proposed for the phenol-type herbicides (Trebst and Draber, 1979, 1986, Trebst *et al.*, 1989b) probably contribute the major energy of the total binding affinity, but the substituents are also involved and this will change in a tolerant mutant. The essential element of one group of herbicides does not bind to the same amino acid as that of another group, as is again clear from the cross resistance experiments. This was known from early QSAR studies, as emphasized by Draber (1987). The QSAR data can now be equated specifically with interactions to certain amino acids in the three-dimensional folding of the D-1 protein in the binding domain (Fig. 1).

Table 2 summarizes some of the data of Wildner *et al.* (1989b) with different inhibitors in five *Chlamydomonas* mutants, as important details for the modelling can be obtained from such data. Some specific points from these and the many other cross sensitivity experiments in the papers quoted here:

a) Usually the classical triazine/urea-type inhibitors show good cross-resistance among themselves, but there are definite specificities.

154

b) As discussed, phenol-type inhibitors bind to the same binding niche, but are different in structure and inhibitory pattern. This difference is even more apparent in cross resistance: the phenol-type herbicides do not show tolerance in the mutations leading to triazine/urea tolerance. Actually the mutants are supersensitive to the phenol type of inhibitors. The ketonitriles and the quinolines are the extremes. The latter will be further discussed below. Ioxynil is between the two inhibitor groups, tolerance to this herbicide is shown in the val 219 and the asn 266 change. This is another indication of the quite different binding of the triazine/urea family vs the phenol-type family (Trebst and Draber, 1986). The functional inhibitory pattern (Trebst and Draber, 1979) indicated the difference between the two families (reviewed by Oettmeier *et al.*, 1983). The two groups have been recently grouped as a serine (264) and the histidine (215) family according to their orientation in the binding niche (Trebst, 1987a), following the X-ray structure of inhibitor binding in the purple bacteria reaction centre (Michel *et al.*, 1986).

c) Also classical inhibitors show increasing sensitivity, like certain triazinones in the mutant with the serine 264 change in higher plants (Oettmeier *et al.*, 1985) or ureas in the asn 266 change in *Synechocystis* (Ajlani *et al.*, 1989b). The DCMU sensitivity in a terbutryn-resistant mutant of *Rps. viridis* is another example (Sinning *et al.*, 1989b).

d) A second amino acid change can reverse tolerance brought about by the first change. The second mutation of phe 255 to leucine reverses the triazine tolerance in the first mutation of serine 264 to alanine in *Synechocystis* (Ajlani *et al.*, 1989a).

e) The ser 264 to alanine change in *Chlamydomonas* leads to both triazine and urea tolerance, whereas triazine but no urea tolerance is obtained in the ser 264 to glycine change in the higher plants. This is explained by results with site directed mutagenesis with *Synechococcus* (Ohad and Hirschberg, 1989). Both these both changes have been engineered. Clearly, an ala instead of serine at position 264 does not allow ureas to bind to the protein, whereas gly does. This should be compared with the quite similar situation in the ser 223 change in purple bacteria (Table 1).

f) Sensitivity (or tolerance) of an inhibitor in a mutant can be predicted. An increased insensitivity towards certain triazinones has been predicted for metribuzin resistance in higher plants (Oettmeier *et al.*, 1985) as well as the effect of a double mutation where the properties of a fourth mutant of *Synechococcus* was predicted from three experimentally measured mutants (Horovitz *et al.*, 1989).

g) Details of the cross resistance, i.e. analysis of which compound with which substituent becomes tolerant or not in a certain mutation allows for fine tuning of the three-dimensional model of the herbicide binding niche.

The hydroxyquinolines (or quinolones) are very effective inhibitors of the phenol-type (Trebst *et al.*, 1989b), yielding no tolerance in any of the mutations (see Table 2) tested so far (Wildner et al., 1989b). HOQNO (2-N-heptyl-hydroxyquinoline-N-oxide) is less effective in an *Anacystis nidulans* mutant (Golden and Sherman, 1984), due to a ser 264 change to ala (Golden and Haselkorn, 1985). It is likely that the N-oxides of quinolines are turned upside down in the binding niche and become a urea/triazine-type (Trebst, 1987b). Indeed, the lipophilic side chain of the N-oxides is in the two-position likely equivalent to the isoprenyl side chain of plastoquinone that is "up" in the binding niche.

Table 2 <u>Cross resistance of inhibitors of the triazine/urea family vs the phenotype in</u> <u>*Chlamydomonas*</u> <u>mutants (from Wildner et al., 1989b)</u>. R/S change = ratio of the pI_{50} values of resistance vs sensitivity in the wild-type. A value below one is an increase in sensitivity.

Compounds	Val 219	Ala 251	Mutation at: Phe 255	Ser 264	Leu 275
Atrazine	2	25	16	158	1
Diuron	32	8	0.8	200	5
Metribuzin	20	500	0.6	5000	25
Cyanoacrylate[1]	2	16	40	32	1
Ioxynil	50	25	2.5	0.5	0.
Bromonitrothymol	2	8	0.6	0.4	1.6
Ketonitrile[2]	0.6	0.5	0.6	0.6	1.3
Quinoline[3]	0.5	0.6	0.4	1	0.4

[1] = 2-cyano-3-ethyl-3-(4-chlorbenzyl)-aminoacrylate (Phillips & Huppatz, 1984);
[2] = 2-phenylthiazolyl-3-hydroxy-4-phenyl-butenonitrile (Bühmann *et al.*, 1987);
[3] = 3-bromo-2,6-di-trifluoromethyl-hydroxyquinoline (Trebst *et al.*, 1989b).

A SPECIFIC AMINO ACID SEQUENCE REGULATES THE TURNOVER OF THE HERBICIDE BINDING PROTEIN

The Q_B binding niche that houses the herbicides when they displace the plastoquinone discussed so far is quite similar to that of the purple bacteria, except for the exchange of some amino acids for an equivalent one. However, there are interesting differences; there is an additional alanine just before serine 264 in the D-1 protein. A major difference between the D-1 subunit of photosystem II to the L subunit in purple bacteria is in a stretch of about 14 amino acids between AS 235 to AS 250 that is specific for the D-1 protein and does not exist in the L subunit. One of the tyrosines (tyr 237) identified by photoaffinity labeling (Dostatni *et al.*, 1988) is in this sequence. It has been proposed (Trebst *et al.*, 1988) that this sequence is part of the Q_B binding niche and folds on top of Q_B and the central Fe connecting the D-1 and D-2 subunit (Fig. 2). This way the site is covered quite unlike that in purple bacterial system. There the M subunit and on top of that the H subunit covers the hydrophobic area of Q binding.

This is supported by and in turn explains two experimental observations with photosystem II. A specific trypsin cut at arginine 238 on the D-1 protein is blocked by diuron, but not by phenol-type inhibitors (Trebst *et al.*, 1988). The rapid turnover of the D-1 protein (Marder *et al.*, 1984) is started by a specific degradation at a cleavage site around glu 245 by a protease guided by a PEST sequence (Greenberg *et al.*, 1987 and

1989) (see Fig. 1). This rapid turnover is prevented by diuron (Mattoo *et al.*, 1984).

It is proposed (Trebst *et al.*, 1988) that under physiological conditions plastoquinone in the Q_B binding site guides the conformation of this amino acid sequence from AS 235 to 250 in photoinhibition and rapid turnover. An empty site is degraded, an occupied site is not. Either Q_B or an inhibitor of the urea/triazine, but not of the phenol-type family alters the conformation or the accessibility of the sequence to proteases. As the continuous physiological degradation of the D-1 protein is a prerequisite for the repair of photosystem II in photoinhibition, it is likely that this mechanism is a major part of the mode of action of photosystem II herbicides. There is again the difference in the functional behaviour of the two herbicidal groups that may well be related to the effectiveness of a herbicide *in vivo* (many of the phenol-type inhibitors are very effective *in vitro*, but did not prove useful in the field). Inhibitors of rapid turnover would sustain the herbicidal activity and prevent repair of a damaged D-1 protein in photoinhibition (Mattoo *et al.*, 1986; Trebst et al., 1988). This might be the reason why photosystem II herbicides do not just block photosynthesis, but actually lead to a destruction of the photosynthetic apparatus.

ACKNOWLEDGEMENT

Work at Bochum is supported by Deutsche Forschungsgemeinschaft and by Fonds der chemischen Industrie.

REFERENCES

Aiach, A., Johanningmeier, U. and Ohmann, E. (1989). A *psbA* mutation in diuron resistant *Euglena* leads to a change at ser 264 to thr in the D-1 protein. (In preparation).

Ajlani, G., Kirilovsky, D., Picaud, M. and Astier, C. (1989a). Molecular analysis of *psbA* mutations responsible for various herbicide resistance phenotypes in *Synechocystis* 6714. Plant Molecular Biology **13**, 469-480.

Ajlani, G., Meyer, I., Vernotte, C. and Astier, A. (1989b). Mutation in phenol-type herbicide resistance maps within the *psbA* gene in *Synechocystis* 6714. FEBS Letters **246**, 207-210.

Allen, J.P., Feher, G., Yeates, T.O., Komiya, H. and Rees, D.C. (1987). Structure of the reaction centre from *Rhodobacter sphaeroides* R-26: The cofactors. Proceedings of the National Academy of Sciences USA **84**, 5730-5734.

Alt, H., Morris, J., Westhoff, P. and Herrmann, R.G. (1984). Nucleotide sequence of the clustered genes for the 44 kD chlorophyll a apoprotein and the "32 kD"-like protein of the photosystem II reaction centre in the spinach plastid chromosome. Current Genetics **8**, 597-606.

Astier, C., Meyer, I., Vernotte, C. and Etienne, A.L. (1986). Photosystem II electron transfer in highly herbicide resistant mutants of *Synechocystis* 6714. FEBS Letters **207**, 234-238.

Barber, J., Chapman, D.J. and Telfer, A. (1987). Characterization of a PS II reaction centre isolated from the chloroplasts of *Pisum sativum*, FEBS Letters **220**, 67-73.

Brusslan, J. and Haselkorn, R. (1988). Molecular genetics of herbicide resistance in cyanobacteria. Photosynthesis Research **17**, 115-124.

Büchel, K.-H. (1972). Mechanisms of action and structure activity relations of herbicides that inhibit photosynthesis. Pesticide Science **3**, 89-110.

Bühmann, U., Herrmann, E.C., Kötter, C., Trebst, A., Depka, B. and Wietoska, H. (1987). Inhibition and photoaffinity labeling of photosystem II by thiazolyliden-ketonitriles. Zeitschrift für Naturforschung **42c**, 704-712.

Bylina, E.J., Jovine, R.V.M. and Youvan, D.C. (1989). A genetic system for rapidly assessing herbicides that compete for the quinone binding site of photosynthetic reaction centers. Biotechnology **7**, 69-74.

Bylina, E.J. and Youvan, D.C. (1987). Genetic engineering of herbicide resistance: Saturation mutagenesis of isoleucine 229 of the reaction centre L subunit. Zeitschrift für Naturforschung **42c**, 769-774.

Debus, R.J., Barry, B.A., Babcock, G.T. and McIntosh, L. (1988). Site-directed mutagenesis identifies a tyrosine radical involved in the photosynthetic oxygen-evolving system. Proceedings of the National Academy of Sciences USA **85**, 427-430.

Deisenhofer, J., Epp, O., Miki, K., Huber, R. and Michel, H. (1985). Structure of the protein subunits in the photosynthetic reaction centre of *Rhodopseudomonas viridis* at 3 Å resolution. Nature **318**, 618-624.

Deisenhofer, J. and Michel, H. (1989). Nobel lecture: The photosynthetic reaction centre from the purple bacterium *Rhodopseudomonas viridis*. EMBO Journal **8**, 2149-2170.

Dostatni, R., Meyer, H.E. and Oettmeier, W. (1988). Mapping of two tyrosine residues involved in the quinone-(Q_B)binding site of the D-1 reaction centre polypeptide of photosystem II. FEBS Letters **239**, 207-210.

Draber, W. (1987). Can quantitative structure activity analyses and molecular graphics assist in designing new inhibitors of photosystem II? Zeitschrift für Naturforschung **42c**, 713-717.

Erickson, J.M., Pfister, K., Rahire, M., Togasaki, R.K., Mets, L. and Rochaix, J.-D. (1989). Molecular and biophysical analysis of herbicide-resistant mutants of *Chlamydomonas reinhardtii*: Structure-function relationship of the photosystem II D1 polypeptide. The Plant Cell **1**, 361-371.

Erickson, J.M., Rahire, M., Bennoun, P., Delepelaire, P., Diner, B. and Rochaix, J.-D. (1984a). Herbicide resistance in *Chlamydomonas reinhardtii* results from a mutation in the chloroplast gene for the 32-kilodalton protein of photosystem II. Proceedings of the National Acadademy of Sciences USA **81**, 3617-3621.

Erickson, J.M., Rahire, M. and Rochaix, J.-D. (1984b). *Chlamydomonas reinhardtii* gene for the M_r 32000 protein of photosystem II contains four large introns and is located entirely within the chloroplast inverted repeat. EMBO Journal **3**, 2753-2762.

Erickson, J.M., Rahire, M., Rochaix, J.-D. and Mets, L. (1985). Herbicide resistance and cross-resistance: Changes at three distinct sites in the herbicide-binding protein. Science **228**, 204-207.

Giardi, M.T., Marder, J.B. and Barber, J. (1988). Herbicide binding to the isolated Photosystem II reaction centre. Biochimica and Biophysica Acta **934**, 64-71.

Gilbert, C.W., Williams, J.G.K., Williams, K.A.L. and Arntzen, C.J. (1985). Herbicide action in photosynthetic bacteria. In "Molecular Biology of the Photosynthetic Apparatus" (K.E. Steinback, S. Bonitz and L. Bogorad, eds), pp. 67-77. Cold Spring Harbor Laboratory, Cold Spring Harbor.

Gingrich, J.C., Buzby, J.S., Stirewalt, V.L. and Bryant, D.A. (1988). Genetic analysis of two new mutations resulting in herbicide resistance in the cyanobacterium *Synechococcus* sp. PCC 7002. Photosynthesis Research **16**, 83-99.

Golden, S.S. and Haselkorn, R. (1985). Mutation to herbicide resistance maps within the *psbA* gene of *Anacystis nidulans* R2. Science **229**, 1104-1197.

Golden, S.S. and Sherman, L.A. (1984). Biochemical and biophysical characterization of herbicide-resistant mutants of the unicellular cyanobacterium, *Anacystis nidulans* R2. Biochimica and Biophysica Acta **764**, 239-246.

Greenberg, B.M., Gaba, V. and Mattoo, A.K. (1987). Identification of a primary *in vivo* degradation product of the rapidly-turning-over 32 kd protein of photosystem II. EMBO Journal **6**, 2865-2869.

Greenberg, B.M., Gaba, V. Mattoo, A.K. and Edelman M. (1989). Degradation of the 32 kDa photosystem II reaction center protein in UV, visible and far red light occurs through a common 23.5 kDa intermediate. Zeitschrift für Naturforschung **44c**, 450-452.

Hansch, C. (1969) Photophosphorylation, CO_2 Fixation, Action Mechanisms of Herbicides. In "Progress in Photosynthesis Research", Vol. III (H. Metzner, ed.) pp. 1685-1692. Metzner, Tübingen.

Hirschberg, J. and McIntosh, L. (1983). Molecular basis of herbicide resistance in *Amaranthus hybridus*. Science **222**, 1346-1348.

Hirschberg, J., Bleecker, A., Kyle, D.J., McIntosh, L. and Arntzen, C.J. (1984). The molecular basis of triazine-herbicide resistance in higher-plant chloroplasts. Zeitschrift für Naturforschung **39c**, 412-420.

Hirschberg, J., Ohad, N., Pecker, I. and Rahat, A. (1987). Isolation and characterization of herbicide resistant mutants in the cyanobacterium *Synechococcus* R2. Zeitschrift für Naturforschung **42c**, 758-761.

Horovitz, A., Ohad, N. and Hirschberg, J. (1989). Predicted effects on herbicide binding of amino acid substitutions in the D1 protein of photosystem II. FEBS Letters **243**, 161–164.

Johanningmeier, U., Bodner, U. and Wildner, G.F. (1987). A new mutation in the gene coding for the herbicide-binding protein in *Chlamydomonas*. FEBS Letters **211**, 221–224.

Johanningmeier, U. and Hallick, R.B. (1987). The *psbA* gene of DCMU-resistant *Euglena gracilis* has an amino acid substitution at serine codon 265. Current Genetics **12**, 465–470.

Karabin, G.D., Farley, M. and Hallick, R.B. (1984). Chloroplast gene for M_r 32000 polypeptide of photosystem II in *Euglena gracilis* is interrupted by four introns with conserved boundary sequences. Nucleic Acids Research **12**, 5801–5813.

Kyle, D.J. (1985). The 32 000 Dalton Q_B protein of photosystem II. Photochemistry and Photobiology **41**, 107–116.

Marder, J.B., Goloubinoff, P. and Edelman, M. (1984). Molecular architecture of rapidly metabolized 32-kilodalton protein of photosystem II. Journal of Biological Chemistry **259**, 3900–3908.

Mattoo, A.K. and Edelman, M. (1987). Intramembrane translocation and postranslational palmitoylation of the chloroplast 32-kd herbicide-binding protein. Proceedings of the National Academy of Sciences USA **84**, 1497–1501.

Mattoo, A.K., Hoffmann-Falk, H., Marder, J.B. and Edelman, M. (1984). Regulation of protein metabolism: Coupling of photosynthetic electron transport to *in vivo* degradation of the rapidly metabolized 32-kilodalton protein of the chloroplast membranes. Proceedings of the National Academy of Sciences USA **81**, 1380–1384.

Mattoo, A.K., Marder, J.B. and Edelman, M. (1989). Dynamics of the photosystem II reaction centre. Cell **56**, 241–246.

Mattoo, A.K., Marder, J.B., Gaba, V. and Edelman, M. (1986). Control of 32 kDa thylakoid protein degradation as a consequence of herbicide binding to its receptor. In "Regulation of Chloroplast Differentiation", (G. Akoyunoglou, ed.) pp. 607–613. Alan R. Liss, New York.

Mattoo, A.K., Pick, U., Hoffmann-Falk, H. and Edelman, M. (1981). The rapidly metabolized 32,000-dalton polypeptide of the chloroplast is the 'proteinaceous shield' regulating photosystem II electron transport and mediating diuron herbicide sensitivity. Proceedings of the National Academy of Sciences USA **78**, 1572–1576.

Michel, H. and Deisenhofer, J. (1988). Relevance of the photosynthetic reaction centre from purple bacteria to the structure of photosystem II. Biochemistry **27**, 1–7.

Michel, H., Epp, O. and Deisenhofer, J. (1986). Pigment-protein interactions in the photosynthetic reaction centre from *Rhodopseudomonas viridis*. EMBO Journal **5**, 2445–2451.

Michel, H., Hunt, D.F., Shabanowitz, J. and Bennett, J. (1988). Tandem mass spectrometry reveals that three photosystem II proteins of spinach chloroplasts contain N-acetyl-0-phosphothreonine at their N-termini. Journal of Biological Chemistry **263**, 1123-1130.

Morden, C.W. and Golden, S.S. (1989). *psbA* genes indicate common ancestry of prochlorophytes and chloroplasts. Nature **337**, 382-385.

Moreland, D.E. (1967). Mechanisms of action of herbicides. Annual Reviews of Plant Physiology **18**, 365-386.

Nanba, O. and Satoh, K. (1987). Isolation of a photosystem II reaction centre consisting of D-1 and D-2 polypeptides and cytochrome *b*-559. Proceedings of the National Academy of Sciences USA **84**, 109-112.

Oettmeier, W., Masson, K., Höhfeld, J., Meyer, H.E., Pfister, , and Fischer, H.-P. (1989). [^{125}I]Azido-ioxynil labels Val$_{249}$ of the photosystem II D-1 reaction centre protein. Zeitschrift für Naturfor-schung **44c**, 444-449.

Oettmeier, W., Masson, K., Soll, H.-J. and Olschewski, E. (1985). QSAR of 1,4-benzoquinones in photosynthetic systems, In "QSAR and Strategies in the Design of Bioactive Compounds" (J.K. Seydel, ed.), pp. 238-244. VCH Verlagsgesellschaft, Weinheim.

Oettmeier, W. and Trebst, A. (1983). Inhibitor and plastoquinone binding to photosystem II, In "The Oxygen Evolving System of Photosynthesis" (Y. Inoue, ed.), pp. 411-420. Academic Press Japan, Tokyo.

Ohad, N. and Hirschberg, J. (1989).A similar structure of the herbicide binding site in photosystem II of plants and cyanobacteria is demonstrated by site specific mutagenesis of the *psbA* gene. Photosynthesis Research, (in press).

Ohad, N., Pecker, I. and Hirschberg, J. (1987). Biochemical and Molecular analysis of herbicide resistant mutants in cyanobacteria. In "Progress in Photosynthesis Research", Vol. III. (J. Biggins, ed.), pp 807-810. Martinus Nijhoff Publishers, Dordrecht.

Osiewacz, H.D. and McIntosh, L. (1987). Nucleotide sequence of a member of the *psbA* multigene family from the unicellular cyanobacterium *Synechocystis* 6803. Nucleic Acids Research **15**, 10585.

Paddock, M.L., Williams, J.C., Rongey, S.H., Abresch, E.C., Feher, G. and Okamura, M.Y. (1989). Characterization of three herbicide resistant mutants of *Rhodopseudomonas sphaeroides* 2.4.1.: Structure-function relationship. In "Progress in Photosynthesis Research", Vol. III. (J. Biggins, ed.), pp 775-778. Martinus Nijhoff Publishers, Dordrecht.

Palmer, J.D. (1985). Comparative organization of chloroplast genomes. Annual Reviews of Genetics **19**, 325-354.

Páy, A., Smith, M.A., Nagy, F. and Márton, L. (1988). Sequence of the *psbA* gene from wild type and triazine-resistant *Nicotiana plumbaginifolia*. Nucleic Acids Research **16**, 8176.

Phillips, J. and Huppatz, J. (1984). Cyanoacrylate inhibitors of photosynthetic electron transport. Nature of the interaction with the receptor site. Zeitschrift für Naturforschung 39c, 335-337.

Pfister, K. and Arntzen, C.J. (1979). The mode of action of photosystem II-specific inhibitors in herbicide-resistant weed biotypes. Zeitschrift für Naturforschung 34c, 996-1009.

Pfister, K., Radosevich, S.R. and Arntzen, C.J. (1979). Modification of herbicide binding to photosystem II in two biotypes of Senecio vulgaris L. Plant Physiology 64, 995-999.

Pfister, K., Steinback, K.E., Gardner, G. and Arntzen, C.J. (1981). Photoaffinity labeling of an herbicide receptor protein in chloroplast membranes. Proceedings of the National Academy of Sciences USA 78, 981-985.

Pucheu, N., Oettmeier, W., Heisterkamp, U., Masson, K. and Wildner, G.F. (1984). Metribuzin-resistant mutants of Chlamydomonas reinhardii, Zeitschrift für Naturforschung 39c, 437-439.

Radosevich, S.R. and DeVilliers, O.T. (1976). Studies on the mechanism of s-triazine resistance in common groundsel. Weed Science 24, 229-232.

Reisfeld, A., Mattoo, A.K. and Edelman, M. (1982). Processing of a chloroplast-translated membrane protein in vivo. European Journal of Biochemistry 124, 125-129.

Rochaix, J.D., Dron, M., Rahire, M. and Malnoe, P. (1984). Sequence homology between the 32k dalton and the D2 chloroplast membrane polypeptide. Plant Molecular Biology 3, 363-370.

Rochaix, J.-D. and Erickson, J. (1988). Function and assembly of photosystem II: genetic and molecular analysis. Trends in Biochemical Sciences 13, 56-59.

Sato, F., Shigematsu, Y. and Yamada, Y. (1988). Selection of an atrazine-resistant tobacco cell line having a mutant psbA gene. Molecular Genetics 214, 358-360.

Sinning, I., Michel, H., Mathis, P. and Rutherford, W.A. (1989a). Characterization of four herbicide-resistant mutants of Rhodopseudomonas viridis by genetic analysis, electron paramagnetic resonance, and optical spectroscopy. Biochemistry 28, 5544-5553.

Sinning, I., Michel, H., Mathis, P. and Rutherford, W.A. (1989b). Terbutryn resistance in a purple bacterium can induce sensitivity toward the plant herbicide DCMU. FEBS Letters 256, 192-194.

Stein, R.R., Castellvi, A., Bogacz, J.P. and Wraight, C.A. (1984). Herbicide-quinone competition in the acceptor complex of photosynthetic reaction centers from Rhodopseudomonas sphaeroides: A bacterial model for PS II-herbicide activity in plants. Journal of Cellular Biochemistry 24, 243-259.

Takahashi, M., Shiraishi, T. and Asada, K. (1988). COOH-terminal residues of D1 and the 44 kDa CPa-2 at spinach photosystem II core complex. FEBS Letters **240**, 6-8.

Trebst, A. (1986). The topology of the plastoquinone and herbicide binding peptides of photosystem II in the thylakoid membrane. Zeitschrift für Naturforschung **41c**, 240-245.

Trebst, A. (1987a). The three-dimensional structure of the herbicide binding niche on the reaction centre polypeptides of photosystem II. Zeitschrift für Naturforschung **42c**, 742-750.

Trebst, A. (1987b). Topology of polypeptides and redox components in PS II and the cytochrome b/f-complex, In "Progress in Photosynthesis Research, Vol. II (J. Biggins, ed.), pp.109-112. Martinus Nijhoff Publishers, Dordrecht.

Trebst, A. and Depka, B. (1985). The architecture of photosystem II in plant photosynthesis. Which peptide subunits carry the reaction centre of PS II? In "Antennas and Reaction Centers of Photosynthetic Bacteria. Structure, Interactions, and Dynamics" (M.E. Michel-Beyerle, ed.), pp. 216-224. Springer Series in Chemical Physics 42. Springer-Verlag, Berlin.

Trebst, A., Depka, B. and Kipper, M. (1989a). The topology of the reaction centre polypeptides of photosystem II, In "Current Research in Photosynthesis" (M. Baltscheffsky, ed.), pp. I.2.217- 222. Kluwer Academic Press, Dordrecht.

Trebst, A., Depka, B., Kraft, B. and Johanningmeier, U. (1988). The Q_B site modulates the conformation of the photosystem II reaction centre polypeptides. Photosynthesis Research **18**, 163-177.

Trebst, A. and Draber, W. (1979). Structure activity correlations of recent herbicides in photosynthetic reactions, In "Advances in Pesticide Science", Part 2 (H. Geissbühler, ed.), pp.223-234. Pergamon Press, Oxford.

Trebst, A. and Draber, W. (1986). Inhibitors of photosystem II and the topology of the herbicide and Q_B binding polypeptide in the thylakoid membrane. Photosynthesis Research **10**, 381-392.

Trebst, A., Pittel, B. and Draber, W. (1989b). The modelling of photosystem II inhibitors into the herbicide binding protein: Inhibitory pattern, QSAR and quantum mechanical calculations of new hydroxyquinoline derivatives. In "Probing Bioactive Mechanisms: Proof, QSAR and Prediction" (P.S. Magee, D.R. Henry and J.H. Block, eds), pp. 215-228. American Chemical Society Book, ACS Symposium Series 413.

Velthuys, B.R. (1981). Electron-dependent competition between plastoquinone and inhibitors for binding to photosystem II. FEBS Letters **126**, 277-281.

Vermaas, W.F.J., Rutherford, A.W. and Hansson, O. (1988). Site-directed mutagenesis in photosystem II of the cyanobacterium *Synechocystis* sp. PCC 6803: Donor D is a tyrosine residue in the D2 protein. Proceedings of the National Academy of Sciences USA **85**, 8477-8481.

Wildner, G.F., Heisterkamp, U., Bodner, U., Johanningmeier, U. and Haehnel, W. (1989a). An amino acid substitution in the Q_B-protein causes herbicide resistance without impairing electron transport. <u>Zeitschrift für Naturforschung</u> **44c**, 431-434.

Wildner, G.F., Heisterkamp, U. and Trebst, A. (1989b). Herbicide cross-resistance and mutations of the psb A gene in *Chlamydomonas reinhardii*, <u>Photosynthesis Research</u>, in press.

Williams, J.C., Steiner, L.A., Feher, G. and Simon, M.I. (1984). Primary structure of the L subunit of the reaction centre from *Rhodopseudomonas sphaeroides*. <u>Proceedings of the National Acadademy of Sciences USA</u> **81**, 7303-7307.

Wolber, P.K., Eilmann, M. and Steinback, K.E. (1986). Mapping of the triazine binding site to a highly conserved region of the Q_B-protein. <u>Archives of Biochemistry and Biophysics</u> **248**, 224-233.

Wolfe, K.H. (1989). Compilation of sequences of protein-coding genes in chloroplast DNA including cyanelle and cyanobacterial homologues. <u>Plant Molecular Biology Reporter</u> **7**, 30-48.

Youvan, D.C., Bylina, E.J., Alberti, M., Begusch, H. and Hearst, J.E. (1984). Nucleotide and deduced polypeptide sequences of the photosynthetic reaction centre, B870 antenna, and flanking polypeptides from *R. capsulata*, <u>Cell</u> **37**, 949-957.

MECHANISMS OF PARAQUAT TOLERANCE

Alan D. Dodge

School of Biological Sciences, University of Bath, Bath, BA2 7AY, UK

The bipyridinium herbicide paraquat, and related compounds such as diquat, interact with photosystem I of chloroplast electron transport. This leads to an inhibition of $NADP^+$ reduction and hence carbon dioxide incorporation. The one electron reduction of the herbicide is followed by the generation of superoxide. Unscavenged superoxide yields the toxic hydroxyl radical that instigates membrane damage and cell death. In the last 15 years, a number of monocotyledonous and dicotyledonous biotypes have been identified that tolerate the normally toxic level of paraquat. Apart from *Lolium perenne*, identified in a breeding programme, the other biotypes from Egypt, Hungary, Japan, Australia and UK have arisen as a result of extensive use of paraquat in plantations for up to 10 to 20 years.
The mechanisms involved in paraquat tolerance are discussed in relationship to the evidence for limited uptake or sequestration, to a failure to interact with the photosynthetic electron transport chain, or to an enhanced level of oxygen radical scavenging enzymes. Evidence at present suggests that limited movement or sequestration is the primary mechanism of tolerance in most biotypes.

THE MECHANISM OF PARAQUAT ACTION

The herbicide paraquat and related bipyridiniums such as diquat have been in use for almost 30 years as total-kill compounds. At the present time we have a relatively good understanding of the mechanism of action of these herbicides. After application and uptake into the plant, they are reduced at the terminal end of photosystem I. This is likely to be via electrons donated from the iron-sulphur centres FeS_A or FeS_B with redox potentials of around -590 mV and -530 mV respectively. In normal photosynthesis these centres donate electrons to ferredoxin, thence to $NADP^+$ for carbon dioxide incorporation. With paraquat present, the diversion of electrons leads to a cessation of $NADP^+$ reduction, and hence carbon dioxide incorporation is prevented.

The one electron reduction of the dicationic paraquat gives rise to a potentially stable radical cation. In the presence of oxygen however, the dication is regenerated and the odd electron donated to oxygen, producing superoxide (O_2^-). Superoxide thus generated may be scavenged by chloroplast components such as superoxide dismutase enzymes, ascorbate and other compounds (see below). In fact what is likely to occur is that the excess generation of superoxide overtaxes these protective systems. Hydrogen peroxide that is formed either enzymically or spontaneously from superoxide (equation 1) will react with further superoxide in the presence of traces of iron salts to generate the very toxic hydroxyl radical (OH·, equation 2).

(1) $$O_2^- + O_2^- + 2H^+ \longrightarrow H_2O_2 + O_2$$

(2) $$O_2^- + H_2O_2 \xrightarrow{\text{Fe catalyst}} O_2 + OH^- + OH·$$

Thus paraquat not only causes the indirect inhibition of photosynthetic carbon dioxide incorporation, but promotes the generation of massive amounts of the OH· radical. This reacts indiscriminately with membrane fatty acids, protein amino acids, nucleic acids, aromatic compounds etc. This leads to wholesale cellular disruption, pigment bleaching and consequently plant death.

PARAQUAT TOLERANT PLANTS

It has been known for many years that there are small differences in the response of some plants to paraquat that are probably related to variations in uptake, leaf structure and so on (Bovey and Davis, 1967). The first plants clearly identified as being tolerant to the normally lethal dose rates were biotypes of perennial rye grass (*Lolium perenne*) that had been selected during a breeding programme (Faulkner, 1975). A number of lines were identified that were tolerant to paraquat throughout their life cycle, but the degree of tolerance varied from approximately 3- to 10-fold, depending upon treatment method and growth conditions (Harvey and Harper, 1982).

Although paraquat tolerant lines of *Poa annua* had been identified in the UK in 1979 (LeBaron and Gressel, 1982), the first published account of paraquat tolerance in natural populations was of the annual weed, hairy fleabane (*Conyza bonariensis*) from citrus and vine populations in Egypt (Youngman and Dodge, 1981). Here, continuous paraquat treatments had been in use for nine years. A resistance ratio calculated by Fuerst *et al.* (1985) on the basis of the I_{50} of the resistant and susceptible biotypes varied from around 94 to 150, according to the method of estimation. Tolerance to the related bipyridiniums diquat and triquat was maximally only 38 times, but minimally only 4.6 times. Paraquat tolerance was first identified in Japan in 1982 as shown by *Erigeron philadelphicus* (Watanabe *et al.*, 1982) from a mulberry plantation that had been treated for 10 years. In 1983, paraquat tolerant *Erigeron canadensis* from a vineyard was identified (Kato and Okuda, 1983). Resistance ratios were calculated as being around 100 times more than the susceptible biotypes. In 1987, paraquat and atrazine tolerant populations of the closely related species *Conyza canadensis* were identified in a Hungarian vineyard which had been treated with paraquat three or four times a year for 10 years (Pölös and Mikulás, 1987). In the same year, paraquat tolerant *Epilobium ciliatum* was reported from hop gardens in Kent (Clay, 1987).

Paraquat tolerance in Australian weeds was first identified by Warner and Mackie (1983) for a barley grass (*Hordeum glaucum*) that was growing in lucerne in Victoria. Subsequent work showed that this biotype was around 250 times more tolerant to paraquat than susceptible plants (Powles, 1986). Two further tolerant plants were identified from this same farm in which paraquat and diquat had been used continuously for over 20 years – hare barley (*Hordeum leporinum*) tolerant to paraquat (Tucker and Powles, 1989) and capeweed (*Arctotheca calendula*) tolerant to diquat (Powles *et al.*, 1989).

Further weeds showing tolerance to paraquat are *Conyza sumatrensis* and *Youngia japonica* from Japan (Matsunaka and Ito, these Proceedings).

MECHANISMS OF TOLERANCE

Limited Uptake and Movement

Paraquat is a water soluble dication, and is strongly adsorbed on to leaf surfaces after spray application (Summers, 1980). It is possible that ultra-violet light induced degradation of the molecule could occur on the surface, to yield the non-herbicidal components 4-carboxyl-1-methyl pyridinium chloride and methylamine hydrochloride (Slade, 1966). Movement into the leaf cells is likely to be impeded by cuticular wax (Thrower *et al.*, 1965). Paradoxically, Vaughn and Fuerst (1985) found that a *C. bonariensis* biotype which was paraquat tolerant possessed a thinner leaf cuticle, and this could have accounted for the enhanced uptake of the herbicide into these plants. After entry into mesophyll cells, movement occurs through the leaf and into the xylem, but there appears to be little movement from leaves to roots (Baldwin, 1963; Thrower *et al.*, 1965). Work with radio-labelled paraquat showed no evidence of metabolic breakdown within the plant (Slade and Bell, 1966).

Evidence for the restricted movement of ^{14}C-paraquat in tolerant plants has been obtained for *C. bonariensis* (Fuerst *et al.*, 1985), *E. philadelphicus* and *E. canadensis* (Tanaka *et al.*, 1986) (Fig. 1) and *Hordeum glaucum* (Bishop *et al.*, 1987). In each of these cases the herbicide apparently showed considerably restricted movement into the apoplast, indicating therefore some form of sequestering or immobilization before movement into or through the mesophyll. Although each of these experimental approaches involved the use of ^{14}C-paraquat fed to cut petiole ends, the failure to move into or through the mesophyll would also occur if the herbicide was sprayed on to the leaf surface.

Although paraquat binds readily to cellulose cell walls (Brown and Nix, 1975), Fuerst *et al.* (1985) failed to find any difference in binding between tolerant and susceptible biotypes of *C. bonariensis*. Powles and Cornic (1987) experimented with protoplasts from tolerant and susceptible *H. glaucum* plants. They found no difference in the penetration of paraquat through the plasmalemma or into the chloroplast. With protoplasts from both plants, there was a uniform inhibition of bicarbonate dependent oxygen evolution with an increasing concentration of paraquat.

Although no differences were found in cell wall binding or uptake by protoplasts it is likely that each system is very different from the whole plant state. Fuerst *et al.* (1985) suggested that an important soluble component, possibly associated with ionic binding or cation exchange, could be lost. They found that the divalent cation calcium reduced cell wall binding uniformly between tolerant and susceptible material. Using a tissue slice technique, Ranson and Morrod (unpublished observations, 1974) showed that paraquat with a neutral charge, as the dimethylsulphonate derivative, was taken up to a more limited extent. Furthermore, both potassium and calcium were shown to inhibit uptake of the dication.

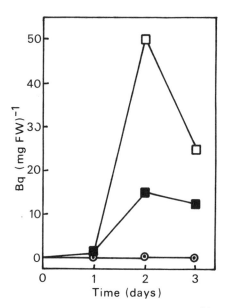

Fig. 1. <u>Time course of the movement of [^{14}C]-paraquat into excised leaves of *Erigeron philadelphicus*</u>. Susceptible biotype incubated in light □ or darkness ■ ; Tolerant biotype incubated in light ○ or darkness ● . Redrawn from Tanaka *et al.* (1986).

Failure to interact with photosynthetic electron transport

It has already been outlined that the effect of an interaction of paraquat with the electron acceptor portion of photosystem I will indirectly lead to a cessation of carbon dioxide incorporation. Carbon dioxide exchange measurements are therefore a convenient and sensitive assay for such interactions. Major differences in inhibition of photosynthetic carbon dioxide uptake were shown by Bishop *et al.* (1987) for susceptible and tolerant *H. glaucum* (Fig. 2). In the experiments of Youngman and Dodge (1981), Shaaltiel and Gressel (1987) and Pölös *et al.* (1988), there was evidence of a transient inhibition of carbon dioxide incorporation by paraquat in tolerant plants, but this was followed by recovery. A similar concentration of paraquat in susceptible plants caused irreversible inhibition.

An alternative method of estimating the interaction of paraquat with the photosynthetic electron transport chain, is to measure whole leaf fluorescence. It is well known that fluorescence emission is related to the ability of photosystem II to donate electrons to photosystem I (Duysens and Sweers, 1963) and that fluorescence will be reduced if an efficient electron acceptor interacts with PSI. Fuerst *et al.* (1985) showed that there were major differences in the concentration of paraquat required to quench the variable fluorescence of susceptible and tolerant

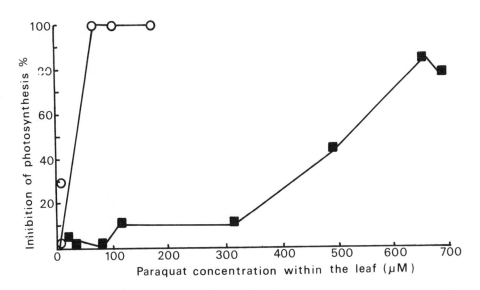

Fig. 2. **The % inhibition of photosynthesis in leaves of tolerant (■)
and susceptible (○)** *Hordeum glaucum* **at different leaf concentrations of
paraquat**. Different concentrations were obtained by placing the leaf in
a paraquat solution for different periods. Redrawn from Bishop *et al.*
(1987).

C. bonariensis.   This was a direct indication of the greatly restricted
movement of paraquat into the tolerant chloroplasts or to a failure to
interact with FeS_A or FeS_B (Fig. 3). Further experiments showed that the
ability of the other bipyridiniums diquat and triquat to quench
fluorescence was not greatly different between tolerant and susceptible
plants (Fuerst *et al.*, 1985).

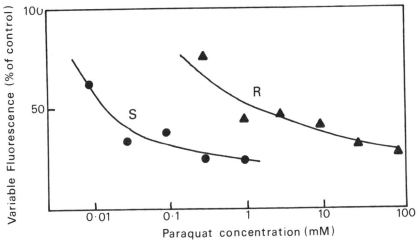

Fig. 3. **Variable fluorescence of leaves of susceptible (S = ●) and
resistant (R = ▲) biotypes of** *Conyza bonariensis* **after incubation with
varying concentrations of paraquat**. Redrawn from Fuerst *et al.* (1985).

There is no doubt therefore that in some paraquat tolerant plants there is a failure of the normal lethal herbicide concentration to interact with PSI. There is evidence however that limited interaction leads to a transient reduction of carbon dioxide incorporation as mentioned above.

It is nevertheless possible that the magnitude of this inhibition was enhanced by the diversion of NADPH from carbon dioxide exchange to maintain the ascorbate-glutathione cycle for superoxide and hydrogen peroxide scavenging (Fig. 4) (Foyer and Halliwell, 1976) and/or that some hydrogen peroxide generated from superoxide led to the transient inhibition of some Calvin cycle enzyme such as fructose bisphosphatase (Law *et al.*, 1983).

Enhanced scavenging bipyridinium produced toxic products

Although the bipyridinium herbicides promote the generation of O_2^- and $OH \cdot$ radicals, the production of O_2^- may occur to a limited extent during "normal" photosynthesis. It is possible that around 10-20% of photosynthetic electron flow could be to oxygen reduction in what is termed an endogenous Mehler reaction (Asada *et al.*, 1977) with reduced ferredoxin acting as the electron flow mediator. The chloroplast is well endowed with scavenging systems to cope with these toxic products. Predominant are the CuZn superoxide dismutase enzymes (SOD) that are partially bound to the chloroplast thylakoids (equation 3). The enzyme has been calculated to maintain the O_2^- concentration at a steady state level of $6.0 \times 10^{-9}M$ (Asada *et al.*, 1977).

$$(3) \quad O_2^- + O_2^- + 2H^+ \xrightarrow{\text{SOD}} H_2O_2 + O_2$$

Hydrogen peroxide is scavenged through a linked series of enzymes, the ascorbate-glutathione reductase, and ferredoxin-NADP$^+$ reductase (Fig. 4). It has been possible to modulate the action of paraquat to an extent by enhancing the superoxide scavenging capacity of leaves. This has been achieved by use of the complex, copper penicillamine (Youngman *et al.*, 1979) or by the use of an ascorbate precursor L-galactono-1,4-lactone (Schmidt and Kunert, 1987). On the other hand, an inhibitor of superoxide dismutase, diethyl dithiocarbamate has been shown to enhance paraquat action (Gillham, 1986, quoted in Dodge, 1989; and Gressel and Shaaltiel, 1988).

It has been a matter of considerable interest and discussion during the last decade as to whether enhanced superoxide and hydrogen peroxide scavenging ability contributes to the mechanism of tolerance. There have been several surveys of the appropriate scavenging enzymes in tolerant and susceptible biotypes. In a detailed survey of 15 *Lolium perenne* lines, Harper and Harvey (1978) found a limited increase in the activity of superoxide dismutase, catalase and peroxidase in the tolerant lines. However, in extracts from isolated chloroplasts they found no relationships between an increase in superoxide dismutase activity and paraquat tolerance. Work with *Conyza* biotypes has been particularly confusing. Youngman and Dodge (1981) showed that a paraquat tolerant biotype had enhanced superoxide dismutatase activity but later work by Vaughn and Fuerst (1985) failed to demonstrate this. Shaaltiel and Gressel (1986) using isolated cloroplasts found that superoxide dismutase

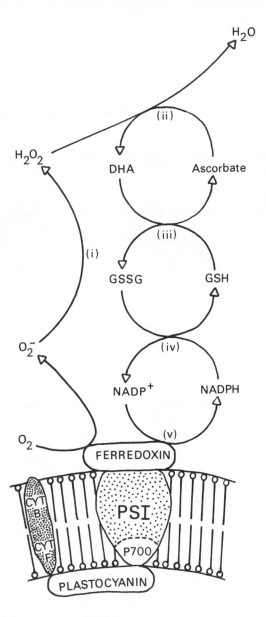

Fig. 4.   Chloroplast scavenging enzymes for the control of superoxide (O_2^-). (i) superoxide dismutase, (ii) ascorbate peroxidase, (iii) dehydroascorbate reductase, (iv) glutathione reductase, (v) ferredoxin-NADP$^+$ reductase.   Enzymes (ii)-(iv) represent the "ascorbate-glutathione" cycle. Abbreviations: DHA, dehydroascorbate; GSH, reduced glutathione; GSSG, oxidized glutathione.

was elevated by around 60%, ascorbate peroxidase by 150% and glutathione reductase by 192%. Work by Pölös *et al.* (1988) failed to find enhanced scavenging enzymes in their *Conyza* biotype, as did Powles and Cornic (1987) for *H. glaucum*.   The most recent report of enhanced scavenging enzyme activities is by Kawaguchi (Matsunaka and Ito, these Proceedings). His experiments with isolated chloroplasts of tolerant and susceptible

E. canadensis showed a 69% increase in superoxide dismutase, 79% ascorbate peroxidase, 94% dehydroascorbate reductase and 61% glutathione reductase.

It is evident that considerable care must be taken in the analysis and interpretation of data relating to scavenging enzymes. Firstly, an increase in superoxide dismutase activity is likely to enhance hydroxyl radical generation unless limited by enhanced hydrogen peroxide scavenging. Secondly, there are likely to be differences in activity between different biotypes, leaf ages and after different treatment conditions. Asada *et al.* (1977) showed major variants in the superoxide dismutatase activity of maize leaves throughout their life. Gillham and Dodge (1987) showed changes in the level of ascorbate peroxidase, ascorbate and glutathione reductase during the course of a year, and an enhanced level of these components when pea leaves were illuminated with higher light levels. Another possibility is that superoxide itself is leading to an enhanced expression of superoxide dismutase. This was demonstrated by Matters and Scandalios (1986) in maize treated with low levels of paraquat. After 12 hours there was a 40% increase in enzyme activity, and this was a consequence of O_2^- induced SOD gene expression. Gillham and Dodge (1984) found that subtoxic paraquat levels also enhanced the activity of some enzymes of the ascorbate-glutathione cycle, while Pölös *et al.* (1988) showed some enhancement of scavenging activity in both tolerant and susceptible biotypes of *Conyza*.

CONCLUSIONS

Asada *et al.* (1977) calculated that a normal dose of bipyridinium herbicide would elevate the level of O_2^- within the chloroplast by 10- to 20-fold. The evidence at present does not support the idea that increased activity of scavenging enzymes could cope with this problem. The most convincing evidence at this time suggests that there is a failure of the herbicide to reach the chloroplast, or to interact with the electron transport chain (Vaughn *et al.*, 1989). There is a necessity to establish in detail the exact reasons for restricted movement or sequestration, and the intriguing feature that paraquat tolerant plants are not necessarily tolerant to diquat. The establishment of a precise discriminating mechanism, possibly involved with molecular structure or redox properties, remains to be elucidated.

REFERENCES

Asada, K., Takahashi, M., Tanaka, K. and Nakano, Y. (1977). Formation of active oxygen and its fate in chloroplasts. In "Biochemical and Medical Aspects of Active Oxygen" (O. Hayaishi and K. Asada, eds), pp. 45-63. Japan Scientific Societies Press, Tokyo.

Baldwin B.S. (1963). Translocation of diquat in plants. Nature **198**, 872-873.

Bishop, T., Powles, S.B. and Cornic, G. (1987). Mechanism of paraquat resistance in *Hordeum glaucum*: 2. Studies in paraquat uptake and translocation. Australian Journal of Plant Physiology **14**, 539-547.

Bovey, R.W. and Davis, F.S. (1967). Factors affecting the phytotoxicity of paraquat. Weed Research **7**, 281-289.

Brown, C.L. and Nix, L.E. (1975). Uptake and transport of paraquat in slash pine. Forestry Science 21, 359-364.

Clay, D.V. (1987). The response of simazine-resistant and susceptible biotypes of *Chamomilla suaveolens*, *Epilobium ciliatum* and *Senecio vulgaris* to other herbicides. British Crop Protection Conference - Weeds 925-932.

Dodge, A.D. (1989). Herbicides interacting with photosystem I. In "Herbicides and Plant Metabolism" (A.D. Dodge, ed.), pp. 37-50. Cambridge University Press, Cambridge.

Duysens, L.W.M. and Sweers, H.E. (1963). Mechanism of two photochemical reactions in algae as studied by means of fluorescence. In "Microalgae and Photosynthetic Bacteria" (Special issue of Plant and Cell Physiology), pp. 353-372. University of Tokyo Press, Tokyo.

Faulkner, J.S. (1975). A paraquat tolerant line of *Lolium perenne*. In "Symposium on Status, Biology and Control of Grass Weeds in Europe" Vol. I, pp. 349-359. European Weed Research Society.

Foyer, C.H. and Halliwell, B. (1976). The presence of glutathione and glutathione reductase in chloroplasts: a proposed role for ascorbic acid metabolism. Planta 133, 21-25.

Fuerst, E.P., Nakatani, H.Y., Dodge, A.D., Penner, D. and Arntzen, C.J. (1985). Paraquat resistance in *Conyza*. Plant Physiology 77, 984-989.

Gillham, D.J. (1986). Aspects of chloroplast protection against photo-oxidative damage. Ph.D. Thesis, University of Bath.

Gillham, D.J. and Dodge, A.D. (1984). Plant Physiology 75 Suppl., 51.

Gillham, D.J. and Dodge, A.D. (1987). Chloroplastic superoxide and hydrogen peroxide scavenging systems from pea leaves: Seasonal variations. Plant Science 50, 105-109.

Gressel, J. and Shaaltiel, Y. (1988). Biorational herbicide synergists. In "Biotechnology for Crop Protection" (P.A. Hedin, J.J. Menn and R.M. Hollingworth, eds), pp. 4-24. A.C.S. Washington.

Harper, D.B. and Harvey, B.M.R. (1978). Mechanism of paraquat tolerance in perennial ryegrass. II. Role of superoxide dismutase, catalase and peroxidase. Plant, Cell and Environment 1, 211-215.

Harvey, B.M.R. and Harper, D.B. (1982). Tolerance of bipyridinium herbicides. In "Herbicide Resistance in Plants" (H.M. LeBaron and J. Gressel, eds), pp. 215-233. John Wiley, New York.

Kato, A. and Okuda, Y. (1983). Paraquat resistance in *Erigeron canadensis*. Weed Research (Tokyo) 28, 54-56.

Law, M.Y., Charles, A. and Halliwell, B. (1983). Glutathione and ascorbic acid in spinach chloroplasts. Biochemical Journal 210, 899-903.

LeBaron, H.M. and Gressel, J. (1982). "Herbicide Resistance in Plants". John Wiley, New York.

Matters, G.L. and Scandalios, J.G. (1986). Effect of the free radical-generating herbicide paraquat on the expression of the superoxide dismutase (Sod) genes in maize. Biochemica et Biophysica Acta **882**, 29-38.

Pölös, E. and Mikulás, J. (1987). Cross-resistance to paraquat and atrazine in *Conyza canadensis*. British Crop Protection Conference - Weeds **8**, 909-916.

Pölös, E., Mikulás, J., Szigetti, Z., Matkovics, B., Quy Hai, D., Párducz, A. and Lehoczki, E. (1988). Paraquat and atrazine co-resistance in *Conyza canadensis*. Pesticide Biochemistry and Physiology **30**, 142-154.

Powles, S.B. (1986). Appearance of a biotype of the weed *Hordeum glaucum* resistant to the herbicide paraquat. Weed Research **26**, 167-172.

Powles, S.B. and Cornic, G. (1987). Mechanisms of paraquat resistance in *Hordeum glaucum*. I. Studies with isolated organelles and enzymes. Australian Journal of Plant Physiology **14**, 81-89.

Powles, S.B., Tucker, E.S. and Morgan, T.W. (1989). A capeweed (*Arctotheca calendula*) biotype in Australia resistant to bipyridyl herbicides. Weed Science **37**, 60-62.

Schmidt, A. and Kunert, K.J. (1987). Antioxidative systems: Defence against oxidative damage in plants. In "Molecular Strategies for Crop Protection" (C.J. Arntzen and C. Ryan, eds), pp. 401-413. Alan R. Liss, New York.

Shaaltiel, Y. and Gressel, J. (1987). Kinetic analysis of resistance to paraquat in *Conyza*. Plant Physiology **85**, 869-871.

Slade, P. (1966). The fate of paraquat applied to plants. Weed Research **6**, 158-167.

Slade, P. and Bell, E.G. (1966). The movement of paraquat in plants. Weed Research **6**, 267-274.

Summers, L.A. (1980). " The Bipyridinium Herbicides". Academic Press, London.

Tanaka, Y., Chisaka, H. and Saka, M. (1986). Movement of paraquat in resistant and susceptible biotypes of *Erigeron philadelphicus* and *E. canadensis*. Physiologia Plantarum **66**, 605-608.

Thrower, A.L., Hallam, N.D. and Thrower, L.B. (1965). Movement of diquat in leguminous plants. Annals of Applied Biology **55**, 253-260.

Tucker, E.S. and Powles, S.B. (1989). A biotype of hare barley (*Hordeum leporinum*) in Australia resistant to the bipyridyl herbicides paraquat and diquat. Weed Science (in press).

Vaughn, K.C. and Fuerst, E.P. (1985). Structural and physiological studies of paraquat-resistant *Conyza*. Pesticide Biochemistry and Physiology **24**, 86-94.

Vaughn, K.C., Vaughan, M.A. and Camilleri, P. (1989). Lack of cross-resistance of paraquat-resistant hairy fleabane (*Conyza bonariensis*) to other toxic oxygen generators indicates enzymatic protection is not the resistance mechanism. Weed Science **37**, 5-11.

Warner, R.B. and Mackie, W.B.C. (1983). A barley grass (*Hordeum leporinum*) spp. *glaucum* population tolerant to paraquat (Gramoxone). Australian Weed Research Newsletter **31**, 16,

Watanabe, Y., Honina, T., Ito, K. and Miyrashara, M. (1982). Paraquat resistance in *Erigeron philadelphicus* L. Weed Research (Japan) **27**, 49-54.

Youngman, R.J., Dodge, A.D., Lengfelder, E. and Elstner, E.F. (1979). Inhibition of paraquat phytotoxicity by a novel copper chelate with superoxide dismutating activity. Experientia **35**, 1295-1296.

Youngman,. R.J. and Dodge, A.D. (1981). On the mechanism of paraquat resistance in *Conyza* sp. In "Photosynthesis and Plant Productivity, Photosynthesis and Environment" (G. Akoyunoglou, ed.), pp. 537-544. Balaban, Philadelphia.

DINITROANILINE RESISTANCE IN *ELEUSINE INDICA* MAY BE DUE TO HYPER-STABILIZED MICROTUBULES[1]

Kevin C. Vaughn[2] and Martin A. Vaughan[3]

USDA/ARS, Southern Weed Science Laboratory, P.O. Box 350, Stoneville, MS 38776 USA

Previous investigations in our laboratory have shown that the dinitroaniline-resistant (R) biotype of *Eleusine indica* Gaertn. is due to an alteration in tubulin, but the mechanism by which this alteration confers resistance is not understood. Treatment of the dinitroaniline-susceptible (S) and R biotypes with solutions of the microtubule-stabilizing drug, taxol, reveals a number of microtubule alterations (cross linking, lateral associations, abnormal configurations, microtubules in the interphase nucleus) and consequences of these alterations (incomplete cell plates, abnormally-oriented walls, multiple nuclei). Some effects were observed in both biotypes but the numbers of abnormalities and their severity was much greater in the R biotype. The S biotype required 10-100 X more taxol to induce the same effects as on the R biotype. Treatment of the S biotype with 1.0 µM taxol produced a phenocopy of the small wall alterations found in the untreated R biotype and, like the R biotype, these taxol-treated S biotypes are resistant to further treatment with dinitroaniline herbicides. These data indicate that the microtubules of the R biotype are hyper-stabilized and that this hyper-stabilization may be responsible for the observed resistance to dinitroaniline herbicides.

INTRODUCTION

Mutants of animal cell lines and fungi that are resistant to microtubule disrupting compounds such as colchicine or benomyl have been described for many years (see Oakley, 1985, for a review). In these systems, the mutants owe their resistance to one of these mechanisms:

---

[1]Supported by a Competitive Research Grant No. 86-CRCR-1-1933 to K.C. Vaughn.

[2]Correspondence and reprints.

[3]Present address: Biology Dept., Rochester Institute of Technology, Rochester, NY 14623.

(1) restricted uptake of the disrupter;
(2) alteration in binding of the disrupter to tubulin;
(3) hyper-stabilization of the tubulin in the microtubule form.

A characteristic of mutants of the third category is that all are hypersensitive to the microtubule-stabilizing agent taxol. Mutants with the other kinds of resistance mechanisms do not share this hypersensitivity to taxol (Cabral *et al.*, 1986).

At present, only one higher plant mutant resistant to a microtubule disrupter has been described, the dinitroaniline-resistant (R) biotype of *Eleusine indica* (Mudge *et al.*, 1984; Vaughn, 1986; Vaughn *et al.*, 1987; Vaughn and Vaughan, 1989). A dinitroaniline-<u>tolerant</u> (or low level resistance) accession of *Setaria viridis* has recently been described by Morrison and Beckie (these Proceedings). The R mutant of *Eleusine* is 1,000–10,000x more resistant than the dinitroaniline-susceptible (S) biotype based upon either changes in the mitotic index (Vaughn, 1986) or growth (Vaughn *et al.*, 1987). This is a much greater level of resistance than many of the other mutants isolated to this point (Oakley, 1985; James *et al.*, 1988; Cabral *et al.*, 1986). Biochemical investigations in our laboratory indicate that differences in tubulin are responsible for the alteration: tubulin from R biotype is able to polymerize in the presence of oryzalin *in vitro*, and Western blots of tubulin revealed a novel beta tubulin isoform (Vaughn and Vaughan 1986, 1989). These data strongly indicate that an alteration in tubulin is responsible for the resistance.

In this report, we examine the sensitivity of the R and S biotype to the microtubule stabilizer taxol to determine if the R biotype has hyperstabilized microtubules. Our data indicate that the R biotype is almost 100x more sensitive to taxol than than the S biotype, indicating that hyperstabilization may be the alteration responsible for dinitroaniline resistance in *Eleusine*.

MATERIALS AND METHODS

Plant material

Seeds of the R and S *Eleusine indica* biotypes were germinated in the dark on moistened Whatman #1 paper in Petri dishes. After 4 days of growth seedlings were transferred to dishes with filter paper saturated with taxol solutions (0.10, 1.0, and 10 μM) for 24 h in the dark. The solutions were prepared from a 10 μM taxol stock in 100% dimethylsulfoxide stored at 20°C. After treatment, the plants were sampled for microscopy. Another group of S *Eleusine* seedlings were treated with 1 μM taxol and subsequently treated with 1.0 μM trifluralin or oryzalin for an additional 24 h. These herbicide levels are 100x the level that are required to cause root tip swelling in the S biotype (Vaughn and Koskinen, 1987). These plants were then sampled for microscopy.

Electron microscopy

Roots were fixed and processed for electron microscopy using the procedures of Vaughn *et al.*, (1987). At least six root tips of each treatment were examined. Low magnification survey photos of the root meristem area (original magnifications 1,250x) were used to compare the number of cells with abnormal wall orientations and multiple nuclei.

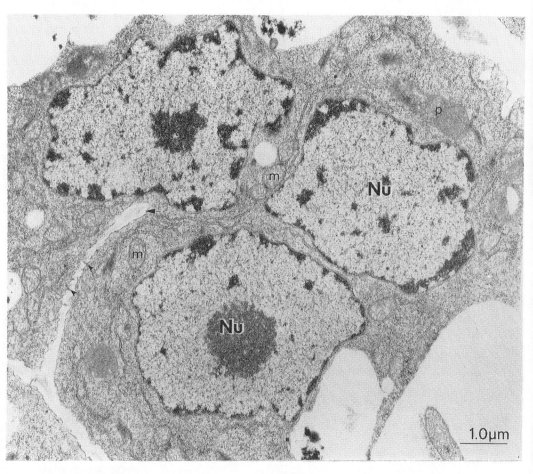

Fig. 1. Cell of R biotype of *Eleusine* treated with 10 μM taxol for 24 h. Three nuclei (Nu) have apparently resulted from a multipolar division. A cell wall is partly formed (arrow) and contains plasmodesmata (small arrows). m= mitochondrion; p= proplastid.

Immunofluorescence microscopy

Roots of R and S *Eleusine*, treated with taxol as above, were processed for immunofluorescence microscopy using the procedures of Wick *et al.* (1981) except that 1% (w/v) Driselase (Fluka Chem. Co.) was used for the digestion steps. Micrographs were taken with an Olympus epifluorescence microscope.

RESULTS AND DISCUSSION

Cell plate and cell wall effects

Dramatic effects are noted in the vicinity of the cell wall in all cell types in the root tip after treatment of both biotypes with 10 μM taxol. Cell plates are often incomplete, leaving the two or more (in a multipolar mitosis) cells that should have resulted from a mitotic division still as one (Fig. 1). Despite the abnormal, incomplete nature of the walls, certain features of walls such as their interspersion with plasmadesmata, are retained. Similar effects of taxol have been noted by other workers examining the effects of taxol on other systems. Microtubules are found abundantly along the incomplete cell walls. Frequently, microtubules in both the cell plate and cell walls occur as doublets or joined structures (Fig. 2). Such associations between microtubules are rarely observed in control preparations, indicating that the taxol may be increasing the cross-bridging (and hence stability) of the microtubule arrays. Along the cell walls of the outer root cap cells, many parallel arrays of microtubules with corresponding parallel wall abnormalities are noted (Fig. 2). The stabilization of the microtubules may have set a pattern for cellulose microfibril deposition that results in an uneven distribution of new wall material (Falconer and Seagull, 1985). Although these abnormal cell walls are not noted in the untreated S biotype, similar but less severe abnormalities are noted (>1% of the cells) in the walls of the untreated R biotype grown on H_2O (Vaughn, 1986), indicating that the alterations of wall structure in the R biotype might also be due to hyper-stabilized tubulin.

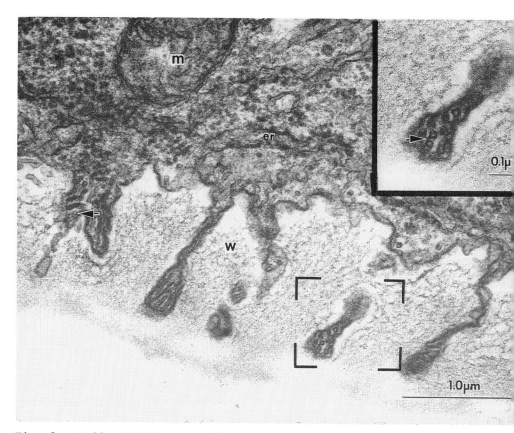

Fig. 2. Cell of R biotype of *Eleusine* treated with 1 μM taxol for 24 h. The surface of the cell wall (w) is undulated and bunches of microtubules are present along the wall. Some of the microtubule profiles show cross-bridging (arrows in both Figure and inset).

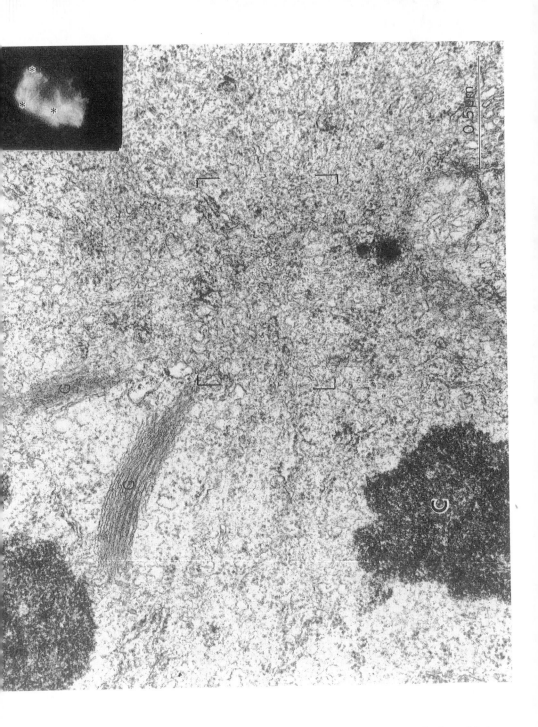

Fig. 3. Cell of R bioytpe of *Eleusine* treated with 10 μM taxol for 24 h. A zone (marked with brackets) contains abundant microtubules and vesicles. This is a zone where microtubules appear to aggregate. Profiles of microtubules in both cross-section and longi-section may be noted. Inset shows an immunofluorescence micrograph of a similar cell with multiple poles (*). C= chromosome; G= golgi body (dictyosome).

Spindle apparatus effects

Many cells were in some type of mitotic stage and, although some appear normal, especially in the S biotype, evidence of microtubules oriented toward several poles was frequently observed (Fig. 3). Similarly, immunofluorescence of spindle microtubules in some of these treated cells shows this 3-dimensional arrangement of microtubules in an apparent multipolar mitosis (Fig. 3 inset). Aster-like structures were observed in *Zinnia* and vetch cell cultures treated with taxol (Weerdenberg *et al.*, 1986); these were similar to the foci of fluorescence observed in the R biotype. When the multipolar division is completed, the cell contains multiple nuclei (e.g. Fig. 1). In the R biotype, these reformed nuclei retain small remnants of the microtubules of the spindle apparatus (Fig. 4). These intranuclear microtubules may represent stabilization of microtubules at nucleation centers, such as the kinetochores of condensed chromatin in the R biotype.

Table 1. <u>Comparison of the sensitivity of S and R biotypes of *Eleusine* to various concentrations of taxol</u>.

Taxol (μM)	S	R
10	Some abnormal walls, rare multipolar division.	Most cells with abnormal wall configurations, micro-tubules in nuclei.
1.0	3 to 5% of cells with abnormal wall configurations, no multi-polar division.	>80% of cells with abnormal cell walls some multipolar division.
0.1	More or less normal, occasional wall pro-tuberance.	Wall protuberances are common in >50% of the cells, rare multipolar division.

Biotype comparisons

Although, at 10 μM taxol treatment the R and S biotypes are both affected, the R biotype is more affected, as noted above in the discussion of spindle apparatus effects. Longitudinal sections through 12 different R and S biotype roots at both 1.0 and 10 μM taxol concentrations were examined for the percentage of cells exhibiting wall and/or mitotic abnormalities (Table 1). At both concentrations, the S biotype is much less affected. After 1.0 μM treatment, the S biotype looks nearly normal except for some small cell wall alterations. These data, on cell wall and mitotic effects and the persistence of spindle and/or kinetochore microtubules in the reformed nucleus of the R biotype (e.g. Fig 4), indicate that the microtubules of the R biotype are more sensitive to taxol and may be inherently more stabilized than the microtubules of the S biotype (Cabral *et al.*, 1986).

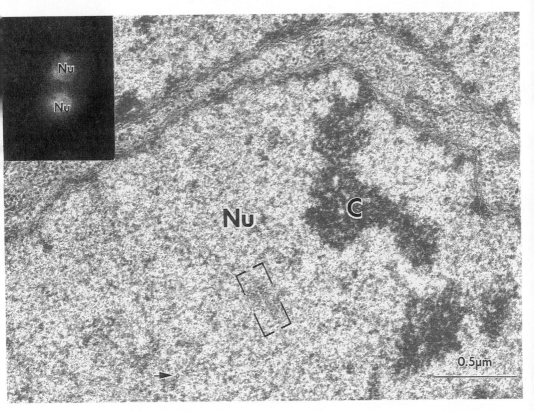

Fig. 4. Cell of R biotype of _Eleusine_ treated with 10 μM taxol. The cell
has two large nuclei (Nu) of irregular shape, presumably reformed after
an abnormal division. Microtubules in profile (brackets) and in cross-
section are observed in the nucleus. Inset shows an immunofluorescence
micrograph of a similar cell with scattered particles of
immunofluorescence within each of the nuclei (Nu).

Protection of the S biotype by taxol

Treatment of the S biotype with 1.0 μM taxol results in cells with
ultrastructural features of the R biotype when grown on H_2O (Vaughn,
1986). Thus, we wondered if taxol treatment of the S biotype would also
confer resistance to dinitroaniline herbicides (i.e. a phenocopy of the
R biotype). To test these plants for potential resistance, we treated
roots of the S biotype with 1.0 μM taxol for 24 h and then transferred
them for a second day to 1.0 μM trifluralin or oryzalin solutions.
Although· S biotypes treated with the same concentrations of DMSO as
contained in the taxol treatments (0.01%) were still sensitive to
treatment with the two dinitroaniline herbicides, the microtubules of the
taxol-treated S biotype appear to be unaffected by either dinitroaniline
herbicide. Treatment of the taxol-protected S biotype with 10 μM
oryzalin did cause some mitotic disruption, indicating that the phenocopy
of the R biotype induced by taxol is not absolute.

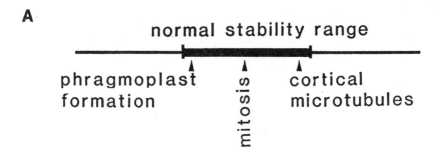

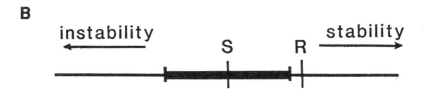

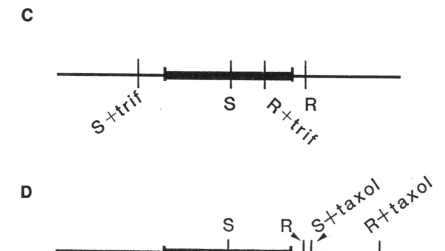

Fig. 5. <u>Model to explain the behaviour of R and S biotypes of *Eleusine* to dinitroaniline and taxol treatments</u> (After Cabral *et al.*, 1986).
A. Normal stability range and microtubule configurations of the plant cell. B. Relation of the R and S biotype microtubules to the normal range of microtubule stability. C. Effect of trifluralin (trif) treatment on microtubule stability of R and S biotype. D. Effect of taxol treatment on microtubule stability of R and S biotype.

184

Model for resistance

Cabral *et al.* (1986) proposed that, in normal cells, there was a balance between microtubules in a stable, polymerized state and an unstabilized, readily mobile, free tubulin pool. This allows for the polymerization of new microtubules and the depolymerization of old microtubules. A model of this concept as it applies to the present results is shown in Fig. 5. Within a certain stability/instability range, the cell is able to produce all of the microtubule arrays (cortical, preprophase, spindle, phragmoplast), perhaps by calling on other factors: calcium, calmodulin, microtubule associated proteins, or other (de)stabilizing factors to modify the stability of microtubules at a certain developmental stage (Fig. 5A). Based on the heightened sensitivity to taxol, we conclude that the R biotype shifts this normal range to a tubulin that is more stabilized in the polymerized form than the tubulin of the S biotype (Fig. 5B). Agents that destabilize microtubules, such as the dinitroanilines, probably do not cause enough destabilization to cause abnormalities in the R biotype (Fig. 5C). Although some level of dinitroaniline herbicide might destabilize the microtubules of the R biotype, the inherently low solubilities of these herbicides preclude a cellular concentration high enough to produce any pronounced effect. Long-term growth of seedlings on very high concentrations (1–10 µM) of oryzalin does cause some reduction of growth, even in the R biotype (Vaughn *et al.* 1987). The R biotype, because of the enhanced stability of its microtubules, is ultrasensitive to agents that stabilize microtubules such as taxol (Fig. 5D). Because of this stabilization, the R biotype appears to be altered in forming normal phragmoplast microtubules and hence cell plates. Although taxol protects the S biotype from dinitroaniline herbicides and, after 1 µM treatments, creates an ultrastructural phenocopy of the R biotype, the stabilization provided by taxol and that provided by the mutation are not the same. Taxol protects against nearly all mitotic disrupters (e.g. steganacin, colchicine, colcemid, dinitroanilines and amiprophosmethyl (Manfredi and Horwitz, 1984). The R biotype is resistant only to the dinitroanilines and amiprophosmethyl (Vaughn *et al.*, 1987). It is possible that the mutation in tubulin found in the R biotype, although increasing the stability of the microtubules, is a pleiotropic effect of this tubulin alteration (Vaughn and Vaughan, 1989), and not the sole mechanism of resistance. Another mutant of *Eleusine* with much lower levels of resistance to dinitroaniline herbicides (Vaughn, Vaughan, and Gossett, submitted) is not a tubulin mutant and appears to have no increased sensitivity to taxol, compared to the S biotype.

The sensitivity of cell plate formation to taxol, or the mutation to herbicide resistance, indicates that the phragmoplast microtubular array is relatively much more dynamic than the other microtubular configurations. The phragmoplast microtubules are relatively short and are associated with thousands of Golgi-derived vesicles during the brief period of cell plate formation. It is likely that the phragmoplast microtubules must undergo more rapid polymerization/depolymerization changes than other microtubular arrays. Thus, the stabilized phragmoplast microtubules in the taxol-treated plants (and in the occasional deviant walls in the untreated R biotype) may be so stabilized that the movements required for their normal functioning are not possible. Because of this, abnormalities of cell plate formation are noted in both the untreated R biotype and in taxol-treated plants of *Eleusine* and other species (Weerdenberg *et al.*, 1986).

185

ACKNOWLEDGEMENTS

Thanks are extended to Ms R.H. Jones for technical assistance, to Dr M. Suffness, NIH, for taxol used in this study and to J. Schepper-Vaughan and R. Stryker for typing the manuscript.

REFERENCES

Cabral, F.R., Brady, R.C., and Schibler, M.J. (1986). A mechanism of cellular resistance to drugs that interfere with microtubule assembly. Annals of the New York Academy of Sciences **466**, 745-756.

Falconer, M.M. and Seagull, R.W. (1985). Xylogenesis in tissue cultures: taxol effects on microtubule reorientation and lateral association in differentiating cells. Protoplasma **128**, 157-166.

James, S.W., Ranum, L.P.W., Silflow, C.D., and Lefebre, P.A. (1988). Mutants resistant to antimicrotubule herbicides map to a locus on the uni linkage group in Chlamydomonas reinhardi. Genetics **118**, 141-147.

Manfredi, J.J. and Horwitz, S.B. (1984). Taxol: an antimitotic agent with a new mechanism of action. Pharmacology and Therapeutics **25**, 83-125.

Mudge, L.C., Gossett, B.J., and Murphy, T.R. (1984). Resistance of goosegrass (Eleusine indica) to dinitroaniline herbicides. Weed Science **32**, 591-594.

Oakley, B.R. (1985). Microtubule mutants. Canadian Journal of Biochemistry and Cellular Biology **33**, 478-488.

Vaughn, K.C. (1986). Cytological studies of dinitroaniline-resistant Eleusine. Pesticide Biochemistry and Physiology **26**, 66-74.

Vaughn, K.C. and Koskinen, W.C. (1987). Effect of trifluralin metabolites on goosegrass (Eleusine indica) root meristems. Weed Science **35**, 36-44.

Vaughn, K.C., Marks, M.D., and Weeks, D.P. (1987). A dinitroaninline-resistant mutant of Eleusine indica exhibits cross-resistance and supersensitivity to antimicrotubule herbicides and drugs. Plant Physiology **83**, 956-964.

Vaughn, K.C. and Vaughan, M.A. (1986). Dinitroaniline resistance in Eleusine is due to altered tubulin. Plant Physiology (Suppl.) **80s**, 67.

Vaughn, K.C. and Vaughan, M.A. (1989). Structural and biochemical characterization of dinitroaniline-resistant Eleusine. American Chemical Society Symposium. In press.

Weerdenberg, C., Falconer, M.M., Setterfield, G. and Seagull, R.W. (1986). Effects of taxol on microtubule arrays in cultured higher plant cells. Cell Motility Cytoskel **6**, 469-478.

Wick, S.M., Seagull, R.W., Osborn, M., Weber, K., and Gunning, B.E.S. (1981). Immunofluorescence microscopy of organized microtubule arrays in structurally stabilized meristematic plant cells. Journal of Cell Biology **89**, 685-690.

MECHANISMS OF RESISTANCE TO ACETOLACTATE SYNTHASE/ACETOHYDROXYACID SYNTHASE INHIBITORS

Dale L. Shaner

American Cyanamid Co., Princeton, NJ 08540

Acetolactate synthase/acetohydroxyacid synthase (ALS) inhibitors can be very potent, broad spectrum herbicides. Two classes of inhibitors, the imidazolinones and the sulfonylureas, have been developed as herbicides and are becoming widely used. Resistant plant populations to these two chemical classes have been isolated through deliberate selection in the laboratory and through continuous use of the sulfonylureas in the field. Mutations have been selected in corn, soybeans, flax, tobacco, *Chlamydomonas*, *Datura*, *Kochia*, and *Arabidopsis* which are resistant to either the sulfonylureas, the imidazolinones or both. In fact, some mutations result in resistance to one subclass within sulfonylureas or imidazolinones but not other subclasses within the same chemical groups. Studies on the mechanism of resistance of these populations to these ALS inhibitors have shown that most of the resistant plant populations have a mutant ALS enzyme that is no longer sensitive to the herbicides. These studies have also demonstrated that although the imidazolinones and sulfonylureas inhibit the same enzyme, they do not appear to bind with the enzyme in exactly the same manner. The reason for these differences lies in the fact that multiple mutations can occur within the ALS gene that give rise to different levels of resistance. These different mutations must affect the way the imidazolinones and sulfonylureas bind to the ALS enzyme.

INTRODUCTION

In the last few years two new classes of highly potent herbicides have been introduced into the weed control market place. These are the imidazolinones (Los, 1987) and sulfonylureas (Beyer *et al.*, 1988). They represent a new generation of herbicides which are used at low rates (in the grams per hectare range) and have extremely low mammalian toxicity. They control a broad spectrum of monocots and dicots and are registered or being developed for use in most of the major agronomic crops including cereals, rice, legumes, cotton and maize (Los, 1987; Beyer *et al.*, 1988).

Both imidazolinones and sulfonylureas kill plants by inhibiting the enzyme that catalyzes the first step in branched chain amino acid biosynthesis, acetolactate synthase (ALS; also known as acetohydroxyacid synthase) (Shaner *et al.*, 1984; Chaleff and Mauvais, 1984). Studies with plants selected for resistance to both classes of herbicides have clearly demonstrated that ALS is the sole site of action of these herbicides (Chaleff and Mauvais, 1984; Newhouse *et al.*, 1989).

Early in the development of the imidazolinones and sulfonylureas, attempts to select for resistance through various selection protocols were successful in a wide range of species including maize (Anderson *et al.*, 1984), tobacco (Chaleff and Ray, 1984), *Arabidopsis* (Haughn and Somerville, 1986), canola (Swanson *et al.*, 1988), flax (Jordan and McHughen, 1987), *Datura innoxia* (Saxena and King, 1988), carrot (Watanabe *et al.*, 1988), and soybeans (Sebastian and Chaleff, 1987).

The most widely used of the first registered sulfonylureas were chlorsulfuron and metsulfuron. These are both very potent herbicides with a use rate of 4 to 20 g a.i. ha^{-1} (Beyer *et al.*, 1988). These two herbicides give greater than 95% control of susceptible weeds and they persist in the soil for several months, particularly in soils with a pH greater than 7.5 (Beyer *et al.*, 1988). In continuous wheat areas, both chlorsulfuron and metsulfuron were used on an annual basis, being applied at 7 to 12 month intervals (Thill *et al.*, 1989).
This use pattern was tailor made for the selection of sulfonylurea resistant weed populations, and in 1987 the first sulfonylurea resistant weed population of *Lactuca serriola* was reported in Lewiston, Idaho (Thill *et al.*, 1989). In 1988, sulfonylurea resistant populations of *Kochia scoparia*, *Salsola iberica* and *Stellaria media* in the dryland wheat belt of the United States and Canada were discovered (Thill *et al.*, 1989). More details on the discovery, distribution, biology and mechanism of sulfonylurea herbicide-resistant weeds are given by Thill *et al.* (these Proceedings).

The purpose of this chapter is to discuss the mechanisms of resistance in plants to ALS inhibitors. There are at least three mechanisms by which plants have become resistant to formerly phytotoxic herbicides. These include (1) a modification of the site of action of the herbicide; (2) a change in the absorption and translocation of the herbicide so that it does not reach the site of action; (3) a change in the ability of the plant to metabolize the herbicide.

Two of these mechanisms appear to account for the resistance of weeds to ALS inhibitors. These are changes at the site of action, and changes in the ability of the plants to metabolize the herbicides. The predominant mechanism, however, has been a change in ALS so that it is no longer sensitive to the inhibitors. Since the second mechanism of resistance (i.e. a change in the ability of a plant to metabolize the ALS inhibitor), is extensively covered in other papers, this chapter will focus on the changes in the ALS enzyme. Furthermore, since ALS inhibitor-resistant weed populations have only recently been discovered, most of the work that has been done on this latter mechanism has been with plant populations that were deliberately selected in the laboratory. Nonetheless, preliminary work on the resistant weed populations has shown that the same principles are working in the weeds as in the laboratory selected species (Thill *et al.*, 1989).

The primary objectives of this paper are to discuss the types of changes in ALS that have occurred in resistant plant populations, to consider the phenomenon of cross resistance in plant populations between different ALS inhibitors, and finally to speculate on why this particular mechanism of resistance has been the predominant one selected and not others such as over-expression of the target site or metabolism.

MECHANISM OF RESISTANCE TO ALS INHIBITORS

Mechanism of Inhibition of ALS by Imidazolinones and Sulfonylureas

ALS is required for the biosynthesis of two acetohydroxyacids, acetolactate and acetohydroxybutyrate (Bryan, 1985), which are key intermediates in the synthesis of the branched chain amino acids, valine, leucine and isoleucine. ALS is one of the three regulation sites within the branched chain amino acid pathway (Bryan, 1985) and is co-operatively feedback regulated by valine and leucine (Miflin and Cave, 1972).

Although there are up to six isozymes of ALS in micro-organisms which differ in their sensitivity to feedback inhibition by amino acids and sulfonylureas and imidazolinones (Schloss *et al.*, 1988), there appears to be only one type of ALS in plants. Chaleff and Bascomb (1987) showed that there are two genes for ALS in *Nicotiana tabacum* that are encoded by two unlinked loci. However, they were unable to separate two forms of the enzyme from plants by various chromatographic techniques and suggested that the two isozymes may be very similar and are expressed at similar levels within the plant. Lee *et al.* (1988) confirmed this speculation when they were able to isolate the two genes for ALS in

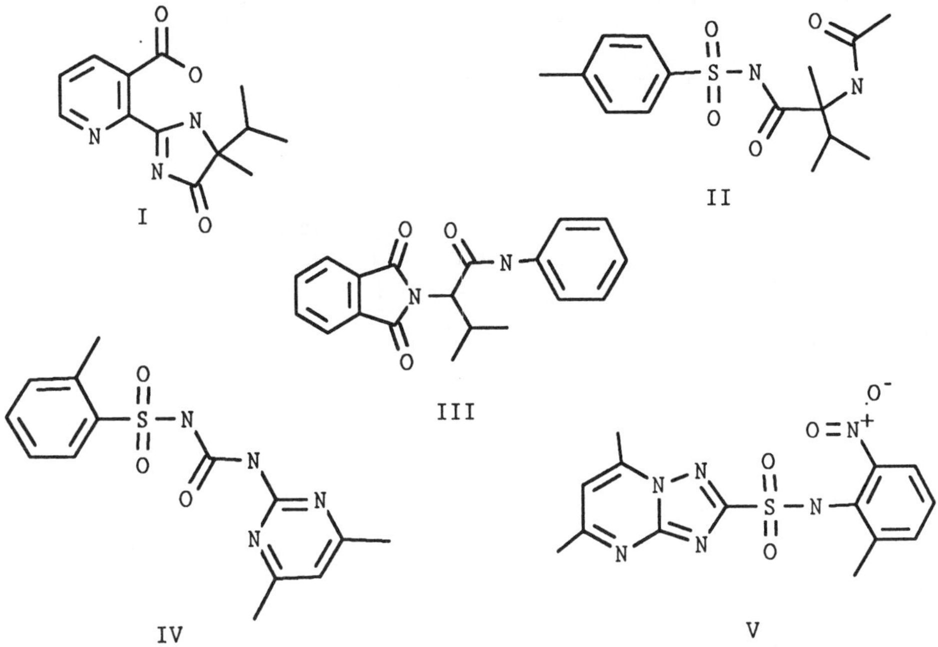

Fig. 1   Examples of chemicals that inhibit plant ALS. The classes represented are: I = imidazolinones; II = sulfonylcarboxamides; III = N-phthalyl-L-valine anilides; IV = sulfonylureas; V = triazolopyrimidines.

tobacco. They found that the two gene products only diverged 0.7% at the amino acid level which explains why Chaleff and Bascomb (1987) were not able to separate the two isozymes from one another. Swanson *et al.* (1988) reported that there are at least two ALS genes in canola, while Haughn and Somerville (1987) found only one ALS gene in *Arabidopsis*.

There are at least five different classes of ALS inhibitors (Fig. 1). These include imidazolinones (Shaner *et al.*, 1984), sulfonylureas (Schloss *et al.*, 1988), triazolopyrimidines (Schloss *et al.*, 1988), sulfonylcarboxamides (Crews *et al.*, 1989), and N-phthalyl-L-valine anilides (Huppatz and Casida, 1985). The triazolopyrimidines appear to interact with plant ALS in a manner similar to the sulfonylureas (Schloss *et al.*, 1988). There is no reported work on the other two classes of inhibitors with regard to ALS-inhibitor resistant plants. Therefore, the majority of the work discussed below will be on the imidazo- linones and sulfonylureas.

Both imidazolinones and sulfonylureas are slow, tight-binding, uncompetitive inhibitors of plant ALS (Muhitch *et al.*, 1987; Ray, 1986). Schloss *et al.*, (1988) found that imidazolinones and sulfonylureas compete for the same site on ALS extracted from bacteria. Shaner *et al.* (1989) also has evidence to demonstrate competition for the same site in plant ALS. However, the two classes of herbicides also interact somewhat differently with plant ALS based on the different types of resistant plants that have been selected in various laboratories. This aspect will be discussed more fully below.

Mechanism of Resistance to ALS Inhibitors

Many different research groups have successfully selected for either imidazolinone or sulfonylurea-resistant plant lines. In all cases where the mechanism of resistance was determined, the resistant plant populations had an altered ALS that was no longer sensitive to the inhibitors (Table 1).

ALS inhibitor resistance is inherited as a single, semi-dominant trait in maize (Newhouse *et al.*, 1989), tobacco (Chaleff and Bascomb, 1987), canola (Swanson *et al.*, 1988), *Arabidopsis* (Haughn and Somerville, 1986), and *Chlamydomonas* (Winder and Spalding, 1988). Homozygous resistant plants are much more resistant to the herbicides than heterozygous plants.

In vitro characterization of ALS from resistant plants has shown that the enzyme is generally expressed at the same level as in sensitive plants (Newhouse, *et al.*, 1989; Chaleff and Bascomb, 1987). Furthermore, the resistant enzyme is still feedback regulated by the branched chain amino acids (Fig. 2). These results indicate that these ALS inhibitors are not interacting with the enzyme at the site of feedback regulation of valine and leucine. Furthermore, the enzyme appears to function normally within the plant and there has been no report that the resistant lines are less fit than susceptible lines.

Lee *et al.* (1988) have reported on the isolation and characterization of the ALS gene from two different sulfonylurea resistant tobacco lines. They were able to show that one of the lines, C3, had one mutation in position 196 of the enzyme from a Pro to a Gln which was responsible for

Table 1. <u>Instances and Mechanisms of Resistance to ALS Inhibitors</u>

Species	Selection Protocol	Selection Agent[a]	Mechanism of Resistance	Reference
Alopecurus mysosuroides	Field selection	CH	Increased metabolism	Moss, 1987
Arabidopsis thaliana	Seed mutagenesis	SU	Altered ALS	Haughn and Somerville, 1986
Canola	Microspore selection	SU	Altered ALS	Swanson *et al.*, 1988
Carrot	Cell culture selection	SU	Altered ALS	Watanabe *et al.*, 1988
Chlamydomonas rheinhardtii	Cell mutagenesis	SU/IM	Altered ALS	Winder and Spalding, 1988
Datura innoxia	Cell culture selection	SU	Altered ALS	Saxena and King, 1988
Flax	Cell culture selection	SU	ND	Jordan and McHughen, 1987
Kochia scoparia	Field selection	SU	Altered ALS	Thill *et al.*, 1989
Lactuca serriola	Fild selection	SU	Altered ALS	Thill *et al.*, 1989
Lolium rigidum	Field selection	DM	Increased metabolism	Heap and Knight, 1986
Maize	Cell culture selection	IM	Altered ALS	Shaner and Anderson, 1985
Salsola iberica	Field selection	SU	Altered ALS	Thill *et al.*, 1989
Soybean	Seed mutagenesis	SU	ND	Sebastian and Chaleff, 1987
	Seed mutagenesis	SU	Altered ALS	Sebastian *et al.*, 1989
Tobacco	Cell culture selection	SU	Altered ALS	Chaleff and Ray, 1984

[a] Abbreviations used: CH = chlorotoluron; SU = sulfonylurea; IM = imidazolinone; DM = diclofop methyl; ND = not determined.

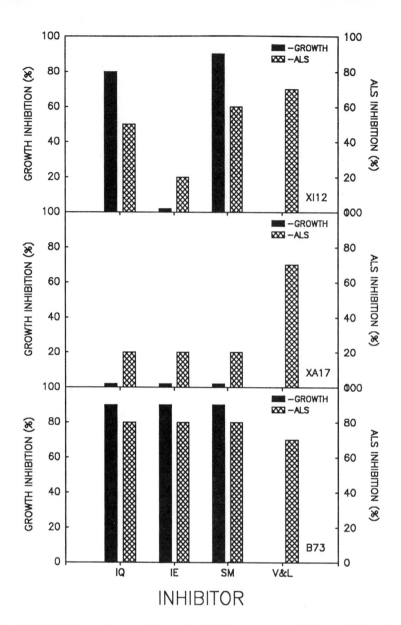

Fig. 2 Inhibition of growth and ALS in susceptible and imadazolinone-resistant maize lines by imazaquin (IQ), imazethapyr (IE), sulfometuron-methyl (SM), and valine and leucine (V & L). The concentrations used for growth inhibition (applied post-emergent) were: IQ - 140 g ha^{-1}; IE - 100 g ha^{-1}; SM - 20 g ha^{-1}. The concentrations used for *in vitro* ALS inhibition were: IQ - 100 μM; IE - 100 μM; SM - 100 nM; V & L - 1 mM; B73 = susceptible inbred line; XA17 and XI12 = imidazolinone-resistant maize lines backcrossed into B73.

the resistance. In a highly resistant line, S4-Hra, there were two mutations, one at position 196 with a Pro to Ala change and the other at position 574 with a Trp to Leu substitution. More information on the molecular biology aspects of these changes are presented by Hartnett *et al.* (these Proceedings).

Cross–Resistance Between ALS Inhibitors

The imidazolinones and sulfonylureas inhibited the same enzyme and there is data to suggest that they may bind to the same site on ALS (Schloss *et al.*, 1988; Shaner *et al.*, 1989). Thus, it is logical to hypothesize that any change in binding of one of these chemicals to ALS will also affect the binding of the other. However, the data on the different imidazolinone and sulfonylurea resistant plant populations show that cross resistance is not very predictable.

When either sulfonylurea or imidazolinone selection of various species was done in the laboratory, all combinations of cross resistance were found. Newhouse *et al.* (1989) reported that imidazolinone resistant maize lines showed not only differences in resistance between imidazolinones and sulfonylureas, but also differences among imidazolinones. Line XA17 was highly resistant to all imidazolinones and sulfonylureas tested, while line XI12 was only highly resistant to imazethapyr with lesser resistance to imazaquin and no resistance to sulfometuron-methyl (Figure 2).

Winder and Spalding (1988) reported that among *Chlamydomonas reinhardtii* mutants selected on imazaquin cross tolerance to chlorsulfuron was rare. But among chlorsulfuron-selected mutants low levels of cross tolerance to imazaquin were common. However, the ALS from these chlorsulfuron resistant lines was not highly resistant to imazaquin *in vitro*. They also found one line that was highly resistant to imazaquin and hypersensitive to chlorsulfuron.

Just the opposite case was reported by Saxena and King (1988) in *Datura innoxia* where they selected for a sulfonylurea-resistant line which was hypersensitive to imazaquin. Other *Datura* lines selected on either chlorsulfuron or sulfometuron-methyl were either not cross resistant to imazaquin or imazapyr or showed intermediate resistance. One line, SMR1, was only moderately tolerant to chlorsulfuron, sulfometuron-methyl and imazaquin, but was highly tolerant to imazapyr.

Thill *et al.* (1989) reported that ALS from sulfonylurea resistant weed populations of *Lactuca serriola* and *Kochia scoparia* are also resistant to the imidazolinones. The degree of this cross resistance, however, varied with the plant population. DeCastro and Youmans (pers. comm.) found that the cross resistance of 4 different chlorsulfuron-resistant populations of *Kochia scoparia* to use rates of 4 different sulfonylureas and 3 different imidazolinones was quite variable (Table 2). One population was resistant to all sulfonylureas and to imazethapyr and imazaquin, while all of the populations were still controlled by a use rate of imazapyr. All of the populations were resistant to chlorsulfuron, which was the selective agent.

These results clearly show that one cannot assume *a priori* that a plant population that is resistant to one class of ALS inhibitors will also be resistant to another class of inhibitors. The reason for this lies in the type of mutation that occurs in the ALS gene. Bedbrook *et al.* (1986) have identified at least six different mutation sites within the ALS gene that can confer resistance to ALS inhibitors including sulfonylureas, imidazolinones and triazolopyrimidines. At least one of the mutations at position 197 of a Pro to Ser substitution results in an ALS that is resistant to chlorsulfuron and sulfometuron-methyl but not to imazapyr. Presumably the other types of mutations selected in other species are the result of different substitutions. These results also show that although the imidazolinones and sulfonylureas share some common binding sites on ALS, there also must be enough differences in their binding to allow this type of discrimination among mutated ALS enzymes.

Table 2. Cross Resistance of five *Kochia scoparia* Populations to Sulfonylureas and Imidazolinones (Data from DeCastro and Youmans, American Cyanamid)

Population	Control Rating[a]						
	Herbicide[b]						
	CS	AL	SM	MS	IM	IE	IQ
I[c]	9	9	9	9	9	9	9
II	0	0	6	5	9	8.3	9
III	0.7	0	9	6.3	9	9	8.7
IV	0	0.3	4	2	9	2.7	6.7
V	2	0	5.3	2.7	9	8.7	6.7

[a]Control rating based on a 0 to 9 scale. 0 = no control; 9 = complete control. Plants were treated in the greenhouse and injury assessed 4 weeks after application.
[b]Herbicide abbreviations and use rates (all herbicides were sprayed at their recommended use rate in g ha^{-1}): CS = 26.3 g ha^{-1} chlorsulfuron; AL = 9 g ha^{-1} DPX-M6316; SM = 52.2 g ha^{-1} sulfometuron-methyl; MS = 4.2 g ha^{-1} metsulfuron; IM = 560 g ha^{-1} imazapyr; IE = 70 g ha^{-1} imazethapyr; IQ = 140 g ha^{-1} imazaquin.
[c]Location of populations of susceptible and sulfonylurea-resistant *K. scoparia*: I = susceptible greenhouse line; II = Wheat field in Bottineu, ND; III = Wheat field in Westhope, ND; IV = Wheat field in Maxbass, ND; V = Railroad right of way in Palmer County, TX.

Why An Altered ALS?

In all of the work done to date on ALS inhibitor resistant plant populations selected on either imidazolinones or sulfonylureas, the only stably inherited type of resistance has been through an altered ALS. Xiao *et al.* (1987) had evidence to suggest that they were able to select a chlorsulfuron-resistant *Datura innoxia* line that had increased levels of ALS. However, this trait disappeared after 2 months of subculturing and chlorsulfuron resistance appeared to be due to an altered ALS. The question that arises is why is an altered ALS the predominant mechanism that has been selected?

Cell culture selection for resistance to glyphosate, phosphinothricin, and Hoe 704, herbicides that also inhibit the biosynthesis of amino acids (Botterman and Leemans, 1988; Schulz *et al.*, 1988), has resulted in plant lines which over-produce the target site (Botterman and Leemans, 1988; Schulz *et al.*, 1988). All of these herbicides inhibited their respective target sites by competing with the substrate for the catalytic site. In these cases, overproduction of the target enzyme may give protection by tying up the herbicide and leaving enough residual enzyme activity for the plant's needs.

In the case of ALS inhibitors, over-production of the target site does not appear to give much protection from the inhibitor. Haughn *et al.* (1988) were able to transform tobacco with an altered ALS gene from *Arabidopsis* which produced a chlorsulfuron resistant ALS enzyme. The transformed tobacco did show increased resistance to chlorsulfuron and approximately 20% of the ALS activity appeared to be resistant to the herbicide. However, in the absence of the herbicide, the amount of ALS activity in the transgenic tissue was consistently lower than in wild type tissue. These results indicate that over-expression of the mutant ALS gene was not contributing to herbicide resistance.

There has also been no case of increased resistance to ALS inhibitors due to altered uptake and translocation or altered metabolism after selection on an ALS inhibitor. There are *Lolium rigidum* and *Alopecurus myosuroides* populations that are resistant to chlorsulfuron and imazamethabenz-methyl presumably through increased metabolism of the herbicides (Heap and Knight, 1986; Moss, 1987; Kemp and Caseley, 1987), but these populations were not selected by either one of the ALS inhibitors. The former was selected after continued use of diclofop-methyl (Heap and Knight, 1986) and the latter with chlorotoluron (Moss, 1987).

Selection for altered metabolism may occur with ALS inhibitors, but probably in populations that already can metabolize the herbicide at least to a limited extent. The frequency of mutations that can give rise to an altered ALS must be higher than that for altered metabolism and will probably account for the mechanism of resistance in the majority of plant populations.

UNANSWERED QUESTIONS

Some of the questions that still wait to be answered are (1) What determines the level of cross resistance in an altered ALS enzyme? How is this related to the mechanism of binding of inhibitors to the enzyme? (2) What is the frequency of genes in weed populations for altered ALS? (3) If ALS can be altered in so many different sites without apparently affecting the catalytic or regulatory sites of the enzyme, why are there not more naturally tolerant ALS enzymes? Are these resistant populations less fit in some unknown manner? The molecular biology studies on ALS from bacteria, yeast and plants indicates a very high degree of homology. There must be some reason for this preservation of the amino acid sequence. What is that reason?

CONCLUSION

The predominant mechanism of resistance to ALS inhibitors is the selection for an altered ALS that is no longer inhibited by the herbicides. There are at least six different sites in the ALS genome that, if altered, can result in herbicide resistance. Given this many sites and all of the different amino acid changes that can result in resistance, then the frequency of resistant genes in any plant population may be very high. This possibility is borne out by the ease with which ALS inhibitor resistant populations have been selected both in the laboratory and in the field.

The accumulation of knowledge on ALS and its interaction with various inhibitors has occurred at an astounding rate. The use of resistant ALS enzymes that are selected either through various selection protocols or through genetic engineering make it possible for us to gain a detailed understanding in the functioning of this important enzyme and in the interactions of various inhibitors with the protein. This knowledge will be useful in the further understanding of how to make new inhibitors.

REFERENCES

Anderson, P.L., Georgeson, M. and Hibberd, K.A. (1984). Selection and characterisation of imidazolinone-tolerant mutants of maize. Agronomy Abstracts 76, 56.

Bedbrook, J., Chaleff, R., Falco, S.C., Mazur, B., Somerville, C. and Yadav, N. (1986). Nucleic acid fragment encoding herbicide resistant plant acetolactate synthase. European Patent Application 257-993-A. 79p.

Beyer, Jr. E.M., Duffy, M.J., Hay, J.V. and Schlueter, D.D. (1988). Sulfonylureas. In "Herbicides: Chemistry, Degradation, and Mode of Action" (P.C. Kearney and D.D. Kaufman, eds), pp. 117-189. Marcel Dekker, Inc., New York.

Botterman, J. and Leemans, J. (1988). Engineering herbicide resistance in plants. Trends in Genetics 4, 219-222.

Bryan, J.K. (1985). Synthesis of the aspartate family and branched chain amino acids. In "The Biochemistry of Plants: A Comprehensive Treatise" (B.J. Miflin, P.K. Stumpf and E.E. Conn, eds) Volume 5. pp. 403-453. Academic Press, New York, NY.

Chaleff, R.S. and Bascomb, N.F. (1987). Genetic and biochemical evidence for multiple forms of acetolactate synthase in Nicotiana tabacum. Molecular and General Genetics 210, 33-38.

Chaleff, R.S. and Mauvais, C.J. (1984). Acetolactate Synthase is the site of action of two sulfonylurea herbicides in higher plants. Science 224, 1443-1445.

Chaleff, R.S. and Ray, T.B. (1984). Herbicide-resistant mutants for tobacco cell cultures. Science 223, 1148-1151.

Crews, A.D., Alvarado, S.I., Brady, T.E., Doehner, R.F., Gange, D.M., Little, D.L. and Wepplo, P.J. (1989). Discovery of a new class of herbicides: sulfonylcarboxamides. Abstracts of 198th ACS National Meeting #39.

Haughn, G.W., Smith, J., Mazur, B. and Somerville, C. (1988). Transformation with a mutant *Arabidopsis* acetolactate synthase gene renders tobacco resistant to sulfonylurea herbicides. <u>Molecular and General Genetics</u> **211**, 266–271.

Haughn, G.W. and Somerville, C.R. (1986). Sulfonylurea-resistant mutants of *Arabidopsis thaliana*. <u>Molecular and General Genetics</u> **204**, 430–434.

Haughn, G.W. and Somerville, C.R. (1987). Selection for herbicide resistance at the whole-plant level. In "Biotechnology in Agricultural Chemistry" (H.M. LeBaron, R.O. Mumma, R.C. Honeycutt and J.H. Duesing, eds). pp. 98–107. American Chemical Society, Washington, D.C.

Heap, I. and Knight, R. (1986). The occurrence of herbicide cross-resistance in a population of annual ryegrass, *Lolium rigidum*, resistant to diclofop-methyl. <u>Australian Journal of Agricultural Research</u> **37**, 149–156.

Huppatz, J.L. and Casida, J.E. (1985). Acetohydroxyacid synthase inhibitors: N-phthalyl-L-valine anilide and related compounds. <u>Zeitschrift für Naturforschung</u> **40**, 652–656.

Jordan, M.C. and McHughen, A. (1987). Selection for chlorsulfuron resistance in flax (*Linum usitatissimum*) cell culture. <u>Journal of Plant Physiology</u> **131**, 333–338.

Kemp, M.S. and Caseley, J.C. (1987). Synergistic effects of 1-amino-benzotriazole on the phytotoxicity of chlorotoluron and isoproturon in a resistant population of black-grass (*Alopecurus myosuroides*). <u>British Crop Protection Conference - Weeds</u> **3**, 895–899.

Lee, K.Y., Townsend, J., Tepperman, J., Black, M., Chui, C.F., Mazur, B., Dunsmuir, P. and Bedbrook, J. (1988). The molecular basis of sulfonylurea herbicide resistance in tobacco. <u>EMBO Journal</u> **7**, 1241–1248.

Los, M. (1987). Synthesis and biology of the imidazolinone herbicides. In "Pesticides and Biotechnology" (R. Greenhalgh and T.R. Roberts, eds). pp. 35–42. Blackwell Scientific, Boston, MA.

Miflin, B.J. and Cave, P.R. (1972). The control of leucine, isoleucine, and valine biosynthesis in a range of higher plants. <u>Journal of Experimental Botany</u> **23**, 511–516.

Moss, S.R. (1987). Herbicide resistance in black-grass (*Alopecurus myosuroides*). <u>British Crop Protection Conference - Weeds</u> **3**, 879–887.

Muhitch, M.J., Shaner, D.L. and Stidham, M.A. (1987). Imidazolinones and acetohydroxyacid synthase from higher plants: Properties of the enzyme from maize suspension culture cells and evidence for the binding of imazapyr to acetohydroxyacid synthase *in vivo*. <u>Plant Physiology</u> **83**, 451–456.

Newhouse, K., Shaner, D., Wang, T. and Fincher, R. (1989). Genetic modification of crop responses to imidazolinone herbicides. American Chemical Society Symposium Series In press.

Ray, T.B. (1986). Sulfonylurea herbicides as inhibitors of amino acid biosynthesis in plants. <u>TIBS</u> **11**, 180–183.

Saxena, P.K. and King, J. (1988).   Herbicide resistance in *Datura innoxia*:
Cross-resistance of sulfonylurea-resistant cell lines to imidazolinones.
Plant Physiology **86**, 863-867.

Schloss, J.V., Ciskanik, L.M. and VanDyk, D.E. (1988). Origin of the
herbicide binding site of acetolactate synthase. Nature **331**, 360-362.

Schulz, A., Sponemann, P., Kocher, H. and Wengenmayer, F. (1988). The
herbicidally active experimental compound Hoe 704 is a potent inhibitor
of the enzyme acetolactate reductoisomerase. FEBS Letters **238**, 375-378.

Sebastian, S.A. and Chaleff, R.S. (1987). Soybean mutants with increased
tolerance for sulfonylurea herbicides. Crop Science **27**, 948-952.

Sebastian, S.A., Fader, G.M., Ulrich, J.F. and Forney, D.R. (1989).
Dominant soybean mutations for resistance to sulfonylurea herbicides.
World Soybean Research Conference IV Abstract 342.

Shaner, D.L., Anderson, P.C. and Stidham, M.A. (1984). Imidazolinones:
Potent inhibitors of acetohydroxyacid synthase. Plant Physiology **76**,
545-546.

Shaner, D.L. and Anderson, P.C. (1985). Mechanism of action of the
imidazolinones and cell culture selection of tolerant maize.   In
"Biotechnology in Plant Science" (M.   Zaitlin,   P.R.   Day and A.
Hollaender, eds.). pp. 287-299. Academic Press, Orlando, FL.

Shaner, D.L., Singh, B.K. and Stidham, M.A. (1989).   Interaction of
imidazolinones with plant acetohydroxyacid synthase: Evidence for *in vitro*
binding and competition with sulfometuron methyl.   Journal of
Agricultural and Food Chemistry submitted.

Swanson, E.B., Coumans, M.P., Brown, G.L., Patel, J.D. and Beversdorf,
W.D. (1988). The characterization of herbicide tolerant plants in *Brassica
napus* L. after *in vitro* selection of microspores and protoplasts. Plant
Cell Reports **7**, 83-87.

Swanson, E.B. (1988). Sulfonylurea and imidazolinone resistant canola
selected through microspore culture selection. Second Canadian Plant
Tissue Culture and Genetic Engineering Workshop, Ottawa, Canada, May,
1988   Canadian Section, International Association for Plant Tissue
Culture.

Thill, D.C., Mallory, C.A., Saari, L.L., Cotterman, J.C. and Primiani,
M.M. (1989). Sulfonylurea resistance-mechanism of resistance and cross
resistance. Weed Science Society of America, Abstracts **29**, 132.

Watanabe, H., Hisamitsu, H. and Ishizuka, K. (1988). Selection of carrot
cells tolerant to bensulfuron methyl and their acetolactate synthase
response. Weed Research (Tokyo) **33**, 285-292.

Winder, T. and Spalding, M.H. (1988).   Imazaquin and chlorsulfuron
resistance and cross resistance in mutants of *Chlamydomonas reinhardtii*.
Molecular and General Genetics **213**, 394-399.

Xiao, W., Saxena, P.K. and Rank, G.H. (1987). A transient duplication
of the acetolactate synthase gene in a cell culture of *Datura innoxia*.
Theoretical and Applied Genetics **74**, 417-422.

DIFFERENTIAL INHIBITION OF PLANT ACETYL COA CARBOXYLASE – THE BIOCHEMICAL BASIS FOR THE SELECTIVITY OF THE ARYLOXY-PHENOXYPROPANOATE AND CYCLOHEXANEDIONE HERBICIDES

W.J. Owen

Dow Elanco Limited, Letcombe Laboratory, Letcombe Regis, Wantage, Oxon, OX12 9JT.

Aryloxyphenoxypropanoates and cyclohexanediones are two classes of herbicides that control a broad spectrum of monocotyledonous weeds, though dicotyledonous plants are generally resistant. Although very different in structure, it has been suggested for some time that both groups of compounds may have a similar mode of action due to their similarity in selectivity and symptomology.

Early work implicated *de novo* fatty acid biosynthesis as the biochemical pathway inhibited and these observations were subsequently reinforced by the emergence of a good correlation between herbicidal efficacy and degree of inhibition of ^{14}C-acetate incorporation into lipids. Until recently the precise enzymic step of fatty acid biosynthesis which constitutes the specific target site for the aryloxyphenoxypropanoates and cyclohexanediones had eluded identification. This resulted largely from an unfortunate coincidence of data from ^{14}C-acetate and ^{14}C-malonate incorporations into fatty acids which implicated fatty acid synthetase as the target enzyme.

However, recent work in our own laboratory and elsewhere has led to the discovery that it is the acetyl CoA carboxylase (acetyl-coenzyme A: bicarbonate ligase [ATP], EC 6.4.1.2) activity of the susceptible plant species which is inhibited by these herbicides.

In this paper the evidence leading to this conclusion will be reviewed along with some of the properties of higher plant acetyl CoA carboxylases.

INTRODUCTION

Aryloxyphenoxypropanoates and cyclohexanediones are two classes of herbicides which selectively control a broad spectrum of monocotyledonous weed species, though dicotyledonous plants are generally very resistant. Although very different in structure (Fig.1) it has been speculated for some time that due to similarities in symptomology and selective properties, both groups of compounds may have the same or similar mode of action.

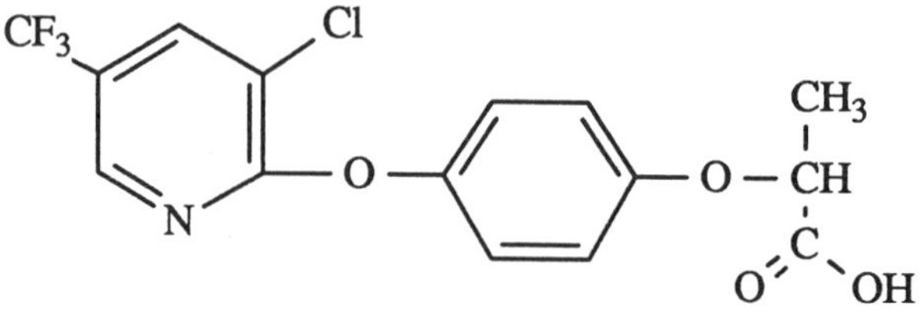

HALOXYFOP

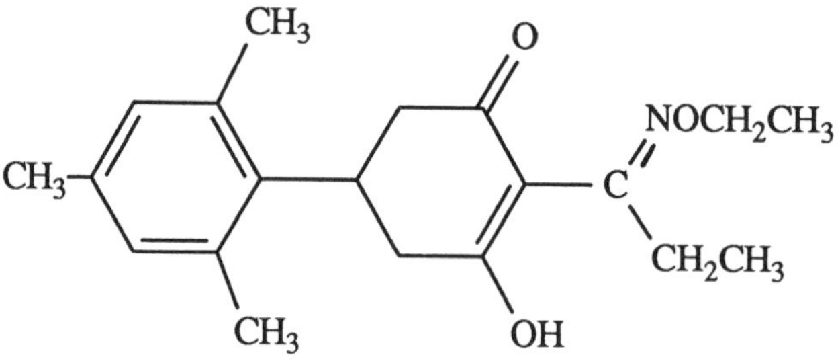

TRALKOXYDIM

Fig. 1.   Structures of haloxyfop and tralkoxydim.

Recent studies in our laboratory (Secor and Cseke, 1988; Secor *et al.*,1989) and elsewhere (Burton *et al.*, 1987; Kobek *et al.*, 1988; Rendina and Felts, 1988; Walker *et al.*, 1988) have provided confirmation for this view and have identified acetyl CoA carboxylase (acetyl-coenzyme A: bicarbonate ligase [ATP], EC 6.4.1.2) as the common target enzyme. The present paper summarises the experimental data that has led to the conclusion that aryloxyphenoxypropanoates and cyclohexanediones exert their herbicidal effects in monocotyledonous plants through inhibition of acetyl CoA carboxylase, which represents the first committed step of fatty acid biosynthesis.

MODE OF ACTION

Studies on the uptake, translocation and metabolic fate of various aryloxyphenoxypropanoate and cyclohexanedione herbicides in susceptible and tolerant plant species have indicated that these factors are generally of little importance with respect to the selectivity of these herbicides (Campbell and Penner, 1981; Swisher and Corbin, 1982; Veerasekaran and Catchpole, 1982; Hendley *et al.*, 1985). These results suggested that the tolerance of broadleaf crops and weeds is probably based on insensitivity of the target site to these compounds. However, despite many investigations aimed at determining the mode of action of aryloxyphenoxypropanoates and, to a lesser extent, cyclohexanediones no specific target site had been identified for either group of compounds. Neither class of compounds interfered with photosynthesis, respiration, protein biosynthesis or nucleic acid biosynthesis (Hoppe, 1980, 1981;

Hosaka and Takagi, 1986). Physiological processes observed to be disrupted by both aryloxyphenoxypropanoates and cyclohexanediones include growth and development, maintenance of membrane integrity, auxin induced growth and lipid metabolism (Shimabukuro *et al.*, 1978; Hoppe, 1980, 1981; Gronwald, 1986). Aryloxyphenoxypropanoates have, additionally, been reported to depolarise membrane potentials (Lucas *et al.*, 1984; Wright and Shimabukuro, 1987). However, with the exception of lipid metabolism, none of these other effects have been correlated with observed herbicidal activity and thus, to date, have not provided an explanation for herbicidal selectivity.

The observation that the aryloxyphenoxypropanoate herbicide clofop-isobutyl was active as a hypolipidemic drug, reducing serum cholesterol and triglycerides in animals (Granzer and Nahm, 1973), prompted Hoppe and colleagues (Hoppe, 1985; Hoppe and Zacher, 1985) to study the effect of related herbicides, principally diclofop-methyl, on plant lipid metabolism. Experiments with diclofop-methyl revealed that this compound inhibited fatty acid biosynthesis in root tips and leaves of sensitive plant species. Fatty acid biosynthesis in plants occurs exclusively in chloroplasts or plastids and Hoppe and co-workers went on to show that a 60 min incubation with diclofop-methyl inhibited acetate incorporation into fatty acids by chloroplasts of susceptible but not tolerant dicotyledonous species. Work by Lichtenthaler and co-workers (Lichtenthaler and Meier, 1984; Lichtenthaler *et al.*, 1987) also established that the cyclohexanedione sethoxydim also altered lipid metabolism by inhibiting acetate incorporation into fatty acids. These studies strongly suggested that *de novo* fatty acid biosynthesis was inhibited by both groups of graminicides though the precise site of inhibition remained to be identified.

ACETYL COA CARBOXYLASE AS TARGET SITE

Identification of the primary physiological lesion occurring in response to aryloxyphenoxypropanoates and cyclohexanediones required that the site of action must be affected both rapidly and by sub-micromolar concentrations of the herbicides. In addition, sensitivity of the target site should also be correlated to herbicidal efficacy at the whole plant level so as to account for selectivity where this cannot be attributed to uptake, translocation or metabolism. Of the various modes of action proposed by other researchers, only disruption of lipid metabolism appeared to meet most of these criteria, and consequently was identified by a number of research groups as an area for more extensive investigation aimed at elucidating the primary site of action of these compounds.

Studies in our laboratory indicated that the aryloxyphenoxypropanoate haloxyfop (Fig.1) rapidly inhibited fatty acid biosynthesis in leaf discs of maize. Within 30 min of application, 1.4 µM of either the free fatty acid or the methyl ester of haloxyfop inhibited ^{14}C-acetate incorporation into organic phase cell constituents by about 50% (Fig.2) but had little or no effect on incorporation into the aqueous fraction.

Further studies indicated that these effects were also discernible within 15 min of herbicide application thus constituting, to our knowledge, the fastest physiological response reported for aryloxyphenoxypropanoates *in situ*. In order to facilitate uptake into

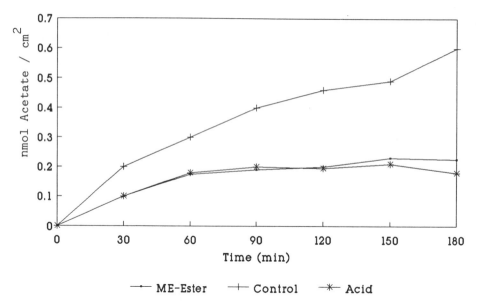

Fig.2. Effect of haloxyfop on acetate incorporation into lipids in maize leaf discs.

plant tissues these herbicides are applied as esters which are generally considered to be inactive but are rapidly de-esterified in the plant to the herbicidal acids. The fact that the methyl ester was as active as the free acid in inhibiting lipid biosynthesis implied the presence of an active esterase in the maize leaf discs. The cyclohexanedione, tralkoxydim (Fig.1) had a similar effect to haloxyfop in reducing acetate incorporation into lipids (Fig.3).

In agreement with the earlier observations of Hoppe and Zacher (1985), a comparison between susceptible barley and tolerant soybean indicated that haloxyfop was effective in inhibiting lipid biosynthesis in the monocotyledonous species only (Table 1). Such differences between

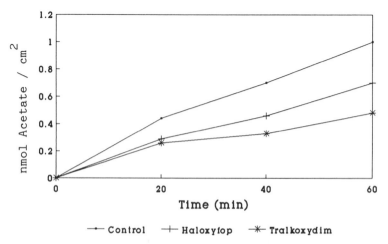

Fig.3. Effect of haloxyfop and tralkoxydim on acetate incorporation into lipids in maize leaf discs.

monocotyledonous and dicotyledonous species have been confirmed in recent publications by other researchers and have also been demonstrated for cyclohexanediones (Burton *et al.*, 1987; Kobek *et al.*, 1988).

In order to confirm that *de novo* fatty acid biosynthesis was inhibited by these herbicides we examined their effect on ^{14}C-acetate incorporation into stearic (16:0) acid, the initial product of fatty acid biosynthesis, again using maize leaf discs. The results obtained (Table 2) indeed indicated that over a 40 min incubation period there was a pronounced inhibition of incorporation of radioactivity into stearate suggesting that both classes of compounds affected an early step in lipid biosynthesis.

Table 1. <u>Effect of haloxyfop on acetate incorporation</u>[*] <u>(in nmol cm^{-2})</u> <u>into lipids in soybean and barley leaf discs</u>.

Treatment	Barley	Soybean
Control	3.10	1.41
1.4 µM haloxyfop	2.04	1.63

[*] after 40 min incubation period

Table 2. <u>The effect of tralkoxydim and haloxyfop on</u> ^{14}C-acetate <u>incorporation into stearic acid in maize leaf discs</u>.

Treatment	^{14}C Incorporation (dpm ± sem)
Control	619 ± 95
1.4 µM Tralkoxydim	144 ± 13
1.4 µM Haloxyfop	297 ± 49

Fatty acid biosynthesis in plants is thought to occur exclusively in chloroplasts or plastids where pyruvate or acetate are the *in vivo* precursors, being converted to acetyl CoA by the chloroplastidic pyruvate dehydrogenase complex and the stromal acetyl CoA synthetase, respectively. Subsequently, the acetyl CoA is carboxylated by acetyl CoA carboxylase (ACCase) to form malonyl CoA. Fatty acid synthetase, a multienzyme complex of seven enzymes, then catalyses the condensation of two carbon units from malonyl CoA to form fatty acids (Stumpf, 1980).

The observation that clofop-butyl acted as a hypolipidemic agent in animals (Granzer and Nahm, 1973) and that some hypolipidemics inhibit mammalian ACCase (Beyen and Geelen, 1982) suggested to us that aryloxyphenoxypropanoates might act in a similar manner by inhibiting ACCase in plants. Based on their observations that diclofop inhibited malonate incorporation into polar lipids in maize root tips, Hoppe and Zacher (1982) had concluded earlier that ACCase did not constitute the site of action.   The conversion of malonate to malonyl CoA prior to entry into the fatty acid biosynthetic pathway would require a thiokinase which, in many plant species, is less active than a malonate decarboxylase (Hatch and Stumpf, 1962).   For these reasons incorporation of malonate into lipids may proceed, at least in part, via acetate requiring the early enzymic steps of fatty acid biosynthesis. Consequently, whether or not ACCase might constitute the target site for aryloxyphenoxypropanoates remained an open question which we sought to answer unequivocally.

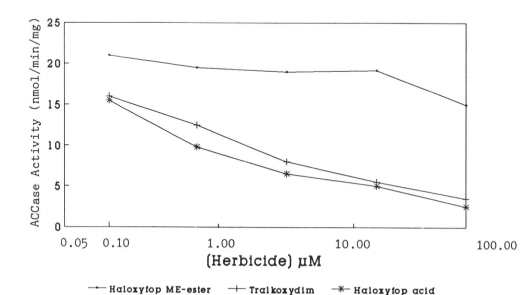

— Haloxyfop ME-ester   + Tralkoxydim   ※ Haloxyfop acid

Fig.4.   Effect of haloxyfop – methyl ester, haloxyfop acid and tralkoxydim on maize ACCase activity.

The activity of ACCase from maize was inhibited by both haloxyfop acid and tralkoxydim in a concentration dependent manner (Fig.4), the I_{50} value for both compounds being about 1 µM.   Similar values have been reported recently for the same and related compounds (Burton *et al.*, 1987; Rendina *et al.*, 1988; Rendina and Felts, 1988; Kobek *et al.*, 1988; Walker *et al.*, 1988).   The methyl ester of haloxyfop was more than 100-fold less active than was the free acid, which is consistent with the methyl ester being de-esterified in the plant.   Since haloxyfop shows stereospecificity in its herbicidal action we tested purified preparations of the R and S stereoisomers with respect to the 2 position of the propionate moiety.   ACCase activity was inhibited by the herbicidally active R(+) (98% enantiomeric excess) haloxyfop acid but not by the S(−) (94% enantiomeric excess) enantiomer (Fig.5).

The much reduced inhibition that was caused by the S(−) enantiomer could be accounted for by the 3% contamination of this preparation by the R(+) enantiomer.   Walker *et al.* (1988) have recently reported a similar distinction with respect to the R and S enantiomers of fluazifop and Rendina and Felts (1988) recorded a lower I_{50} value for the R(+) enantiomer of trifop compared to the racemate.   Such extrapolation of differing enantiomeric efficacies from whole plant to enzymic level provides convincing supportive evidence for ACCase being

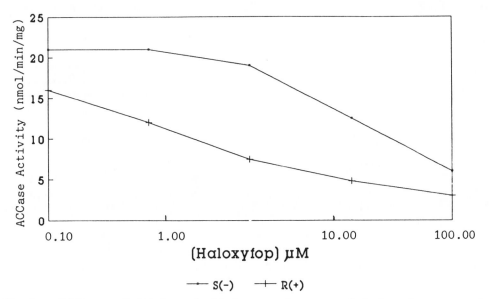

Fig.5 Effect of R(+) and S(-) enantiomers of haloxyfop on maize ACCase activity.

the primary target site of aryloxyphenoxypropanoates.

A comparison of the aryloxyphenoxypropanoates quizalofop, fenoxaprop, diclofop, haloxyfop and fluazifop indicated that they are very effective inhibitors of maize ACCase with I_{50} values of 0.02, 0.11, 0.13, 0.31 and 2.82 M respectively. However, I_{50} values are not necessarily well correlated with whole plant activity. Thus, diclofop has a higher recommended field rate than does haloxyfop, yet diclofop is some 3-fold more active at the enzyme level. This apparent discrepancy is accounted for by oxidative metabolism of diclofop to form less active plant metabolites (Gorecka *et al.*, 1981).

Inhibition of plant ACCases by haloxyfop and tralkoxydim is apparently reversible. Thus activity was restored to inhibited enzyme preparations by passage through a Sephadex G-25 desalting column indicating that neither herbicide was covalently bound.

SELECTIVITY

Since the selectivity of aryloxyphenoxypropanoates and cyclohexanediones is not generally considered to be related to uptake, translocation or metabolism it was considered that any difference in sensitivity to these herbicides between ACCases from tolerant and susceptible species would lend further support to the view that ACCase is the target enzyme. Consequently we examined the effects of haloxyfop and tralkoxydim on the enzyme from soybean (a tolerant dicotyledonous species) and the monocotyledonous species red fescue (*Festuca rubra*) (tolerant to both herbicides), tall fescue (*Festuca arundinacea*) (susceptible to both herbicides), wheat (tolerant to tralkoxydim, susceptible to haloxyfop) and maize (susceptible to both herbicides). The results obtained (I_{50} values) are compared (Table 3) with the efficacy of the two herbicides in reducing growth of the

Table 3. **Effect of haloxyfop and tralkoxydim on whole plant growth and ACCase.**

Species	GR$_{50}$ (μM)		I$_{50}$ (μM)	
	Haloxyfop	Tralkoxydim	Haloxyfop	Tralkoxydim
Maize	19	18	0.50	0.52
Wheat	83	>760	1.22	0.91
Tall fescue	133	225	0.94	0.40
Red fescue	1250	>6000	23.32	13.83
Soybean	>6000	>6000	138.50	516.72

various plants tested by 50% (GR$_{50}$ values) assessed 2 weeks after application.

These data revealed a high degree of parallelism between ACCase I$_{50}$ data and response at the whole plant level. Soybean was most tolerant to both herbicides at whole plant and enzyme levels whereas maize was the most susceptible. Similar findings have been recorded recently by other research groups using ACCase preparations from other monocotyledonous and dicotyledonous species (Burton *et al.*, 1987, 1989; Rendina and Felts, 1988; Kobek *et al.*, 1988; Walker *et al.*, 1988). Though monocotyledonous species are generally sensitive to both groups of compounds, *Festuca rubra* and *Poa annua* (Kobek *et al.*, 1988) have been identified as being tolerant. Data presented in Table 3 would indicate that the red fescue ACCase has reduced sensitivity to both haloxyfop and tralkoxydim which probably accounts for the cross-resistance of the species to the two classes of graminicides. We have not tested the ACCase of *Poa annua* but resistance to cyclohexanediones in the case of this species is apparently not due to differences in uptake (Struve *et al.*, 1987) or sensitivity of fatty acid biosynthesis (Kobek *et al.*, 1988).

The only notable discrepancy between ACCase I$_{50}$ values and whole plant response in our study was for tralkoxydim and wheat (Table 3). As expected, wheat was tolerant to tralkoxydim whereas its ACCase activity was sensitive to inhibition, leading us to conclude that tolerance of wheat to this cyclohexanedione has a basis in other phenomena. The selective control of wild oat (*Avena fatua*) in wheat by the aryloxy-phenoxypropanoate diclofop has been attributed to rapid detoxification in the crop by ring hydroxylation. This view is supported by the data presented in Table 4 which indicates that the diclofop sensitivity of the wheat ACCase is typical to that of other monocots.

Table 4. **Inhibition of acetyl CoA carboxylase activity of various species by diclofop.**

Species	Diclofop I$_{50}$ (nM)
Maize	80
Oat	101
Wheat	143
Avena fatua	440

OUTSTANDING QUESTIONS

Many fundamental questions concerning the interaction of aryloxy-phenoxypropanoates and cyclohexanediones with plant ACCases are still to be addressed. For example: What are the kinetics of inhibition? Do both classes of herbicide bind at the same site? What are the molecular determinants for inhibition? How do ACCases from susceptible and tolerant species differ? Partially purified (40 - 100 fold) ACCase may not be sufficiently pure to answer many of these questions. We have observed that an extract purified on a Sephacryl S-300 column to a specific activity of 300 - 400 nmol min^{-1} mg^{-1} will catalyse the carboxylation of short chain acetyl CoAs other than acetyl CoA. Though both herbicides inhibited carboxylation regardless of whether acetyl or propionyl CoA was used as substrate (data not shown) it is not clear whether ACCase has a fairly broad substrate specificity or whether a propionyl CoA carboxylase, also sensitive to inhibition by these herbicides, was also present in these preparations.

The covalently bound biotin in ACCase acts as the carboxyl carrier between an ATP-dependent biotin carboxylation site and a carboxyl-transferase site. In characterising the inhibition of barley chloroplast ACCase by cyclohexanedione herbicides Rendina and Felts (1988) found that inhibition patterns versus any of the substrates, MgATP, HCO_3^-, or acetyl CoA both fitted an equation for linear non-competitive inhibition. However, since the apparent inhibition constant was most sensitive to the level of acetyl CoA rather than MgATP or HCO_3^-, these workers suggested that cyclohexanediones interfere with the transcarboxylation step. Additional experiments are needed to further characterise the binding site(s) of both classes of herbicide and to determine whether the catalytic functions are directly affected or whether regulatory sites are involved.

There has been a certain amount of confusion in the literature as to the molecular size of native plant ACCases. Early attempts at purifying the plant enzyme yielded preparations apparently containing 3-6 biotin-containing sub-units, as detected by conjugation with avidin after SDS-PAGE. Such complexity may have resulted from a susceptibility of ACCase to degradation by endogenous proteases and inclusion of the protease inhibitor PMSF in extraction buffers subsequently gave rise to extracts consisting predominantly of high-molecular mass (240 kDa and 210 kDa) polypeptides (Egin-Buhler *et al.*, 1980). A further improvement resulted from a more rapid and extensive purification procedure utilising monomeric avidin affinity chromato-graphy. This gave a molecular mass of about 420 kDa for the native enzyme and about 210 kDa for the enzyme sub-unit (Egin-Buhler and Ebel, 1983). Use of the avidin affinity chromatography has also led to the recent purification of ACCases from seeds and leaf tissue of oilseed rape (Hellyer *et al.*, 1986) and from developing soybean seeds (Charles and Cherry, 1986) with polypeptide molecular masses in the range 220-240 kDa. Though the more recent results obtained using affinity chromatography might suggest that all plant ACCases are high-molecular-mass multifunctional proteins these latter authors failed to detect a 240 kDa sub-unit in similarly obtained preparations from soybean leaf tissue. They further argued that since ACCase in leaves is required in a number of different cell types to supply malonyl CoA for at least six different biosynthetic pathways, and ACCase activities from different tissues apparently differ in their molecular organisation, there is the possibility that these activities represent isoenzymes of ACCase.

Identification of bands conjugating with avidin or streptavidin on Western blots as sub-units of ACCase is also fraught with difficulties. Though it has frequently been quoted in the literature (see e.g., Hawke and Leech, 1987) that ACCase is the only biotin-containing protein in plants this is almost certainly not the case and propionyl CoA carboxylase, pyruvate carboxylase and methyl crotonyl CoA carboxylase activities have all been demonstrated in leaf extracts. Until such time as biotin-containing bands detected by the SDS-PAGE/Western blotting can be unequivocally assigned and the uncertainties concerning the sub-unit structure of leaf ACCases finally resolved one can only speculate as to whether the marked differences in sensitivity between ACCases of tolerant and susceptible species results from different molecular organisations of the enzymes from monocotyledonous and dicotyledonous species.

REFERENCES

Beyen, A.C. and Geelen, M.J.H. (1982) Short-term inhibition of fatty acid biosynthesis in isolated hepatocytes by mono-aromatic compounds. Toxicology **24**, 183-197.

Burton, J.D., Gronwald, J.W., Somers, D.A., Connelly, J.A., Gengenbach, B.G. and Wyse, D.L. (1987). Inhibition of plant acetyl coenzyme A Carboxylase by the herbicides sethoxydim and haloxyfop. Biochemical and Biophysical Research Communications **148**, 1039-1044.

Burton, J.D., Gronwald, J.W., Somers, D.A., Gengenbach, B.G. and Wyse, D.L. (1989). Inhibition of corn acetyl CoA carboxylase by cyclohexanedione and aryloxyphenoxypropanoate herbicides. Pesticide Biochemistry and Physiology **34**, 76-85.

Campbell, J.R. and Penner, D. (1981). Metabolism of BAS9052 by susceptible and tolerant plant species. Proceedings of the North Central Weed Control Conference: East Lansing **36**, 33-34.

Charles, D.J. and Cherry, J.H. (1986). Purification and characterisation of acetyl CoA carboxylase from developing soybean seeds. Phytochemistry **25**, 1067-1071.

Egin-Buhler, B. and Ebel, J. (1983). Improved purification and further characterisation of acetyl CoA carboxylase from cultured cells of parsley (*Petroselinum hortense*). European Journal of Biochemistry **133**, 335-339.

Egin-Buhler, B., Loyal, R. and Ebel, J. (1980). Comparison of acetyl CoA carboxylase from parsley cell cultures and wheat germ. Archives of Biochemistry and Biophysics **203**, 90-100.

Gorecka, K., Shimabukuro, R.H. and Walsh, W.C. (1981). Aryl hydroxylation: a selective mechanism for the herbicides diclofop-methyl and clofop-isobutyl in gramineous species. Physiologia Plantarum **53**, 55-63.

Granzer, E. and Nahm, H. (1973). HCG004, a new highly potent hypolipidaemic drug. Arzneimittel Forschung 23, 1353-1354.

Gronwald, J.W. (1986). Effect of haloxyfop and haloxyfop-methyl on elongation and respiration of corn (*Zea mays*) and soybean (*Glycine max*) roots. Weed Science 34, 196-202.

Hatch, M.D. and Stumpf, P.K. (1962). Fat metabolism in higher plants. XVII. Metabolism of malonic acid and its α-substituted derivatives in plants. Plant Physiology 37, 121-126.

Hawke, J.C. and Leech, R.M. (1987). Acetyl CoA carboxylase activity in normally developing wheat leaves. Planta 171, 489-495.

Hellyer, A., Bambridge, H.E. and Slabas, A.R. (1986). Plant acetyl CoA carboxylase. Biochemical Society Transactions 14, 565-568.

Hendley, P., Dicks, J.W., Monaco, T.J., Slyfield, S.M., Tumman, O.J. and Barrett, J.C. (1985). Translocation and metabolism of pyridinyl-oxyphenoxypropionate herbicides in rhizomatous quackgrass (*Agropyron repens*). Weed Science 33, 11-16.

Hoppe, H.H. (1980) Veranderungen der membranpermeabilitat, des Kohlenhydratgehaltes, des Lipidgehaltes und der Lipidzusammensetzung in Keimwurzelspitzen von *Zea mays* L. nach behandlung mit diclofop-methyl. Zeitschrift für Pflanzenphysiologie 100, 415-426.

Hoppe, H.H. (1981). Einfluss von diclofop-methyl auf die Protein-Nukleinsaure-und Lipidbiosynthese der Keimwurzelspitzen von *Zea mays* L. Zeitschrift für Pflanzenphysiologie 102, 189-197.

Hoppe, H.H. (1985). Differential effect of diclofop-methyl on fatty acid biosynthesis in leaves of sensitive and tolerant plant species. Pesticide Biochemistry and Physiology 23, 297-308.

Hoppe, H.H. and Zacher, H. (1982). Hemmung der fettsaeure-biosynthese durch diclofop-methyl in Keimwurzelspitzen von *Zea mays*. Zeitschrift für Pflanzenphysiologie 106, 287-298.

Hoppe, H.H. and Zacher, H. (1985). Inhibition of fatty acid biosynthesis in isolated bean and maize chloroplasts by herbicidal phenoxy-propionic acid derivatives and structurally related compounds. Pesticide Biochemistry and Physiology 24, 298-305.

Hosaka, H. and Takagi, M. (1987). Biochemical effects of sethoxydim in excised root tips of corn (*Zea mays*). Weed Science 35, 612-618.

Kobek, K., Focke, M. and Lichtenthaler, H.K. (1988). Fatty acid biosynthesis and acetyl-CoA carboxylase as a target of diclofop, fenoxaprop and other aryloxyphenoxypropionic acid herbicides. Zeitschrift für Naturforschung 43c, 47-54.

Kobek, K., Focke, M., Lichtenthaler, H.K., Retzlaff, G. and Wurzer, B. (1988). Inhibition of fatty acid biosynthesis in isolated chloroplasts of cycloxydim and other cyclohexane-1,3-diones. Physiologia Plantarum 72, 492-498.

Lichtenthaler, H.K. and Meier, D. (1984). Inhibition by sethoxydim of chloroplast biogenesis, development and replication in barley seedlings. Zeitschrift für Naturforschung. **39c**, 115–122.

Lichtenthaler, H.K., Kobek, K. and Ishii, K. (1987). Inhibition by sethoxydim of pigment accumulation and fatty acid biosynthesis in chloroplasts of *Avena* seedlings. Zeitschrift für Naturforschung **42c**, 1275–1279.

Lucas, W.J., Wilson, C and Wright, J.P. (1984). Perturbation of *Chara* plasmalemma transport function by 2(4(2,4-dichlorophenoxy)phenoxy)-propionic acid. Plant Physiology **74**, 61–66.

Rendina, A.R. and Felts, J.M. (1988). Cyclohexanedione herbicides are selective and potent inhibitors of acetyl CoA carboxylase from grasses. Plant Physiology **86**, 983–986.

Rendina, A.R., Felts, J.M., Beaudoin, J.D., Carig-Kennard, A.C., Look, L.L., Paraskos, S.L. and Hagenah, J.A. (1988). Kinetic characterisation, stereoselectivity and species selectivity of the inhibition of plant acetyl CoA carboxylase by the aryloxyphenoxypropionic acid grass herbicides. Archives of Biochemistry and Biophysics **265**, 219–225.

Secor, J. and Cseke, C. (1988). Inhibition of acetyl CoA carboxylase activity by haloxyfop and tralkoxydim. Plant Physiology **86**, 10–12.

Secor, J., Cseke, C. and Owen, W.J. (1989). Aryloxyphenoxypropanoate and cyclohexanedione herbicides. Inhibition of acetyl coenzyme A carboxylase. In 'Biocatalysis in Agricultural Biotechnology' (J.R.Whitaker and P.E.Sonnet, eds), pp. 265–276. American Chemical Society, Washington, DC.

Shimabukuro, M.A., Shimabukuro, R.H., Nord, W.S. and Hoerauf, R.A. (1978). Physiological effects of methyl-2-(4-(2,4-dichlorophenoxy)phenoxy) propanoate on oat, wild oat and wheat. Pesticide Biochemistry and Physiology **8**, 199–207.

Struve, I., Golle, B. and Luttge, U. (1987). Sethoxydim uptake by leaf slices of sethoxydim resistant and sensitive grasses. Zeitschrift für Naturforschung **42c**, 279–282.

Stumpf, P.K. (1980). Biosynthesis of saturated and unsaturated fatty acids. In 'Biochemistry of Plants' (P.K.Stumpf and E.E.Conn, eds), Vol IV, pp. 177–204. Academic Press, New York.

Swisher, B.A. and Corbin, F.T. (1982). Behaviour of BAS-9052 OH in soybean (*Glycine max*) and johnson grass (*Sorghum halepense*) plant and cell cultures. Weed Science **30**, 640–650.

Veerasekaran, P., and Catchpole, A.H. (1982). Studies of the selectivity of alloxydim-sodium in plants. Pesticide Science **13**, 452–462.

Walker, K.A., Ridley, S.M., Lewis, T. and Harwood, J.L. (1988). Fluazifop, a grass-selective herbicide which inhibits acetyl-CoA carboxylase in sensitive plant species. Biochemical Journal **254**, 307–310.

Wright, J.P. and Shimabukuro, R.H. (1987). Effects of diclofop and diclofop-methyl on the membrane potentials of wheat and oat coleoptiles. Plant Physiology **85**, 188–193.

CYTOCHROME P$_{450}$ AND HERBICIDE RESISTANCE

O.T.G. JONES

Biochemistry Department, Bristol University, Bristol, BS8 1TD.

Microsomes derived from smooth endoplasmic reticulum of animal and plant cells contain a mixed function oxidase which oxygenates a variety of xenobiotics. In higher plants this can lead to the inactivation of a wide range of herbicides. NADPH and oxygen are required for the process together with cytochrome P$_{450}$ and, perhaps, cytochrome $\underline{b}_5$. Flavoprotein reductases transfer electrons from NAD(P)H to the cytochromes. The xenobiotic binds close to the haem of cytochrome P$_{450}$ and oxygen binds to the iron of its haem, where it is activated by reduction with electrons transferred from NADPH.

The oxygenase activity of plant microsomes varies with the age of the plant and can be increased by treating plants with inducers, such as phenobarbitone, *trans*-cinnamic acid, geraniol or the herbicides 2,4-D and monuron. Herbicide tolerance in some plants is associated with increased cytochrome P$_{450}$ content and oxygenase activity.

Cytochrome P$_{450}$ dependent metabolism of herbicides can be prevented by treatment with diagnostic reagents, including CO and 1-aminobenzotriazole, which form complexes with the haem of cytochrome P$_{450}$.

INTRODUCTION

The smooth endoplasmic reticulum of plant and animal cells contains an electron transport system distinct in composition and function from that found in mitochondria and chloroplasts. It is best characterised in animal cells, where it is known to have a variety of functions. It is concerned in the oxygenation reactions leading to the formation of steroid sex hormones, bile acids, active forms of vitamin D and of some hydroxylated fatty acids. The oxygenase reaction requires the participation of oxygen and of NADPH as electron donor. It can be summarised as follows, where RH is the substrate for oxygenation:-

$$NADPH + H^+ + O_2 + RH \xrightarrow{\text{oxygenase}} NADPH^+ + H_2O + ROH$$

Its characteristic is that one atom of the molecule of oxygen contributes to the production of the water molecule and the second to the oxygenated product. This gives rise to another name for this electron transport system, "a mixed function oxidase".

In some mammalian tissues, particularly liver, lung and kidney, but also the gastrointestinal tract and nasal epithelium, (Paine, 1981), this oxygenase serves an additional important function in contributing to the detoxification of foreign molecules (xenobiotics) particularly hydrophobic molecules. Such molecules may be drugs administered for therapeutic reasons or ingested by chance from the environment. The latter group includes solvents, hydrocarbons from petrol, cosmetics, and natural products present in normal food stuffs. Hydrophobic molecules present problems for the usual mammalian excretory mechanisms *via* bile or urine and have first to be made water soluble. This process takes place in two stages; Phase 1 is an activation step, commonly catalysed by an oxygenase, causing the formation of an -OH group, followed by Phase 2 in which a water soluble substituent is added, usually to the newly formed -OH group. Examples of such substituents are glucuronic acid, glycine, cysteine and glutathione. The conjugate is much more soluble than the parent xenobiotic and is safely excreted (Gibson and Skett, 1986, summarise this metabolism). If, for some reason, the xenobiotic is resistant to this detoxification process it is likely to accumulate in the body lipids. A somewhat similar system is present in higher plants, where there is no route for excretion to the outside; products are accumulated in the large plant vacuoles or become attached to structural polymers of the plant (Shimabukoro, 1985). Conjugation of the activated xenobiotic is important in plants and can involve reduced glutathione (GSH), glucose and amino acids.

COMPONENTS OF THE ENDOPLASMIC RETICULUM MONOOXYGENASE SYSTEM

Cytochromes P_{450}

In plants and animals the monooxygenase systems are similar. They involved a group of isoenzymes (perhaps about twenty in a mammal) called cytochromes P_{450} which are the product of a multigene family (Nebert and Gonzales, 1987). These cytochomes have been extensively studied in mammalian systems and as a consequence there is detailed information about their reactivity, primary sequence and genetics (see Ortiz de Montellano (1986) for collection of papers on these topics). Cytochromes P_{450} are hydrophobic proteins intimately associated with the smooth endoplasmic reticulum. The cytochrome has a protohaem prosthetic group and one of the axial ligands to the haem iron is the thiollate anion of cysteine : this has a pronounced effect upon the cytochrome absorption spectrum. The reduced cytochrome P_{450} binds CO to yield a complex with a strong absorption maximum at 450nm. This can be used to test for the presence of cytochrome P_{450} and sensitivity of xenobiotic metabolism to CO is an indicator, but not proof, that a cytochrome P_{450} is involved. CO and O_2 compete for attachment to the iron of haem in cytochrome c oxidase of mitochondria and in haemoglobin : it is probable that the same is true of cytochromes P_{450}.

Cytochromes P_{450} have a molecular weight around 56,000 daltons (see Gabriac *et al.* (1985) for Jerusalem artichoke and O'Keefe and Leto (1989) for avocado cytochrome P_{450}) and can exist in a variety of configurations. The haem iron may be Fe^{2+} or Fe^{3+} and may be either high or low spin. Different spin states are stabilised by the addition of different groups of molecules which interact with protein adjacent to the haem. Nucleophiles, such as amines favour the low spin state, which has an absorption maximum around 420nm whereas aromatic hydrocarbons favour the high spin state (absorption maximum around 390nm). The binding of xenobiotics to cytochrome P_{450} often produces spectra characteristic of high spin (Type 1) or low spin (Type 2) states, which are sketched in Figure 1. Such spectra are a useful indication of interaction of substrates with cytochrome P_{450}.

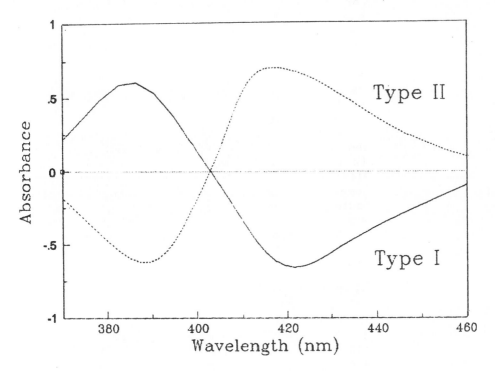

Fig. 1. <u>Diagram of difference spectra produced by addition of Type I or Type II substrates to a suspension of microsomes.</u> The baseline (at 0) is obtained when untreated microsomes are present in both reference and sample cuvette (modified from Jefcoate, 1978).

The way in which cytochrome P_{450} participates in the oxygenation of xenobiotics is shown in Figure 2. It is suggested that the xenobiotic, RH, binds to P_{450} protein adjacent to the haem, but not liganded to its irons, which must remain accessible to oxygen. P_{450} haem is then reduced by an electron from NADPH, delivered *via* a flavoprotein, NADPH-cytochrome P_{450} reductase. O_2 binds to the reduced P_{450}, giving another spectroscopic change (maximum at 440nm) and the complex receives a second electron delivered from either NADPH-cytochrome P_{450} reductase or from a quite different donor, cytochrome $\underline{b}_5$. Cytochrome $\underline{b}_5$ is reduced by the flavoprotein NADH-cytochrome $\underline{b}_5$ reductase. The redox state of the O_2 liganded to cytochrome P_{450} haem is equivalent to that of bound superoxide anion O_2^-. In this complex, oxygen-oxygen bonds are cleaved with formation of water leaving a reactive haem iron-oxene complex (ferryl iron) which oxygenates the adjacent xenobiotic, RH, to form ROH with the regeneration of oxidised cytochrome P_{450}. Cytochrome P_{450} can function as an oxygenase in the absence of reductases and other proteins if supplied with activated oxygen; a convenient donor is cumene hydroperoxide (Ortiz de Montellano, 1986). Hydrogen peroxide and superoxide anion are commonly side products of the activity of the monooxygenase system. One or two of the cytochrome P_{450} isoenzymes are preferentially involved in the metabolism of only one group of xenobiotics: others may be quite ineffective.

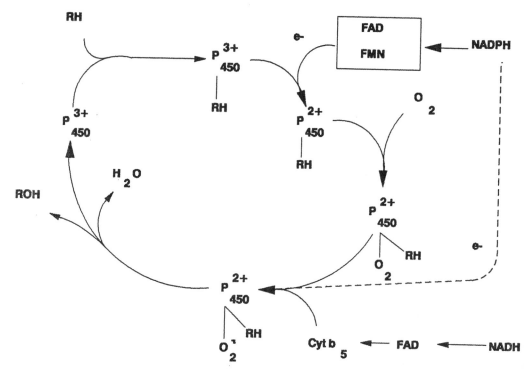

Fig. 2. <u>Representation of suggested (Estabrook *et al.* 1971) electron transport system involved in the metabolism of a xenobiotic (RH) by the cytochrome P_{450} system</u>. P_{450}^{2+} and P_{450}^{3+} represent the cytochrome with the haem iron in these two oxidation states. RH is believed to bind adjacent to haem iron and O_2 to the haem iron.

Flavoproteins of the monooxygenase system

NADPH-Cytochrome P_{450} reductase in animals and plants (M_r around 76,000 (Madhyastha and Coscia, 1979; Benveniste *et al.*, 1986) is an unusual reductase which contains 1 mole of both FAD and FMN per mole protein as prosthetic groups. FAD acts as a $2e^-$ acceptor from NADPH and FMN as a $1e^-$ donor to cytochrome P_{450}. Although this flavoprotein will participate in the transfer of up to four electrons, in steady state electron flow, it appears to cycle between incomplete reduction and incomplete oxidation (one electron and three electron reduction, Iyanagi *et al.*, 1981). The protein is firmly associated with the endoplasmic reticulum and is much less abundant than cytochrome P_{450}. One reductase is able to service several cytochromes P_{450}.

NADH cytochrome $\underline{b}_5$ reductase is also a membrane protein, of $M_r = 33,000$, but has only FAD as a prosthetic group. It accepts two electrons from NADPH and donates them in successive $1e^-$ steps to cytochrome $\underline{b}_5$ via the flavin semiquinone intermediate (Iyanagi *et al.*, 1984). It may function in plants in fatty acid desaturation which does not require the participation of cytochrome P_{450} (Smith *et al.*, 1990).

Cytochrome b$_5$

This cytochrome (M_r = 16,000) has spectroscopic properties characteristic of the low spin cytochromes. It is membrane associated and has an E_{m7} of around 0mV. This is significant because the E_{m7} of cytochromes P$_{450}$ is much lower (-330mV) and consequently it can only be reduced at a significant rate by cytochrome b$_5$ when it is complexed with the xenobiotic and O$_2$. This shifts the E_{m7} of cytochrome P$_{450}$ to a more positive value (around +50mV, Guengerich, 1983).

Cytochrome b$_5$ can also be reduced by NADPH-cytochrome P$_{450}$ reductase, but less effectively than by NADH cytochrome b$_5$ reductase. There is considerable evidence that cytochrome b$_5$ is not obligatory for the successful metabolism of all xenobiotics by the cytochrome P$_{450}$ system (Lu and Levin, 1974). Electrons from NADPH cytochrome P$_{450}$ reductase may donate at the two sites indicated in Figure 2. Thus in a reconstituted digitoxin hydroxylase assembled from purified cytochrome P$_{450}$ of foxglove together with its NADPH-cytochrome P$_{450}$ reductase, the addition of cytochrome b$_5$ and its reductase was unnecessary in order to sustain activity (Peterson and Seitz, 1988). There was no synergism when NADH was added as well as NADPH. In plants the role of cytochrome b$_5$ in xenobiotic metabolism is less established than it is in mammalian systems: indeed there is little evidence of its involvement.

NORMAL FUNCTIONS OF PLANT ENDOPLASMIC RETICULUM OXYGENASES

This topic has been well reviewed by Hendry (1986) who points out that what are xenobiotics for animals are synthesized by plants as secondary metabolites. Compounds in this category include coumarins, polyphenols, alkaloids and plant steroids; these are isolated in plant vacuoles and may serve as defence against herbivores. Also products of plant cytochrome P$_{450}$ are some hydroxy fatty acids intermediate in cutein synthesis and oxidised terpene metabolites, some of great importance to growth, such as the gibberellin family of plant hormones and abscisic acid. The 4-hydroxylation of *trans*-cinnamate (3-phenylpropenoic acid), catalysed by cytochrome P$_{450}$, is an essential step in the biosynthesis of phenolic derivatives such as anthocyanins and quinones as well as flavonoids and lignins. This activity is likely to be present in all plants.

The tissue distribution of plant cytochromes P$_{450}$ is not clear. They are difficult to detect spectroscopically in green tissue in the presence of strongly absorbing carotenoids and chlorophylls, although enzymic activities have been found in such samples (see the early report by Frear *et al.*, 1969). Most accurate measurements have been made on etiolated seedlings or from bulbs or tubers and buds or fruit, where they are relatively abundant, although present at less than 10% of their concentration in liver. However, soon after germination the cytochrome P$_{450}$ content of both dark-grown and light-grown mung beans declines rapidly (Hendry *et al.*, 1981), (Fig. 3).

ACTION OF OXYGENASES ON XENOBIOTICS

The microsomal oxygenases of plants and animals catalyse a wide variety of reactions, some of which are listed in Figure 4. All of these reactions will have importance in the metabolism of herbicides. The N-

217

dealkylation and 0-demethylation (reactions 6 and 7 in Figure 4) almost certainly proceed following initial oxygenation of the alkyl group.

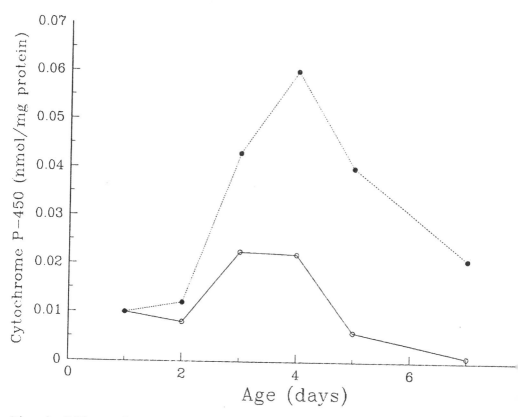

Fig. 3. <u>Effect of age on cytochrome</u> P_{450} <u>content of microsomes from mung beans germinated in the dark</u> (●) <u>or light</u> (o); <u>re-drawn from Hendry</u> <u>*et al.* (1981)</u>.

INDUCTION OF OXYGENASE ACTIVITY

The activity of the P_{450} system in the liver is not fixed: it is dependent upon age, sex, stress and most importantly, to exposure to inducing molecules. These last are substrates for the P_{450} system and treatment of an animal with an inducing substrate will cause a great increase in the synthesis of mRNA for one or more isozymes of cytochrome P_{450}, leading to increased (sometimes more than a hundred-fold) content of specific isozymes of cytochrome P_{450}. In plants a similar induction occurs and some of the known inducers are listed in Table 1. Their effects are not so pronounced as those seen in liver cells, but show clearly that the plant can respond to treatment with foreign molecules, including herbicides, and to tissue damage and ageing. It is possible that metal ions produce a non-specific damaging effect. Phenobarbital is an excellent inducer of liver as well as plant cytochrome P_{450}. Some 24 h of incubation with the inducer is necessary to produce the enhanced levels of plant cytochrome P_{450} and there is evidence of specificity: phenobarbital is a better inducer of laurathydroxylase in Jerusalem artichoke tubers than Mn^{2+} ions, but the reverse is true of induction of

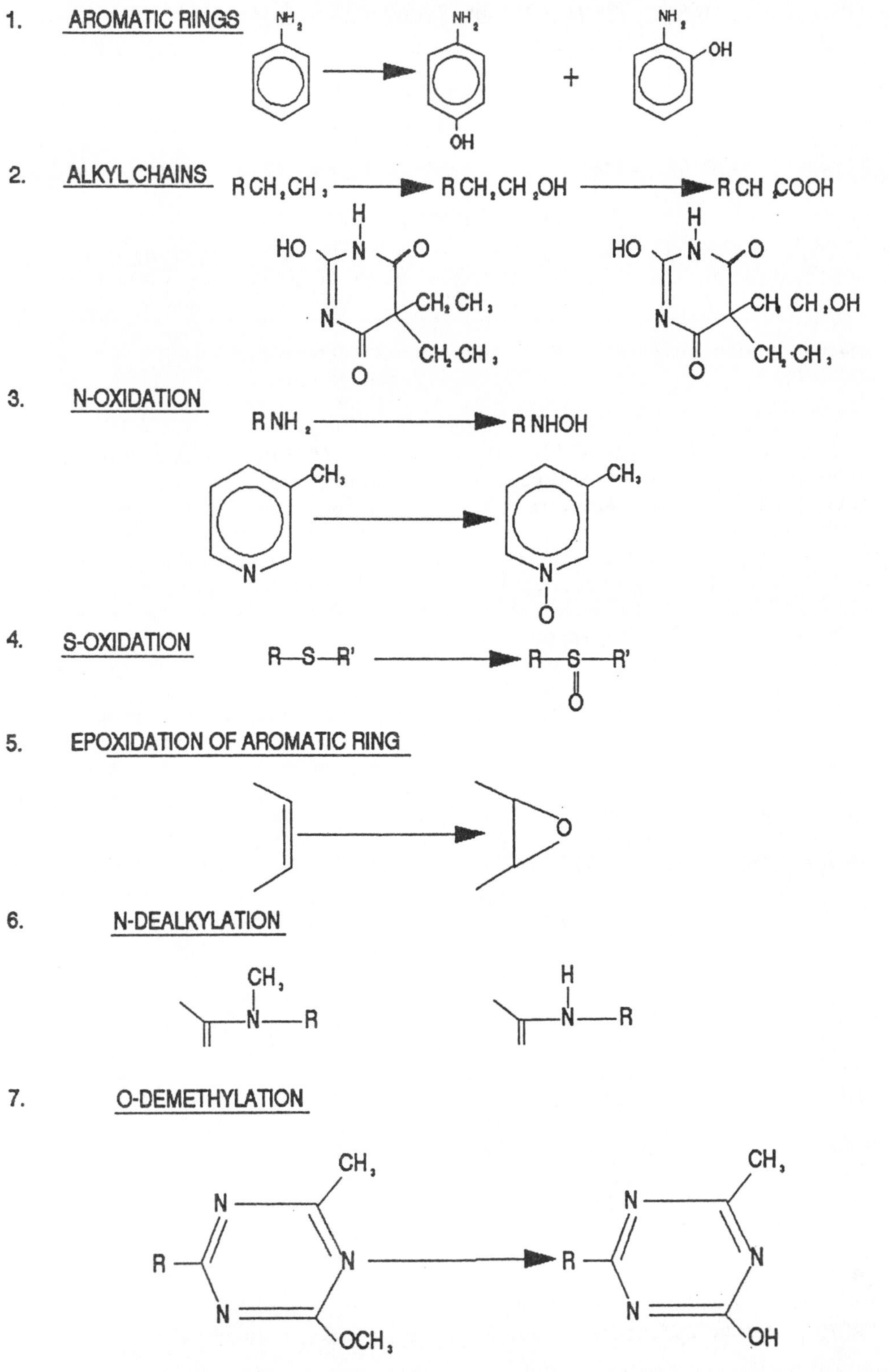

1. **AROMATIC RINGS**

2. **ALKYL CHAINS** $R CH_2 CH_3 \longrightarrow R CH_2 CH_2 OH \longrightarrow R CH_2 COOH$

3. **N-OXIDATION** $R NH_2 \longrightarrow R NHOH$

4. **S-OXIDATION** $R—S—R' \longrightarrow R—S—R'$

5. **EPOXIDATION OF AROMATIC RING**

6. **N-DEALKYLATION**

7. **O-DEMETHYLATION**

Fig. 4. Some reactions catalysed by microsomal mixed function oxidase systems.

Table 1. Inducers of cytochromes P_{450} on higher plants.

Inducer	Reference
Tissue damage to tubes	Benveniste *et al.* (1977) ; Rich and Lamb (1977)
Mn^{2+}	Reichhart *et al.* (1980)
Ethanol	Reichhart *et al.* (1980)
Phenobarbital	Reichhart *et al.* (1980)
Monuron	Reichhart *et al.* (1980)
2,4-D	Reichhart *et al.* (1980)
trans-Cinnamic acid	Hendry and Jones (1984)
geraniol	Hendry and Jones (1984)

the enzyme cinnamic acid 4-hydroxylase (Salaun *et al.*, 1981). The hypolipidaemic drug clofibrate induced lauric acid-hydroxylase activity some 20-fold in *Vicia sativa*, broad bean and soya bean seedlings, with much less effect on total cytochrome P_{450} or cinnamate 4-hydroxylase. In contrast, in Jerusalem artichoke clofibrate induced a 7-fold increase in cinnamic acid 4-hydroxylase with no effect on lauric acid or hydroxylase (Salaun *et al.*, 1986). The effects of 2,4-D and clofibrate (ethyl 2-[4-chlorophenoxy]-2-methylpropanoate) are very similar with respect to the microsomal hydroxylation of tubers and legumes and they are also similar in structure. They may owe their action to auxin-like effects. Hendry and Jones (1984) noted that the absorption maximum of the cytochromes P_{450} induced by geraniol (at 449nm) differed from that induced by cinnamate (451nm). In summary, there is good evidence that a family of cytochromes P_{450} exists in higher plants, one or more of which respond specifically to an inducing stimulus.

METABOLISM OF HERBICIDES

This topic has been reviewed in some detail by Hatzios and Penner (1982), Cole (1983), Shimabukuro (1985) and O'Keefe *et al.* (1987) and only some illustrative examples will be discussed here, where there is clear evidence for the involvement of the cytochrome P_{450} system. The reactions involved in the metabolism include aliphatic hydroxylation, O-dealklylation, N-dealkylation, ring hydroxylation, N-oxide and S-oxide formation. In some cases these modifications (Phase I) are clearly followed by conjugation with a sugar or amino acid (Phase 2).

Frear *et al.* (1969) showed that a microsomal fraction from etiolated cotton contained a N-demethylase which required NADPH or NADH as a cofactor to demethylate monuron (Fig. 4). An unstable hydroxymethyl intermediate was formed which then lost formaldehyde. In the case of 2,4-D a microsomal system catalysed its 4-hydroxylation through a "NIH shift" (Makeev *et al.*, 1977). The 4-Cl substituent was displaced to the 3- or 5- position (Fig. 5). This activity was found in pea and cucumber microsomes, required NADPH and was inhibited by CO in a light-sensitive fashion.

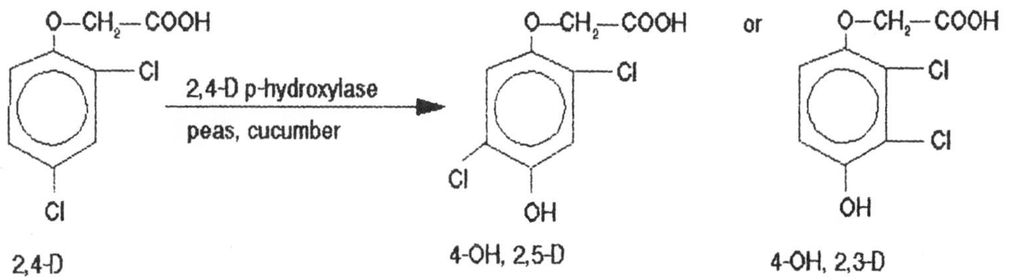

Fig. 5. Products of 2,4-D hydroxylation by plant microsomes, demonstrating the NIH shift.

The N-demethylation of chlorotoluron is carried out by wheat and *Alopecurus myosuroides* by different pathways. In wheat the 4-methyl group is first oxidised to hydroxymethyl before the N-demethylation. In both species the 4-methyl is eventually oxidised to 4-COOH (Ryan *et al.*, 1981).

Zimmerlin and Durst (these Proceedings) have shown that a microsomal preparation from wheat catalysed the aryl hydroxylation of diclofop. The reaction required O_2 and NADPH and was blocked by CO and other inhibitors of cytochrome P_{450}.

A comprehensive list of herbicides which are believed to be metabolized by plant cytochrome P_{450} systems is given by Hatzios and Penner (1982) and Owen (1987). They include metamitron, dicamba, atrazine, trifluralin, prometryne and many others. The evidence to confirm the involvement of a mixed function oxygenase is not always complete.

HERBICIDE-TOLERANT PLANTS

Repeated treatment of crops with a herbicide can lead to the selection of weed species which are resistant to that herbicide (these Proceedings). This process appears to be one of selection rather than induction as described previously in this article. We have examined the chlorotoluron-tolerant strain of *Alopecurus myosuroides* (Moss, 1987) where there is evidence for the involvement of cytochrome P_{450} in the resistance process. We found that suspension cell cultures derived from chlorotoluron-resistant strains of *A. myosuroides* retained the capacity to degrade chlorotoluron and that microsomes prepared from them were also active, if supplied with NADPH. Further, these microsomes had enhanced content of cytochrome P_{450} when compared with microsomes from susceptible strains (Caseley *et al.*, 1990). The microsomes from the chlorotoluron-tolerant strain gave a Type I binding curve when treated with chlorotoluron: this was not observed when microsomes from a susceptible strain were treated similarly. Our evidence suggests that the enhanced metabolism of chlorotoluron found in the resistant strains arises from an increased content of cytochrome P_{450}. It is noteworthy that 1-aminobenzotriazole , an inhibitor of cytochrome P_{450} function, diminishes the chlorotoluron-degradative activity of the tolerant *A. myosuroides* in cell cultures (Kemp and Caseley, 1987) and in excised leaves and shoots of resistant *Galium aparine, Veronica persica, Bromus sterilis* and wheat (Gonneau *et al.*, 1987).

INHIBITORS OF CYTOCHROME P$_{450}$ FUNCTION

The compound 1-aminobenzotriazole and some of its analogues inhibit the activity of cytochromes P$_{450}$ by alkylating its protohaem prosthetic group (Ortiz de Montellano *et al.*, 1984). This suicide substrate inhibition is useful, both as a test of cytochrome P$_{450}$ involvement in a metabolic process and possibly as a way of maintaining the effectiveness of herbicides on resistant weed species.

Carbon monoxide inhibits by attaching to the haem iron of cytochrome P$_{450}$ in competition with oxygen. This inhibition is sensitive to light. Light-reversal of CO-inhibition has the action spectrum of the CO-complex of cytochrome P$_{450}$ and this is considered to be a convincing way to demonstrate the involvement of cytochrome P$_{450}$ in any process. Other haem proteins (such as peroxidases) which also are CO-sensitive and may be involved in herbicide metabolism can be distinguished by their different action spectrum.

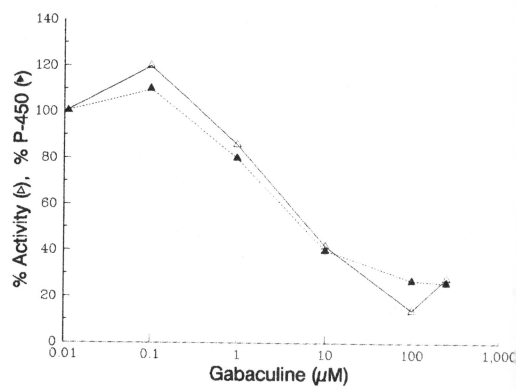

Fig. 6. <u>Effect of gabaculine on cytochrome P$_{450}$ induction (▲) and *trans*-cinnamic acid 4-hydroxylase (Δ) in artichoke tissues aged in the dark.</u> Tissues were induced by treatment with 25 mM MnCl$_2$ in the dark for 48 h; gabaculine, an inhibitor of the synthesis of 5-aminolaevulinate, a haem and chlorophyll precursor, was added to induction medium at the concentrations shown. (Re-drawn from Werck-Reichhart *et al.*, 1988).

The cytochrome P_{450} content of plant tissues can be modified by using gabaculine, which is an inhibitor of pyridoxal phosphate-linked transaminases (Rando, 1977). Gabaculine inhibits chlorophyll biosynthesis because it is a potent inhibitor of the higher plant pathway for 5-aminolevulinate (ALA) biosynthesis which uses glutamate as a substrate (the C5 pathway). Since ALA is required in plants for the synthesis of both haem and chlorophylls gabaculine also blocked cytochrome P_{450} induction (Werck-Reichhart *et al.*, 1988).

The content of cytochrome P_{450} in induced artichoke tubers declined to about one-fifth over 72 h of gabaculine treatment, which suggests that, as in animals, cytochrome P_{450} has a relatively short half-life. The half-life of liver cytochrome P_{450} haem is reported as 7-34 h (Gasser *et al.*, 1982; Sadano and Omura, 1983). The decline in cytochrome P_{450} in gabaculine-treated tubers was closely matched by decline in *trans*-cinnamic acid 4-hydroxylase activity (Fig. 6).

CONCLUDING REMARKS

Cytochrome P_{450} in higher plants is involved in the metabolism of many herbicides. This activity is important in influencing resistance to herbicides in both weeds and crop plants and thus in determining selectivity. The capacity of plants to degrade pesticides to non-toxic products is likely to be relevant in attempting to reduce pesticide residues in food and in the soil. It is possible to change the activity of the plant microsomal oxygenase detoxifying system by the application of specific inhibitors and inducers. The availability of methods for purifying plant cytochrome P_{450} will lead to information about the regulation of its synthesis and the possibility of transfer of herbicide metabolising genes into crop plants.

REFERENCES

Benveniste, I., Gabriac, B. and Durst, F. (1986). Purification and characterisation of NADPH cytochrome P_{450} reductase from higher plant microsomes. Biochemical Journal **235**, 365-373.

Benveniste I., Salaun, J.P. and Durst, F. (1977). Wounding of induced cinnamic acid hydroxylase in Jerusalem artichoke tuber. Phytochemistry **16**, 69-73.

Caseley, J.C., Kueh, J., Jones, O.T.G., Hedden, P. and Cross, A.R. (1990). Mechanism of chlortoluron resistance in *Alopecurus myosuroides*. Abstracts 7th International Congress of Pesticide Chemistry **1**, (H. Frehse, E. Kesseler-Schmitz and S. Conway, eds), p. 417. IUPAC, Hamburg, Germany.

Cole, D.J. (1983). Oxidation of xenobiotics in plants. In "Progress in Pesticide Biochemistry and Toxicology", Vol 3. (D.H. Hutson and T.R. Roberts, eds.), pp. 199-254. John Wiley & Sons, Chichester.

Estabrook, R.W., Hildebrandt, A.G., Baron, G., Netter, N.J. and Leibman, K. (1971). A new spectral intermediate associated with cytochrome P_{450} function in liver microsomes. Biochemical and Biophysical Research Communications **42**, 132-139.

Frear, D.S., Swanson, H.R. and Tanaka, F. S. (1969). N-demethylation of substituted 3-(phenyl)-1-methyl urea: isolation and characterisation of microsomal mixed function oxidase from cotton. Phytochemistry **8**, 2157-2169.

Gasser, R., Hauri, H.P. and Meyer, V.A. (1982). The turnover of cytochrome P_{450b}. FEBS Letters **147**, 239-242.

Gibson, G.G. and Skett, P. (1986). "Introduction to Drug Metabolism". Chapman & Hall, London.

Gabriac, B., Benveniste, I. and Durst F. (1985). Isolation and characterisation of cytochrome P_{450} from higher plants. Comptes Rendus des Seances de l'Academie des Sciences Serie 3, Sciences de la Vie. **301**, 753-758.

Gonneau, M., Pasquette, B., Cabanne, F., Scalla, R. and Loughman, B.C. (1987). Transformation of phenoxyacetic acid and chlorotoluron in wheat, barren brome, cleavers and speedwell: Effects of an inactivator of monooxygenases. British Crop Protection Conference - Weeds 329-336.

Guengerich, F.P. (1983). Oxidation-reduction properties of rat liver cytochromes P_{450} and NADPH-cytochrome P_{450} reductase related to catalysis in reconstituted systems. Biochemistry **22**, 2811-18.

Hatzios K.K. and Penner D. (1982). In "Metabolism of Herbicides in Higher Plants". CEPCO Division, Burgess Publ. Co., Minneapolis. pp. 15-122.

Hendry, G.A.F. (1986). Why do plants have cytochrome P_{450}? Detoxidation versus defense. New Phytology **102**, 239-247.

Hendry, G.A.F. and Jones, O.T.G. (1984). Induction of cytochrome P_{450} in intact mung beans. New Phytology **96**, 153-159.

Hendry, G.A.F., Houghton, J.D. and Jones, O.T.G. (1981). The cytochromes in microsomal fractions of germinating mung beans. Biochemical Journal **194**, 743-751.

Iyanagi, T., Makino, R. and Koichi Anan, F. (1981). Studies on the microsomal mixed function oxidase system: Mechanism of action of hepatic NADPH cytochrome P_{450} reductase. Biochemistry **20**, 1722-1730.

Iyanagi, T., Watanabe, S. and Anan, K.F. (1984). One-electron oxidation-reduction properties of hepatic NADH-cytochrome b5 reductase. Biochemistry **23**, 1418-1425.

Jefcoate, C.R. (1978). Measurement of substrate and inhibitor binding to microsomal cytochrome P_{450} by optical difference spectroscopy. Methods in Enzymology **52**, 258-279.

Kemp, M.S. and Caseley, J.C. (1987). Synergistic effects of 1-aminobenzotriazole of the phytotoxicity of chlorotoluron and isoproturon in a resistant strain of black grass (*Alopecurus myosuroides*). British Crop Protection Conference - Weeds 895-899.

Lu, A.Y.H. and Levin, W. (1974). Liver microsomal electron transport systems: Involvement of cytochrome b in the NADPH supported cytochrome P_{450} dependent hydroxylation of chlorobenzene. Biochemical and Biophysical Research Communications 61, 1348-1355.

Madhyastha, K.N. and Coscia, C.J. (1979). Detergent solubilisation of NADPH cytochrome c (P_{450}) reductase from the higher plant Cathranthus roseus. Journal of Biological Chemistry 254, 2419-2427.

Makeev, A.M., Maloveichuk, A.Y. and Chkanikov, D.I. (1977). Microsomal hydroxylation of 2,4-D in plants. Doklady Akademia Nauk SSSR. 233, 1222-1225.

Moss, S.R. (1987). Herbicide resistance in black-grass (Alopecurus myosuroides). British Crop Protection Conference - Weeds 879-886.

Nebert, D.W. and Gonzales, F.J. (1987). P_{450} genes. Structure, evolution and regulation. Annual Review of Biochemistry 56, 945-953.

O'Keefe, D.P. and Leto, K.J. (1989). Cytochrome P_{450} from the mesocarp of Avocado. Plant Physiology 89, 1141-1149.

O'Keefe, D.P., Romesser, J.A. and Leto, K.J. (1987). Plant and bacterial cytochromes P_{450}. Involvement in herbicide metabolism. In "Phytochemical Effects of Environmental Compounds". (J.A. Saunders, L. Kosak-Channing and E.E. Conns, eds), pp. 151-173. Plenum Press, New York.

Ortiz de Montellano, P.R. (1986). "Cytochrome P_{450}". Plenum Press, New York.

Ortiz de Montellano, P.R., Mathews, J.M. and Langry, K.C. (1984). Autocatalytic inactivation of cytochrome P_{450} and chloroperoxidase by 1-aminobenzotriazole and other aryne precursors. Tetrahedron 40, 511-519.

Owen, W.J. (1987). Herbicide detoxification & selectivity. British Crop Protection Conference - Weeds 309-317.

Paine, A.J. 1981. Hepatic cytochrome P_{450}. In "Essays in Biochemistry" (P.N. Campbell & R.D. Marshall, eds), pp. 85-126. Biochemical Society, London.

Petersen, M. and Seitz, H.U. (1988). Reconstitution of cytochrome P_{450}-dependent digitoxin 12B-hydroxylase from cell cultures of foxglove. Biochemical Journal 252, 537-43.

Rando, R.R., (1977). Mechanism of the irreversible inhibition of γ-aminobutyric acid -α-ketoglutaric acid transaminase by the neurotoxin gabaculine. Biochemistry 16, 4604-4610.

Reichhart D, Salaun, J.P., Benveniste I. and Durst, F. (1980). Time course of induction of cytochrome P_{450}, NADPH-cytochrome c reductase and cinnamic acid hydroxylase by phenobarbital, ethanol, herbicides and manganese in higher plant microsomes. Plant Physiology 66, 600-604.

Rich, P.R. and Lamb, C.J. (1977). Biophysical and enzymological studies upon the interaction of trans-cinnamic acid with higher plant microsomal cytochrome P_{450}. European Journal of Biochemistry **72**, 353-360.

Ryan, P.J., Gross, D., Owen, W.J. & Laanio, T.L. (1981). The metabolism of chlortoluron, diuron and CGA43057 in tolerant and susceptible plants. Pesticide Biochemistry and Physiology **16**, 213-221.

Sadano, H. and Omura, T. (1983). Reversible transfer of heme between different molecular species of microsome-bound cytochrome P_{450} in rat liver. Biochemical and Biophysical Research Communications **116**, 1013-1019.

Salaun, J.P., Simon, A. and Durst, F. (1986). Specific induction of lauric acid w-hydroxylase by Clofibrate, diethylhexyl-phthalate and 2,4-Dichloro-phenoxyacetic acid in higher plants. Lipids **21**, 776-779.

Salaun, J.P., Benveniste, I., Reichhard, D. and Durst, F. (1981). Induction and specificity of a cytochrome P_{450} dependent laurate in-chain hydroxylase from higher plant microsomes. European Journal of Biochemistry **119**, 651-655.

Shimabukuro, R.H. (1985). Detoxification of herbicides. In "Weed Physiology, Vol.2, Herbicide Physiology". (S.O. Duke, ed.), pp. 215-240. CRC Press, Boca Raton.

Smith, M.A., Cross, A.R., Jones, O.T.G., Griffiths, W.T., Stymne, S. and Stobart, K. (1990). Electron transport components of the 1-acyl-2-oleoyl-*sn*-glycero-3-phosphocholine Δ^{12}-desaturase [Δ^{12}-desaturase] in microsomal preparations from developing safflower (*Carthamus tinctorius* L.) cotyledons. Biochemical Journal **272**, 23-29.

Werck-Reichhart, D., Jones, O.T.G. and Durst, F. (1988). Haem synthesis during cytochrome P_{450} induction in higher plants. Biochemical Journal **249**, 473-480.

GLUTATHIONE AND GLUCOSIDE CONJUGATION IN HERBICIDE SELECTIVITY

Gerald L. Lamoureux, Richard H. Shimabukuro and D. Stuart Frear

Biosciences Research Laboratory, United States Department of Agriculture, Agricultural Research Service, P.O. Box 5674, State University Station, Fargo, ND, 58105, USA

The conjugation of herbicides with glutathione or glucose are frequently species specific reactions that result in herbicide detoxification. Therefore, glutathione and glucoside conjugation play an important role in herbicide detoxification and selectivity. Phase 1 activation reactions are sometimes necessary before glutathione or glucoside conjugation can occur. In these cases, the Phase 1 reaction rather than conjugation may be responsible for herbicide selectivity. This appears to be particularly true in the metabolism of herbicides to O-glucoside conjugates. The glutathione-S-transferases and glucosyl-transferases that catalyze these conjugation reactions, the role that these reactions play in the selectivity of specific herbicides, factors that affect these reactions, and the secondary metabolism of glutathione and glucoside conjugates are reviewed.

INTRODUCTION

Differences in herbicide metabolism between plant biotypes is one of the most common factors responsible for herbicide selectivity, and any consideration of the mechanisms involved in the development of herbicide resistance in weeds should take into account the potential role of herbicide metabolism. Listed in approximate order of occurrence are the most commonly observed reactions in the metabolism of herbicides in plants: oxidation, hydrolysis, glucoside conjugation, glutathione (GSH) conjugation, malonyl conjugation, amino acid conjugation, and reduction (Lamoureux and Rusness, 1986b). The order of importance of these reactions to the selective detoxification of herbicides is difficult to establish because these reactions frequently occur sequentially, i.e., oxidation (hydroxylation) followed by glucoside conjugation. Oxidative metabolism appears to be the most important reaction involved in herbicide selectivity, and is discussed by Jones (these Proceedings). Following oxidative metabolism, GSH conjugation and glucoside conjugation are clearly next in importance in the metabolic detoxification and selectivity of herbicides.

GLUTATHIONE AND HOMOGLUTATHIONE CONJUGATION

Requirements for glutathione conjugation of a herbicide

In order to undergo GSH conjugation, a herbicide must contain an electrophilic site that is susceptible to a nucelophilic attack by GSH or it must undergo an activation reaction to produce such an electrophilic site. The electrophilic site is usually an alkyl or aryl carbon, but it also can be a carbonyl carbon or a hetero atom. Glutathione conjugation can occur by nucleophilic displacement or by nucleophilic addition, Reactions 1 and 2.

ATRAZINE

TRIDIPHANE

Many GSH conjugates of herbicides are formed by the nucleophilic displacement of a halogen or an alkyl sulfoxide (Lamoureux and Rusness, 1986b). Although some GSH conjugation reactions occur non-enzymatically, especially with highly activated substrates (Frear and Swanson, 1970), the presence of an appropriate glutathione-S-transferase (GST) (E.C. 2.5.1.18) is usually required (Lamoureux and Rusness, 1989b). Even in those cases where an enzyme is not necessary, the reaction may proceed much faster in the presence of a GST. The concentration of GSH in the plant must be sufficient to allow the reaction to proceed to completion at a reasonable rate. For these reasons, both the level of GSH and the nature of the GST enzymes should be taken into account in considering the role of GSH conjugation in herbicide metabolism and selectivity.

Gluthathione and homoglutathione in plants

Glutathione (γ-glutamylcysteinylglycine, GSH) is the major form of free thiol in most plant species, including some leguminous species such as pea and peanut; however, homoglutathione (γ-glutamylcysteinyl-β-alanine, HomoGSH) is the major form of free thiol in other leguminous species such as soybean and mung bean (Rennenberg and Lamoureux, 1990). The signficance of HomoGSH instead of GSH in species such as soybean and mung bean is not understood, but HomoGSH may serve the same functions as GSH. For example, plant species that contain HomoGSH form HomoGSH conjugates instead of GSH conjugates (Lamoureux and Rusness, 1986b). There is evidence that GSH and HomoGSH can be used interchangeably with GSTs from corn and soybean, but there is no specific information in the literature regarding this (Lamoureux and Rusness, 1989b). Glutathione and HomoGSH are synthesized in the chloroplast and the cytoplasm by the same sequence of reactions utilized in the synthesis of GSH in mammals (Rennenberg and Lamoureux, 1990). Within the plant cell, GSH levels appear to be higher in the chloroplast than in the cytoplast or vacuole. The chloroplasts may contain up to 76% of the GSH (Rennenberg, 1982). Glutathione exists primarily in the thiol form and the levels tend to be higher in the shoots than in the roots; however, the highest levels of glutathione occur in the seed where it exists primarily as the disulfide dimer (GSSG) (Rennenberg and Lamoureux, 1990). The concentration of GSH and HomoGSH ranged from 180 to 590 nmol g fr. wt^{-1} in the seedlings of 12 plant species (Breaux *et al.*, 1987). In some organisms, GSH levels vary depending upon the age and state of development (Lamoureux and Bakke, 1984). Large differences in GSH levels were observed in the seed and developing seedling of *Setaria faberii* (giant foxtail) and corn, Table 1. Giant foxtail is susceptible and corn is tolerant to chloroacetamide herbicides during the early seedling stage and a relationship has been observed between GSH levels in 12 plant species and tolerance to the chloroacetamide herbicide, acetachlor (Breaux *et al.*, 1987). Since these herbicides are metabolized by GSH conjugation, GSH levels could be an important factor in tolerance.

Table 1. <u>Amounts of GSH (in nmol g fr. wt^{-1}) in corn and giant foxtail</u>.

Tissue	Corn (resistant)	Giant foxtail (susceptible)
Dry seed	380	38
Leaves, 2 cm long	1800	380
Leaves, 5 cm long	600	300
Leaves, 10 cm long	300	300

(from Lamoureux *et al.*, 1990).

Glutathione–S–transferase enzymes

Although some herbicides react non-enzymatically with GSH at an appreciable rate, rapid GSH conjugation of many herbicides, such as atrazine (Reaction 1), occurs only in the presence of an appropriate GST. Glutathione-S-transferase activity has been detected in many crop, weed, and tree species (Frear and Swanson, 1970, 1973; Burkholder, 1977; Lamoureux and Rusness, 1980; Lamoureux *et al.*, 1981; Balabaskaran and Muniandy, 1984; Schroder, *et al.*, 1990). The GST enzymes from plants appear to be a family of isozymes with varying degrees of substrate specificity, comparable to the GST isozymes in animals. The first plant GST to be studied was isolated from corn and was characterised with atrazine as a substrate. This crude enzyme system had a pH optimum of 6.7 and apparent K_m values with atrazine and GSH of 3.7×10^{-5}M and 2.4×10^{-3}M, respectively. It was inhibited by atrazine analogues and by sulfobromophthalein (Frear and Swanson, 1970). The second plant GST to be studied was from pea. The GST from pea was characterised with fluorodifen as a substrate, had a pH optimum of approximately 9.4, and apparent Km values for fluoridifen and GSH of 1.2×10^{-5}M and 7.4×10^{-4}M, respectively. The enzyme from pea did not utilize atrazine as a substrate and differed from the GST in corn in response to inhibitors (Frear and Swanson, 1973). Recent studies have shown that multiple forms of GST activity are present in both corn and pea (Diesperger and Sandermann, 1979; Mozer *et al.*, 1983). Three GST isozymes from corn have been purified and studied intensively (Mozer *et al.*, 1983; Moore *et al.*, 1986; Shah *et al.*, 1986; Wiegand *et al.*, 1986; Edwards and Owen, 1988; Grove *et al.*, 1988; Wosnick *et al.*, 1989). However, corn may contain more than three GST isozymes and some differences in GST isozyme composition apparently exist among different corn lines (Shimabukuro *et al.*, 1978). The three GST isozymes from corn that have been studied have molecular weights of approximately 50 KDa and appear to be homodimeric and heterodimeric proteins (Mozer *et al.*, 1983; Moore *et al.*, 1986; Shah *et al.*, 1986; Edwards and Owen, 1988; Grove *et al.*, 1988; Wosnick *et al.*, 1989). GST I from corn is active with chloroacetamide herbicides, chloro-*s*-triazine herbicides, and 1-chloro-2,4-dinitrobenzene (CDNB), and it appears to be a homodimeric protein composed of two identical 25 KDa subunits (214 amino acid) (Mozer *et al.*, 1983; Shah *et al.*, 1986; Wosnick *et al.*, 1989). Treatment of corn with herbicide safeners causes a 4- to 9-fold increase in the steady-state level of mRNA that encodes for GST. This results in a 2-fold increase in the level of GST I and also causes the production of a second GST (GST II) (Wiegand *et al.*, 1986; Edwards and Owen, 1988). GST II is heterodimeric and consists of the subunit present in GST I and a second subunit with a lower molecular weight (Mozer *et al.*, 1983) . GST I and II are equally effective with alachlor as a substrate, but GST II is less effective with CDNB (Mozer *et al.*, 1983). Less information has been published regarding the substrate specificity of GST III. The amino acid sequences of GST I and GST III were determined. Approximately 45% conservation of homology was observed, but differences in sequence have been reported by different laboratories (Moore *et al.*, 1986; Grove *et al.*, 1988). Some conservation of homology also was observed between GST from corn and rat (Shah *et al.*, 1986; Grove *et al.*, 1988; Wosnick *et al.*, 1989). An antiserum for GST has been prepared (Wiegand *et al.*, 1986; Edwards and Owen, 1988; Grove *et al.*, 1988), the cDNAs for GST I and GST III have been cloned, and GST activity for CDNB and atrazine has been introduced into yeast and *E. coli* (Mozer *et al.*, 1983; Moore *et al.*, 1986; Wiegand *et al.*, 1986; Grove *et al.*, 1988; Wosnick *et al.*, 1989). Thus, it may soon be possible to produce transgenic plants that are triazine and chloroacetamide resistant.

Glutathione-*S*-transferase activity can be induced in corn by certain herbicide safeners and this induction may also occur in sorghum and rice (Mozer *et al.*, 1983; Gronwald *et al.*, 1987; Komives and Dutka, 1989a). Safeners also may cause post-translational changes in GST and the substrate specificity of GST may change as a function of tissue differentiation (Edwards and Owen, 1988).

The GST isozymes are generally considered to be soluble proteins, but microsomal GST activity has been detected in both pea and corn (Diesperger and Sandermann, 1979; Komives *et al.*, 1985). The soluble GST isozymes are major proteins in corn. Two laboratories, using different methods, have reported that GST isozymes account for approximately 1% of the soluble protein in corn (Mozer *et al.*, 1983; Edwards and Owen, 1988). Unfortunately, there is little information regarding the natural function of these enzymes. The GST from pea was reported to catalyze the GSH conjugation of cinnamic acid (Diesperger and Sandermann, 1979) and it was also reported to have hydroperoxidase activity with linoleic acid hydroperoxide (Williamson and Beverley, 1987). The GST isozymes may have two distinct functions: the repair of peroxidative damage to membrane lipids that occurs during photosynthesis and the GSH conjugation/detoxification of toxins. If this is the case, the GST isozymes could play an extremely important role in protecting plants from herbicides that inhibit or catalyze the destruction of membrane lipids and that are also metabolized by GSH conjugation.

GSH and HomoGSH in herbicide metabolism and selectivity

S-Triazine herbicide GSH/HomoGSH conjugation and selectivity

The 2-chloro-s-triazine herbicides inhibit electron transport in Photosystem II (Steinback et al., 1982) and are used commonly to control weeds in crops such as corn, sorghum and sugarcane (Best, 1983). Atrazine, the most commonly used of these herbicides, is metabolized in plants by three different routes: non-enzymatic hydrolysis of the 2-chloro group, N-dealkylation, and GSH conjugation (Shimabukuro et al., 1978). The selectivity of the 2-chloro-s-triazines between resistant crop and susceptible weed species is usually related to the rate of detoxification (Shimabukuro et al., 1978), but resistance due to an altered photosynthetic electron transport system has been detected in some weed species (Steinback et al., 1982; De Prado et al., 1989).

The primary route of metabolism of the 2-chloro-s-triazine herbicides in most resistant crop species is conjugation with GSH, Reaction 1, (Lamoureux et al., 1972). The GST that catalyzes this reaction is present in tolerant sorghum, corn, sugarcane, Sorghum halepense and Sudan grass, but is absent in susceptible pea, oats, wheat, barley and Amaranthus retroflexus (Frear and Swanson, 1970). Studies with an atrazine-tolerant and an atrazine-susceptible inbred corn line showed that atrazine was metabolized by GSH conjugation in the tolerant, but not in the susceptible corn line. Inhibition of photosynthesis was overcome by metabolism of atrazine in the tolerant, but not in the susceptible corn line. Non-enzymatic hydroxylation of atrazine, which occured in both corn lines, was not sufficient to prevent atrazine toxicity in the susceptible line. Glutathione-S-transferase activity in eight tolerant inbred corn lines ranged from 1.62 to 3.62 nmol GSH-atrazine mg protein^{-1} h^{-1}. Glutathione conjugation appeared to be the critical factor in tolerance (Shimabukuro et al., 1971). The same relationship between GSH conjugation and recovery of photosynthesis was observed in tolerant sorghum and susceptible pea (Shimabukuro et al., 1978) and in 53 grass species (Jensen et al., 1977). Recent investigations have confirmed the relationship between atrazine metabolism and selectivity, and metabolism by GSH conjugation appears to be the most common basis for tolerance to the 2-chloro-s-triazines in crop species (Weimer et al., 1988).

Resistance to the 2-chloro-S-triazine herbicides has developed in a number of weed species as a result of the repeated use of these herbicides. In many cases, the mechanism of resistance is due to an altered electron transport system in Photosystem II (Steinback et al., 1982; De Prado et al., 1989). However, resistance due to an elevated rate of GSH conjugation may also occur (Gronwald et al., 1989).

Chloroacetamide herbicide GSH/HomoGSH conjugation and selectivity

The 2-chloroacetamide herbicides, such as propachlor, are used to control weeds in crops such as corn, sorghum and soybean. Conjugation with GSH was shown to be the initial step in the metabolism of propachlor in both resistant and susceptible species (Lamoureux et al.,1971), Reaction 3.

$$GSH + \text{(aryl)}-N-\overset{Y}{\underset{O}{C}}-CH_2Cl \xrightarrow{GST} \text{(aryl)}-N-\overset{Y}{\underset{O}{C}}-CH_2\text{-}SG + HCl$$

PROPACHLOR

Additional studies with propachlor (Lamoureux and Rusness, 1989a), acetochlor (Breaux, 1987), metazachlor (Ezra et al., 1986), metolachlor (Khalifia and Lamoureux, 1989), dimetachlor, and pretilachlor (Blattmann et al., 1986) have failed to disclose the presence of other major reactions in the initial stages of chloroacetamide metabolism in various plant species. During the development of the chloroacetamide herbicides, resistant corn and soybean were shown to metabolize prototype chloroacetamide herbicides faster than susceptible oat and cucumber (Jaworski, 1969). More recently, a positive relationship was observed between resistance and metabolism of acetochlor in corn, sorghum, *Sorghum bicolor*, *Echinochloa crus-galli*, soybean, *Ipomoea lacunosa*, *Abutilon theophrasti* and alfalfa (Breaux, 1987). Differences in metabolism between tolerant and susceptible species were greatest at high treatment levels. In other studies, metolachlor was metabolized faster in tolerant corn than in susceptible *Cyperus esculentus* (Dixon and Stoller, 1982), and acetochlor was metabolized faster in resistant corn than in susceptible wheat (Jablonkai and Dutka, 1985). Propachlor and CDAA undergo rapid non-enzymatic conjugation with GSH, and based on early studies it was not clear whether the metabolism of these herbicides was enzymatic (Frear and Swanson, 1970). However, recent studies with alachlor and metolachlor have established the role of GST in the metabolism of the chloroacetamides (O'Connell et al., 1988). Therefore, a high titre of the necessary GST isozymes may be involved in the selectivity of these herbicides.

Chloroacetamide herbicides can injure crops such as corn and sorghum, and safeners can reduce this injury. Safeners effective against metolachlor injury to sorghum cause an elevation in GST activity and an increased rate of GSH conjugation with metolachlor (Gronwald, 1988). Metazachlor, which is toxic to corn, can be safened by dichlormid and BAS 145 138. Safening is associated with an accelerated rate of metazachlor metabolism that dramatically reduces the level of metazachlor in the developing leaves (Fuerst et al., 1988; Lamoureux et al., 1990). Although the mechanism of safener action is not fully understood, an increased rate of herbicide metabolism appears to be a major factor. This is consistent with the hypothesis that rapid GSH conjugation is a major factor responsible for tolerance to these herbicides.

Chlorfenprop-methyl GSH conjugation and selectivity

Chlorfenprop-methyl has some chemical similarity to the 2-chloroacetamide herbicides, but its biological activity resembles that

232

of the phenoxy-phenoxy herbicides. A cysteine conjugate, apparently formed by nucleophilic displacement of chlorine, was the major metabolite of chlorfenprop-methyl in wheat and oat, and hydrolysis was a minor route of metabolism (Collet and Pont, 1978; Pont and Collet, 1980). Although the direct enzymatic formation of cysteine conjugates has been reported (Rusness and Still, 1977), the first product of chlorfenprop-methyl metabolism actually may have been the GSH conjugate. The cysteine conjugate could have then been produced by catabolism of the GSH conjugate. The cysteine conjugate of chlorfenprop-methyl was formed more rapidly in resistant wheat than in susceptible oat. At low doses, these differences in metabolism were small, but at higher doses, metabolism was more extensive in resistant wheat. Injury and death were observed in susceptible oat treated with the higher doses, so it is uncertain whether metabolism is the basis of chlorfenprop-methyl selectivity (Pont and Collet, 1980).

Thiocarbamate herbicide GSH/HomoGSH conjugation and selectivity

The primary route of thiocarbamate herbicide metabolism in resistant and susceptible species appears to involve a transitory sulfoxide that is rapidly metabolized to a GSH or HomoGSH conjugate (Lay and Casida, 1976; Hubbell and Casida, 1977; Ikeda *et al.*, 1986; Unai *et al.*, 1986), Reaction 4.

There is some evidence that tolerance to thiocarbamate herbicides is related to the rate of metabolism. EPTC controls proso millet in corn and is metabolized approximately two times faster in the roots of corn than proso millet. Glutathione-S-transferase activity was 10x greater in the foliage and 3x greater in the roots of corn than in proso millet (Ezra and Stephenson, 1985). Orbencarb was metabolized faster to GSH/HomoGSH related metabolites in resistant soybean, wheat, and corn (19%, 12%, and 10%, respectively) than in susceptible *Digitaria sanguinalis* (3%); however, the interpretation of these data was complicated by the high percentage of orbencarb that remained: soybean (44%), wheat (32%), corn (37%), and *D. sanguinalis* (54%) (Ikeda *et al.*, 1986).

In corn, thiocarbamate injury can be reduced by safeners such as dichlormid, which elevates GSH and GST levels and accelerates the rate of EPTC metabolism in corn (Lay and Casida, 1976). These data are consistent with the hypothesis that thiocarbamate selectivity is related to the rate of metabolism. However, it must be recognized that oxidative enzyme activity also can be elevated by safeners and it is uncertain which is more important to thiocarbamate tolerance, oxidative enzymes or GST (Sweetser, 1985; Komives and Dutka, 1989b).

Metribuzin, dimethametryn, and terbutryne GSH/HomoGSH conjugation and selectivity

Metribuzin inhibits photosynthesis and is used to selectively control weeds in soybean, potatoes and tomatoes. Metribuzin is metabolized rapidly in tomato seedlings to a β-D-(malonyl)-N-glucoside conjugate (81% in 24 h), but it is metabolized more slowly and in a more complex fashion in soybean (Frear *et al.*, 1983b; Frear *et al.*, 1985), Reaction 5.

Only 40–50% metabolism was observed after 48 h in a resistant cultivar of soybean. From 12 to 20% of the residue was in the form of polar metabolites and 20 to 30% of the residue was non-extractable. A HomoGSH conjugate, presumably formed after oxidation, was the most abundant metabolite. Enzymes capable of catalyzing the HomoGSH conjugation of metribuzin were not isolated from soybean, but an NADPH/GSH-dependent microsomal system isolated from rat catalyzed the formation of the GSH conjugate. When the microsomal reaction was run in the absence of GSH, unstable oxidation products were formed. It was not determined whether a GST was required for GSH conjugation of metribuzin. Other soluble metabolites identified in the metabolism of metribuzin in soybean were the N-glucoside, an N-(malonyl)glucoside, and a malonylamido conjugate. These metabolites were observed in the following ratio: HomoGSH conjugate, 100; N-(malonyl)glucoside, 43; malonylamido, 10; and the N-glucoside, 7. Thus, at least three competing pathways are present in the metabolism of metribuzin to soluble metabolites in soybean, Reaction 5.

The selectivity of metribuzin between soybean and *Sesbania exaltata* has been attributed to differences in absorption, translocation, and metabolism (Hardgroder and Rogers, 1974); however, even some soybean cultivars are sensitive to metribuzin. An analysis of the uptake, translocation, and metabolism of metribuzin in tolerant and susceptible soybean cultivars indicated that tolerance was primarily due to metabolism (Mangeot *et al.*, 1979). The concentration of water-soluble metribuzin metabolites was 4.5x greater in the shoots of the tolerant cultivar than in the susceptible cultivar. Tetraploid soybean cells and plants were more tolerant to metribuzin than diploid cells and plants. Differences in tolerance were due primarily to metabolism, but some differences in uptake and translocation were also observed. After four days, water-soluble metabolites accounted for about 2% of the dose in the roots and shoots of the diploid plants, but they accounted for 85% and 54% of the radioactivity in the roots and shoots of the tetraploid plants (Abusteit *et al.*, 1985). The water-soluble metabolites were not identified, so it is uncertain whether

HomoGSH or glucoside conjugation was the predominant route of metabolism. Metabolism as the basis for tolerance to metribuzin has also been investigated in potatoes, barley, winter wheat and *Bromus tectorum* (Gawronski *et al.*, 1986; Devlin *et al.*, 1987; Gawronski *et al.*, 1987).

Dimethametryn, a methylthio- substituted *s*-triazine herbicide, is metabolized slowly in rice to a GSH conjugate that is subsequently metabolized to other products (Mayer *et al.*, 1981). Initial metabolism probably involves the oxidation of the methylsulfide group which is then displaced by GSH. It is uncertain whether GSH conjugation occurs enzymatically and no information was presented regarding the basis of selectivity. A closely related herbicide, terbutryne, was not metabolized by GSH conjugation in potato, a species that slowly metabolizes atrazine to a GSH conjugate. In both potato and wheat, terbutryne was metabolized to hydroxylated derivatives (Edwards and Owen, 1989). Additional studies are needed to determine if GSH conjugation plays a major role in the metabolism and tolerance of methylthio-substituted triazine herbicides and to determine if these GSH/HomoGSH conjugation reactions are enzymatic.

Acifluorfen and fluorodifen GSH/HomoGSH conjugation and selectivity
Data on the mechanism of selectivity of the diphenylether herbicides is not conclusive, but there is evidence that differential uptake and metabolism are important factors (Ritter and Coble, 1981; Higgins *et al.*, 1988). Fluorodifen is metabolized in pea and peanut by a nucelophilic displacement reaction that yields an aryl GSH conjugate and a phenol (Shimabukuro *et al.*, 1973), Reaction 6.

Fluorodifen

This reaction is catalyzed by a GST enzyme characterized from pea (Frear and Swanson, 1973). Higher levels of this GST were detected in resistant cotton, corn, peanut, pea, soybean and okra than in susceptible tomato, cucumber and squash. Glutathione conjugation was observed in excised roots, hypocotyls, epicotyls, leaves and callus. The substrate specificity of the GST from pea was very different from the GST from corn.

Acifluorfen, a structural analogue of fluorodifen, was metabolized rapidly to a HomoGSH conjugate in soybean (Frear *et al.*, 1983a). A small amount of GSH conjugate (< 5%) was also detected. This was the first HomoGSH conjugate of a herbicide to be reported. The phenol, formed as a result of the diphenyl ether cleavage, was metabolized further by glucoside conjugation. Interestingly, the GST enzyme from pea that catalyzed GSH conjugation of fluorodifen was not active with acifluorfen as a substrate. This indicates either a high degree of substrate specificity for some plant GST isozymes, or that conjugation of acifluorfen is more complex than it appears.

Based on limited studies, differential uptake and metabolism appear to be involved in acifluorfen selectivity. More rapid uptake and slower metabolism of ^{14}C-acifluorfen was observed in susceptible *Xanthium*

235

pensylvanicum and *Ambrosia artemisiifolia* than in tolerant soybean. After 7 days, polar metabolites were approximately 2x higher in tolerant soybean than in the susceptible *X. pensylvanicum* and *A. artemisiifolia* (Ritter and Coble, 1981). Differences in acifluorfen tolerance among different types of *Ipomoea lacunosa* have been observed, but these differences appear to be related to uptake since only 11% metabolism was observed 96 h after treatment (Higgins *et al.*, 1988). Under slightly different conditions, 85 to 95% metabolism of acifluorfen to a HomoGSH conjugate was observed within 24 h in soybean (Frear *et al.*, 1983a).

Chlorimuron ethyl HomoGSH conjugation and selectivity

Chlorimuron ethyl is a sulfonylurea herbicide used for the control of a broad spectrum of weeds in soybean. The herbicidal activity of the sulfonylurea herbicides is due to inhibition of acetolactate synthase, an enzyme required for valine and isoleucine biosynthesis (LaRossa and Falco, 1984; Ray, 1986). In soybean, chlorimuron ethyl is metabolized to a HomoGSH conjugate, and to a lesser extent to the free acid. The reaction involves the nucelophilic displacement of chlorine from a heterocyclic ring. Although the reaction appears to be enzymatic, the enzyme that catalyzes this reaction was not isolated. Metabolism of chlorimuron ethyl results in the loss of toxicity as measured by inhibition of acetolactate synthase. The I_{50} for chlorimuron ethyl was 6.2 nM while the I_{50} values for the HomoGSH conjugate and the free acid were 13,000 nM and 7,100 nM, respectively. The half-life of chlorimuron ethyl in tolerant soybean was 1-3 h while the half-life in susceptible *Xanthium strumarium* and *Amaranthus retroflexus* was greater than 30 h (Brown and Neighbors, 1987). A further indication of the role of metabolism in selectivity was obtained in research on safeners. Corn, which is moderately sensitive to chlorimuron ethyl, can be safened by BAS 145 138. Increased tolerance to chlorimuron ethyl was associated with an increased rate of metabolism by several routes, including glutathione conjugation and hydroxylation/glucoside conjugation (Lamoureux and Rusness, 1989c).

Formation of toxic or biologically active products by GSH conjugation

Glutathione conjugation is usually a detoxification process, but toxic or biologically active products have been produced as a result of GSH conjugation. Examples of this are the production of a mutagenic alkylating agent by GSH conjugation of ethylene dibromide in mammals and the release of HCN by GSH conjugation of thiocyanate insecticides in insects (Ohkawa and Casida, 1971; van Bladeren *et al.*, 1980). In plants, a potentially toxic halogenating agent is produced by GSH conjugation of the fungicide, dichlofluanide (Schuphan *et al.*, 1981), Reaction 7.

DICHLOFLUANID

The initial GSH conjugates of the thiocyanate insecticides and dichlofluanide are transitory and difficult to detect.

Tridiphane, a herbicide synergist, is metabolized to a GSH conjugate in plants and insects, Reaction 2. This GSH conjugate inhibits GSH conjugation of atrazine in *Setaria faberii* GSH conjugation of diazinon in the housefly, and GSH reductase from spinach (Lamoureux and Rusness, 1986a, 1987a, 1989b). Tridiphane synergizes atrazine toxicity in *S. faberii* and diazinon toxicity in the housefly. This synergistic effect may be due to inhibition of metabolism. The herbicide safener, fluorazole, is metabolized to a GSH conjugate in corn and it is hypothesised that this GSH conjugate stimulates GSH synthesis (Breaux *et al.*, 1989). These examples indicate the possibility that biologically active GSH conjugates may be formed in the metabolism of herbicides by GSH conjugation.

Metabolism of GSH conjugates to other products

Detailed studies of the metabolism of GSH conjugates in plants have been conducted with atrazine in sorghum, PCNB in peanut and onion, propachlor in soybean and corn, and EPTC and orbencarb in soybean and corn (Lamoureux *et al.*, 1973, Rusness and Lamoureux, 1980; Lamoureux and Rusness, 1980; Lamoureux *et al.*, 1981; Ikeda *et al.*, 1986; Unai, 1986; Lamoureux and Rusness, 1987b, 1989a; Khalifia and Lamoureux, 1989). In most cases, these GSH conjugates are metabolized rapidly to cysteine conjugates, but in the metabolism of tridiphane by *S. faberii* signficant concentrations of the GSH conjugate were detected up to 8 days following treatment (Lamoureux *et al.*, 1990).

Cysteine conjugates (shown as 1 on Fig. 1) frequently undergo metabolism to other products and they represent the most common branch point in metabolism. Rearrangement of the *S*-cysteine conjugate to the *N*-cysteine conjugate (2) is a major process in the metabolism of atrazine in sorghum, but this rearrangement has not been commonly reported in the metabolism of other GSH conjugates (Lamoureux *et al.*, 1973). In sorghum, the *N*-cysteine conjugate was metabolized to a lanthionine conjugate (3) (Lamoureux *et al.*, 1973). Thiolactic acid conjugates (5) are major metabolites of the chloroacetamide and thiocarbamate herbicides in corn, cotton and soybean (Blattmann *et al.*, 1986; Lamoureux *et al.*, 1990). They are probably formed from cysteine conjugates (1) through thiopyruvate intermediates (4), but thiopyruvate derivatives have not been commonly observed. A malonylthiolactic acid conjugate (6) was a major metabolite of EPTC in corn and cotton, and a malonylthiolactic acid conjugate was a major transitory metabolite of propachlor in corn (Lamoureux and Rusness, 1989a; Khalifia and Lamoureux, 1989). The malonylthiolactic acid conjugate of propachlor appeared to be in equilibrium with the thiolactic acid conjugate (5) which was slowly oxidized to the sulfoxide (7) and hydrolyzed to the free hydroxylated derivative of the herbicide (8). Although the free hydroxy derivative of propachlor was not observed, it is the apparent precursor of the O-malonyl-glucoside (9). A thioacetic acid conjugate (10) was a minor metabolite of PCNB in peanut (Rusness and Lamoureux, 1980).

Malonylcysteine conjugates (11) are major residues of pesticides in a number of plant species and the formation of malonylcysteine conjugates is in competition with the formation of thiolactic acid derivatives (Lamoureux and Rusness, 1983). Methylsulfides (14), methylsulfoxides (15) and methylsulfones (16) are metabolites of PCNB in onion and other species and are formed through a thiophenol (13) intermediate (Lamoureux and Rusness, 1980; Rusness and Lamoureux,

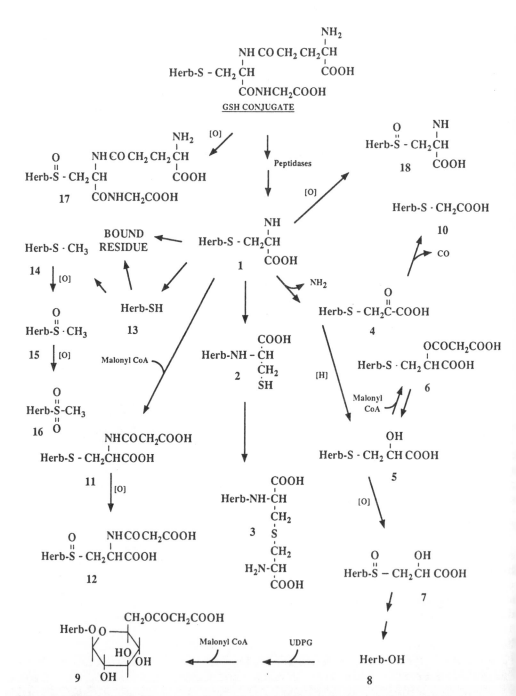

Fig. 1. <u>The metabolism of glutathione and homoglutathion conjugates in higher plants.</u>

1980; Lamoureux and Rusness, 1981). Methylsulfide, methylsulfoxide and methylsulfone derivatives do not appear to be major metabolites of herbicides in plants, but they may be volatile or have properties similar to the parent herbicide. Precautions may be needed if they are not to be overlooked. Because the methylsulfide and methylsulfoxide derivatives frequently resemble the original herbicide in physical and chemical properties, they should be examined for herbicidal properties. Methylsulfoxide derivatives are common substrates of GST enzymes and these metabolites can be metabolized to GSH conjugates (Lamoureux and Bakke, 1984).

High levels of bound residues can be produced from pesticides that are metabolized by GSH conjugation. Thiophenols are also excellent precursors of bound residues, but it is not known if thiophenols are key intermediates in the formation of bound residues of these pesticides (Lamoureux and Rusness, 1981). Sulfoxides can be formed from many of the intermediates shown in Figure 1, including GSH (17), cysteine (18) and malonylcysteine (12) conjugates. However, care must be taken to prevent the formation of oxidation artifacts of metabolites during purification (Lamoureux and Rusness, 1989a).

GLUCOSIDE CONJUGATION

Herbicides that contain HO-, HN-, HON-, HOOC-, or HS- functional groups, or herbicides that can be metabolized to derivatives that contain these functional groups, are frequently metabolized to glucoside conjugates. Therefore, glucosides are among the most commonly reported herbicide conjugates in plants (Lamoureux and Rusness, 1986b). The majority of these conjugates are β-1-0-glucosides, *N*-glucosides and glucose esters; but several *N*-0-glucosides have also been observed (Lamoureux and Rusness, 1986b; Celorio *et al.*, 1987). Two factors are responsible for the common occurrence of glucoside conjugates of herbicides: (A) the ability of plants to metabolize herbicides to derivatives with HO-, HN-, and HOOC- functional groups (Phase 1 metabolism), and (B) the well developed capacity of plants to form glucosides. A strong relationship has been observed between herbicide tolerance and metabolism of herbicides to glucosides. However, Phase 1 metabolism frequently precedes glucoside conjugation and the relative importance these two steps to herbicide selectivity and detoxification has rarely been established.

0-glucosyltransferases in herbicide metabolism

Although the enzymatic synthesis of several glucoside conjugates of herbicides have been demonstrated, the glucosyltransferases that catalyze these reactions have not been studied extensively. Herbicide metabolism by glucoside conjugation or by glucosyltransferases are discussed in several reviews (Lamoureux and Frear, 1979; Edwards *et al.*, 1982; Schmitt and Sandermann, 1984; Lamoureux and Rusness, 1986b). Most 0-glucoside conjugates of phenols and alkyl alcohols are formed by glucosyltransferases that utilize UDP-glucose as the glucose donor, Reaction 8.

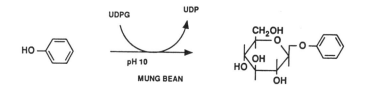

Many of the transferases that catalyze glucoside conjugation of phenolic substrates are soluble enzymes with MWs of 44 to 62 KDa, but membrane bound forms of these enzymes are important in the synthesis of sterol glucosides (Lamoureux and Frear, 1979; Hino *et al.*, 1982; Kalinowska and Wojciechowski, 1987; Dixon *et al.*, 1989). The glucosyltransferases are frequently specific for UDP-glucose as the glucosyl donor, but their specificity for the acceptor substrate varies considerably (Lamoureux and Frear, 1979; Keil and Schreier, 1989). The pH optima of these enzymes varies widely (6.5-10), and enzyme activity is frequently stimulated by divalent ions such as Mg^{++}, Mn^{++}, or Ca^{++} (Lamoureux and Frear, 1979; Keil and Schreier, 1989). Considerable variation in enzyme stability has been observed (Storm and Hassid, 1974; Hino *et al.*, 1982; Frear *et al.*, 1983b; Keil and Schreier, 1989). Glucosyltransferase activity has been increased in response to chemical treatments. For example, the ability of corn to form a glucoside conjugate of a hydroxylated form of chlorimuron ethyl was increased by treatment with the safener, BASF 145-138 (Lamoureux and Rusness, 1989c), and the specific activity of a purified glucosyltransferase from tobacco tissue culture was increased 10-fold by pretreatment of the cell cultures with 2,4-D (Hino *et al.*, 1982). Increased enzyme activity appeared to be due to an energy-dependent activation of pre-exisiting glucosyltransferase (Hino *et al.*, 1982). The UDP-glucose:glucosyltransferase from germinating mung bean that utilized a variety of phenolic and alkyl alcohols as substrates is an excellent example of the broad substrate specificity of some of the plant O-glucosyltransferases (Storm and Hassid, 1974).

Several recent studies with O-glucosyltransferases that utilized substrates with structures similar to herbicides have been reported. A UDP-glucose:phenol-β-*D*-glucosyltransferase from papaya fruit was purified to electrophoretic homogeneity. The 56 KDa glucosyl transferase appeared to be a dimeric protein made up of two 28 KDa subunits. This enzyme had a pH optimum of 7.5, a temperature optimum of 50°C, and K_m values of 0.17 mM and 0.07 mM for UDP-glucose and p-nitrophenol, respectively. The enzyme was active with p-nitrophenol, phenol, benzyl alcohol, 2-phenyl-ethanol, and 4-hydroxyphenyl-ethanol; but it was not active with o-nitrophenol, *m*-nitrophenol, 2,4-dinitrophenol, linalool, nerol, or geraniol. The enzyme had a high degree of specificity for UDP-glucose and activity was stimulated by Mg^{++}, Mn^{++}, and Ca^{++} (Keil and Schreier, 1989). A 45 KDa O-glucosyl-transferase that catalyzed the conjugation of scopoletin was purified 200-fold in 10% yield from tobacco tissue culture. One band of protein was observed upon electrophoresis (Hino *et al.*, 1982). A soluble 44 KDa O-glucosyltransferase was purified 2,500-fold in 6% yield from *Phaseolus lunatus*. The enzyme displayed a high degree of substrate specificity towards *trans*-zeatin (K_m 28 μM), but utilized both UDP-glucose and UDP-xylose as donors (K_m values of 0.2 and 2.7 mM, respectively) (Dixon *et*

240

al., 1989). A soluble UDP-Glucose: glucosyltransferase that utilized pentachlorophenol as a substrate was purified from wheat and soybean cell suspension cultures. This glucosyltrasferase was contaminated with β-glucosidase and it was assayed in the presence of 2.5 mM 4-nitrophenyl -β-glucoside and 2.5 mM salicin. Pentachlorophenol was converted to the corresponding glucoside in 1 to 4% yield (Schmitt *et al.*, 1985).

N-glucosyltransferases in herbicide metabolism

The *N*-glucosides of pyridate, chloramben and metribuzin, have been synthesized *in vitro* with UDP-glucose:glucosyltransferases isolated from peanut callus, soybean and tomato, respectively. The glucosyltransferase from soybean had a pH optimum of 7.5, utilized either UDP-glucose (K_m = 1.88 mM) or TDP-glucose as the donor substrate, and utilized a broad range of arylamines, including 3,4-dichloroaniline (K_m = 0.563 mM) and chloramben as acceptors (Frear, 1968). The enzyme was heat labile, but after purification it could be stored frozen. There was some evidence that more than one glucosyltransferase was present in the crude extracts from soybean. In contrast, metribuzin and ethyl metribuzin were good substrates for an *N*-glucosyl transferase from tomato shoots, but chloramben, picloram and pyrazon were not. This enyzme was present in the foliar tissue of tomato, but not in the roots. Tolerant tomato cultivars contained higher titres of the enzyme than susceptible cultivars. The *N*-glucosyl-transferase from tomato has a MW of 36 to 40 KDa, a pH optimum of 7.5 to 8.5, and is quite specific for UDP-glucose as the donor substrate (Frear *et al.*, 1983b; Frear and Swanson, 1987). An *N*-glucosyl-transferase from peanut callus that was active with a hydrolysis product of pyridate had a pH optimum of 7.0 and was purified 50-fold in 10% yield. It was necessary to stabilize the enzyme with 0.20 mM UDP-glucose and 0.01% Tween 80. This enzyme utilized UDP-glucose as the donor and the hydroxy-pyridate as the acceptor in the formation of an *N*-glucoside (Kroath *et al.*, 1988). Since *N*-glucosyltransferases are involved in the selective detoxification of both chloramben and metribuzin, and an *N*-glucosyltransferase was also involved in the critical second step in the metabolism of pyridate, these enzymes could be involved in the development of herbicide resistance in weeds and they may be useful for the development of herbicide-resistant transgenic plants.

The enzymatic syntheses of herbicide-glucose esters

The enzymatic synthesis of 1-0-glucose esters from herbicides that contain carboxylic acid or ester groups have not been studied intensively even though herbicides such as chloramben, diclofop methyl, and 2,4-D are commonly metabolized to glucose esters (Lamoureux and Frear, 1979; Lamoureux and Rusness, 1986b). The enzymatic syntheses of the 1-0-glucose esters of a variety of hydroxy-cinnamic and hydroxy-benzoic acid derivatives have been studied with crude preparations from geranium leaves and the the enzymatic syntheses of the 1-0-glucose esters of anthranilic acid and IAA have been studied using enzyme preparations from lentils and corn endosperm, respectively (Lamoureux and Frear, 1979). These enzymes required UDP-glucose as the glucosyl donor, but varied considerably in specificity towards the acceptor substrate. The crude enzyme from corn that utilizes IAA as a substrate is stable, but there is evidence that this enzyme requires an unknown co-factor that can be removed upon purification. The enzymatic synthesis of the 1-0-glucose ester of IAA is highly reversible and equilibrium favors free IAA and UDPG (Leznicki and Bandurski, 1988a, 1988b). This is consistent with data discussed later that suggests that the glucose ester of the herbicide, diclofop, is biologically active (Jacobson *et al.*, 1985).

0-Glucosides in herbicide metabolism and selectivity

The most common class of herbicide glucoside conjugates in plants are β-0-*D*-glucosides (Lamoureux and Rusness, 1986b). As previously discussed, the formation of 0-glucoside conjugates usually involves UDPG, a glucosyltransferase, and the hydroxylated substrate. Since very few herbicides contain a free hydroxy group (Best, 1983), these herbicide conjugates are usually formed in a multi-step process that includes oxidation as the initial step in metabolism. Exceptions include cases where hydroxylated derivatives are formed as a result of GSH conjugation, ester hydrolysis, or isomerization (Lamoureux and Rusness, 1986b). However, oxidation is often the first step in metabolism and there is considerable evidence that oxidation and glucoside conjugation results in detoxification and serves as the basis for selectivity (Hatzios and Penner, 1982; Lamoureux and Rusness, 1986b). The free hydroxylated derivatives of herbicides are not commonly reported as major metabolites; therefore, it is likely that the oxidative step is most commonly the rate-limiting step in metabolism. Thus, even if glucoside conjugation is necessary to complete herbicide detoxification, glucoside conjugation may not play a major role in herbicide selectivity if the necessary glucosyl-transferase enzymes are present in most crop and weed species. Exceptions to this are the metabolism of chlorsulfuron in soybean (Hutchison *et al.*, 1984) and the metabolism of chlorimuron ethyl in corn (Lamoureux and Rusness, 1989c). In both of these cases, hydroxylated derivatives are formed and exist as major metabolites. Very few hydroxylated herbicide derivatives have been tested *in vitro* with the target site enzyme systems, so the extent to which hydroxylation results in herbicide detoxification has not been well established. Nor does it appear that many hydroxylated herbicide derivatives have been tested *in vivo* and *in vitro* to determine if they are substrates for glucosyltransferase enzymes in both resistant and susceptible plant species. These lines of investigation would be useful in establishing the importance of 0-glucoside conjugation to herbicide selectivity and whether the 0-glucosyltransferases should be considered for the development of herbicide resistance through genetic engineering. Since oxidation rather than 0-glucoside conjugation appears more likely to be the critical factor in the selectivity of herbicides metabolized to 0-glucosides in plants, only three examples of 0-glucoside conjugation in herbicide selectivity are discussed.

0-Glucoside conjugation in bentazon metabolism and selectivity
Bentazon is an inhibitor of Photosystem II and is typical of a number of selective herbicides that are metabolized by aryl-hydroxylation and glucoside conjugation. Tolerant plant biotypes are able to detoxify bentazon before it causes irreversible damage, but susceptible plant biotypes are not (Mine *et al.*, 1975; Mine and Matsunaka, 1975; Retzlaff and Hamm, 1976). The most commonly observed metabolite of bentazon is the 6-0-β-D-glucoside, but the 8-0-β-D-glucoside is also produced in tolerant soybean (Otto *et al.*, 1978), Reaction 9.

BENTAZON

Up to 80% of the bentazon is metabolized within 24 h in tolerant rice, but only 25–50% of the bentazon is metabolized after 7 days in susceptible *Cyperus serotinus*. Differential uptake of bentazon was not involved in selectivity and the major metabolite was the 6–0–β–D–glucoside (Mine *et al.*, 1975). Severe inhibition of photosynthesis was observed in tolerant and susceptible species, but complete recovery of photosynthesis was observed within 5 days in the tolerant species (Mine and Matsunaka, 1975). More recently, metabolism as the basis for bentazon selectivity was verified with tolerant rice, susceptible wild rice (*Zizania paulustris*) and susceptible giant bur–reed (*Sparganium eurycarpum*). After 24 h, bentazon underwent 98% metabolism to a polar conjugate in rice as compared to less than 2% metabolism in wild rice or bur–reed (Clay and Oelke, 1988). Bentazon was also metabolized rapidly in tolerant soybean hypocotyls and cell suspension cultures, 62% and 79% in 6 h, respectively. No metabolism was observed in hypocotyls and cell suspension cultures of susceptible *Abutilon theophrasti* (velvetleaf). The free hydroxylated derivatives of bentazon were not observed (Sterling and Balke, 1988). Cell suspension cultures of other tolerant plant species also metabolize bentazon rapidly to the 6–0–β–D–glucoside (Sterling and Balke, 1989). Soybean genotypes vary in their tolerance to bentazon and tolerance has been related to a single recessive gene (Bernard and Wax, 1975). In the tolerant genotypes, up to 80% of the bentazon was metabolized in 24 h to a mixture of the 6–0 and 8–0 glucosides, but the susceptible genotypes were unable to metabolize bentazon (Connelly *et al.*, 1988). Based on these reports, it appears that a critical factor for metabolic detoxification of bentazon is the presence of the mono–oxygenase enzyme(s) that catalyze the formation of the 6– and 8–hydroxy derivatives of bentazon.

0–Glucoside conjugation in terbacil metabolism and selectivity

The substituted uracil herbicide, terbacil, is an inhibitor of the Hill reaction of photosynthesis. Terbacil selectivity has been attributed to several factors; however, there is considerable evidence that tolerance in orange seedlings, alfalfa, strawberry and *Solidago canadensis* is at least partially due to oxidation and glucoside conjugation (Jordan *et al.*, 1975; Rhodes, 1977; Genez and Monaco, 1983). Terbacil is metabolized more rapidly in the roots of tolerant strawberry (65% in 12 h and 85% in 48 h) than in the roots of sensitive cucumber (20% in 12 h and 30% in 48 h). The major metabolite of terbacil was characterized as the glucoside of 3–*tert*-butyl–5–chloro–6–hydroxymethyl uracil. The free 6–hydroxymethyl derivative of terbacil was also detected, but it accounted for only 0–6% of the residue. It would appear that oxidation rather than glucoside conjugation is the rate–limiting factor in the selective detoxification of terbacil. Restricted translocation of terbacil from the roots to the shoots may also be involved in the selectivity of this herbicide, but restricted translocation actually may be due to

243

metabolism in the roots and a failure of the resulting glucoside conjugates to be translocated (Genez and Monaco, 1982). Although *S. canadensis* is more tolerant to terbacil than strawberry, terbacil is metabolized more rapidly in strawberry. Therefore, factors in addition to metabolism may be involved in selectivity (Genez and Monaco, 1983).

O–Glucoside conjugation in chlorsulfuron metabolism and selectivity

The sulfonylurea herbicides, such as chlorsulfuron, are inhibitors of the acetolactate synthetase step in branched chain amino acid biosynthesis (Ray, 1986). Tolerance to chlorsulfuron is due to rapid detoxification by hydroxylation and glucoside conjugation (Sweetser *et al.*, 1982; Hutchison *et al.*, 1984). In the leaves of tolerant wheat, barley, *Avena fatua* and *Poa annua* chlorsulfuron accounted for less than 10% of the residue 24 h following treatment. In susceptible cotton, soybean, mustard and sugar beet, chlorsulfuron accounted for 78 to 100% of the residue. The major metabolite of chlorsulfuron in wheat was the 5–O–β–D–glucoside, Reaction 10.

CHLORSULFURON

This metabolite was not detected in susceptible sugarbeet, but it accounted for 42 to 60% of the of the residue after 24 h in the five resistant species. The free 5–hydroxy derivative of chlorsulfuron was not present in significant levels in either wheat or sugarbeet (Sweetser *et al.*, 1982). Hydroxylation appears to be the rate–limiting step in the detoxification of chlorsulfuron in these species. An alternative route of chlorsulfuron metabolism and detoxification occurs in tolerant flax (*Linum usitatissimum* L.) and black nightshade (*Solanum nigrum* L.). The major metabolite in these species was the 4–hydroxymethyl derivative, Reaction 10. The glucoside of this metabolite was observed, but it was present at lower concentrations. The 5–hydroxy derivative has some biological activity, but it is less toxic than chlorsulfuron (Hutchison *et al.*, 1984). The formation of the β–D–glucoside conjugate appears to be the rate–limiting step in chlorsulfuron metabolism in flax and black nightshade, but does not appear to be necessary for tolerance. Recently, chlorimuron ethyl was shown to be metabolized in corn by hydroxylation of the 5–position of the pyrimidine ring and this metabolite was only partially converted to the corresponding glucoside (Lamoureux and Rusness, 1989c). Therefore, the glucosyltransferase enzymes that catalyze glucoside formation have considerable substrate specificity and in some cases glucoside conjugation may be a limiting factor in herbicide detoxification.

Chlorsulfuron and other sulfonylurea herbicides can be safened with dichlormid and naphthalic anhydride. An increased rate of metabolism has been associated with this safening effect, but it is not clear whether an increased rate of metabolism is the primary basis for the protective effect of these safeners (Barrett, 1989). The sulfonylurea herbicides have been the subject of two recent reviews (Blair and Martin, 1988; Barrett, 1989).

Formation of *N*-glucosides in plants

N-Glucoside conjugates have been reported as metabolites of several herbicides in higher plants, including chloramben, dinoben, metribuzin, picloram, propanil and pyrazon (Hatzios and Penner, 1982; Lamoureux and Rusness, 1986b). *N*-glucosides frequently are formed as the first step in the metabolism of a herbicide (chloramben, metribuzin, picloram and pyrazon), but some are formed after an initial Phase 1 reaction, such as the reduction of a nitro group in the metabolism of dinoben or the hydrolysis of an ester group in the metabolism of pyridate (Colby, 1966; Kroath *et al.*, 1988). In plants, the primary metabolic pathways that compete with *N*-glucoside conjugation of herbicides appear to be the formation of bound residues and malonyl derivatives (Lamoureux and Rusness, 1986b). Tolerance to several herbicides may be related to rapid metabolism of these herbicides to *N*-glucosides. Plant tolerance to chloramben and metribuzin has been attributed to the rate at which these herbicides are metabolized to *N*-glucoside conjugates.

N-Glucoside conjugation in chloramben metabolism and selectivity

Chloramben is metabolized more rapidly in tolerant soybean than in susceptible barley and metabolism was proposed as the basis for selectivity (Colby, 1965). A good correlation between chloramben tolerance and metabolism to an *N*-glucoside was observed in tissue sections exposed to 10^{-5}M [^{14}C]chloramben for 7 h, Reaction 11.

The following percentages of metabolism were observed: tolerant *I. lacunosa* (76%) squash (84%), snapbean (67%), soybean (62%), intermediately susceptible corn and cucumber (28% and 33%, respectively), and susceptible *A. theophrasti* (23%), barley (19%), *S. faberii* (15%) and *E. crus-galli* (12%) (Frear *et al.*, 1978). Chloramben is translocated readily from the roots of susceptible cucumber, but not from the roots of tolerant squash. Based on reciprocal root grafts with squash and cucumber, resistance appeared to be due to a lack of translocation from the roots (Baker and Warren, 1962). However, lack of translocation may have been due to the metabolism of chloramben to an immobile *N*-glucoside. Only 3% of the chloramben remained in squash roots after 7 h while 22% of the chloramben remained in cucumber roots (Frear *et al.*, 1978). Chloramben is also metabolized partially to a glucose ester in several plant species, Reaction 11. The soluble UDPG-glucosyltransferase from soybean that catalyzes the formation of the N-glucoside of chloramben has a broad substrate specificity for different aromatic amines. This would suggest that other herbicides might also be metabolized to N-glucosides in soybean (Frear, 1968). The N-glucoside of chloramben is stable in soybean for at least 50 days and it is assumed that metabolism to the N-glucoside results in detoxification of the herbicide (Swanson *et al.*, 1966). N-Glucoside conjugation is reduced under low light (Swanson *et al.*, 1966).

N-Glucoside conjugation in metribuzin metabolism and selectivity

Tomato cultivars differ in their tolerance to metribuzin during early seedling growth, but most cultivars are sufficiently tolerant that metribuzin can be used safely by the six to seven leaf-stage (Stephenson *et al.*, 1976). Tolerance appeared to be related to metribuzin metabolism in the shoots. Metribuzin was metabolized to the N-glucoside and then to the corresponding N-(6-0-malonyl)glucoside, Reaction 5. The UDPG-glucosyltransferase that catalyzes the first reaction was isolated from tomato leaves. A higher titre of this enzyme was present in the leaves from young tolerant cultivars than from susceptible cultivars. After 4 weeks, when all cultivars were tolerant, no signficant differences in glucosyltransferase activity were observed (Frear *et al.*, 1983b; Frear and Swanson, 1987).

N-Glucoside conjugation in picloram metabolism and selectivity

Picloram is metabolized both to an N-glucoside and to a glucose ester in a variety of plant species. Although it is metabolized primarily to a glucose ester in sensitive species, it is metabolized slowly to the N-glucoside in sensitive sunflower (72% in 7 days). The N-glucoside of picloram appears to be nontoxic and resistant to further metabolism in French bean (Chkanikov *et al.*, 1983). It is uncertain whether differential rates of metabolism to the N-glucoside or differential routes of metabolism to the glucose ester versus the N-glucoside are involved in the selectivity of this herbicide (Hall and Vanden Born, 1988; Frear *et al.*, 1989). Pyrazon is also metabolized to an N-glucoside, but it is uncertain whether metabolism is the basis of selectivity of this herbicide (Ries *et al.*, 1968).

Glucose–ester conjugates

Herbicides that contain a free carboxylic acid or a carboxylic acid ester, such as diclofop methyl, picloram, clopyralid, 2,4-D, MCPA, thiobencarb and chloramben, are usually metabolized in plants to glucose ester conjugates (Hatzios and Penner, 1982; Lamoureux and Rusness, 1986b; Frear *et al.*, 1989). Although glucose ester conjugates are frequently assumed to be β-0-1 esters, chloramben was metabolized to the α-0-1 ester (Frear *et al.*, 1978). Amino acid conjugation is competitive with glucose ester conjugation in the metabolism of some herbicides that contain carboxylic acid groups, but amino acid conjugation is reported much less commonly (Lamoureux and Rusness, 1986b). Herbicides that contain a carboxylic acid/ester group may also contain an aromatic ring or a free amino or nitro group and thus may also be metabolized to N- or 0-glucosides. There is evidence that glucose ester conjugates are readily converted back to the unconjugated form. This can be due to esterases or it can also be due to the reversibility of the reaction (Leznicki and Bandurski, 1988b).

246

These conjugates may isomerize by migration of the carboxylic acid group during purification, ie. the C–1 glucose ester conjugate may be converted to a C–2 or C–3 glucose ester conjugate conjugate during purification. Furthermore, they may undergo transesterification and be converted to methyl esters if improper isolation procedures are employed (Lamoureux and Rusness, 1986b). For these reasons, studies dealing with the detoxification of herbicides via glucose ester conjugation have been complicated and some incorrect conclusions may have been reached.

Glucose – ester conjugation in diclofop–methyl metabolism and selectivity

The best evidence that glucose ester conjugation may not result in herbicide detoxification comes from studies dealing with diclopfop methyl metabolism in resistant wheat and in susceptible oat. Diclofop methyl is metabolized rapidly in both wheat and oat. In wheat, over 80% of the diclofop was metabolized by hydroxylation and O–glucoside conjugation. In 24 h, no significant levels of diclofop or diclofop methyl remained. In susceptible oat, diclofop methyl and free diclofop were still present after 24 h (40%) and the the glucose ester was the major metabolite (60%) (Jacobson and Shimabukuro, 1984), Reaction 12. Toxicity studies showed that the glucose *tetra*-acetate ester of diclofop (1 µM) caused a 58% reduction in root growth of oat as compared to a 70% reduction with 1 µM diclofop acid. The hydroxyolated form of diclofop was not toxic at 10 µM, but it was not reported whether this metabolite underwent glucose conjugation in oat roots during the 48 h assay. Therefore, it is likely that conversion to the glucose ester does not result in detoxification (Jacobson *et al.*, 1985). It can be concluded that a slower rate of metabolism of diclofop methyl to a toxic glucose ester conjugate as compared to rapid metabolism to a nontoxic O–glucoside is the basis of selectivity of diclofop methyl between these species. Chloramben and picloram can both be metabolized slowly to glucose esters, as discussed in the section on *N*-glucosides, and this appears to be the major route of metabolism of these herbicides in susceptible plant species (Frear *et al.*, 1978; Frear *et al.*, 1989).

*N*O– and *S*-Glucosides

*N*O– and *S*-glucosides are formed in the metabolism of various natural products and xenobiotics, including an *N*O–glucoside of the herbicide, phenmedipham, in sugar beet (Lamoureux and Rusness, 1986b; Celorio *et al.*, 1987). These conjugates do not occur with sufficent frequency to be considered an important factor in the metabolic detoxification of most currently used herbicides.

Formation of complex glucoside conjugates

Simple glucoside or glucose ester conjugates of pesticides frequently are metabolized to more complex secondary conjugates as illustrated in Figure 2. When secondary metabolism such as this occurs, it becomes more difficult to determine if glucoside conjugation is involved in the initial detoxification because it may not be possible to use β-glucosidase hydrolysis or simple chromatographic comparisons to determine if a glucoside conjugate was formed. For these reasons, the secondary metabolism of glucoside conjugates of herbicides is a concern, not only to residue chemists, but also to scientists attempting to study herbicide selectivity.

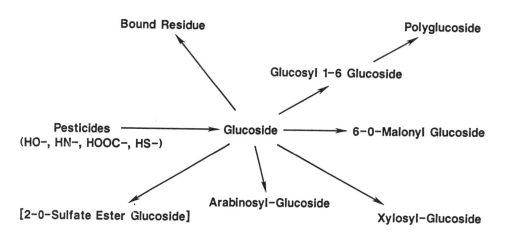

Fig. 2. The secondary metabolism of glucoside conjugates of pesticides in higher plants.

The formation and occurence of complex secondary glucoside conjugates of pesticides in plants has been reviewed (Lamoureux and Rusness, 1986b). O-Glucosides, N-glucosides, NO-glucosides, and glucose ester conjugates can be metabolized to these more complex glucoside conjugates in plants. Based on studies with gentiobiose and malonylglucoside conjugates, it appears that these complex conjugates are formed by the sequential addition of monosaccharide or acyl moieties rather than by the transfer of intact gentiobiose or malonylglucose residues (Matern and Heller, 1983; Schmitt *et al.*, 1985). The sequential addition of these residues may occur so rapidly that the simple glucoside conjugates are not detected or are only detected in low concentrations (Suzuki and Casida, 1981; Frear *et al.*, 1983a).

6-0-Malonylglucoside conjugates of herbicides have been detected in a variety of plant species (Mikami *et al.*, 1985; Lamoureux and Rusness, 1986b). Acifluorfen, diclofop methyl, flamprop, methazole and metribuzin are among the herbicides that are metabolized to malonylglucosides (Lamoureux and Rusness, 1986b). The enzyme that catalyzes the transfer of the malonyl group from malonyl coenzyme A to the 6-0- position of the glucose residue has been isolated from several plant species, including parsley and cell suspension cultures of soybean and wheat (Matern and Heller, 1983; Schmitt *et al.*, 1985). The malonylglucoside of pentachlorophenol was synthesized from the glucoside in 90 to 100% yield with malonyl transferases from soybean and wheat cell cultures (Schmitt *et al.*, 1985). The function of the malonyl transferase reaction may be to prevent hydrolysis of the glucoside by β-glucosidases that frequently are present in plant tissues (Matern and Heller, 1983). The 6-0-malonylglucose bond is labile and may be hydrolyzed by esterases present in plant extracts or by isolation procedures that involve the use of weak bases or some acids (Lamoureux and Rusness, 1986b). The malonyl residue may also be removed during derivatization with diazomethane (Frear *et al.*, 1983b) or by base catalyzed hydrolysis with reagents such as hydroxylamine (Schmitt *et al.*, 1985). Because of their lability, it is likely that some malonylglucoside conjugates of herbicides have been identified incorrectly as simple glucoside conjugates. The malonylglucoside conjugates are resistant to hydrolysis by β-glucosidase from almond, but malonylglucose conjugates have been reported to be hydrolyzed by cellulase from *Aspergillus niger* (Mikami *et al.*, 1984, 1985) and they can be hydrolyzed partially by hesperidinase (Schneider *et al.*, 1984). The malonyl group can also be hydrolyzed by a malonyl esterase from parsley (Matern and Heller, 1983).

Glucosylglucoside conjugates of a number of pesticides have also been reported (Lamoureux and Rusness, 1986b). Diphenamid and maleic hydrazide are metabolized to disaccharide gentiobiose [β-(1-->6) glucosyl beta-(1-->0) glucose] conjugates (Lamoureux and Rusness, 1986b). This pathway is particularly common in tomato, but has been observed in cotton, cabbage, kidney bean, wheat, tobacco, horseradish, gladiolus and *Dalbergia sisso* (Mikami *et al.*, 1984, 1985; Lamoureux and Rusness, 1986b). The gentiobiose conjugates are formed in two steps involving two different UDP-glucosyltransferases (Yamaha and Cardini, 1960). The UDP-glucosyltransferases that catalyzes the addition of glucose to the phenolic glucoside has a broad substrate specificity for phenolic glucosides, but it does not catalyze the addition of glucose to a simple phenolic substrate, nor does it catalyze the addition of glucose to a gentiobiose conjugate in the formation of trisaccharide conjugates (Yamaha and Cardini, 1960). The gentiobiose ester of 3-phenoxybenzoic acid is hydrolyzed by β-glucosidase (Mikami *et al.*, 1984, 1985). Conjugates of cellobiose [β-(1-->4) glucosyl β-(1-->0) glucose], sophorose [β-1-->2 glucosyl β-(1-->0) glucose] and several triglucosyl conjugates have also been reported as pesticide metabolites in cotton and tomato (Mikami *et al.*, 1984, 1985; Lamoureux and Rusness, 1986b). Arabinosylglucose and xylosylglucose conjugates of pesticides have been isolated from grape and cotton (Lamoureux and Rusness, 1986b). These disaccharide conjugates also appear to be formed by stepwise addition of the hexose or pentose residues. A 2-0-sulfate ester of glucose conjugated to a herbicide has also been reported (Celorio *et al.*, 1987).

The stability and the significance of complex glucoside conjugates of herbicides and other pesticides in plants has not been well studied. Some of the simple glucoside and malonylglucoside conjugates are stable within the plant for long periods of time while some complex glucose ester conjugates are known to be turned over rapidly with a half-life of only a few hours (Lamoureux and Rusness, 1986b). The mechanisms and the extent to which these complex glucoside conjugates become incorporated into the soluble and insoluble polymeric fractions of the plant is not well understood.

SUMMARY

Many herbicides that contain an electrophilic site are metabolized to GSH or HomoGSH conjugates in plants. The rate of metabolism varies greatly among different plant biotypes and has frequently been correlated with herbicide tolerance. The concentration and nature of the GST isozymes present is usually the major factor that determines the rate of GSH/HomoGSH conjugation, but under some conditions GSH/HomoGSH levels could be a contributing factor. Unfortunately, very little research has been conducted on GSH/HomoGSH levels and the effect of modulating these levels in susceptible plant species. In at least one case, the development of resistance in weed species has been attributed to an accelerated rate of GSH conjugation. Glutathione-S-transferase isozymes from corn have been studied extensively and it appears possible to produce transgenic plants tolerant to herbicides because of elevated levels of GST isozymes. Some GST isozymes can be induced and GST activity can also be inhibited. Consequently, herbicide metabolism by GSH conjugation can be modulated chemically. The extent to which the modulation of GST activity can be used to deal with herbicide resistant weeds needs to be explored more fully.

In plants, most GSH conjugates of herbicides are metabolized to a variety of other metabolites. These metabolites are usually considered to be biologically inactive, but toxic metabolites can be produced as a result of GSH conjugation, and the possibility that biologically active metabolites may be formed during herbicide metabolism should be considered. Some GSH conjugates of herbicides are formed after a Phase 1 reaction. In these cases it has not been established whether GSH conjugation plays a major role in herbicide selectivity.

Glucoside conjugation is a major process in the metabolism of many classes of herbicides, but the O-glucosides are usually formed after a Phase 1 reaction that appears to be primarily responsible for herbicide selectivity. The O-glucosyl transferases responsible for herbicide conjugation have not been studied extensively. The formation of N-glucosides appears to be an important factor in the selectivity of several herbicides. At least three UDPG:N-glucosyltransferase enzymes that catalyze herbicide conjugation reactions have been studied. It is uncertain whether plants contain multiple forms of N-glucosyltransferases as appears to be the case with the GST isozymes, but there is evidence that different N-glucosyltransferases are present in different plant species. There is some evidence that glucosyltransferase activity can be elevated *in vivo* by safeners and other chemicals, but it is uncertain whether this can be used to improve weed control. The N-glucosyltransferases should be considered for their potential usefulness in the production of transgenic plants resistant to herbicides. Some studies have been conducted with the UDPG:transferases that catalyze the formation of glucose esters of substrates such as IAA. However, it is questionable whether the formation of glucose esters of herbicides results in detoxification because of the apparent reversibility of the reaction

250

and the ease of enzymatic hydrolysis.

REFERENCES

Abusteit, E.O., Corbin, F.T., Schmitt, D.P., Burton, J.W., Worsham, A. and Thompson L. Jr. (1985). Absorption translocation and metabolism of metribuzin in diploid and tetraploid soybean (*Glycine max*) plants and cell cultures. Weed Science **33**, 618-628.

Baker, R.S. and Warren, G.F. (1962). Selective herbicidal action of amiben on cucumber and squash. Weeds **10**, 219-224.

Balabaskaran, S. and Muniandy, N. (1984). Glutathione S-transferase from *Hevea brasiliensis*. Phytochemistry **23**, 251-256.

Barrett, M. (1989). Protection of grass crops from sulfonylurea and imidazolinone toxicity. In "Crop safeners for herbicides" (K.K. Hatzios and R.E. Hoagland, eds), pp. 195-220. Academic Press, Inc., New York.

Bernard, R.L. and Wax, L.M. (1975). Inheritance of a sensitive reaction to bentazon herbicide: Soybean. Genetics Newsletter **2**, 46-47.

Best, C.E. (1983). In "Herbicide Handbook of the Weed Science Society of America", pp. 30-35. Weed Science Society of America, Champaign, Illinois.

Blair, A.M. and Martin, T.D. (1988). A review of the activity, fate and mode of action of sulfonylurea herbicides. Pesticide Science **22**, 195-219.

Blattmann, P., Gross, D., Kriemler, H.P., and Ramsteiner, K. (1986). Identification of thiolactic acid type conjugates as major degradation products in glutathione dependent metabolism of the 2-chloroacetamide herbicides metolachlor (Dual), dimetachlor (Teridox) and Pretilachlor (Kifit). The Sixth International Congress of Pesticide Chemistry (IUPAC), Ottawa, Canada (Abstract 7A-02).

Breaux, E.J. (1987). Initial metabolism of acetochlor in tolerant and susceptible seedlings. Weed Science **35**, 463-468.

Breaux, E.J., Hoobler, M.A., Patanella, J.E. and Leyes, G.A. (1989). Mechanisms of action of thiazole safeners. In "Crop Safeners for Herbicides" (K.K. Hatzios and R.E. Hoagland, eds), pp. 163-174. Academic Press Inc., New York.

Breaux, E.J., Patanella, J.E. and Sanders, E.F. (1987). Chloro-acetanilide herbicide selectivity: Analysis of glutathione and homoglutathione in tolerant, susceptible, and safened seedlings. Journal of Agricultural and Food Chemistry **35**, 474-478.

Brown, H.M. and Neighbors, S.M. (1987). Soybean metabolism of chlorimuron ethyl: Physiological basis for soybean selectivity. Pesticide Biochemistry and Physiology **29**, 112-120.

Burkholder, R.R.S. (1977). "A Survey of Several Important Agronomic Plant Species for the Presence of Glutathione-S-Transferase Activity". Thesis for Master of Science in Chemistry, North Dakota State University, College of Chemistry, Fargo, North Dakota.

Celorio, J.-I., Hoyer, G.-A., Iwan, J. and Kolsch, L. (1987). Metabolism of phenmedipham in the sugar beet (*Beta vulgaris* L.). In "Pesticide Science and Biotechnology" (R. Greenhalgh and T.R. Roberts, eds), pp. 495–498. Blackwell Scientific Publications.

Chkanikov, D.I., Makeev, A.M., Pavlova, N.N. and Nazarova, T.A. (1983). Formation of picloram N-glucoside in plants. Fiziologiya Rastenii **30**, 95–101.

Clay, S.A. and Oelke, E.A. (1988). Basis for differential susceptibility of rice (*Oryza sativa*), wild rice (*Zizania paulstris*), and Giant Bur-reed (*Sparganium eurycarpum*) to Bentazon. Weed Science **36**, 301–304.

Colby, S.R. (1965). Herbicide metabolism: N-glycoside of amiben isolated from soybean plants. Science **150**, 619–620.

Colby, S.R. (1966). The mechanism of selectivity of amiben. Weeds **14**, 197–201.

Collet, G.F. and Pont, V. (1978). Physiologie Vegetale.–Le role de la cysteine dans la detoxification d'un herbicide. Comptes Rendus Hebdomadaires de Seances de L'Academie des Sciences D: Sciences Naturelles t. 286, 681–684.

Connelly, J.A., Johnson, M.D., Gronwald, J.W. and Wyse, D.L. (1988). Bentazon metabolism in tolerant and susceptible soybean (*Glycine max*) genotypes. Weed Science **36**, 417–423.

De Prado, R., Dominguez, C. and Tena, M. (1989). Characterization of triazine-resistant biotypes of common lambsquarters (*Chenopodium album*), hairy fleabane (*Conyza bonariensis*), and yellow foxtail (*Setaria glauca*) found in Spain. Weed Science **37**, 1–4.

Devlin, D.L., Gealy, D.R. and Morrow, L.A. (1987). Differential metabolism of metribuzin by downy brome (*Bromus tectorum*) and winter wheat (*Triticum aestivum*). Weed Science **35**, 741–745.

Diesperger, H. and Sandermann, H. (1979). Soluble and microsomal glutathione S-transferase activities in pea seedlings. Planta **146**, 643–648.

Dixon, G.A. and Stoller, E.W. (1982). Differential toxicity, absorption, translocation and metabolism of metolachlor in corn, *Zea mays* and yellow nutsedge, *Cyperus esculentus*. Weed Science **30**, 225–230.

Dixon, S.C., Martin, R. C., Mok, M. C., Shaw, G. and Mok, D.W.S. (1989). Zeatin glycosylation enzymes in *Phaseolus*. Plant Physiology **90**, 1316–1321.

252

Edwards, R. and Owen, W.J. (1988). Regulation of glutathione-S-transferase of *Zea mays* in plants and cell cultures. Planta **175**, 99-106.

Edwards, R. and Owen, W.J. (1989). The comparative metabolism of the s-triazine herbicides atrazine and terbutryne in suspension cultures of potato and wheat. Pesticide Biochemistry and Physiology **34**, 246-254.

Edwards, V.T., McMinn, A.L. and Wright, A.N. (1982). Sugar conjugates of pesticides and their metabolites in plants – current status. In "Progress in Pesticide Biochemistry, Vol. 2" (D.H. Hutson and T.R. Roberts, eds), pp. 71-125. John Wiley and Sons, Ltd.

Ezra, G. and Stephenson, G.R. (1985). Comparative metabolism of atrazine and EPTC in Proso Millet (*Panicum miliaceum* L.) and corn. Pesticide Biochemistry and Physiology **24**, 207-212.

Ezra, G., Stephenson, G.R. and Lamoureux, G.L. (1986). Metabolism of ^{14}C metazachlor and action of safener 145 138 in corn. 26th Meeting of the Weed Science Society of America, (Abstract # 250).

Frear, D.S. (1968). Herbicide metabolism in plants. 1. Purification and properties of UDP-glucose:arylamine N-glucosyl-transferase from soybean. Phytochemistry **7**, 381-390.

Frear, D.S. and Swanson, H.R. (1970). Biosynthesis of S-(4-ethylamino-6-isopropylamino-2-s-triazino)glutathione. Partial purification and properties of glutathione-S-transferase from corn. Phytochemistry **9**, 2123-2132.

Frear, D.S. and Swanson, H.R. (1973). Metabolism of substituted diphenylether herbicides in plants. I. Enzymatic cleavage of fluorodifen in peas (*Pisum sativum* L.). Pesticide Biochemistry and Physiology **3**, 473-482.

Frear, D.S. and Swanson, H.R. (1987). Purification and properties of UDP-glucose: metribuzin N-glucosyltransferase from tomato. National Meeting of the American Chemical Society, New Orleans, LA, August, (Abstract).

Frear, D.S., Swanson, H.R., and Mansager, E.R. (1983a). Acifluorfen metabolism in soybean: diphenylether bond cleavage and the formation of homoglutathione, cysteine, and glucose conjugates. Pesticide Biochemistry and Physiology **20**, 299-316.

Frear, D.S., Swanson, H.R., and Mansager, E.R. (1985). Alternate pathways of metribuzin metabolism in soybean: Formation of N-glucoside and homoglutathione conjugates. Pesticide Biochemistry and Physiology **23**, 56-65.

Frear, D.S., Swanson, H.R. and Mansager, E.R. (1989). Picloram metabolism in leafy spurge: Isolation and identification of glucose and gentiobiose conjugates. Journal of Agricultural and Food Chemistry **37**, 1408-1411.

Frear, D.S., Swanson, H.R., Mansager, E.R. and Tanaka, F.S. (1983b). Metribuzin metabolism in tomato: Isolation and identification of N-glucoside conjugates. Pesticide Biochemistry and Physiology **19**, 270-281.

Frear, D.S., Swanson, H.R., Mansager, E.R. and Wien, R.G. (1978). Chloramben metabolism in plants: Isolation and identification of glucose ester. Journal of Agricultural and Food Chemistry **26**, 1347-1351.

Fuerst, E.P., Ahrens, W.H. and Lamoureux, G.L. (1988). The role of corn seedling anatomy in dichloroacetamide antidote activity. 28th Meeting of the Weed Science Society of America, Las Vegas (Abstract 199).

Gawronski, S.W., Haderlie, L.C., Callihan, R.H. and Gawronski, H. (1986). Mechanism of metribuzin tolerance herbicide metabolism as a basis for tolerance in potatoes (*Solanum tuberosum*). Weed Research **26**, 307-314.

Gawronski, S.W., Haderlie, L.C. and Stark, J.C. (1987). Metribuzin metabolism as the basis for tolerance in barley (*Hordeum vulgare* L.). Weed Research **27**, 49-56.

Genez, A.L. and Monaco, T.J. (1982). Uptake and translocation of terbacil in strawberry (*Fragaria* x *ananassa*) and goldenrod (*Solidago fistulosa*). Weed Science **31**, 56-62.

Genez, A.L. and Monaco, T.J. (1983). Metabolism of terbacil in strawberry (*Fragaria* x *ananassa*) and Goldenrod (*Solidago fistulosa*), Weed Science **31**, 221-225.

Gronwald, J. W. (1988). Influence of herbicide safeners on herbicide metabolism In "Crop Safeners for Herbicides" (K.K. Hatzios and R.E. Hoagland, eds), pp. 103-128. Academic Press Inc., New York.

Gronwald, J.W., Andersen, R.N. and Yee, C. (1989). Atrazine-resistance in velvetleaf (*Abutilon theophrasti*) due to enhanced atrazine detoxification. Pesticide Biochemistry and Physiology **34**, 149-163.

Gronwald, J.W., Fuerst, E.P., Eberlein, C.V. and Egli, M.A. (1987). Effect of herbicide antidotes on glutathione content and glutathione-S-transferase activity of sorghum shoots. Pesticide Biochemistry and Physiology **29**, 66-76.

Grove, G., Zarlengo, R.P., Timmerman, K.P., Li, Nan-qian, Tam, M.F. and Tu, C-P. (1988). Characterization and heterospecific expression of cDNA clones of genes in the maize GSH S-transferase multigene family. Nucleic Acid Research **16**, 425-438.

Hall, J.C. and Vanden Born, W.H. (1988). The absence of a role of absorption, translocation or metabolism in the selectivity of picloram and clopyralid in two plant species. Weed Science **36**, 9-14.

Hardgroder, T.G. and Rogers, R.L. (1974). Behavior and fate of metribuzin in soybeans and hemp sesbania. Weed Science 22, 238-245.

Hatzios, K.K. and Penner, D. (1982). "Metabolism of Herbicides in Higher Plants", Burgess Publishing Company, Minneapolis, Minn., USA.

Higgins, J.M., Whitwell, T., Corbin, F.T., Carter, G.E. Jr, and Hill, H.S. Jr. (1988). Absorption, translocation, and metabolism of acifluorfen in pitted morningglory (Ipomoea lacunosa) and ivyleaf morningglory (Ipomoea hederacea). Weed Science 36, 141-145.

Hino, F., Okazaki, M., and Miura, Y. (1982). Effect of 2,4-dichlorophenoxyacetic acid on glucosylation of scopoletin to scopolin in tobacco tissue culture. Plant Physiology 69, 810-813.

Hubbell, J.P. and Casida, J.E. (1977). Metabolic fate of the N,N-dialkylcarbamoyl moiety of thiocarbamate herbicides in rats and corn. Journal of Agricultural and Food Chemistry 25, 404-413.

Hutchison, J.M., Shapiro, R. and Sweetser, P.B. (1984). Metabolism of chlorsulfuron by tolerant broadleaves. Pesticide Biochemistry and Physiology 22, 243-247.

Ikeda, M., Unai, T. and Tomizawa, C. (1986). Metabolism of the herbicide orbencarb in soybean, wheat, corn and crabgrass seedlings. Weed Research 31, 238-243.

Jablonkai, I. and Dutka, F. (1985). Metabolism of acetochlor in tolerant and sensitive plant species. Journal of Radioanalytical and Nuclear Chemistry Letters 94, 271-280.

Jacobson, A. and Shimabukuro, R.H. (1984). Metabolism of diclofop-methyl in root-treated wheat and oat seedlings. Journal of Agricultural and Food Chemistry 32, 742-746.

Jacobson, A., Shimabukuro, R.H. and McMichael, C. (1985). Response of wheat and oat seedlings to root-applied diclofop-methyl and 2,4-dichlorophenoxyacetic acid. Pesticide Biochemistry and Physiology 24, 61-67.

Jaworski, E.G. (1969). Analysis of the mode of action of herbicidal α-chloroacetamides. Journal of Agricultural and Food Chemistry 17, 165-169.

Jensen, K.I.N., Stephenson, G.R., and Hunt, L.A. (1977). Detoxification of atrazine in three gramineae subfamilies. Weed Science 25, 212-220.

Jordan, L.S., Zurqiyah, A.A., Clerx, W.A. and Leasch, J.G. (1975). Metabolism of terbacil in orange seedlings. Archives of Environmental Contamination and Toxicology 3, 268-277.

Kalinowska, M. and Wojciechowski, Z.A. (1987). Subcellular localization of UDPG:nuatingenin glucosyltransferases in oat leaves. Phytochemistry 26, 353-357.

Keil, U. and Schreier, P. (1989). Purification and partial characterization of UDP-glucose:phenol-β–D-glucosyltransferase from papaya fruit. <u>Phytochemistry</u> **28**, 2281-2284.

Khalifia, M. and Lamoureux, G.L. (1989). The effect of BASF 145 138 safener on the metabolism of propachlor and metolachlor in corn. <u>Pesticide Biochemistry and Physiology</u> (in preparation).

Komives, T. and Dutka, F. (1989a). Biochemical mode of action of the antidote fenclorim in rice. <u>Biochemie und Physiologie der Pflanzen</u> **184**, 475-477.

Komives, T. and Dutka, F. (1989b). Effects of herbicide safeners on levels and activity of cytochrome P-450 and other enzymes in corn. <u>In</u> "Crop Safeners for Herbicides" (K.K. Hatzios and R.E. Hoagland, eds), pp. 129-145. Academic Press Inc., New York.

Komives, T., Komives, V.A., Balazs, M. and Dutka, F. (1985). Role of glutathione-related enzymes in the mode of action of herbicide antidotes. <u>British Crop Protection Conference – Weeds</u> **3**, 1155-1160.

Kroath, H., Susani, M. and Zohner, A. (1988). Genetic engineering of resistance to the phenylpyridazine herbicide, pyridate. <u>In</u> "Factors Affecting Herbicidal Activity and Selectivity". <u>Proceedings EWRS Symposium</u>.

Lamoureux, G.L. and Bakke, J.E. (1984). Formation and metabolism of xenobiotic glutathione conjugates in various life forms. <u>In</u> "Foreign Compound Metabolism" (J. Caldwell and G.D. Paulson, eds), pp. 185-199. Taylor and Francis, London.

Lamoureux, G.L. and Frear, D.S. (1979). Pesticide metabolism in higher plants: *In vitro* enzyme studies. <u>In</u> "Xenobiotic Metabolism: *In vitro* Methods" (G.D. Paulson, D.S. Frear and E.P. Marks, eds), pp. 77-128. American Chemical Society, Washington, D.C.

Lamoureux, G.L., Fuerst, E.P. and Rusness, D.G. (1990). Selected aspects of glutathione conjugation research in herbicide metabolism and selectivity. <u>In</u> "Sulfur Nutrition and Assimilation in Higher Plants" (H. Rennenberg, ed.) SPB Academic Publishing, The Hague. (in press).

Lamoureux, G.L., Gouot, J.-M., Davis, D.G. and Rusness, D.G. (1981). Pentachloronitrobenzene metabolism in peanut. 3. Metabolism in peanut cell suspension cultures. <u>Journal of Agricultural and Food Chemistry</u> **29**, 996-1002.

Lamoureux, G.L. and Rusness, D.G. (1980). *In vitro* metabolism of pentachloronitrobenzene to pentachloromethylthiobenzene by onion: Characterization of glutathione-*S*-transferase, cysteine C-S lyase, and *S*-adenosylmethionine methyl transferase activities. <u>Pesticide Biochemistry and Physiology</u> **14**, 50-61.

Lamoureux, G.L. and Rusness, D.G. (1981). Catabolism of glutathione conjugates of pesticides in higher plants. In "Sulfur in Pesticide Action and Metabolism" (J.D. Rosen, P.S. Magee and J.E. Casida, eds), pp. 133–164. American Chemical Society, Washington, D.C.

Lamoureux, G.L. and Rusness, D.G. (1983). Malonylcysteine conjugates as end-products of glutathione conjugate metabolism in plants. In "Pesticide Chemistry Human Welfare and the Environment" (Miyamoto, J. and Kearney, P.C., eds), pp. 295–300. Pergamon Press, New York.

Lamoureux, G.L. and Rusness, D.G. (1986a). Tridiphane [2-(3,5-dichlorophenyl)-2-(2,2,2-trichloroethyl)oxirane] an atrazine synergist: enzymatic conversion to a potent glutathione S-transferase inhibitor. Pesticide Biochemistry and Physiology 26, 323–342.

Lamoureux, G.L. and Rusness, D.G. (1986b). Xenobiotic conjugation in higher plants. In "Xenobiotic Conjugation Chemistry" (G.D. Paulson, J. Caldwell, D.H. Hutson and J.J. Menn, eds), pp. 62–105. American Chemical Society, Washington, D.C.

Lamoureux, G.L. and Rusness, D.G. (1987a). Synergism of diazinon toxicity and inhibition of diazinon metabolism in the house fly by tridiphane: inhibition of glutathione-S-transferase activity. Pesticide Biochemistry and Physiology 27, 318–329.

Lamoureux, G.L. and Rusness, D.G. (1987b). EPTC metabolism in corn, cotton, and soybean: Identification of a novel metabolite derived from the metabolism of a glutathione conjugate. Journal of Agricultural and Food Chemistry 35, 1–7.

Lamoureux, G.L. and Rusness, D.G. (1989a). Propachlor metabolism in soybean plants, excised soybean tissues, and soil. Pesticide Biochemistry and Physiology 34, 187–204.

Lamoureux, G.L. and Rusness, D.G. (1989b). The role of glutathione and glutathione S-transferases in pesticide metabolism, selectivity, and mode of action in plants and insects. In "Coenzymes and Cofactors, Vol IIIB, Glutathione", (D. Dolphin, R. Poulson and O. Avramovic, eds), pp. 153–196. John Wiley and Sons, New York.

Lamoureux, G.L. and Rusness, D.G. (1989c). Unpublished research.

Lamoureux, G.L., Stafford, L.E. and Tanaka, F.S. (1971). Metabolism of 2-chloro-N-isopropylacetanilide (propachlor) in the leaves of corn, sorghum, sugarcane, and barley. Journal of Agricultural and Food Chemistry 19, 346–350.

Lamoureux, G.L., Stafford, L.E. and Shimabukuro, R.H. (1972). Conjugation of 2-chloro-4,6-bis(alkylamino)-s-triazines in higher plants. Journal of Agricultural and Food Chemistry 20, 1004–1009.

Lamoureux, G.L., Stafford, L.E., Shimabukuro, R.H. and Zaylskie, R.G. (1973). Atrazine metabolism in sorghum: catabolism of the glutathione conjugate of atrazine. Journal of Agricultural and Food Chemistry 21, 1020–1030.

LaRossa, R.A. and Falco, S.C. (1984). Amino acid biosynthetic enzymes as targets of herbicide action. Trends in Biotechnology 2, 158-161.

Lay, M.-M. and Casida, J.E. (1976). Dichloroacetamide antidotes enhance thiocarbamate sulfoxide detoxification by elevating corn root glutathione content and glutathione-S-transferase activity. Pesticide Biochemistry and Physiology 6, 442-456.

Leznicki, A.J. and Bandurski, R.S. (1988a). Enzymatic synthesis of indole-3-acetyl-1-0-β-D-glucose. I. Partial purification and characterization of the enzyme from Zea mays. Plant Physiology 88, 1474-1480.

Leznicki, A.J. and Bandurski, R.S. (1988b). Enzymatic synthesis of indole-3-acetyl-1-0-β-D-glucose. II. Metabolic characteristics of the enzyme. Plant Physiology 88, 1481-1485.

Mangeot, B.L., Slife, F.E. and Rieck, C.E. (1979). Differential metabolism of metribuzin by two soybean (Glycine max) cultivars. Weed Science 27, 267-269.

Matern, U. and Heller, W. (1983). Conformational changes of apigenin 7-0-(6-0-malonylglucoside), a vacuolar pigment from parsley, with solvent composition and proton concentration. European Journal of Biochemistry 133, 439-448.

Mayer, P., Kriemler, H.-P., and Laanio, T.L. (1981). Metabolism of N-(1',2'-dimethylpropyl)-N'-ethyl-6-methylthio-1,3,5-triazine-2,4-diamine (C 18 898) in paddy rice. Agricultural Biological Chemistry 45, 361-368.

Mikami, N., Wakabayashi, N., Yamada, H. and Miyamoto, J. (1984). New conjugated metabolites of 3-phenoxybenzoic acid in plants. Pesticide Science 15, 531-542.

Mikami, N., Wakabayashi, N., Yamada, H. and Miyamoto, J. (1985). The metabolism of fenvalerate in plants: the conjugation of the acid moiety. Pesticide Science 16, 46-58.

Mine, M. and Matsunaka, S. (1975). Mode of action of bentazon: Effect on photosynthesis. Pesticide Biochemistry and Physiology 5, 440-450.

Mine, M., Miyakado, M. and Matsunaka, S. (1975). The mechanism of bentazon selectivity. Pesticide Biochemistry and Physiology 5, 566-574.

Moore, R.E., Davies, M.S., O'Connell, K.M., Harding, E.I., Wiegand, R.C. and Tiemeier, D.C. (1986). Cloning and expression of a cDNA encoding a maize glutathione S-transferase in E. coli. Nucleic Acid Research 14, 7227-7235.

Mozer, T.J., Teimeier, D.C. and Jaworski, E.G. (1983). Purification and characterization of corn glutathione S-transferase. Biochemistry 22, 1068-1072.

O'Connell, K.M., Breaux, E.J. and Fraley, R.T. (1988). Different rates of metabolism of two chloroacetanilide herbicides in Pioneer 3320 corn. Plant Physiology **86**, 359-363.

Ohkawa, H. and Casida. J.E. (1971). Glutathione-S-transferase liberates hydrogen cyanide from organic thiocyanates. Biochemical Pharmacology **20**, 1708.

Otto, S., Beutel, P., Decker, N. and Huber, R. (1978). Investigations into the degradation of bentazon in plant and soil. Advances in Pesticide Science **3**, 551-556.

Pont, V. and Collet, G.F. (1980). Metabolism du chloro-s (p-chloro-phenyl)-3 propionate de methyle et probleme de selectivite. Phytochemistry **19**, 1361-1363.

Ray, T.R. (1986). Sulfonylurea herbicides as inhibitors of amino acid biosynthesis. TIBS **11**, 180-183.

Rennenberg, H. (1982). Glutathione metabolism and possible biological roles in higher plants. Phytochemistry **21**, 2771-2781.

Rennenberg, H. and Lamoureux, G.L. (1990). Physiological processes that modulate the concentration of glutathione in plant cells. In "Sulfur Nutrition and Assimilation in Higher Plants" (H. Rennenberg, ed.). SPB Academic Publishing, The Hague, The Netherlands. (in press).

Retzlaff, G. and Hamm, R. (1976). The relationship between CO_2 assimilation and the metabolism of bentazon in wheat plants. Weed Research **16**, 263-266.

Rhodes, R.C. (1977). Metabolism of $(2-^{14}C)$terbacil in alfalfa. Journal of Agricultural and Food Chemistry **25**, 1066-1068.

Ries, S.K., Zabik, M.J., Stephenson, G.R. and Chen, T.M. (1968). N-glucosyl metabolite of pyrazon in red beets. Weed Science **16**, 40-41.

Ritter, R.L. and Coble, H.D. (1981). Penetration, translocation, and metabolism of acifluorfen in soybean (*Glycine max*), common ragweed (*Ambrosia artemisiifolia*), and common cocklebur (*Xanthium pensylvanicum*). Weed Science **29**, 474-480.

Rusness, D.G. and Lamoureux, G.L. (1980). Pentachloronitrobenzene metabolism in peanut. 2. Characterization of chloroform-soluble metabolites produced *in vivo*. Journal of Agricultural and Food Chemistry **28**, 1070-1077.

Rusness, D.G. and Still, G.G. (1977). Partial purification and properties of S-cysteinyl-hydroxychlorpropham transferase from oat (*Avena sativa* L.). Pesticide Biochemistry and Physiology **7**, 220-231.

Schmitt, R., Kaul, J., Trenck, T., Schaller, E. and Sandermann, H. Jr. (1985). β-glucosyl and O-malonyl-β-D-glucosyl conjugates of pentachlorophenol in soybean and wheat: Identification and enzymatic synthesis. Pesticide Biochemistry and Physiology **24**, 77-85.

Schmitt, R. and Sandermann, H. Jr. (1984). Herbicide resistance through gene transfer. Purification of a glucosyltransferase for the detoxification of pentachlorophenol. Hoppe-Seyler's Zeitschrift für Physiologie Chemie **365**, 1058.

Schneider, B., Schutte, H.R., Tewes. A. (1984). Comparative investigations on the metabolism of 2-(2,4-dichlorophenoxy)isobutyric acid in plants and cell suspensions cultures of *Lycopersicon esculentum*. Plant Physiology **76**, 989-992.

Schroder, P., Lamoureux, G.L., Rusness, D.G. and Rennenberg, H. (1990). Glutathione S-transferase in spruce trees. Pesticide Biochemistry and Physiology **37**, 211-218.

Schuphan, I., Westphal, D. and Ebing, W. (1981). Biological and chemical behavior of perhalogenmethylmercapto fungicides: metabolism and *in vitro* reactions of dichlofluanid in comparison with captan. In "Sulfur in Pesticide Action and Metabolism" (J.D. Rosen, P.S. Magee and J.E. Casida, eds), pp. 85-96. American Chemical Society, Washington, D.C.

Shah, D.M., Hironaka, C.M., Wiegand, R.C., Harding E.I., Krivi, G.G. and Tiemeier, D.C. (1986). Structural analysis of a maize gene coding for glutathione-S-transferase involved in herbicide detoxification. Plant Molecular Biology **6**, 203-211.

Shimabukuro, R.H., Frear, D.S., Swanson, H.R. and Walsh, W.C. (1971). Glutathione conjugation an enzymatic basis for atrazine resistance in corn. Plant Physiology **47**, 10-14.

Shimabukuro, R.H., Lamoureux, G.L., and Frear, D.S. (1978). Glutathione conjugation: A mechanism for herbicide detoxification and selectivity in plants. In "Chemistry and Action of Herbicide Antidotes" (F.M. Pallos and J.E. Casida, eds), pp. 133-149. Academic Press Inc., New York.

Shimabukuro, R.H., Lamoureux, G.L., Swanson, H.R., Walsh, W.C., Stafford, L.E. and Frear, D.S. (1973). Metabolism of substituted diphenylether herbicides in plants. II. Identification of a new fluorodifen metabolite, S-(2-nitro-4-trifluoromethylphenyl)-glutathione in peanut. Pesticide Biochemistry and Physiology **3**, 483-494.

Steinback, K.E., Pfister, K. and Arntzen, C.J. (1982). Identification of the receptor site for triazine herbicides in chloroplast thylakoid membranes. In "Biochemical Responses induced by Herbicides", (D.E. Moreland, J.B. St. John and F.D Hess, eds), pp.37-55. American Chemical Society, Washington, DC.

Stephenson, G.R., McLeod, J.E. and Phatak, S.C. (1976). Differential tolerance of tomato cultivars to metribuzin. Weed Science **24**, 161-165.

Sterling, T.M. and Balke, N.E. (1988). Use of soybean (*Glycine max*) and velvetleaf (*Abutilon theophrasti*) suspension-cultured cells to study bentazon metabolism. Weed Science **36**, 558-565.

Sterling, T.M. and Balke, N.E. (1989). Differential bentazon metabolism and retention of bentazon metabolites by plant cell cultures. Pesticide Biochemistry and Physiology **34**, 39–48.

Storm, D.L. and Hassid, W.Z. (1974). Partial purification and properties of a β-glucosyltransferase occurring in germinating *Phaseolus aureus* seeds. Plant Physiology **54**, 840–845.

Suzuki, T. and Casida, J.E. (1981). Metabolites of diuron, linuron, and methazole formed by liver microsomal enzymes in spinach plants. Journal of Agricultural and Food Chemistry **29**, 1027–1033.

Swanson, C.R., Hodgson, R.H., Kadunce, R.E. and Swanson, H.R. (1966). Amiben metabolism in plants II. Physiological factors in *N*-glucosyl amiben formation. Weeds **14**, 323–327.

Sweetser, P.B. (1985). Safening of sulfonylurea herbicides to cereal crops, mode of herbicide action. British Crop Protection Conference – Weeds, 1147–1153.

Sweetser, P.B., Schow, G.S. and Hutchison, J.M. (1982). Metabolism of chlorsulfuron by plants: biological basis for selectivity of a new herbicide for cereals. Pesticide Biochemistry and Physiology **17**, 18–23.

Unai, T., Ikeda, M., and Tomizawa, C. (1986). Metabolic fate of the chlorobenzyl moiety of orbencarb sulfoxide and benthiocarb sulfoxide in soybean seedlings. Weed Research **31**, 228–237.

van Bladeren, P.J., Breimer, D.D., Rotteveel-Smijs, G.M.T., DeJong, R.A.W., Buijs, W., Van Der Gen, A. and Mohn, G.R. (1980). The role of glutathione conjugation in the mutagenicity of 1,2-dibromoethane. Biochemical Pharmacology **29**, 2975–2982.

Weimer, M.R., Swisher, B.A., and Vogel, K.P. (1988). Metabolism as a basis for differential atrazine tolerance in warm-season forage grasses. Weed Science **36**, 436–440.

Wiegand, R.C., Shah, D.M., Mozer, T. J., Harding, E. I., Diaz-Collier, J., Saunders, C., Jaworski, E. G. and Teimeier, D.C. (1986). Messenger RNA encoding a glutathione-S-transferase responsible for herbicide tolerance in maize is induced in response to safener treatment. Plant Molecular Biology **7**, 235–243.

Williamson, G. and Beverley, M.C. (1987). The purification of acidic glutathione-S-transferase from pea seeds and their activity with lipid peroxidation products. Biochemical Society Transactions **15**, 1103–1104.

Wosnick, M.A., Barnett, R.W. and Carlson, J.E. (1989). Total chemical synthesis and expression in *Escherichia coli* of a maize glutathione-transferase (GST) gene. Gene **76**, 153–160.

Yamaha, T. and Cardini. C.E. (1960). The biosynthesis of plant glycosides. II. Gentiobiosides. Archives of Biochemistry and Biophysics **86**, 133–137.

THE ROLE OF COMPARTMENTATION OF HERBICIDES AND THEIR METABOLITES IN RESISTANCE MECHANISMS

David Coupland

Department of Agricultural Sciences, University of Bristol,
AFRC Institute of Arable Crops Research, Long Ashton Research Station,
Bristol, BS18 9AF, UK.

Compartmentation is defined as a process, or a
series of processes, which physically
separates xenobiotics and their toxic
metabolites, from entering into the main
biochemical reactions of the plant cell. In
some cases, this cellular sequestration has
been shown to limit herbicide transport and
prevent the active ingredient reaching its
target site. This review will describe the
biochemical and physiological processes
involved in achieving this physical separation
between the herbicide and its 'site of
action'. Emphasis is placed on describing
those studies which have shown a link between
compartmentation and herbicide resistance. As
many of these studies have implicated the cell
vacuole as the likely site, or compartment, of
these xenobiotics or their metabolites, its
form and function is briefly discussed.
Similarly, as it is generally assumed that
xenobiotic conjugates need to be formed prior
to compartmentation, their biosynthesis is
also briefly reviewed.

INTRODUCTION

Herbicide resistance may result from differences in the rates of
absorption, translocation, detoxification or sub-cellular distribution
of active ingredient (a.i.) between resistant and susceptible biotypes.
Variations in sensitivity to the herbicide at a target site would also
influence susceptibility to the herbicide. Differences in absorption and
translocation as bases for herbicide selectivity have been reviewed by
Hess (1985). Gressel (1985) also considered these mechanisms, together
with altered sensitivity at the site of action, as explanations for
resistance to photosystem II herbicides.

This review summarises the literature on herbicide compartmentation and
describes its role in herbicide resistance mechanisms. Compartmentation
is defined here as a process, or series of processes, which physically
separates xenobiotics and their phytotoxic metabolites from entering into
the main biochemical reactions of the cell. Herbicide compartmentation,
therefore, involves metabolism of the a.i. and the subsequent 'storage'
of metabolites, probably within the vacuole. Accordingly, the mechanisms

of herbicide detoxification in plants, and the structure and function of the vacuole are also briefly reviewed.

HERBICIDE METABOLISM IN PLANTS

Herbicides are metabolised in plants in many ways. Although these reactions have been reviewed previously (Lamoureux and Frear, 1979; Shimabukuro, 1985; Cole *et al.*, 1987) a brief description of the main biochemical mechanisms whereby herbicides are detoxified in plants is necessary as a background for this review (Fig. 1). Phase I reactions may be considered as the primary metabolism of the herbicide, leading to the formation of compounds which can then undergo further chemical changes in Phases II and III. This sequence of Phase I followed by II then III is common, particularly where metabolism confers herbicide selectivity. For example, Shimabukuro *et al.* (1979) showed that selective control of *Avena fatua* in wheat by diclofop-methyl is due to greater detoxification of the herbicide in the crop, as shown in Figure 2. It should be realised that certain compounds can enter into Phase II or Phase III reactions directly without prior metabolism.

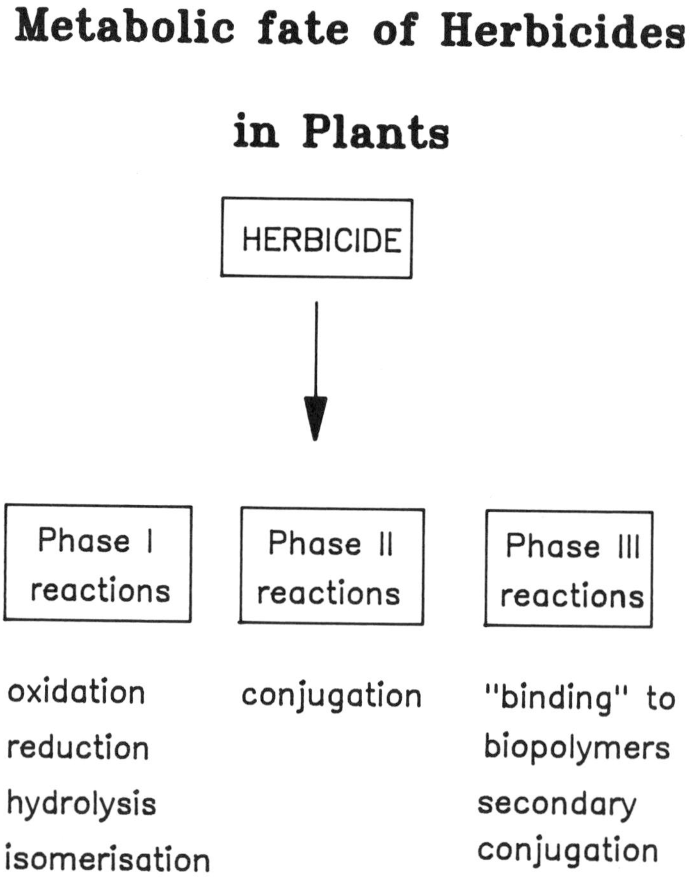

Metabolic fate of Herbicides

in Plants

HERBICIDE

Phase I reactions	Phase II reactions	Phase III reactions
oxidation	conjugation	"binding" to
reduction		biopolymers
hydrolysis		secondary
isomerisation		conjugation

Fig. 1. Herbicide metabolism in plants

Phase II reactions can be classified as 'conjugation metabolism' occurring within the cytoplasm, covalently bonding the herbicide (or metabolites thereof) to naturally occurring compounds such as sugars, amino acids, malonic acid, glutathione, and certain lipophilic compounds, including fatty acids and glycerol (Lamoureux and Rusness, 1986). By far the most common reactions are those involving sugars, in particular glucose (Edwards *et al.*, 1982) as shown in Figure 3. Simple (mono-glycosides) and complex (homo- and heterogeneous) oligo-glycosides may be formed (Lamoureux and Rusness, 1986).

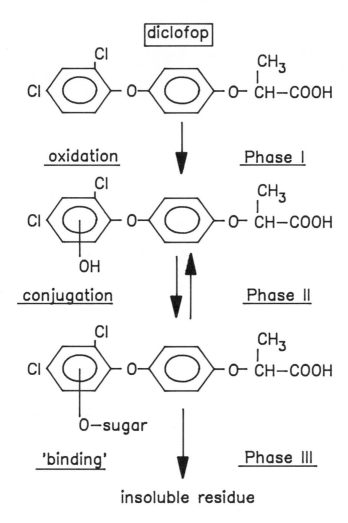

Fig. 2. <u>Metabolic pathway of diclofop in wheat</u> (after Shimabukuro *et al.*, 1979).

Phase III reactions principally involve the incorporation of herbicides and their metabolites into biopolymers such as lignin, cellulose and proteins (Hatzios and Penner, 1982), although Shimabukuro *et al.* (1982) also describe secondary conjugation reactions as Phase III. Compounds that contain, or can be metabolised to form, certain functional groups, particularly $-OH$, $-COOH$, $-NH_2$ and $-SH$, frequently form bound residues.

Glycoside Conjugation

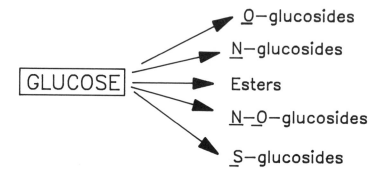

Mono-glycosides ➤ Oligo-glycosides

eg. — Glu–Glu– etc

— arabinose
(or xylose)

— malonyl glucoside

Fig. 3. <u>Phase II conjugation reactions involving sugars</u>.

Whatever the metabolic fate of a compound, Phase II and III reactions have two features in common, even though their products are chemically dissimilar. Firstly, these products are immobile in the plant (cf. the active ingredient); and secondly, they are relatively non-toxic.

The products of Phase III metabolism are immobile and non-toxic due to their insolubility. They are often so tightly bound to biopolymers that they are able to resist the action of hydrolytic enzymes (Pillmoor *et al.*, 1984) and 'extreme' solubilizing techniques, such as boiling for 16 h with strong acid, are often needed for their extraction and analysis (Still *et al.*, 1976).

It has been proposed that some Phase II herbicide metabolites are non-toxic and immobile because they are stored inside the cell vacuole. Once inside this organelle, the metabolites do not diffuse out to any extent, nor are they actively transported back into the cytoplasm (Nakamura *et al.*, 1985; Sterling and Balke, 1989). This immobility of the xenobiotic within the cell, and lack of primary metabolic function within these organelles, effectively renders these herbicide metabolites non-phytotoxic.

STRUCTURE AND FUNCTION OF THE CELL VACUOLE

A brief description of vacuole physiology is necessary in order to appreciate the implications of xenobiotic storage in this organelle. Typically, vacuoles constitute more than 90% of the mature cell volume and it is not surprising, therefore, that these organelles have a vital role in osmoregulation (Keller *et al.*, 1985). Vacuoles are also known to store intermediary metabolites such as organic acids (MacLennan *et al.*, 1963), secondary plant products, e.g. dhurrin (Saunders and Conn, 1978) and carbohydrates (Frehner *et al.*, 1984). As well as these dynamic aspects of vacuolar function, the classical role of the vacuole as a 'static compartment' is also true. There are many examples of substances such as tannins (Matile, 1978) and calcium oxalate (Franceschi and Horner, 1980) permanently deposited within the vacuole.

The vacuolar compartment is enclosed by a cell membrane, the tonoplast, thought to be derived from the Golgi system (Marty, 1978). Transport of substances across the tonoplast is a vital element of vacuole physiology. Isolated vacuoles, have been found to possess transport systems for many substances including organic compounds and inorganic ions (Doll *et al.*, 1979; Buser-Suter *et al.*, 1982; Rasi-Caldogno *et al.*, 1982). Many of these transport systems for organic compounds have been shown to be substrate-specific (Deus-Neumann and Zenk, 1984). In view of the known properties of tonoplast membranes, how can herbicides or their metabolites be transported into the vacuole? It seems probable that herbicides are conjugated with natural metabolites in the cytosol, since this is where the transferases (e.g. glucosyl, malonyl, etc.) which are necessary for these conjugation reactions, are located. (Hrazdina *et al.*, 1982; Werner and Matile, 1985). Thus, herbicides are converted to their conjugate metabolites prior to entering the vacuole. In view of the water solubility of these metabolites, simple diffusion across the lipophilic tonoplast membrane is considered unlikely, or at best, slow. If so, then naturally-occurring permease systems may be responsible for the transport of herbicide conjugates into the vacuole, perhaps 'recognising' the natural metabolite part of the molecule. This function of acting as a 'signal' for deposition of the conjugate in the vacuole has been suggested for the malonyl group – a common feature of many herbicide conjugates (Schmitt *et al.*, 1985; Lamoureux and Rusness, 1986).

Vacuolar sap is acidic, maintained by an active ATP-dependent proton pump (Mandala and Taiz, 1985). Thus, there is the potential for accumulating weak bases in the vacuole by an ion-trap mechanism. Once inside, the protonated cations are unable to diffuse back through the tonoplast. Compounds may also be accumulated if they become associated with charged, high molecular weight substances inside the vacuole, such as tannins, which then act as ion exchangers. However, to the author's knowledge, neither of these two mechanisms is responsible for the vacuolar compartmentation of xenobiotics.

VACUOLAR COMPARTMENTATION OF HERBICIDE METABOLITES

If the conjugates that are formed as a result of Phase II metabolism are inherently non-toxic and cannot be re-converted to the a.i., then they will not be phytotoxic and do not need to be compartmentalised. Gronwald *et al.*, (1989) found that a biotype of *Abutilon theophrasti*, which was resistant to atrazine, converted two to five times more herbicide to a glutathione conjugate than did a susceptible biotype. Conjugation with glutathione is known to occur rapidly in many species (Lamoureux *et al.*, 1971), producing metabolites which undergo a variety of enzymic or non-enzymic reactions, generally involving hydrolysis, to form cystine conjugates. These are then *N*-acylated with malonic acid forming non-toxic end-products (Lamoureux and Rusness, 1986). Recently, Winkler and Sandermann (1989) have shown that N-malonyl conjugates of chlorinated aniline herbicides are not compartmentalised in cell cultures of soybean.

For those herbicides whose metabolites are toxic or readily converted to the parent herbicide, compartmentation in the cell is critical if the plant is to remain unaffected. For example, the amino acid conjugates of phenoxyacid herbicides have been shown to be biologically active (Feung *et al.*, 1974). Glucose esters of acidic herbicides are prone to hydrolysis and are inherently unstable in most plant cells, readily forming the parent herbicide (Cole *et al.*, 1987; Owen 1987).

Despite many years of research, there is only one report in which the specific vacuolar localisation of a herbicide conjugate has been found. Schmitt and Sandermann (1982) incubated cell suspension cultures of soybean with ^{14}C 2,4-D and converted the cells to protoplasts. A vacuole preparation was made from these protoplasts by mild osmotic shock. The purified vacuoles contained β-D-gluosides of 2,4-D, no amino acid conjugates and only minor, perhaps contaminant traces of free 2,4-D.

The reason for the paucity of information on this subject is undoubtedly the technical difficulties of isolating cell vacuoles and the problems of using organic compounds which undergo extensive and often rapid metabolism. Hitherto there have been three main techniques used to determine herbicide localisation in plants: micro-autoradiography, sub-cellular fractionation and efflux analysis. None of these techniques is easy to perform and all have problems in interpretation. Table 1 outlines these problems and gives examples of research in which these techniques have been used to study herbicide localisation in plants. A more recent technique uses an immunological procedure, also detailed in Table 1.

COMPARTMENTATION AND HERBICIDE RESISTANCE

There are no publications that unequivocally relate herbicide resistance with the vacuolar compartmentation of the herbicide or its toxic metabolites. However, there are several reports which demonstrate differences in herbicide metabolism between resistant and susceptible biotypes and *imply* that vacuolar compartmentation is partly or wholly responsible for the lack of phyto-toxicity in the resistant plants. These studies are summarised in Table 2.

Table 1. Techniques used to determine herbicide localisation in plants

Technique/Problems	Example[*]
Micro-autoradiography	
- Leakage and redistribution of water-soluble radioactive compounds during tissue preparation, sectioning and exposure. - Inability to distinguish the chemical identity of radioactivity. - At best, only semi-quantitative. - Poor resolution, even with low energy isotopes. - Reduction in resolution with higher energy isotopes. - Long exposure times required.	2,4-D and dalapon[1] Trifluralin[2] Diuron[3] Picloram[4] Atrazine[5]
Sub-cellular fractionation	
- Diffusion of soluble metabolites out of organelles during isolation and/or centrifugation. - Inability to obtain pure fractions. - Inability to distinguish between soluble metabolites from cell wall, cytosol or vacuole.	2,4-D[6] Diphenamid[7]
Efflux analysis	
- An indirect measurement, assumes efflux is determined and limited by diffusion across cell membranes. - The herbicides or metabolites in the intra-cellular compartments are assumed to have the same specific activity. - Assumes that the compartmentation of herbicide has reached a steady-state at the end of the loading period. - Loading, compartmentation and efflux are all affected by herbicide metabolism. - Herbicides, and many of their formulation compounds, are known to directly affect membrane permeability.	2,4-D[8] Flamprop[9] 2,4-D[10]
Immunofluorescence	
- Facilities are required for antibody production and purification. - The 'target' herbicide needs to be immunogenic, if not, a derivative needs to be synthesised. - Antibodies to the immunogenic herbicide derivative may not be specific only to the parent herbicide.	Atrazine[11]

[*1] Pickering, 1966; [2] Strang and Rogers, 1971a; [3] Strang and Rogers, 1971b; [4] O'Donovan amd Vanden Born, 1981; [5] Norris and Fong, 1983; [6]

Hallam and Sargent, 1970; [7] Yaklich and Scott, 1973; [8] Shone *et al.*, 1974; [9] Pillmoor *et al.*, 1982; [10] Young, 1986; [11] Huber and Sautter, 1986.

Table 2. Studies correlating herbicide resistance with compartmentation of active ingredient or toxic metabolites.

Herbicide	Plant species	Main metabolites formed	Reference
Metribuzin	Soybean (*Glycine max*)	N-glycoside	Smith and Wilkinson, 1974
2,4-D	Tobacco (*Nicotiana tabacum*)	β-glycoside*	Nakamura *et al.*, 1985
Metribuzin	Potato (*Solanum tuberosum*)	Unspecified conjugates	Gawronski *et al*, 1985
Bentazon	Soybean (*Glycine max*)	6-0-glucoside 8-0-glucoside	Sterling and Balke, 1989
Mecoprop	Chickweed (*Stellaria media*)	Unspecified ester conjugates	Coupland *et al.*, 1990

*Compartmentation of parent 2,4-D rather than the glucoside was suggested.

SEQUESTRATION AND HERBICIDE RESISTANCE

Compartmentation ought to be considered as distinct from sequestration. The former implies specific 'boxes' or compartments into which, in this case, herbicides or their metabolites are stored, whereas sequestration is less specific and, simply implies that the herbicide is kept remote from its site of action. There are several examples in the literature of herbicide sequestration, all of which concern paraquat and all show a lack of movement of the a.i. in the resistant species due to a binding of the herbicide in tissues remote from the sites of action. As such, these may be classified as Phase III-type reactions as described earlier in this review. A characteristic of this sequestration is the apparent absence of herbicide metabolism. Table 3 summarises those studies where sequestration has resulted in resistance to paraquat.

CONCLUDING REMARKS

There is an increasing number of reports which suggest that herbicide compartmentation may be responsible, at least in part, for herbicide resistance in some plant species. The two ways of achieving this compartmentation are: storage of the a.i. or its toxic metabolites in the cell vacuole; and the sequestration of the xenobiotic in cells or tissues

Table 3. Paraquat sequestration and subsequent resistance.

Plant species	Location of a.i.	Reference
Conyza bonariensis	Restricted to the leaf vacular tissues.	Fuerst *et al.*, 1985
"	Restricted to cut edge of leaf disc.	Vaughn and Fuerst, 1985
"	Not determined but inability of a.i. to reach the site of action was demonstrated.	Vaughn *et al.*, 1989
Erigeron spp.	Restricted to lower portion of petiole. Less movement in light than dark.	Tanaka *et al.*, 1986
Hordeum glaucum	In or around the leaf vascular tissues. Sequestered in the apoplast?	Bishop *et al.*, 1987

remote from the site of action. The mechanisms of herbicide compartmentation, however, are not well defined and there are problems associated with both the vacuolar 'storage' and the sequestration of herbicides in plants. First, many herbicides and their formulation components, such as surfactants, are known to disrupt cell membranes and affect their physiological function (Table 4). Thus, certain herbicides are likely to inhibit their own compartmentation within vesicles. Second, herbicide metabolites contained within vacuoles are likely to be metabolised, especially if they are prone to hydrolysis, since vacuoles are known to contain a range of hydrolytic enzymes (Boller and Wiemken, 1986). Paradoxically, it is precisely herbicide ester, amide and malonyl conjugates that are thought to be compartmentised in vacuoles because they are so prone to hydrolysis (in the cytoplasm) forming the parent, phytotoxic herbicide. There are two possible solutions to this paradox. First, the hydrolases in the vacuole may be substrate-specific, and therefore cannot hydrolyse xenobiotic conjugates. Alternatively, β-glucosidase and carboxylic acid esterases, the enzymes that would degrade the most common types of conjugate, may be absent from the vacuoles of higher plants (Butcher *et al.*, 1977). Third, the sequestration of paraquat in resistant biotypes, in particular of *Erigeron* spp. and *Conyza bonariensis*, has only been observed after application through cut stems or leaves. Shaaltiel and Gressel (1987) have suggested that this method of application produces artefacts. Furthermore, they concluded that as the sequestration of ^{14}C (from ^{14}C-paraquat) is only apparent after 4 h or more, it is a secondary effect, and perhaps is due to the re-incorporation of ^{14}C-labelled products of ^{14}C-paraquat photo-degradation or metabolism. Gressel and co-workers believe that paraquat resistance in *Conyza bonariensis* is due to increases in the activities of enzymes responsible for the detoxification of paraquat-generated toxic oxygen species (Shaatiel and Gressel, 1986; Jansen *et al.*, in press).

Table 4. **Disruption of membrane structure and function by herbicides and surfactants.**

Substance	Effect	Reference
Herbicides		
EPTC	Membrane disruption	Wilkinson, 1985
Paraquat	Membrane disruption due to lipid peroxidation	Riely *et al.*, 1974
Oxyflurofen	Increased membrane permeability	Prendeville and Warren, 1977
Diclofop	Perturbation of membrane potential affecting transport function	Lucas *et al.*, 1984
Chlorsulfuron	Inhibition of K^+/H^+ transport	Agazio and Giardina, 1987

MCPA)
 Dinitramine)
 Chloropropham)
 Fluazifop-)
   butyl)
 Flamprop-)
   isopropyl)
 Acifluorfen)
 Alachlor)
	Various inhibitory effects on proton gradient development and maintenance by plasmalemma and tonoplast vesicles.	Ratterman and Balke, 1988

Surfactants		

Silwet L-77)
   (organo-)
   silicone))
 Triton X-45)
   (non-ionic))
 Hyspray)
   (cationic))
	Plasmalemma and tonoplast disruption	Coupland *et al.*, 1989

Tween 20) (non-ionic)) DAXAD 21) (anionic))	Increased membrane permability	St. John *et al.*, 1974

Vaughn and colleagues, on the other hand, believe that sequestration of paraquat, preventing it reaching the active site in the chloroplast, is the main mechanism of resistance (Vaughn and Fuerst, 1985; Vaughn *et al.*, 1989). They base this conclusion on several pieces of evidence, the latest of which (above reference) describes the lack of cross-resistance to other toxic oxygen generators such as morfamquat, which has the same mode of action as paraquat (Brian, 1972). Assuming this compound was

272

taken up to the same extent as paraquat, the difference in susceptibility between the resistant and susceptible *Conyza* biotypes to these two herbicides indicates that sequestration may be specific for the paraquat molecule in this species.

Clearly there is a long way to go before the role of herbicide compartmentation in resistance mechanisms can be defined accurately. What is required is an integrated approach, studying herbicide uptake, movement, metabolism and subsequent location of metabolites at the tissue and cell level. Omission of any one of these processes will result in some doubt as to the precise mechanism of resistance to a particular herbicide.

REFERENCES

Agazio, M. de and Giardina, M.C. (1987). Inhibition of fusicoccin stimulated K^+/H^+ transport in root tips from maize seedlings pretreated with chlorsulfuron. Plant, Cell and Environment **10**, 229-232.

Bishop, T., Powles, S.B. and Cornic, G. (1987). Mechanism of paraquat resistance in *Hordeum glaucum*. II. Paraquat uptake and translocation. Australian Journal of Plant Physiology **14**, 539-547.

Boller, T. and Wiemken, A. (1986). Dynamics of vacuolar compartmentation. Annual Review of Plant Physiology **37**, 137-164.

Brian, R.C. (1972). Observations on the physiological basis for the selectivity of morfamquat to graminaceous plants. Pesticide Science **3**, 409-414.

Buser-Suter, C., Wiemken, A. and Matile, P. (1982). A malic acid permease in isolated vacuoles of a crassulacean acid metabolism plant. Plant Physiology **69**, 456-459.

Butcher, H.C., Wagner, G.J. and Siegelman, H.W. (1977). Localization of acid hydrolases in protoplasts. Plant Physiology **59**, 1098-1103.

Cole, D.J., Edwards, R. and Owen, W.J. (1987). The role of metabolism in herbicide selectivity. In "Progress in Pesticide Biochemistry and Toxicology" Vol. 6, (D.H. Hutson and T.R. Roberts, eds), pp. 57-104. John Wiley & Sons, Chichester, England.

Coupland, D., Lutman, P.J.W. and Heath, C. (1990). Uptake, translocation, and metabolism of mecoprop in a sensitive and a resistant biotype of *Stellaria media*. Pesticide Biochemistry and Physiology **36**, 61-67.

Coupland, D., Zabkiewicz, J.A. and Ede, F.J. (1989). Evaluation of three techniques used to determine surfactant phytotoxicity. Annals of Applied Biology **115**, 147-156.

Deus-Neumann, B. and Zenk, M.H. (1984). A highly selective alkaloid uptake system in vacuoles of higher plants. Planta **162**, 250-260.

Doll, S., Rodier, F. and Willenbrink, J. (1979). Accumulation of sucrose in vacuoles isolated from red beet tissue. Planta **144**, 407-411.

Edwards, V.T., McMinn, A.L. and Wright, A.N. (1982). Sugar conjugates of pesticides and their metabolites in plants - current status. In "Progress in Pesticide Biochemistry and Toxicology" Vol. 2, (D.H. Hutson and T.R. Roberts, eds), pp. 71-125. John Wiley & Sons, Chichester, England.

Feung, C-s., Mumma, R.O. and Hamilton, R.H. (1974). Metabolism of 2,4-dichlorophenoxyacetic acid. VI. Biological properties of amino acid conjugates. Journal of Agricultural and Food Chemistry 22, 307-309.

Franceschi, V.R. and Horner, H.T. (1980). Calcium oxalate crystals in plants. Botanical Review 46, 361-427.

Frehner, M., Keller, F. and Wiemken, A. (1984). Localization of fructan metabolism in the vacuoles isolated from protoplasts of Jerusalem-artichoke tubers (Helianthus tuberosis L.). Journal of Plant Physiology 116, 197-208.

Fuerst, E.P., Nakatani, H.Y., Dodge, A.D., Penner, D. and Arntzen, C.J. (1985). Paraquat resistance in Conyza. Plant Physiology 77, 984-989.

Gawronski, S.W., Haderlie, L.C., Callihan, R.H. and Dwelle, R.B. (1985). Metribuzin absorption, translocation and distribution in two potato (Solanum tuberosum) cultivars. Weed Science 33, 629-634.

Gressel, J. (1985). Herbicide tolerance and resistance: alteration of site of activity. In "Weed Physiology, Vol. II, Herbicide Physiology" (S.O. Duke, ed.), pp. 159-189. CRC Press, Boca Raton, USA.

Gronwald, J.W., Andersen, R.N. and Yee, C. (1989). Atrazine resistance in velvetleaf (Abutilon theophrasti) due to enhanced atrazine detoxification. Pesticide Biochemistry and Physiology 34, 149-163.

Hallam, N.D. and Sargent, J.A. (1970). The localization of 2,4-D in leaf tissue. Planta 94, 291-295.

Hatzios, K.K. and Penner, D. (1982). Terminal residues of herbicides and herbicide binding in plants. In "Metabolism of Herbicides in Higher Plants" (K.K. Hatzios and D. Penner, eds), pp. 75-82. Burgess Publishing Company, Minneapolis.

Hess, F.D. (1985). Herbicide absorption and translocation and their relationship to plant tolerances and susceptibility. In "Weed Physiology, Vol. II, Herbicide Physiology" (S.O. Duke, ed.), pp. 191-214. CRC Press, Boca Raton, USA.

Hrazdina, G., Marx, G.A. and Hoch, H.C. (1982). Distribution of secondary plant metabolites and their biosynthetic enzymes in pea (Pisum sativum L.) leaves. Plant Physiology 70, 745-748.

Huber, S.J. and Sautter, C. (1986). Immunofluorescence localisation of conjugated atrazine in leaf pieces of corn Zea mays. Zeitschrift für Pflanzenkrankheiten und Pflanzenschutz 93, 108-113.

Jansen, M.A.K., Shaaltiel, Y., Kazzes, D., Canaani, O., Malkin, S. and Gressel, J. Increased tolerance to photoinhibitory light in paraquat-resistant *Conyza bonariensis* measured by photoacoustic spectroscopy and $^{14}CO_2$-fixation. Plant Physiology **91**, 1174-1178.

Keller, F., Matile, P. and Wiemken, A. (1985). Distribution of saccharides between cytoplasm and vacuole in protoplasts. In "Physiological Properties of Plant Protoplasts" (P.E. Pilet, ed.), pp. 116-121. Springer-Verlag, Berlin.

Lamoureux, G.L. and Frear, D.S. (1979). Pesticide metabolism in higher plants: *in vitro* enzyme studies. In "Xenobiotic Metabolism – *In Vitro* Methods". American Chemical Society Symposium Series No. 97, (G.D. Paulson, D.S. Frear and E.P. Marks, eds), pp. 72-128. American Chemical Society, Washington D.C.

Lamoureux, G.L. and Rusness, D.G. (1986). Xenobiotic conjugation in higher plants. In "Xenobiotic Conjugation Chemistry", American Chemical Society Symposium Series No. 299, (G.D. Paulson, J. Caldwell, D.H. Hutson and J.J. Menn, eds), pp. 62-105. American Chemical Society, Washington D.C.

Lamoureux, G.L., Stafford, L.E., Tanaka, F.S. (1971). Metabolism of 2-chloro-*N*-isopropylacetanilide (Propachlor) in the leaves of corn, sorghum, sugar cane and barley. Journal of Agricultural and Food Chemistry **19**, 346-350.

Lucas, W.J., Wilson, C. and Wright, J.P. (1984). Perturbation of Chara plasmalemma transport function by 2(4(2'4'-dichlorophenoxy) phenoxy) propionic acid. Plant Physiology **74**, 61-66.

MacLennan, D.H., Beevers, H. and Harley, J.L. (1963). 'Compartmentation' of acids in plant tissues. Biochemistry Journal **89**, 316-327.

Mandala, S. and Taiz, L. (1985). Proton transport in isolated vacuoles from corn coleoptiles. Plant Physiology **78**, 104-109.

Marty, F. (1978). Cytochemical studies on GERL, provacuoles, and vacuoles in root meristematic cells of *Euphorbia*. Proceedings National Academy of Science USA **75**, 852-856.

Matile, P. (1978). Biochemistry and function of vacuoles. Annual Review of Plant Physiology **29**, 193-213.

Nakamura, C., Nakata, M., Shioji, M. and Ono, H. (1985). 2,4-D resistance in a tobacco cell culture variant: Cross-resistance to auxins and uptake, efflux and metabolism of 2,4-D. Plant and Cell Physiology **26**, 271-280.

Norris, R.F. and Fong, I.E. (1983). Localization of atrazine in corn (*Zea mays*), oat (*Avena sativa*), and kidney bean (*Phaseolus vulgaris*) leaf cells. Weed Science **31**, 664-671.

O'Donovan, J.T. and Vanden Born, W.H. (1981). A microradioautographic study of ^{14}C-labelled picloram distribution in soybean following root uptake. Canadian Journal of Botany **59**, 1928-1931.

Owen, W.J. (1987). Herbicide detoxification and selectivity. British Crop Protection Conference - Weeds 309-318.

Pickering, E.R. (1966). Autoradiography of mobile, ^{14}C-labelled herbicides in sections of leaf tissue. Stain Technology 41, 131-137.

Pillmoor, J., Gaunt, J. and Roberts, T. (1984). Examination of bound (non-extractable) residues of MCPA and flamprop in wheat straw. Pesticide Science 15, 375-381.

Pillmoor, J.B., Roberts, T.R. and Gaunt, J.K. (1982). Compartmentation and mobility of flamprop and its metabolites in leaf slices of oats. Journal of Experimental Botany 33, 919-928.

Prendeville, G.N. and Warren, G.F. (1977). Effect of four herbicides and two oils on leaf-cell membrane permeability. Weed Research 17, 251-258.

Rasi-Caldogno, F., DeMichelis, M.I. and Pugliarello, M.C. (1982). Active transport of Ca^{2+} in membrane vesicles from pea. Evidence for a H^+/Ca^{2+} antiport. Biochimica et Biophysica Acta 693, 287-295.

Ratterman, D.M. and Balke, N.E. (1988). Herbicidal disruption of proton gradient development and maintenance by plasmalemma and tonoplast vesicles from oat root. Pesticide Biochemistry and Physiology 31, 221-236.

Riely, C.A., Cohen, G. and Lieberman, M. (1974). Ethane evolution: a new index of lipid peroxidation. Science 183, 208-210.

Saunders, J.A. and Conn, E.E. (1978). Presence of the cyanogenic glucoside dhurrin in isolated vacuoles from sorghum. Plant Physiology 61, 154-157.

Schmitt, R., Kaul, J., Trenk, T.V.D., Schaller, E. and Sandermann, H. (1985). β-D-Glucosyl and O-malonyl-β-D-glycosyl conjugates of pentachlorophenol in soybean and wheat: Identification and enzymic synthesis. Pesticide Biochemistry and Physiology 24, 77-85.

Schmitt, R. and Sandermann, H. (1982). Specific localization of β-D-glucoside conjugates of 2,4-dichlorophenoxyacetic acid in soybean vacuoles. Zeitschrift für Naturforschung 37c, 772-777.

Shaaltiel, Y. and Gressel, J. (1986). Multienzyme oxygen radical detoxifying system correlated with paraquat resistance in Conyza bonariensis. Pesticide Biochemistry and Physiology 26, 22-28.

Shaaltiel, Y. and Gressel, J. (1987). Kinetic analysis of resistance to paraquat in Conyza. Plant Physiology 85, 869-871.

Shimabukuro, R.H. (1985). Detoxification of herbicides. In "Weed Physiology. Vol. 2 Herbicide Physiology" (S.O. Duke, ed.), pp. 215-140. CRC Press, Boca Raton, USA.

Shimabukuro, R.H., Lamoureux, G.L., and Frear, D.S. (1982). Pesticide metabolism in plants. Reactions and mechanisms. In "Biodegradation of Pesticides (F. Matsumura and G.R. Krishna Murti, eds), pp. 21-66. Plenum Press, New York.

Shimabukuro, R.H., Walsh, W.C. and Hoerauf, R.A. (1979). Metabolism and selectivity of diclofop-methyl in wild oat and wheat. Journal of Agricultural and Food Chemistry 27, 615-623.

Shone, M.G.T., Bartlett, B.O. and Wood, A.V. (1974). A comparison of the uptake and translocation of some organic herbicides and a systemic fungicide by barley. II. Ralationship between uptake and translocation to shoots. Journal of Experimental Botany 25, 401-409.

Smith, A.E. and Wilkinson, R.E. (1974). Differential absorption, translocation and metabolism of metribuzin [4-amino-6-tert-butyl-3-(methylthio)-as-triazine-5(4H)one] by soybean cultivars. Physiologia Plantarum 32, 253-257.

St John, J.B., Bartels, P.G. and Hilton, J.L. (1974). Surfactant effects on isolated plant cells. Weed Science 22, 233-237.

Sterling, T.M. and Balke, N.E. (1989). Differential bentazon metabolism and retention of bentazon metabolites by plant cell cultures. Pesticide Biochemistry and Physiology 34, 39-48.

Still, G.G., Norris, R.F. and Iwan, J. (1976). Solubilization of bound residues from 3,4-dichloroaniline -^{14}C and propanil-phenyl-^{14}C treated rice root tissues. In "Bound and Conjugated Pesticide Residues". American Chemical Society Symposium Series No. 29, (D.D. Kaufman, G.G. Still, G.D. Paulson and S.K. Bandal, eds), pp. 156-165. American Chemical Society, Washington, D.C.

Strang, R.H. and Rogers, R.L. (1971a). A microradioautographic study of ^{14}C-trifluralin absorption. Weed Science 19, 363-369.

Strang, R.H. and Rogers, R.L. (1971b). A microradioautographic study of ^{14}C-diuron absorption by cotton. Weed Science 19, 355-362.

Tanaka, Y., Chisaka, H. and Saka, H. (1986). Movement of paraquat in resistant and susceptible biotypes of Erigeron philadelphicus and E. canadensis. Physiologia Plantarum 66, 605-608.

Vaughn, K.C. and Fuerst, E.P. (1985). Structural and physiological studies of paraquat-resistant Conyza. Pesticide Biochemistry and Physiology 24, 86-94.

Vaughn, K.C., Vaughan, M.A. and Camilleri, P. (1989). Lack of cross-resistance of paraquat-resistant Hairy Fleabane (Conyza bonariensis) to other toxic oxygen generators indicates enzymic protection is not the resistance mechanism. Weed Science 37, 5-11.

Werner, C. and Matile, P. (1985). Accumulation of coumaroylglucosides in vacuoles of barley mesophyll protoplasts. Journal of Plant Physiology 118, 237-249.

Wilkinson, R.E. (1985). Gibberellin influence on membrane disruption by thiocarbamate herbicides. Pesticide Biochemistry and Physiology **24**, 103-111.

Winkler, R. and Sandermann, H. (1989). Plant metabolism of chlorinated anilines: Isolation and identification of N-glucosyl and N-malonyl conjugates. Pesticide Biochemistry and Physiology **33**, 239-248.

Yaklich, R.W. and Scott, E.G. (1973). The intra-cellular localization of the herbicide diphenamid in corn root tip. Physiologia Plantarum **28**, 447-451.

Young, G.J. (1986). "The Stability and Localisation of Herbicide Metabolites in Plants", 266 pp. Ph.D. thesis, University College of North Wales, Bangor.

SYNERGISTS TO COMBAT HERBICIDE RESISTANCE

M.S. Kemp and J.C. Caseley

Department of Agricultural Sciences, University of Bristol, AFRC Institute of Arable Crops Research, Long Ashton Research Station, Long Ashton, Bristol, BS18 9AF, U.K.

Synergism is defined as "the co-operative action of two components of a mixture such that the total effect is greater than the sum of the effects of the components used independently". Mechanisms known to be involved in herbicide synergism are increased uptake and mobility and decreased detoxification. These offer direct approaches to overcoming some types of herbicide resistance, notably that due to enhanced degradation. Weeds with this latter form of resistance, such as chlorotoluron-resistant *Alopecurus myosuroides*, are often cross-resistant to herbicides with other modes of action, and this seriously limits treatment options.

In *A. myosuroides*, resistance to the phenylureas is due to rapid ring hydroxylation and N-demethylation. This detoxification can be overcome by treatment with inhibitors of cytochrome P_{450} mixed function oxidase such as aminobenzotriazole.

Interspecific variations in oxidative herbicide catabolism account for the selectivity of several herbicides. A greater knowledge of the inter- and intra-specific variations of the oxygenases involved in crops and weeds could provide a rational basis for the design of synergists and herbicide molecules with improved activity and selectivity and the combined capacity to control some types of herbicide-resistant weeds.

INTRODUCTION

Strategies for controlling herbicide resistance in weeds depend currently upon the use of herbicide rotations and mixtures, and changes in agricultural practices including modifications of cultural methods and the rotation of crops (Parochetti *et al.*, 1982; Slife, 1986; Powles, 1987; Gressel and Segel, 1990; LeBaron and McFarland, 1990; Gressel, these Proceedings). These methods are mainly preventative and are designed to delay the development of resistant weed populations by decreasing, varying or breaking herbicide selection pressure and decreasing the seed-bank. Once resistance is established, alternative herbicides have to be used and this may involve growing other crops. However, advances in our understanding of the modes of action of herbicides (Dodge, 1989), their metabolism and mechanism of selectivity (Hatzios and Penner, 1982; Shimabukuro, 1985; Lamoureux and Rusness, 1986a; Cole *et al.*, 1987) and synergistic interactions with adjuvants (Hatzios and Penner, 1985;

Gressel, 1990) indicate that it may be possible to combat some forms of herbicide resistance more directly by the use of synergists. Such an approach has potential economic and environmental advantages.

Synergy

Tammes (1964) defined synergy as "The co-operative action of two components of a mixture such that the total effect is greater or more prolonged than the sum of the effects of the compounds taken independently". Several mathematical models of varying complexity have been developed to measure synergy (Colby, 1967; Morse, 1978; Hatzios and Penner, 1985), but for the present purposes only substantial synergistic interactions are of interest and the simple model illustrated in Figure suffices.

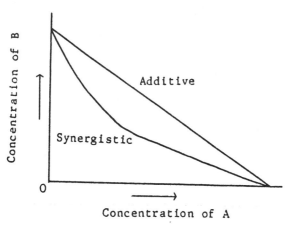

Fig. 1. Isoboles of additive and synergistic interactions.

In the case of additive interactions, isoboles of, for example, I_{50} values are represented by a straight line joining I_{50} values for 100% A and 100% B. Isoboles of synergistic interactions are curves lying below this line, the greater the synergy, the deeper the curve.

MECHANISMS OF SYNERGISTIC ACTION

Numerous examples have been reported of synergistic interactions between herbicides and other agrochemicals including herbicides, fungicides, insecticides, fertilisers and adjuvants (Hatzios and Penner, 1985), and some of these synergistic herbicide/herbicide and herbicide/adjuvant combinations are used as commercial formulations (Table 1).

In many cases the mechanisms of these synergistic interactions are not known, but in those which have been elucidated, synergism of herbicide activity is attributed to increased amounts of herbicide reaching the site of action through increased absorption and mobility and/or through decreased degradation and detoxification. Synergism arising from interactions at the physiological and biochemical site of action may also occur, but clear examples of this mechanism have yet to be demonstrated.

Table 1.   Synergistic   combinations   of   herbicide/herbicide   and
herbicide/adjuvant used in commercial herbicide formulations (after Polge,
1989).

Herbicide	Herbicide/ adjuvant	Product (Company)	Weed/ Reference
Amitrole	ammonium thiocyanate	Radaxone TL (SOPRA) Weedazol TL (Union Carbide)	*Agropyron repens*[1]
Amitrole	paraquat	Groundhog (ICI)	*Agropyron*
Simazine	paraquat	Gramazine (ICI) Terraklene (SOPRA)	*repens*[2]
Paraquat	diuron	Dexuron (Chipman) Totacol (SOPRA)	*Paspalum conjugatum*[3]
Simetryn	MCPA	Grakill (Hokko Chemical)	*Echinochloa crus-galli*[4]
Simetryn	thiobencarb	Saturn (Hokko Chemical)	
Alachlor	atrazine	Lasso (Monsanto) Rambo (Siapa)	*Echinochloa crus-galli*[5]
Atrazine	tridiphane	Tandem (Dow)	*Panicum milaceum*[6] *Setaria faberii*[7]

[1] Donnaley and Ries (1964); [2] Putnam and Ries (1967); [3] Headford (1970); [4] Hagimoto and Yoshikawa (1972); [5] Akobundu *et al.* (1975); [6] Ezra *et al.* (1985); [7] Zorner (1983).

Surfactants, oils and emulsifiers may increase foliar uptake of herbicides (Hodgson, 1982; Holloway and Stock, 1990) and the addition of fertiliser salts including ammonium ions, and plant growth regulators including phenoxyalkanoic acids have been reported to increase apoplastic movement (McWhorter, 1971; Devlin and Yaklich, 1971, 1972; Blair, 1975; Suwanketnikom and Penner, 1978; Hatzios and Penner, 1985). Ethephon and the morphactins promote symplastic movement by the activation of dormant buds on roots and rhizomes thus increasing sink strength (Binning *et al.*, 1971; Caseley, 1972; Carson and Bandeen, 1975; Baradari *et al.*, 1980).

An early example of synergism resulting from the inhibition of herbicide degradation and detoxification was seen when propanil was used to control *Echinochloa crus-galli* (L.) Beauv. in rice treated with carbamate insecticides. This interaction was memorable, not so much for the improved control of *E. crus-galli* but rather for the devastating effect it had upon the crop. Carbamate and organophosphate insecticides inhibit the hydrolysis of propanil by aryl acrylamidase which is the major detoxification pathway conferring tolerance to propanil in rice (Fig. 2), (Frear and Still, 1968; Matsunaka, 1968, 1971; Still and Herrett, 1976).

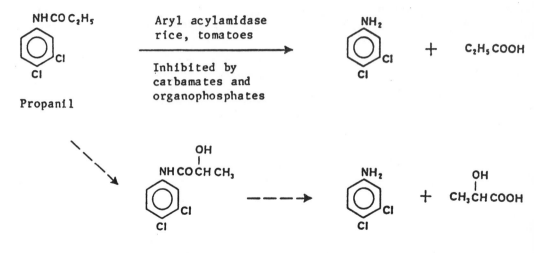

Fig. 2. **Inhibition of propanil detoxification by carbamates and organophosphates**.

A more desirable example of herbicide synergy resulting from inhibited detoxification is that of amitrole with ammonium thiocyanate. As previously mentioned ammonium ions may promote herbicide uptake and transport, but in this case the synergism is also associated with the inhibited condensation of amitrole with serine (Fig. 3) (Racusen, 1958; Carter, 1975; Cook and Duncan, 1979; Cook *et al.*, 1981). The practical advantages of this synergy are realised in commercial formulations such as Amitrole-T (Table 1).

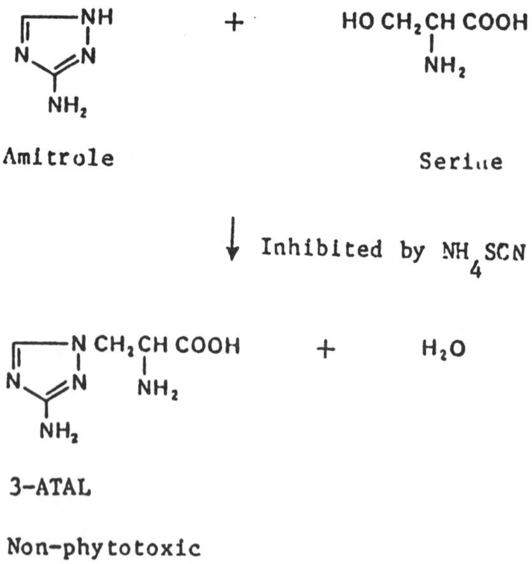

Amitrole   +   HO CH₂CH COOH
                                                                NH₂

Amitrole   Serine

↓ Inhibited by NH₄SCN

3-ATAL   +   H₂O

3-ATAL

Non-phytotoxic

Fig. 3. **Inhibition of amitrole detoxification by ammonium thiocyanate**.

SYNERGY AND TYPES OF HERBICIDE RESISTANCE

How can synergistic interactions be used most efficiently to combat herbicide resistance? This question can only be answered by considering the mechanisms known to confer resistance. These include – modifications to, and over–abundance of, the target binding site and increased herbicide detoxification by metabolism and sequestration.

It is difficult to envisage how synergists could overcome resistance due to modification of the target site, for example, chloroplastic resistance to PS II inhibitors (van Oorschot, these Proceedings; Trebst, these Proceedings) and insensitivity to sulphonylurea and imidazolinone herbicides in mutated acetolactate synthase (Thill *et al.*, these Proceedings; Shaner, these Proceedings).

Greater success might be expected where resistance is attributable to an increased amount of the target enzyme. The synergistic promotion of herbicide availability at the site of action could increase phytotoxicity, but the beneficial effects would depend on the abundance of the target enzyme. This mechanism of resistance has not been widely demonstrated in weeds; however, over–expression of 5–enolpyruvyl-shikimate–3–phosphate synthase activity has been shown to confer resistance in glyphosate–resistant *Convolvulus arvensis* (Kosinski and Weller, 1989). Bocion (1986) has reported strong synergy between glyphosate and imazapyr in several species and studies with this herbicide combination should be considered for controlling glyphosate–resistant *C. arvensis*.

The most effective applications of synergism to the control of resistance should arise where the synergistic mechanism is carefully matched and chosen to eliminate or reduce the mechanism of resistance. Those instances where resistance results from enzyme–mediated herbicide degradation and detoxification appear most suited to this approach. Such mechanisms are associated with the particularly problematical cross–resistance that has developed in *Alopecurus myosuroides* in the UK (Kemp and Caseley, 1987; Kemp *et al.*, 1990). Herbicide selectivity is also dependent upon differences in herbicide detoxification, but there are indications that there may be sufficient interspecific variation to enable herbicide phytotoxicity to be synergised in resistant weeds without adversely affecting crop tolerance.

The pathways of S–triazine degradation in higher plants (Fig. 4) are species–dependent (Hatzios and Penner, 1982; Shimabukuro, 1985; Lamoureux and Rusness, 1986a). In maize (*Zea mays*), hydroxylation is a major degradation pathway in root tissues (Willard and Penner, 1976), but glutathione conjugation is a significant contributor to degradation in leaf tissues (Shimabukuro *et al.*, 1971). In giant foxtail (*Setaria faberi*), degradation via glutathione conjugation predominates (Thompson, 1972; Jensen *et al.*, 1977). These degradative and detoxifying processes confer resistance to the S–triazine atrazine in both maize and *S. faberi*. Tridiphane synergises atrazine phytotoxicity by inhibiting glutathione conjugation in *S. faberi*, but it has little effect upon the tolerance of maize to this herbicide (Boydston and Slife, 1986). Lamoureux and Rusness (1986b) have shown that tridiphane forms glutathione conjugates (Fig. 5) which are active inhibitors of glutathione–*S*–transferases in *S. faberi* but are less inhibitory to glutathione–*S*–transferases from maize. Furthermore, the tridiphane–glutathione conjugates are quickly degraded in maize. Resistant populations of velvetleaf (*Abutilon theophrasti*) which also detoxify atrazine via glutathione conjugation (Gronwald *et al.*, 1989) might be controlled by a similar selective synergism.

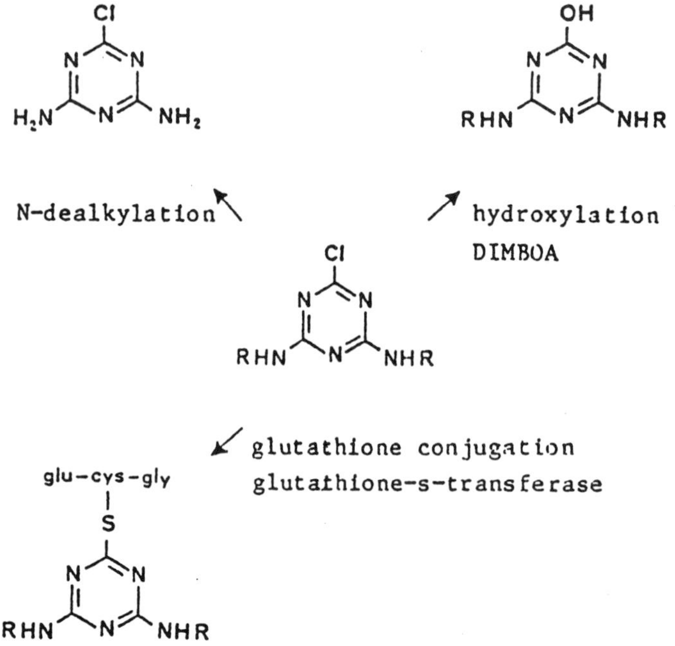

N-dealkylation   hydroxylation
        DIMBOA

glutathione conjugation
glutathione-s-transferase

glu-cys-gly

Fig. 4. Metabolism of chloro-S-triazines in higher plants.

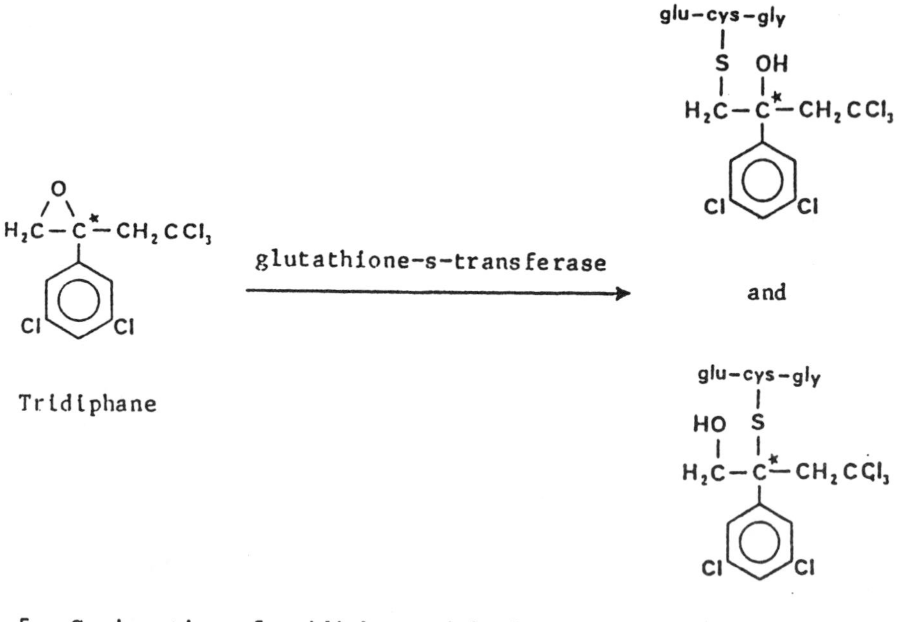

glu-cys-gly

glutathione-s-transferase

and

Tridiphane

glu-cys-gly

Fig. 5. Conjugation of tridiphane with glutathione (after Lamoureux and Rusness, 1986b).

In addition to its conjugation with xenobiotics, glutathione also

284

interacts with and quenches free radicals, and represents one of several routes by which these highly damaging moieties can be sequestered. Dionigi and Dekker (1988) have reported that pre-treatment with tridiphane synergised paraquat phytotoxicity in soybean (*Glycine max* (L.) Merr.), possibly as a result of the removal of free glutathione through its conjugation with tridiphane and a subsequent reduction in quenching of the free radicals generated by paraquat.

Other synergistic interactions of tridiphane may be associated with the inhibition of cytochrome P_{450} mixed-function oxygenases. The degradation of metalochlor by a microsomal fraction prepared from shoots of grain sorghum (*Sorghum bicolor* (L.) Moench.) required NADPH plus oxygen and was inhibited by tridiphane and other known inhibitors of cytochrome P_{450} including carbon monoxide, piperonyl butoxide and tetcyclacis (Moreland *et al.*, 1989a; Moreland *et al.*, 1989b).

Combinations of tridiphane with several herbicides including isoproturon and tri-allate were often synergistic enabling control of herbicide-resistant *A. mysosuroides* (Caseley *et al.*, these Proceedings). In view of the metabolic profiles of these herbicides it is likely that tridiphane synergism of isoproturon involves inhibition of cytochrome P_{450} and tridiphane synergism of tri-allate involves inhibition of glutathione-*S*-transferases. Another synergist which limits degradation is MZH2091 (picolinic acid t-butyl) which increases the activity of metribuzin against *Ipomoea hederacea* by inhibiting deamination of the herbicide (Klamroth *et al.*, 1989).

A more widely reported inhibitor of cytochrome P_{450} is 1-aminobenzotriazole (ABT), which is a suicide substrate of the haem centre. Oxidation of the 1-amino group in ABT results in the generation of benzyne which binds covalently to the adjacent nitrogen atoms in the porphyrin ring and displaces the iron at the haem centre (Ortiz de Montellano and Mathews, 1981). The phytotoxicity of chlorotoluron and isoproturon is synergised by 1-aminobenzotriazole in resistant *A. mysosuroides*, but not in susceptible populations (Kemp and Caseley, 1987).

This observation suggests that the resistance is attributable to rapid herbicide degradation since the oxidative degradation of phenylureas via N-dealkylation and ring alkyloxidation (Ryan *et al.*, 1981) has been associated with cytochrome P_{450} (Frear *et al.*, 1969; Gonneau *et al.*, 1988) and is inhibited in several plant species by 1-aminotriazole (Gaillardon *et al.*, 1985; Cole and Owen, 1987; Cabanne *et al.*, 1987). Metabolic studies confirm that herbicide metabolism is more rapid in resistant *A. mysosuroides* and is inhibited by 1-aminobenzotriazole (Kemp *et al.*, 1990 and Fig. 6). Suspension cell cultures and microsomes of resistant compared with susceptible *A. mysosuroides* degraded ^{14}C-chlorotoluron more rapidly (x2) and addition of ABT reduced herbicide metabolism in both systems. The CO difference spectra of the resistant biotype showed a substantially larger peak at 450 nm providing further evidence that chlorotoluron resistance is associated with elevated levels of cytochrome P_{450} and blockers of this enzyme are potential synergists (Caseley *et al.*, 1990). However, a range of other inhibitors of cytochrome P_{450} were found to differ widely in their ability to synergise chlorotoluron activity (Kemp *et al.*, 1988; Kemp *et al.*, 1990).

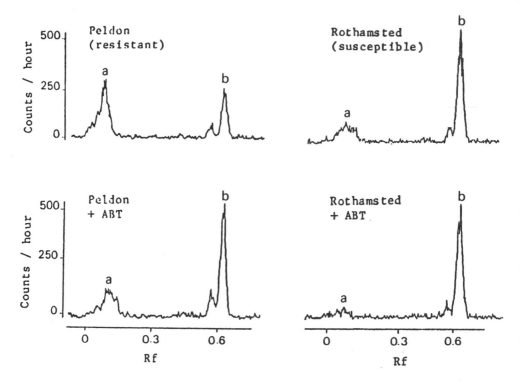

Fig. 6.   Inhibition of ^{14}C-chlorotoluron degradation in foliage of _Alopecurus myosuroides_ by 1-aminobenzotriazole (ABT). Chlorotoluron (0.2 mg l^{-1}) and ABT (7.5 mg l^{-1}) were incorporated in the liquid medium of hydroponically-grown plants. After one day the plants were transferred to chlorotoluron-free medium, and after two days the foliage was extracted with 80% methanol/water (unpublished data).
a = conjugated metabolites; b = chlorotoluron.

FUTURE DEVELOPMENTS

Pre-requisites for successful use of synergists to control herbicide-resistant weeds are that the mechanism of resistance is understood, and that the mode of action of the synergist is appropriate for counteracting it.   Thus, synergists offer little chance of negating resistance resulting from a modified site of action, have limited potential for combating over-expression, but provide considerable opportunities for overcoming resistance due to enhanced herbicide metabolism.

Cytochrome P_{450} mixed-function oxidases have been associated with herbicide degradation in plants (Frear _et al._, 1969; Makeev _et al._, 1977), and further examples provide evidence of inter- and intraspecific variations in their occurrence and specificity (Fonne-Pfister _et al._, 1988; Sterling and Balke, 1989; McFadden _et al._, 1989;   Zimmerlin and Durst, these Proceedings). Furthermore there is evidence of selectivity in the activity of plant cytochrome P_{450} inhibitors (Reichhart _et al._, 1982; Rademacher _et al._, 1987, Burden _et al._, 1987). Numerous examples of these potential herbicide synergists are available including insecticide synergists e.g. piperonyl butoxide (Raffa and Priester, 1985), the ergosterol biosynthesis inhibiting fungicides (Henry and Sisler, 1984) and the kaurene-oxidase inhibiting plant growth regulators (Rademacher *et al.*, 1987).

Cytochrome P_{450} oxidases are widespread in living organisms (Schuster, 1989). Relatively little is known of their activity in plants, but they have been studied more intensively in animal systems. Their importance in drug metabolism was recognised at an early stage and with the aid of iron-porphyrin model systems the chemistry and mechanisms of cytochrome P_{450}-substrate interactions are now being studied to improve knowledge of drug metabolism and toxicology (Mansuy *et al.*, 1989). By studying the inter- and intraspecific variations of oxygenase and glutathione-*S*-transferase enzyme systems in weed and crop plants, and the stereochemical selectivity of their inhibitors we may be able to gain the knowledge required to design more environmentally acceptable selective herbicides and synergistic mixtures with improved selectivity and reduced problems of resistance.

REFERENCES

Akobundu, I.O., Sweet, R.D. and Duke, W.B. (1975). A method of evaluating herbicide combinations and determining herbicide synergism. Weed Science 23, 20-25.

Baradari, M.R., Haderlie, L.C. and Wilson, R.G. (1980). Chlorflurenol effects on absorption and translocation of dicamba in Canada thistle (*Cirsium arvense*). Weed Science 28, 197-200.

Binning, L.K., Penner, D. and Meggitt, W.F. (1971). The effect of 2-chloroethyl phosphonic acid on dicamba translocation in wild garlic (*Allium vincale* L.). Weed Science 19, 73-75.

Blair, A.M. (1975). The addition of ammonium salts of phosphate esters to herbicides to control *Agropyron repens* (L.) Beauv. Weed Research 15, 101-105.

Bocion, P. (1986). Synergistic herbicidal compositions containing glyphosate. European Patent EP 234, 379, 31 pp.

Boydston, R.A. and Slife, F.W. (1986). Alteration of atrazine uptake and metabolism by tridiphane in giant foxtail (*Setaria faberii*) and corn (*Zea mays*). Weed Science 34, 850-858.

Burden, R.S., Carter, G.A., Clark, T., Cooke, D.T., Croker, S.J., Deas, A.H.B., Hedden, P., James, C.S. and Lenton, J.R. (1987). Comparative activity of the enantiomers of triadimenol and paclobutrazol as inhibitors of fungal growth and plant sterol and gibberellin biosynthesis. Pesticide Science 21, 253-267.

Cabanne, F., Huby, D., Gaillardon, P., Scalla, R. and Durst, F. (1987). Effect of the cytochrome P_{450} inactivator 1-aminobenzotriazole on the metabolism of chlorotoluron and isoproturon in wheat. Pesticide Biochemistry and Physiology 28, 371-380.

Carson, A.G. and Bandeen, J.D. (1975). Influence of ethephon on absorption and translocation of herbicides in Canada thistle. Canadian Journal of Plant Science 55, 795-800.

Carter, M.C. (1975). Amitrole. In "Herbicides – Chemistry Degradation and Mode of Action" 2nd Ed. Volume 1. (P.C. Kearney and D.D. Kaufman, eds), pp. 377–398. Marcel Dekker Inc., New York.

Caseley, J. (1972). The use of plant growth regulators in the control of *Agropyron repens*. 11th British Weed Control Conference, 736–743.

Caseley, J.C. Kueh, J., Jones, O.T.G., Hedden P. and Cross, A.R. (1990). Mechanism of chlorotoluron resistance in *Alopecurus myosuroides*. Abstracts 7th International Congress of Pesticide Chemistry 1, (H. Frehse, E. Kesseler-Schmitz and S. Conway, eds), p. 417. IUPAC, Hamburg.

Colby, S.R. (1967). Calculating synergistic and antagonistic responses of herbicide combinations. Weeds 15, 20–22.

Cole, D.J. and Owen, W.J. (1987). Influence of mono-oxygenase inhibitors on the metabolism of the herbicides chlorotoluron and metolachlor in cell suspension cultures. Plant Science 50, 13–20.

Cole D.J., Edwards, R. and Owen, W.J. (1987). The role of metabolism in herbicide selectivity. In "Progress in Pesticide Biochemistry and Toxicology", Volume 6, Herbicides, (D.H. Hutson and T.R. Roberts, eds), pp. 57–104. John Wiley & Sons, Chichester.

Cook, G.T., and Duncan, H.J. (1979). Mode of action of thiocyanates and iodides in aminotriazole formulations. Pesticide Science 10, 281–290.

Cook, G.T., Stephen, N.H. and Duncan, H.J. (1981). Influence of ammonium thiocyanate on scorching and control of bracken (*Pteridium aquilinum*) by amitrole. Weed Science 29, 196–200.

Devlin, R.M. and Yaklich, R.W. (1971). Influence of GA on uptake and accumulation of naptalan by bean plants. Weed Science 19, 135–137.

Devlin, R.M. and Yaklich, R.W. (1972). Influence of two phenoxy growth regulators on the uptake and accumulation of naptalan by bean plants. Physiologia Plantarum 27, 317–320.

Dionigi, C.P. and Dekker, J.H. (1988). The effects of tridiphane pretreatment on response of soybeans to paraquat-induced oxidative stress. Weed Science Society of America Abstracts 28, 201.

Dodge, A. (1989). "Herbicides and Plant Metabolism". Society for Experimental Biology Seminar Series 38, Cambridge University Press, Cambridge. pp. 277.

Donnaley, W.F. and Ries, S.K. (1964). Amitrole translocation in *Agropyron repens* increased by the addition of ammonium thiocyanate. Science 145, 497–498.

Ezra, G., Dekker, J.H. and Stephenson, G.R. (1985). Tridiphane as a synergist for herbicides in corn (*Zea mays*) and proso millet (*Panicum miliaceum*). Weed Science 33, 287–290.

Fonne-Pfister, R., Simon, A., Salaun, J-P. and Durst, F. (1988). Xenobiotic metabolism in higher plants. Involvement of microsomal cytochrome P_{450} in aminopyrine N-demethylation. Plant Science 55, 9–20.

Frear, D.S. and Still, G.G. (1968). The metabolism of 3,4-dichloropropionanilide in plants. Partial purification and properties of an aryl acylamidase from rice. Phytochemistry 7, 913-920.

Frear, D.S., Swanson, H.R. and Tanaka, F.S. (1969). N-demethylation of substituted 3-(phenyl)-1-methylureas : isolation and characterisation of a microsomal mixed-function oxidase from cotton. Phytochemistry 8, 2157-2169.

Gaillardon, P., Cabanne, F., Scalla, R. and Durst, F. (1985). Effect of mixed-function oxidase inhibitors on the toxicity of chlortoluron and isoproturon to wheat. Weed Research 25, 397-402.

Gonneau, M. Pasquette, B., Cabanne, F. and Scalla, R. (1988). Metabolism of Chlortoluron in tolerant species : possible role of cytochrome P_{450} mono-oxygenases. Weed Research 28, 19-25.

Gressel, J. (1990). Synergising herbicides. Reviews in Weed Science (in press).

Gressel, J. and Segel, L.A. (1990). Herbicide rotations and mixtures, effective strategies to delay resistance. In "Managing Resistance to Agrochemicals: From Fundamental Research to Practical Strategies" (M.B. Green, H.M. LeBaron and W.K. Moberg, eds), pp. 430-458. American Chemical Society, Washington, DC.

Gronwald, J.W., Andersen, R.N. and Yee, C. (1989). Atrazine resistance in velvetleaf (Abutilon theophrasti) due to enhanced atrazine detoxification. Pesticide Biochemistry and Physiology 34, 149-163.

Hagimoto, H. and Yoshikawa, H. (1972). Synergistic interactions between inhibitors of growth and photosynthesis. 1. The "growth-dilution" hypothesis. Weed Research 12, 21-30.

Hatzios, K.K. amd Penner, D. (1982). "Metabolism of Herbicides in Higher Plants". Burgess Publishing Company, Minneapolis.

Hatzios, K.K. amd Penner, D. (1985). Interactions of herbicides with other agrochemicals in higher plants. Reviews of Weed Science 1, 1-63.

Headford, D.W.R. (1970). Influence of light on paraquat activity in the tropics. Pesticide Science 1, 41-42.

Henry, M.J. and Sisler, H.D. (1984). Effects of sterol biosynthesis-inhibiting (SBI) fungicides on cytochrome P_{450} oxygenations in fungi. Pesticide Biochemistry and Physiology 22, 262-275.

Hodgson, R.H. (1982). "Adjuvants for Herbicides". Weed Science Society of America, Champaign.

Holloway, P.J. and Stock, D. (1990). Factors affecting the activation of foliar uptake of agrochemicals by surfactants. In "Industrial Applications of surfactants", Volume II. (d. Karsa, ed.), Royal Society of Chemistry Special Publication 77, pp. 303-337. Cambridge.

Jensen, K.I.N., Stephenson, G.R. and Hunt, C.A. (1977). Detoxification of atrazine in three gramineae subfamilies. Weed Science 25, 212-220.

Kemp, M.S. and Caseley, J.C. (1987). Synergistic effects of 1-amino-benzotriazole on the phytotoxicity of chlorotoluron amd isoproturon in a resistant population of black-grass (*Alopecurus myosuroides*). British Crop Protection Conference - Weeds, 895-899.

Kemp, M.S., Moss, S.R. and Thomas, T.H. (1990). Herbicide resistance in *Alopecurus myosuroides*. In "Managing Resistance to Agrochemicals: From Fundamental Research to Practical Strategies" (M.B. Green, H.M. LeBaron and W.K. Moberg, eds), pp. 376-393. American Chemical Society, Washington, DC.

Kemp, M.S., Newton, L.V. and Caseley, J.C. (1988). Synergistic effects of some P_{450} oxidase inhibitors on the phytotoxicity of chlorotoluron in a resistant population of black-grass (*Alopecurus myosuroides*). Proceedings of the European Weed Research Society Symposium, 1988, "Factors affecting herbicidal activity and selectivity", 121-126. Wageningen.

Klamroth, E.E.., Fedtke, C. and Kühbuch, W.C. (1989). Mechanism of synergism between metribuzin and MZH2091 on ivyleaf morning glory (*Ipomoea hederacea*). Weed Science 37, 517-520.

Kosinski, W. and Weller, S.C. (1989). 5-Enolpyruvyl-shikimate-3-phosphate synthase activity in field bindweed, *Convolvulus arvensis* L., biotypes. Weed Science Society of America Abstract 29, 76-77.

Lamoureux, G.L. and Rusness, D.G. (1986a). Xenobiotic conjugation in higher plants. In "Xenobiotic Conjugation Chemistry", (G.D. Paulson, J. Caldwell, D.H. Hutson and J.J. Menn, eds), pp. 62-105. American Chemical Society, Washington, DC.

Lamoureux, G.L. and Rusness, D.G. (1986b). Tridiphane [2-(3,5-dichloro-phenyl)-2-(2,2,2-trichloroethyl)oxirane] an atrazine synergist : enzymatic conversion to a potent glutathione S-transferase inhibitor. Pesticide Biochemistry and Physiology 26, 323-342.

LeBaron, H.M. and McFarland, J. (1990). Herbicide resistance in weeds and crops: an overview and prognosis. In "Managing Resistance to Agrochemicals: From Fundamental Research to Practical Strategies" (M.B. Green, H.M. LeBaron and W.K. Moberg, eds), pp. 336-352. American Chemical Society, Washington, DC.

Makeev, A.M. Makoviechuk, A.I.U. and Chkanikov, D.C. (1977). Microsomal hydroxylation of 2,4-D in plants. Doklady Akademic Nauk SSSR 233, 1222-1225.

Mansuy, D., Battioni, P. and Battioni, J.P. (1989). Chemical model systems for drug-metabolizing cytochrome P_{450}-dependent mono-oxygenases. European Journal of Biochemistry 184, 267-285.

Matsunaka, S. (1968). Propanil hydrolysis : Inhibition in rice plants by insecticides. Science 160, 1360-1361.

Matsunaka, S. (1971). Metabolism of acylamilide herbicides. In "Pesticide Terminal Residues" (A.S. Tahori, ed.), pp. 343-365. International Union of Pure and Applied Chemistry, Butterworths, London.

McFadden, J.J., Frear, D.S. and Mansager, E.R. (1989). Aryl hydroxylation of diclofop by a cytochrome P_{450} dependent mono-oxygenase from wheat. Pesticide Biochemistry and Physiology 34, 92-100.

McWhorter, C.G. (1971). The effect of alkali metal salts on the toxicity of MSMA and dalapon to johnsongrass. Weed Science Society of America Abstracts, p. 84.

Moreland, D.E., Corbin, F.T. and Novitzky, W.P. (1989a). Metabolism of metolachlor by a microsomal fraction isolated from grain sorghum, Sorghum bicolor (L.) Moench. shoots. Weed Science Society of America Abstracts 29, 70-71.

Moreland, D.E., Novitzky, W.P. and Levi, P.E. (1989b). Selective inhibition of cytochrome P_{450} isozymes by the herbicide synergist tridiphane. Pesticide Biochemistry and Physiology 35, 42-49.

Morse, P.M. (1978). Some comments on the assessment of joint action in herbicide mixtures. Weed Science 26, 58-71.

Ortiz de Montellano, P.R. and Mathews, J.M. (1981). Autocatalytic alkylation of the cytochrome P_{450} prosthetic haem group by 1-aminobenzo-triazole. Biochemical Journal 195, 761-764.

Parochetti, J.V., Schnappinger, M.G., Ryan, G.F. and Collins, H.A. (1982). Practical significance and means of control of herbicide-resistant weeds. In "Herbicide Resistance in Plants" (H.M. LeBaron and J. Gressel, eds), pp. 309-324, John Wiley & Sons, New York.

Polge, N.D. (1989). "Studies into the Modification of Herbicide Activity by Chemical Safeners and Synergists". Ph.D. Thesis, University of Bath.

Powles, S.B. (1987). A review of weeds in Australia resistant to herbicides. Proceedings of the 8th Australian Weeds Conference, pp. 109-113.

Putnam, A.R. and Ries, S.K. (1965). The effects of adjuvants on the activity of herbicides for the control of quackgrass (Agropyron repens L. Beauv.). Proceedings North Central Weed Control Conference 19, 300.

Putnam, A.R. and Ries, S.K. (1967). The synergistic action of herbicide combinations containing paraquat on Agropyron repens (L.) Beauv. Weed Research 7, 191-199.

Racusen, D. (1958). The metabolism and translocation of 3-amino-triazole in plants. Archives of Biochemistry and Biophysics 74, 106-113.

Rademacher, W., Fritsch, H., Graebe, J.E., Sauter, H. and Jung, G. (1987). Tetcyclacis and triazole-type plant growth retardants : their influence on the biosynthesis of gibberellins and other metabolic processes. Pesticide Science 21, 241-252.

Raffa, K.F. and Priester, T.M. (1985). Synergists as research tools and control agents in agriculture. Journal of Agricultural Entomology 2, 27-45.

Reichhart, D., Simon, A. and Durst, F. (1982). Autocatalytic inactivation of plant cytochrome P_{450} enzymes : Selective inactivation of cinnamic acid 4-hydroxylase from *Helianthus tuberosus* by 1-amino-benzotriazole. Archives of Biochemistry and Biophysics 216, 522-529.

Ryan, P.J., Gross, D., Owen W.J. and Laanio, T.L. (1981). The metabolism of chlortoluron, diuron and CGA 43057 in tolerant and susceptible plants. Pesticide Biochemistry and Physiology 16, 213-221.

Schuster, I. (1989). "Cytochrome P_{450} : Biochemistry and Biophysics". Taylor and Francis, London.

Shimabukuro, R.H. (1985). Detoxication of herbicides. In "Weed Physiology", Volume II. Herbicide Physiology, (S.O. Duke, ed.), CRC Press Inc., Boca Raton.

Shimabukuro, R.H., Frear, D.S., Swanson, H.R. and Walsh, W.C. (1971). Glutathione conjugation: an enzymatic basis for atrazine resistance in corn. Plant Physiology 47, 10-14.

Slife, F.W. (1986). Resistance in weeds. In "Pesticide Resistance : Strategies and Tactics for Management", (U.S. National Research Council, Board of Agriculture, eds), pp. 327-334. National Academy Press, Washington, DC.

Sterling, T.N. and Balke, N.E. (1989). Bentazon metabolism by cultured plant cells of several species. Weed Science Society of America Abstracts 29, 70.

Still, G.G. and Herrett, R.A. (1976). Methylcarbamates, carbanilates and acylanilides. In "Herbicides : Chemistry, Degradation and Mode of Action". Volume 2, 2nd edition (P.C. Kearney and D.D. Kaufman, eds), pp. 609-664. Marcel Dekker, New York.

Suwanketnikom, R. and Penner, D. (1978). Effect of ammonium salts on bentazon and glyphosate activity on yellow nutsedge (*Cyperus esculentus*). Weed Science Society of America Abstracts p. 73.

Tammes, P.M.L. (1964). Isoboles, a graphic representation of synergism in pesticides. Netherlands Journal of Plant Pathology 70, 73-80.

Thompson, L.Jr. (1972). Metabolism of chloro-S-triazine herbicides by *Panicum* and *Setaria*. Weed Science 20, 584-587.

Willard, J.I. and Penner, D. (1976). Benzoxazinones : cyclic hydroxamic acids found in plants. Residue Reviews 64, 67-76.

Zorner, P.S. (1983). Physiological interactions between tridiphane and atrazine in panicoid grasses. Proceedings North Central Weed Control Conference 38, 109.

ENHANCEMENT OF CROP TOLERANCE TO HERBICIDES WITH CHEMICAL SAFENERS

Kriton K. Hatzios

Laboratory for Molecular Biology of Plant Stress, Department of Plant Pathology, Physiology and Weed Science, Virginia Polytechnic Institute and State University, Blacksburg, VA 24061-0330, USA

Successful safeners (also known as antidotes) protect grass crops from herbicide injury without reducing weed control. Safeners are applied either to the crop seed prior to planting ("seed safeners") or to the soil together with the herbicide in a single formulation package. Commercialised safeners are chemically diverse and include dichloro-acetamide, oxime ether, thiazole carboxylate, dichlormethyldioxolan, and phenylpyrimidine derivatives. These safeners protect large-seeded grass crops against injury caused by soil-active herbicides such as the carbamo-thioates and chloroacetanilides. Safeners protect plants by acting either as "bio-regulators" influencing the amount of a given herbicide that reaches its target site in an active form or as "antagonists" of herbicidal effects at a common site of action.

INTRODUCTION

Extensive research during the last four decades has resulted in the discovery and development of a plethora of synthetic herbicides which are utilised efficiently in crop production. Some form of selective chemical weed control is currently available for all major crops produced worldwide. However, the weed complexes affecting any given crop are dynamic and new weed problems develop as soon as existing weeds are controlled. At first, chemically-induced shifts in weed populations were documented by observed increases of weeds which were botanically related to crops and exhibited similar physiological tolerance to specific herbicides. More recently, however, the development of weed biotypes which are resistant to specific herbicides such as the triazines, dinitroanilines, aryloxyphenoxypropanoates, bipyridyliums, and sulfonyl-ureas provide us with prime examples of chemically-induced shifts in weed populations (Le Baron and Gressel, 1982; Gressel, 1986).

Traditional approaches for addressing the challenges provided by natural or chemically-induced shifts in weed populations have relied either on the development of new herbicides possessing good crop selectivity and strong activity against new weed problems or on the selective application of currently available herbicides which possess limited crop selectivity (McWhorter and Gebhardt, 1988). These two approaches will continue to be important, and new classes of active and selective herbicides will be

293

undoubtedly discovered and developed in the future, in spite of the ever-increasing cost required for their development and registration.

More recently, however, considerable attention has been given to two alternative approaches which attempt to confer crop tolerance on non-selective herbicides either genetically or chemically. The potential uses of plant breeding techniques as well as of genetic engineering for developing crops resistant to herbicides have been reviewed (Le Baron and Gressel, 1982; Mazur and Falco, 1989). In addition, these Proceedings focus on recent advances in understanding the development of weed resistance to herbicides and in circumventing and/or manipulating such resistance.

The main purpose of this paper is to discuss briefly the concept of using chemical safeners as a promising alternative for enhancing the physiological tolerance of crops to herbicides. General or specific aspects of the development, chemistry, practical applications and mechanisms of action of herbicide safeners have been reviewed previously (Pallos and Casida, 1978; Hatzios, 1983; Parker, 1983; Fedtke and Trebst, 1987; Stephenson and Ezra, 1987; Hatzios and Hoagland, 1989).

BENEFITS OF SAFENER CONCEPT

The idea of manipulating crop tolerance to herbicides with chemical agents was conceived by Otto Hoffman in the late 1940s during his doctoral studies at Iowa State University (Hoffman, 1978). As a result of Hoffman's vision and dedicated research, this concept became a reality in 1962, when Hoffman introduced the compound S-449 (4-chloro-2-hydroxy-iminoacetanilide) as the first commercially developed safener of wheat against injury caused by the herbicide barban (Hoffman, 1962). Subsequent developments included the introduction of naphthalic anhydride (NA, naphthalene-1,8-dicarboxylic acid anhydride) and dichlormid as safeners of maize in the early 70s and the recent introduction of seed safeners for grain sorghum (Hatzios, 1989a). Today, chemical crop safening against herbicide injury is an accepted agricultural practice (Hatzios and Hoagland, 1989).

Manipulation of the physiological tolerance of crop plants to herbicides with the use of chemical safeners is appealing for several reasons (Hatzios, 1989a). A major benefit resulting from the use of herbicide safeners is the selective control of weeds closely related to the crop. Controlling effectively wild oats (*Avena fatua* L.) in cultivated oats, shattercane in grain sorghum (both *Sorghum bicolor* (L.) Moench), red rice in cultivated rice (both *Oryza sativa* L.), itchgrass (*Rottboelia exaltata* L.) in corn (*Zea mays*) and wild mustard (*Brassica kaber* (DC.) L.C. Wheeler) in rapeseed (*Brassica campestris* L.) is still a challenge confronting weed technologists (Smith, 1971; Hoffman, 1978; Parker, 1983). This problem has been confounded in recent years by the appearance of cultivated crops as "volunteer weeds" in selected crop rotation systems. "Volunteer" wheat (*Triticum aestivum* L.) or barley (*Hordeum vulgare* L.) are currently serious weeds in a barley-wheat rotation system used in many European countries and the western United States (Parker, 1983). "Volunteer" maize is often encountered as a weed problem in maize-soybean (*Glycine max* (L.) Merr.) rotations used extensively in the midwestern United States (Paul and Knake, 1984).

Equally important is the fact that safeners improve the efficacy of existing herbicides by allowing the use of increased herbicide rates to obtain a wider spectrum and longer duration of weed control without injuring the crop. Other benefits afforded by the use of safeners include: (a) the counteraction of the residual activity of soil-applied persistent herbicides in crop rotation systems, (b) increased spectrum of herbicides available for use in "minor" crops, (c) expansion of the use of "out-of-patent" herbicides, and (d) a reduction in the cost of weed control by allowing the use of cheaper herbicides. Last but not least, safeners provide us with useful biochemical or molecular probes which could be used for the elucidation of the mechanisms of action of herbicides or their metabolic fate in plants (Hatzios, 1989b).

PRACTICAL APPLICATIONS OF HERBICIDE SAFENERS

Practical success in crop safening against herbicide injury has been achieved mainly with large-seeded grass crops such as maize, grain sorghum and rice and soil-applied herbicides such as carbamothioates and chloroacetanilides. Varying degrees of success have also been achieved in safening other grass crops including oats, barley and wheat and additional herbicides with soil activity such as the sulfonylureas, imidazolinones, aryloxyphenoxypropanoates, cyclohexanediones and isoxazolidinones (Parker, 1983; Hatzios, 1989a).

Dichloroacetamide derivatives such as dichlormid, BAS-145128 (1-dichloro-acetyl-hexahydro-3,3-8α-trimethyl-pyrrolo-[1,2-α]-pyrmidin-6-(2$\underline{H}$)-one, and CGA-154281 [4-(dichloroacetyl)-3,4-dihydro-3-methyl-2$\underline{H}$-1,4-benzoxazine] or dichloromethyldioxolans such as MG-191 (2-dichloromethyl-2-methyl-1,3-dioxolane) have been the most successful commercialized safeners of maize against injury from carbamothioate and chloro-acetanilide herbicides (Table 1). These safeners are applied primarily as prepackaged tank mixtures with the respective herbicides. NA, the first commercially developed safener for protecting maize against carbamothioate herbicide injury is not marketed today. However, NA is still quite useful for identifying potential crop-herbicide-safener interactions (Parker, 1983; Hatzios, 1989a).

Oxime ether derivatives such as cyometrinil, oxabetrinil and CGA-133205 {O-[1,3-dioxolan-2-yl-methyl]-2,2,2-trifluoro-4'-chloroacetophenone-oxime} or substituted thiazole carboxylates such as flurazole are currently marketed as successful safeners of grain sorghum against chloracetanilide herbicide injury (Table 2). These safeners are applied exclusively as seed dressings.

Fenclorim is the first commercially developed safener of rice against injury from the chloroacetanilide herbicide pretilachlor. It is applied as a prepackaged mixture with the herbicide. In early studies by Smith (1971), NA protected rice against injury from the carbamothioate herbicide molinate. Nevertheless, this successful combination was not exploited commercially. Other safeners of rice against carbamothioate and sulfonylurea herbicides, used primarily in Japan, are listed in Table 3 and they have been discussed in more detail by Matsunaka and Wakabayashi (1989). A safener based on triazole chemistry protects wheat against the herbicide fenoxaprop-ethyl and is currently under development by Hoechst.

Table 1. **Herbicide safeners for maize.**

Safener	Herbicide	Herbicide chemistry	Method of safener application
Naphthalic anhydride	EPTC, butylate vernolate	carbamothioate	seed dressing
Dichlormid	EPTC, butylate vernolate	carbamothioate	formulation additive
MG–191	EPTC	carbamothioate	formulation additive
BAS–145138	metazachlor	chloroacetanilide	formulation additive
CGA–154281	metolachlor	chloroacetanilide	formulation additive

Table 2. **Herbicide safeners for grain sorghum.**

Safener	Herbicide	Herbicide chemistry	Method of safener application
flurazole	alachlor, butachlor	chloroacetanilide	seed dressing
cyometrinil	metolachlor	chloroacetanilide	seed dressing
oxabetrinil	metolachlor	chloroacetanilide	seed dressing
CGA–133205	metolachlor	chloroacetanilide	seed dressing

Table 3. **Herbicide safeners for rice.**

Safener	Herbicide	Herbicide chemistry	Method of safener application
fenclorim	pretilachlor	chloroacetanilide	formulation additive
methoxy-phenone	thiobencarb	carbamothioate	formulation additive
BCS[1]	thiobencarb	carbamothioate	formulation additive
thiobencarb	bensulfuron-methyl	sulfonylurea	formulation additive

[1] BCS = 4-bromophenyl chloromethyl sulfone

The application of safeners in the field does not involve any extra operation by the farmer (Hatzios, 1989a). As mentioned earlier, herbicide safeners are marketed either as prepackaged mixtures with herbicides or as treated crop seeds (seed safeners). Safeners formulated as prepackaged mixtures with herbicides should exhibit a high degree of specificity for the desirable crop and not provide any protection to target weeds. In addition, these safeners should be as water-soluble or biodegradable as the herbicide, and compatible with other agrochemicals (e.g. pesticides) applied to the same crop. The effectiveness of safeners formulated as prepackaged mixtures with herbicides is usually expressed by the ratio of safener-to-herbicide doses included in the mixture. For marketed safeners, safener-to-herbicide ratios range from 1:6 to 1:30 (Hatzios, 1989a).

Dressing of crop seeds with safeners provides the ultimate in safener selectivity (Hoffman, 1978). Seed safeners should not be washed away from seeds and they should be compatible with insecticides or fungicides which are also applied as seed dressings. Dressing of crop seeds with safeners is conducted at the manufacturer's or distributor's level rather than at the user's level. Chemical companies developing seed safeners market them through either affiliated or co-operating seed companies.

Intraspecific differential responses of several inbred lines or varieties of hybrid crops to herbicide-safener combinations have been reported (Hatzios, 1989a). In addition, environmental factors such as temperature, light, soil moisture and soil type have all been reported to influence the efficacy of herbicide safeners under field conditions (Hatzios, 1989a).

MECHANISMS OF ACTION OF HERBICIDE SAFENERS

Considerable research aimed at elucidating the mechanisms of action of herbicide safeners has been conducted by numerous groups of investigators in the last two decades. A recent book on this subject has summarised and evaluated the results of most of these studies (Hatzios and Hoagland, 1989). Undoubtedly, we have learned a lot about the interactive biochemistry of plants, herbicides and safeners. However, we have yet to come up with a single theory on the mode of action of herbicide safeners which is unequivocally accepted by all researchers active in this field. Of the mechanisms proposed two have received the most attention and suggest either a safener-induced enhancement of herbicide detoxification or a competitive antagonism of herbicides and safeners at a similar target site (Hatzios, 1989b).

Enhanced herbicide detoxification by safeners

A safener-induced enhancement of the metabolic reactions detoxifying a given herbicide in protected plants is the major mechanism by which safeners may exclude the herbicide from reaching its site of action in an active form. Alternatively, this could also result from a safener-induced alteration in the rate of herbicide uptake and/or translocation in the protected plant, but evidence in support of such a hypothesis is currently limited (Hatzios, 1989b). Safeners may enhance herbicide detoxification by inducing either the synthesis of specific co-factors such as reduced glutathione (GSH) or the activity of key metabolic enzymes involved in herbicide metabolism such as glutathione-

S-transferases (GSTs, EC 2.5.1.18) and mixed function oxidases (MFOs, EC 1.14.14.1) (Gronwald, 1989; Komives and Dutka, 1989).

Elevated levels of GSH in safened plants may result from a direct effect of safeners on GSH synthesis. Dichloroacetamide safeners have been shown to regulate the assimilation of sulfate into "bound sulfide" and cysteine. Adams *et al.* (1983) reported that dichlormid enhanced the activity of maize ATP-sulfurylase (EC 2.7.7.4), the first enzyme in sulfate assimilation, which catalyses the formation of adenosine-5'-sulfate from ATP and sulfate. More recently, Farago and Brunold (1989) argued that dichlormid and CGA-154281 enhanced GSH synthesis in maize by enhancing the activity of adenosine-5'-phospho-sulfate sulfotransferase rather than the activity of ATP-sulfotrans-ferase. Dichlormid has also been reported to increase the activity of GSH synthetase (EC 6.3.2.3), a key enzyme catalysing the synthesis of GSH from γ-glutamylcysteine and glycine (Carringer *et al.*, 1978; Rennenberg, 1987). Seed safeners such as flurazole and oxime ethers can conjugate with GSH in protected grass crops including maize and grain sorghum (Breaux *et al.*, 1989; Yenne and Hatzios, 1990a). Breaux *et al.* (1989) proposed that GS-conjugates of safeners may bind to GSH synthetase enzyme and override the feedback inhibition caused by GSH. Such a mechanism could explain the enhanced levels of GSH seen in safener-treated plants.

Safeners may also elevate GSH levels by inducing the activity of glutathione reductase (GR, EC 1.6.4.2), a NADPH-dependent enzyme which catalyses the reduction of oxidised glutathione (GSSG) to GSH. A safener-induced enhancement of GR activity will maintain a high GSH/GSSG ratio in the cells of safened plants compensating for the GSH used as a reductant in the formation of GS-conjugates of herbicides or in the ascorbate-dehydro- ascorbate redox system in the chloroplast (Rennenberg, 1987). Dichlormid and MG-191 enhanced significantly the activity of GR in maize seedlings (Komives *et al.*, 1985). Oxime ether safeners such as CGA-133205 have also been reported to enhance the activity of GR in grain sorghum seedlings (Yenne and Hatzios, 1990a).

Enhanced metabolism of chloroacetanilide and carbamothioate herbicides by GSH conjugation could result also from a safener-induced increase of the activity of the respective glutathione-S-transferase enzymes (GSTs) which catalyse this reaction. Maize and other plants contain multiple forms of GST enzymes which exhibit a rather high degree of substrate specificity (Timmerman, 1989). A strong correlation between the efficacy of safeners in protecting grain sorghum from chloroacetanilide injury and their ability to increase GST activity has been demonstrated (Gronwald, 1989). Apart from enhancing the activity of constitutive GST isozymes in maize (GST I and GST III), the safener flurazole has been shown also to induce the *de novo* synthesis of a novel GST isozyme (GST II) with greater activity in conjugating chloroacetanilide herbicides with GSH (Mozer *et al.*, 1983). At the molecular level, flurazole induced a 3- to 4-fold increase in the level of mRNA coding for the GST I gene in maize (Wiegand *et al.*, 1986). Thus, it is likely that safeners may activate the genes coding for herbicide-degrading enzymes in protected plants.

Current information on the interactions of herbicide safeners with plant oxidative enzymes is limited. The inherent difficulty in isolating, characterising and assaying such enzymes from plants is partially responsible for the lack of information on this topic (Gronwald, 1989;

Komives and Dutka, 1989). Indirect studies employing the use of mixed function oxidase inhibitors have shown that a safener-induced enhancement of oxidative enzymes is very likely (Fedtke and Trebst, 1987).

Interactions of herbicides and safeners at the site of action

When it is not possible to exclude the herbicide from reaching its site of action, safeners could antagonise the herbicide by competing with it at a common site of action. Support for such a mechanism comes from the apparent structural similarity of dichloroacetamide safeners to carbamothioate and chloroacetanilide herbicides (Stephenson *et al.*, 1978). Computer-assisted molecular modelling (CAMM) studies by Yenne and Hatzios (1990b) have shown that at the molecular level flurazole and the oxime ether safeners are quite similar to their respective herbicides alachlor and metolachlor. Nevertheless, our understanding of the potential interactions of herbicides and safeners at a common site of action is complicated because of the lack of evidence for a specific mode of action of chloroacetanilide or carbamothioate herbicides or the involvement of single target site affected by these herbicides (Fuerst, 1987).

Plant metabolic processes such as synthesis of lipids, terpenoids, lignin, protein and nucleic acid, and membrane function and ion transport can be affected by carbamothioate or chloroacetanilide herbicides (Fuerst, 1987). Consequently, antagonistic interactions between these herbicides and their safeners at these sites of action are possible and the topic has been reviewed (Hatzios, 1989b). Many of these metabolic processes are dependent on acetyl-CoA intermediates and because of that, Fuerst (1987) proposed that the action of carbamothioate and chloroacetanilide herbicides may be related to some aspect of acetyl-CoA metabolism. Recent studies by Wilkinson (1989) and by Yenne and Hatzios (1989) indicate that chloroacetanilide herbicides and their safeners may be affecting the formation of acetyl-CoA rather than the incorporation of acetyl-CoA into fatty acids or other lipids of grain sorghum seedlings.

The safening of grass crops from the phytotoxic effects of chemically diverse groups of herbicides such as the carbamothioates, chloroacet-anilides, sulfonylureas, imidazolinones, aryloxyphenoxypropanoates, cyclo- hexanediones, and isoxazolidinones does not favour the "competitive inhibition" theory of safener action. The potential interactive effects of safeners and herbicides on specific target enzymes such as acetohydroxyacid synthase (AHAS, EC 4.1.3.18) and acetyl-CoA carboxylase (ACCase, EC 6.4.1.2) have been studied (Barrett, 1989; Yenne and Hatzios, 1989). From such studies it is evident that safeners do not protect grass crops against herbicide injury by antagonising herbicide effects on these target enzymes.

CONCLUDING REMARKS

The use of chemical safeners offers a highly effective way for increasing the crop selectivity of certain herbicides. Our present experience in manipulating crop tolerance to herbicides chemically indicates that it is easier to find safeners for herbicides which are somewhat selective to a crop. It is also apparent that it is easier to find safeners for large-seeded grass crops and certain classes of herbicides than others. All of the currently marketed safeners are particularly effective against herbicides that are active when applied to the soil. Success has been very limited in developing chemical safeners for the protection of any crop against photosynthesis-inhibiting herbicides or broad spectrum weed killers such as glyphosate and paraquat. In addition, the interactions of broad-leaved crops with chemical safeners against herbicide injury have not been very promising (Hatzios and Hoagland, 1989).

A safener-induced enhancement of herbicide detoxification in safened plants seems to be the major mechanism involved in the protective action of the currently developed safeners. Most safeners resemble structurally their respective herbicides and they induce the substrates and enzymes needed for their own metabolism as well as that of the antidoted herbicides in safened plants. Although safeners can compete with herbicides at common target sites, such a mechanism seems unlikely. As mentioned earlier, the ratio of safener-to-herbicide doses in prepackaged formulated mixtures of herbicide and safeners ranges from 1:6 to 1:30. Such ratios do not favour the "antagonist" theory of safener action since very little safener will be available at a site of action to compete effectively with its respective herbicide which would be present at considerably higher concentrations. A better understanding of the mechanisms of action of current safeners and herbicides will allow more positive attempts towards increasing the number of situations in which crop safeners for herbicides could be used.

Most of the currently exploited crop-herbicide-safener combinations have been discovered by empirical screening and subsequent chemical optimisation (Hoffman, 1978; Gray *et al.*, 1982). Future advances in the development of crop safeners for herbicides, however, are likely to be more rational, based on the principles of biotechnology and chemistry. Targeted synthesis of crop safeners for herbicides in the future would use knowledge of the characterics of the target site of herbicide action and detoxification mechanisms of herbicides coupled with computer graphics.

REFERENCES

Adams, C.A., Blee, E. and Casida, J.E. (1983). Dichloroacetamide herbicide antidotes enhance sulfate metabolism in corn roots. Pesticide Biochemistry and Physiology **19**, 350–360.

Barrett, M. (1989). Protection of grass crops from sulfonylurea and imidazolinone toxicity. In "Crop Safeners for Herbicides: Development, Use, and Mechanisms of Action" (K.K. Hatzios and R.E. Hoagland, eds), pp. 195–220. Academic Press, San Diego.

Breaux, E.J., Hoobler, M.A., Patanella, J.E. and Leyes, G.A. (1989). Mechanisms of action of thiazole safeners. In "Crop Safeners for Herbicides: Development, Use, and Mechanisms of Action" (K.K. Hatzios and R.E. Hoagland, eds), pp. 163–175. Academic Press, San Diego.

Carringer, R.E., Rieck, C.E. and Bush, L.P. (1978). Effect of R-25788 on EPTC metabolism in corn (*Zea mays*). Weed Science **26**, 167–171.

Farago, S. and Brunold, C. (1989). Regulatory effects on assimilatory sulfate reduction by herbicide antidotes in *Zea mays* L. In "Sulfate Assimilation in Plants" (H. Rennenberg and C. Brunold, eds), in press.

Fedtke, C. and Trebst, A. (1987). Advances in understanding herbicide modes of action. In "Pesticide Science and Biotechnology" (R. Greenhalgh and T.R. Roberts, eds), pp. 161-168. Blackwell, Oxford.

Fuerst, E.P. (1987). Understanding the mode of action of the chloro-acetamide and thiocarbamate herbicides. Weed Technology **1**, 270-277.

Gray, R.A., Green, L.L., Hoch, P.E. and Pallos, F.M. (1982). The evolution of practical crop safeners. British Crop Protection Conference - Weeds 431-437.

Gressel, J. (1986). Modes and genetics of herbicide resistance in plants. In "Pesticide Resistance: Strategies and Tactics for Management" (E.H. Glass, ed.), pp. 54-73. National Academy Press, Washington, DC.

Gronwald, J.W. (1989). Influence of herbicide safeners on herbicide metabolism. In "Crop Safeners for Herbicides: Development, Uses, and Mechanisms of Action" (K.K. Hatzios and R.E. Hoagland, eds). pp. 103-128. Academic Press, San Diego.

Hatzios. K.K. (1983). Herbicide antidotes: Development, chemistry and mode of action. Advances in Agronomy **36**, 265-316.

Hatzios, K.K. (1989a). Development of herbicide safeners: Industrial and academic perspectives. In "Crop Safeners for Herbicides: Development, Uses, and Mechanisms of Action" (K.K. Hatzios and R.E. Hoagland, eds), pp. 3-45. Academic Press, San Diego.

Hatzios, K.K. (1989b). Mechanisms of action of herbicide safeners: An overview. In "Crop Safeners for Herbicides: Development, Uses, and Mechanisms of Action" (K.K. Hatzios and R.E. Hoagland, eds), pp. 65-102. Academic Press, San Diego.

Hatzios, K.K. and Hoagland, R.E. (1989). "Crop Safeners for Herbicides: Development, Uses, and Mechanisms of Action". Academic Press, San Diego.

Hoffman, O.L. (1962). Chemical seed treatments as herbicide antidotes. Weeds **10**, 322-323.

Hoffman,. O.L. (1978). Herbicide antidotes: From concept to practice. In "Chemistry and Action of Herbicide Antidotes" (F.M. Pallos and J.E. Casida, eds), pp. 1-13. Academic Press, New York.

Komives, T. and Dutka, F. (1989). Effects of herbicide safeners on cytochrome P_{450} levels and activity and on other enzymes of corn. In "Crop Safeners for Herbicides: Development, Uses, and Mechanisms of Action" (K.K. Hatzios and R.E. Hoagland, eds), pp. 129-145. Academic Press, San Diego.

Komives, T., Komives, V.A., Balazs, M. and Dutka, F. (1985). Role of glutathione-related enzymes in the mode of action of herbicide antidotes. British Crop Protection Conference - Weeds 1155-1162.

LeBaron, H.M. and Gressel, J. (1982). "Herbicide Resistance in Plants". Wiley, New York.

Matsunaka, S. and Wakabayashi, K. (1989). Crop safening against herbicides in Japan. In "Crop Safeners for Herbicides: Development, Uses, and Mechanisms of Action" (K.K. Hatzios and R.E. Hoagland, eds), pp. 47-62. Academic Press, San Diego.

Mazur, B. and Falco, S.C. (1989). The development of herbicide resistant crops. Annual Review of Plant Physiology and Plant Molecular Biology **40**, 441-470.

McWhorter, C.G. and Gebhardt, M.R. (1988). "Methods of Applying Herbicides". Monograph #4, Weed Science Society of America, Champaign, Illinois.

Mozer, T.J., Tiemeier, D.C. and Jaworski, E.G. (1983). Purification and characterization of corn glutathione-S-transferase. Biochemistry **22**, 1068-1072.

Pallos, F.M. and Casida, J.E. (1978). "Chemistry and Action of Herbicide Antidotes". Academic Press, New York.

Parker, C. (1983). Herbicide antidotes - A review. Pesticide Science **14**, 40-48.

Paul, L.E. and Knake, E.L. (1984). Control of volunteer corn in soybean. Abstracts of the Weed Science Society of America Meeting **24**, 3.

Rennenberg, H. (1987). Aspects of glutathione function and metabolism in plants. In "Plant Molecular Biology" (D. von Wettstein and N-H. Chua, eds), pp. 279-292. NATO Advanced Study Institute, Plenum Press, New York.

Smith, R.J. (1971). Red rice: A problem weed in rice. Weeds Today **2**(2), 12-13.

Stephenson, G.R. and Ezra, G. (1987). Chemical approaches for improving herbicide selectivity and crop tolerance. Weed Science **35** (Suppl. 1), 24-27.

Stephenson, G.R., Bunce, J.J., Makowski, R.I. and Curry, J.C. (1978). Structure-activity relationships for S-ethyl N,N-dipropylthiocarbamate (EPTC) antidotes in corn. Journal of Agricultural and Food Chemistry **26**, 137-140.

Timmerman, K.P. (1989). Molecular charactization of corn glutathione S-transfertase isozymes involved in herbicide detoxication. Physiologia Plantarum (submitted).

Wiegand, R.C., Shah, D.M., Mozer, T.J., Harding, E.I., Diaz-Collier, J., Sounders, C., Jaworski, E.G. and Tiemeier, D.C. (1986). Messenger RNA encoding a glutathione-S-transferase responsible for herbicide tolerance in maize is induced in response to safener treatment. Plant Molecular Biology **7**, 235-243.

Wilkinson, R.E. (1988). Consequences of metolachlor induced inhibition of gibberellin biosynthesis in sorghum seedlings. Pesticide Biochemistry and Physiology **32**, 25-37.

Yenne, S.P. and Hatzios, K.K. (1989). Influence of oxime ether safeners and metolachlor on acetate incorporation into lipids and on acetyl-CoA carboxylase of grain sorghum. Pesticide Biochemistry and Physiology **35**, 146-154.

Yenne, S.P. and Hatzios, K.K. (1990a). Influence of oxime ether safeners on glutathione levels and glutathione-related enzyme activity in seeds and seedlings of grain sorghum. Zeitschrift für Naturforschung **45c**, 96-106.

Yenne, S.P. and Hatzios, K.K. (1990b). Molecular comparisons of selected herbicides and their safeners by Computer-Aided Molecular Modeling. Journal of Agricultural and Food Chemistry **38**, 1950-1956.

THE GENETICAL ANALYSIS AND EXPLOITATION OF DIFFERENTIAL RESPONSES TO HERBICIDES IN CROP SPECIES

John W. Snape[1], D. Leckie[1], B.B. Parker[1] and E. Nevo[2]

[1] AFRC Institute of Plant Science Research, Cambridge Laboratory, Trumpington, Cambridge, CB2 2JB
[2] Institute of Evolution, University of Haifa, Haifa, Israel

Crop species can show differential responses to successful, widely used, herbicides where some varieties are unaffected by application, whilst others show symptoms of damage ranging from a slight reduction in vigour to complete plant death. The elucidation of the genetical control of such responses is important for developing strategies for breeding for herbicide resistance within crop species and also in understanding the modes of action of the herbicides and the evolution of resistance in weed species.

To investigate this phenomenon in cereals detailed studies of the control of responses of wheat to difenzoquat and to three phenylureas, chlorotoluron, metoxuron and isoproturon, have been carried out. These have revealed that the primary differences in response between resistant and susceptible varieties are due to single major genes, although the influence of other "modifier" genes has also been detected. It has also been shown that the responses to chlorotoluron and metoxuron are determined by the same gene. Studies of related wild grass species have indicated that polymorphisms for response also exist in such species. This suggests that the genes for differential responses have evolved prior to the domestication of the cultivated cereals and not in response to the development and use of the chemicals. The importance of these results for developing strategies for the incorporation of herbicide resistance in new varieties of crop species by conventional and non-conventional methods of plant breeding is discussed.

INTRODUCTION

The identification of chemical compounds suitable for use as herbicides relies on a primary screen which tests activity on a small number, or even a single variety, of the crop and weed species of interest. Subsequent progress towards commercialisation depends on a resistant response of one or more crop species combined with a susceptible response of one or more of their associated weed species. If these criteria are satisfied then further progress of promising compounds depends, of course, on the satisfaction of toxicology and production criteria.

305

However, the use of only a single or small number of genotypes for testing responses during these processes can have unforeseen consequences since it effectively ignores any genetical variation that may exist within the crop species for response. On the one hand, this can result in promising compounds being rejected as being phytotoxic to a crop whereas resistance may be common and exist in available alternative commercial varieties. On the other hand, this can result in the commercialisation of compounds which are subsequently found to damage certain cultivars of the target crop species, because such genotypes were not included in the testing regimes. Indeed it is now widely documented, for example in cereals (Tottman *et al.*, 1982; Tottman *et al.*, 1984), that many crop species show differential responses to successful and widely used herbicides. In such cases certain varieties are unaffected by application whilst others show symptoms of damage ranging from a slight reduction in vigour, and hence yield, to severe leaf necrosis leading to plant death.

Undoubtedly, the occurrence of differential responses can cause problems to companies marketing such products. Alternatively, however, such responses can be a valuable tool for the plant scientist. Firstly, they provide the material which enables the genetical control of the responses to be elucidated, leading to the location and identification of specific genes for resistance. This then allows the activity of such genes to be investigated at the physiological, biochemical, and ultimately, the molecular level, leading to an understanding of their mode of action. This information can be useful to the biochemist in designing more effective compounds or in highlighting target sites for new compounds. At the whole plant level, knowledge of the genetical control enables the plant breeder to develop strategies for breeding for herbicide resistance, thereby obviating the problem in future cultivars. In addition, the information obtained can give an insight into the evolution of resistance in weed species.

This paper discusses the methodologies for examining such responses and the significance of the information obtained for breeding herbicide resistant crops. Examples from our own work with cereals, particularly wheat, will be used as illustrations.

GENETICAL ANALYSIS OF DIFFERENTIAL RESPONSES

Types of responses

The type of differential response exhibited by a crop species to a herbicide will be determined by the biological activity of the herbicide. There are two possible aspects of this response. First, whether the differential response is determined by differences acting at the target site of the herbicide within the plant, or, secondly, whether the differential response is determined by differences in the concentration of the herbicide at the target site. The former depends on the mode of action of the herbicide; the latter on the uptake, metabolism and translocation of the herbicide to the target site. Evidence in wheat for example, by Leckie (1989) working with difenzoquat suggests that genetical variation exists within crop species for both uptake and movement. However the evidence also indicates that the most important and significant effects are due to genetical variation for response at the target site (Leckie, 1989).

Differential responses of crop varieties are seen as changes in plant growth or morphology following application of the herbicide. Generally, for herbicides which attack the photosynthetic apparatus, for example the phenylureas, susceptible varieties of the crop exhibit effects which mimic those of the weed species at which the herbicide is targeted. Thus in cereals there is leaf chlorosis and necrosis which progresses through the plant, leading, eventually, to complete leaf senescence and plant death. Resistant varieties are unaffected, and consequently the difference between resistant and susceptible genotypes is qualitative and dramatic. For herbicides which affect the growth and development of the plant through influencing nucleic acid or protein synthesis, for example difenzoquat and chlorsulfuron, respectively, primary responses of susceptible varieties are a retardation of growth which can also, eventually, lead to plant senescence and death. However death need not always occur and differential responses are seen as qualitative or quantitative changes in vigour and growth rate manifested as stunting and/or late flowering. For example, wheat varieties susceptible to difenzoquat exhibit severe stunting and prolific lateral tillering; however plants generally survive to maturity where flowering is delayed, height is shortened and yield is reduced (Snape *et al.*, 1987). Overall, therefore, differential responses within a crop species to any particular herbicide can be qualitative or quantitative and different characters and components can be measured as an indicator of difference in response. The first aspect of studying differential responses, therefore, is to establish the criteria and characters used to evaluate the presence of any differences before genetic analysis can proceed. This is best established by choosing a character where qualitative or quantitative assessment is simple, quick and accurate. For example, in wheat, Snape and Parker (1988) used visual assessments of the extent of leaf senescence, measured on a 1-9 scale, to quantify differences in response to chlorotoluron. Alternatively Bowran and Blacklow (1987) measured the rate of elongation of the third leaf of seedlings to identify differences in wheat cultivar sensitivity to chlorotoluron. Other workers have used flowering time, plant height, or relative reductions in yield or yield components to quantify differential responses.

It is also important, of course, to have an appropriate environment for testing, where these differences are maximally expressed. This commonly requires the establishment, through experimentation, of the most appropriate dose rates of application and conditions of plant growth in terms of temperature, light and moisture. Frequently such conditions are best obtained through the use of controlled environment facilities, although commonly good differentiation can also be achieved under field conditions (Snape *et al.*, 1990).

Genetic analysis

Having established a suitable scale of measurement for a differential response, genetical analysis can be pursued to identify the genes responsible. However, the type of analysis that needs to be performed will depend on whether, first, the variation observed is nuclear or cytoplasmic in origin, and secondly, if nuclear in origin whether the variation exhibited in segregating generations can be considered as discrete or continuous.

Although many herbicides exhibit their effects through actions on cytoplasmic organelles, the chloroplast or mitochondrion, many of the proteins involved are nuclear encoded and it is an easy matter to distinguish whether the origin of the variation is nuclear or cytoplasmic. This is achieved by carrying out reciprocal crosses between varieties of contrasting response and testing the response of the F_1s. Any reciprocal difference will indicate cytoplasmic inheritance whilst an identical reaction of the F_1s combined with segregation for response in their F_2 families will indicate a nuclear encoded effect.

If an effect is established as being due to variation in the nuclear genes, it is then necessary to identify the number and location of the genes involved so that detailed analysis of their effects can proceed. This is achieved by examination of the phenotypic variation in the segregating generations F_2s, F_3s or backcrosses of crosses between varieties of contrasting response. If the responses of individual plants fall into discrete categories which mimic the responses of the parents and F_1s, the segregation of one or a small number of major genes is indicated. If the variation does not fall into discrete categories but a continuous gradual change between extreme phenotypes is observed the segregation of several loci each of small but accumulative effect is presumed. It is also necessary, of course, to measure the environmental variation, to establish the relative influences of genetical and environmental components on response.

Our studies on differential responses in cereals suggest that large qualitative differences between varieties are in fact, determined by single major genes. For example, Figure 1 shows the differential responses of wheat to chlorotoluron exhibited by the resistant variety, Capelle-Desprez and the sensitive variety, Chinese Spring. In this case sensitivity to the chemical is shown by severe leaf necrosis leading to plant death. Table 1 shows the response of a random sample of 60 F_3 families of the cross between these varieties to 2x application of chlorotoluron. Six individual plants of each family were tested and the families fell into three types; those uniformly resistant, those uniformly susceptible and those families with both resistant and susceptible plants. The segregation ratio observed, both between individual families and within the segregating families indicated that the difference in response is due to a single major dominant gene.

Once a major gene segregation is shown the exact chromosomal and within chromosomal locations of such genes can be determined by conventional Mendelian genetics. This requires the establishment of linkage between the locus for herbicide response and other markers, which is facilitated by the availability of comprehensive genetic maps within the species of interest. Currently the development of such maps in many crop species is being rapidly improved through the use of biochemical and molecular marker systems and genetic mapping is becoming a routine matter (Tanksley *et al.*, 1989). Considering the case quoted above for chlorotoluron response in wheat, subsequent analysis by Snape *et al.* (1990) was able to show that the single major gene involved designated <u>Su1</u>, is located on chromosome 6B, closely linked to the locus <u>B2</u> which controls awn production. Similarly, other responses, such as wheat to difenzoquat, have been shown to be determined by single major genes. Overall these results suggest that major qualitative differences in response between varieties of a crop species are most likely due to single major gene differences.

Fig. 1. <u>Differential responses of wheat varieties to 2x dose of chlorotoluron</u>.
a) Cappelle–Desprez, untreated, (b) Cappelle–Desprez, treated,
(c) Chinese Spring, untreated, (d) Chinese Spring, treated.

Table 1. <u>Responses of random F_3 families of the cross of the wheat varieties Cappelle–Desprez x Chinese Spring to 2x application of chlorotoluron</u>.

Type of family observed	Number observed	Expected Seg^n single gene
Uniformly resistant	20	15
Segregating resistant/susceptible	27	30
Uniformly susceptible	13	15

$X^2_{[2]}$ test for single gene segregation = 2.23 NS.

It can be speculated that the allelic differences of these single genes reflect primary changes at the target site for the herbicide. Such changes could come about in three ways, as pointed out by Oxtoby and Hughes (1989). Firstly, there could be modification of target gene products such that in resistant varieties the herbicide is unable to disable the product. Secondly there could be an increased level of the product, either through gene amplification or modified expression so that the herbicide is not at a high enough concentration to disable all the product in resistant varieties. Thirdly, there may be an improved detoxification of the herbicide, again either by higher activity of an altered product, or greater concentration of this product. As yet, no investigations to date have shown which of these mechanisms operates to bring about differential responses, and indeed, different genes may have different mechanisms.

Although these major genes are the most important determinants of differential responses, other quantitative changes in the degrees of herbicide damage are also apparent. An interesting example is the response of wheat varieties to difenzoquat. Here a major gene determining differences in response between resistant and susceptible varieties has been located on chromosome 2B (Snape *et al.*, 1987).

However, some commercial varieties are classified as "indeterminate" in response, which effectively means that they are damaged only under certain environmental conditions but generally not as severely as susceptible varieties. Work by Leckie (1989) has shown that these varieties possess the susceptible allele at the locus on chromosome 2B determining the difference in response between "truly" resistant and "fully" susceptible varieties. The greater resistance of intermediate varieties is possibly due, therefore, to the concentration of the herbicide at the target site. Through the development and use of chromosome substitution lines (Law *et al.*, 1987) further analysis was able to show that the degree of damage to the very susceptible variety Sicco, could be greatly reduced by the introduction of whole chromosomes from a resistant variety, Chinese Spring.

Although the derived lines possessed the allele for susceptibility to difenzoquat damage at the locus on chromosome 2B certain lines suffered significantly less damage than Sicco with quantitative increases in growth rate. Thus, as Table 2 shows, the possession of chromosome 3B from Chinese Spring increased mean fresh weight at 35 days after spraying, a measure of recovery, by 85%. Chromosome 5D appeared to carry similar genes mediating a smaller, but still statistically significant effect. Other lines, however, such as those with chromosomes 6A and 6B of Chinese Spring responded exactly the same as Sicco. Clearly there are genes mediating a quantitative change in response such as those chromosomes 3B and 5D of Chinese Spring. Presumably, therefore, the responses of commercial "intermediate" varieties are the results of the actions of such genes which can be conveniently referred to as "modifier" genes. It would be interesting to establish the mechanisms of such effects to see if these genes influence the uptake and translocation of difenzoquat in the plant. It is also of interest to establish if other differential responses have similar types of genetical differences between varieties.

Table 2. Responses of single chromosome substitution lines in the susceptible variety Sicco to 2x application of difenzoquat. Response is measured by mean fresh weight per plant 35 days after spraying.

Genotype	Plant Mean fr. wt. (g)	Significance of difference from Sicco	% Change
Sicco	35.6	–	
Chinese Spring	185.3	***	+420
Sicco (CS 3B)	66.0	**	+ 85
Sicco (CS 5D)	50.4	*	+ 42
Sicco (CS 6B)	36.7	NS	+ 3
Sicco (CS 6A)	30.1	NS	– 15

Significance levels: NS>0.05, * = 0.05–0.01, ** = 0.01–0.001, ***= >70.00

ORIGINS OF DIFFERENTIAL RESPONSES

The presence of differential responses within a crop species to a herbicide can only be observed once commercialisation has begun. However, these differences must reflect genetical variation which already exists within the species and it is extremely unlikely that they arose by mutation and selection subsequent to commercial use. Two questions arise from these assumptions. Firstly, do wild related and progenitor species also exhibit the genetical variation or did it arise during or after domestication of the crop species? Secondly, does the genetical variation have any adaptive significance either to the wild species or the crop species which could be responsible for maintaining the polymorphism? Answers to these questions can help us to understand the evolution of resistance and also to predict the consequences of breeding for resistance on agronomic performance.

Table 3. Responses to difenzoquat of accessions of wild diploid *Triticeae* species.

Species	Genome	Number of accessions		
		Resistant	Intermediate	Susceptible
Triticum thaoudar	A	0	1	4
T. urartu	A	0	3	2
Aegilops sharonensis	S^1	1	0	0
Ae. longissima	S^1	0	0	2
Ae. searsii	S^s	0	0	1
Ae. speltoides	S	0	1	0
Ae. squarrosa	D	0	0	10
Ae. comosa	M	0	0	1
Ae. umbellulata	U	1	0	0

To investigate these questions in cereals we have examined the responses to phenylureas and difenzoquat of wild *Triticeae* which are related to or are the wild progenitors of the genomes of cultivated bread and durum wheats. For example, Table 3 shows the qualitative responses of wild diploid species to difenzoquat. These were tested in a field experiment by application of a dose twice that recommended for use on bread wheat, and which is known to severely damage susceptible cultivars. Clearly resistance exists in some accessions of these wild species, whilst others are susceptible.

Thus, resistance exists in *Aegilops sharonensis* and *Ae. umbellulata* and it is interesting to note that the species with genomes related to the B genome of cultivated wheat (the S^1, S^s and S genomes) are polymorphic for response whilst species with genomes related to the A(A) genomes or D genomes (D,M) are uniformly susceptible. In cultivated wheat the polymorphism is due to variation at a locus on chromosome 2B, and no homoeoallelic resistance has yet been found on chromosomes 2A or 2D. It is also noteworthy that certain accessions had an intermediate response so that these may contain "modifier" genes as found previously in the cultivated form.

Overall these results show that there is interspecific variation in resistance to difenzoquat in wild species related to the cultivated wheats although the sample sizes for each species were too small to examine if there is likely to be intraspecific variation. However, we have also carried out a much wider survey of intraspecific variation to difenzoquat and to phenylureas of wild accessions of the tetraploid species *Triticum dicoccoides* (wild emmer wheat) which is the progenitor species of the cultivated macaroni wheats. Table 4 shows the responses to chlorotoluron of 113 such families, each derived from a different individual plant collected from 11 different localities in Israel. These were tested in a replicated and randomised field experiment alongside resistant and susceptible bread wheat controls.

Table 4. Responses of wild populations of emmer wheat (*Triticum dicoccoides*) to application of 2x chlorotoluron (1-9 Scale).

Population	No. of families tested	Mean response	Range	Qualitative classification
Amirim	8	7.3	6-9	all S
Daliyya	8	6.2	3-9	4R:4S
Yabed	6	4.5	3-8	all R
Yehudiyya	14	5.8	2-9	7R:7S
Rosh Pinna	10	4.9	2-9	8R:2S
Sanhedriyya	5	4.4	2-9	4R:1S
Beit Meir	6	8.2	6-9	all S
Mt Dov	12	6.3	2-9	5R:7S
Tabigha	12	7.1	3-9	3R:9S
Bat Shelomo	16	7.3	3-9	6R:10S
Kokhav Hashahar	16	7.2	3-9	8R:8S

Clearly this species exhibits intraspecific variation for response to chlorotoluron and this is also repeated for response to metoxuron (Snape *et al.*, 1990). Certain populations were monomorphic for response, such as Amirim for susceptibility and Yabed for resistance, but most populations were polymorphic. Thus, gene(s) for differential responses are present in this species with the alternative morphs at almost equal frequency. Nevo and Beiles (1989) have shown that these populations are also polymorphic for a wide range of morphological and biochemical characteristics and that many of the differences appear to be associated with ecoclimatic differences. It can be argued, therefore, that variation for response to herbicides in natural populations which have never previously been exposed to the herbicides is a byproduct of underlying biochemical differences which may have adaptive significance to the species. Conversely they could be neutral and maintained by genetic drift or "hitchhiking" effects.

These results indicate that, at least in the *Triticeae*, natural populations exhibit both interspecific and intraspecific variation for response to widely used herbicides of the cultivated forms. It is likely that this situation pertains in other Families and Genera. It could also follow that the emergence of weed genotypes resistant to such herbicides is not the result of new mutations for resistance but the selection for alleles which already exist, albeit perhaps at a low frequency in the natural populations. Such alleles could provide a source of resistance for the cultivated forms if they could be transferred by interspecific hybridisation and chromosome engineering technology or by protoplast fusion. This would be particularly useful in situations where the cultivated species is monomorphic for susceptibility to a useful herbicide whilst wild related species are monomorphic or polymorphic for resistance.

EXPLOITING DIFFERENTIAL RESPONSES

Breeding for herbicide resistance

When differential responses occur there is, generally, a necessity for new varieties to incorporate resistance. The location of single major gene loci for response greatly simplifies this process. Thus when the target species is polymorphic for major genes, conventional methods of gene transfer are easily exploited. Ideally, resistance will be available in adapted genotypes of good agronomic performance, which can then be crossed by conventional hybridisation procedures to desirable, but susceptible varieties. Resistant progenies are selected amongst the early filial generations by direct selection following application of the appropriate chemical. Alternatively, if sources of resistance can only be found in unadapted, exotic germplasm, then a programme of backcrossing is necessary, combined with selection for resistance. These procedures are, of course, standard plant breeding practice. If resistance is determined by more than one gene or is quantitative in nature, then more sophisticated procedures may be necessary. For example, if all appropriate loci have been identified by genetic analysis they can be selected, either by phenotypic selection for an appropriate level of resistance, or more conveniently by marker mediated approaches. The latter, in particular are now achieving prominence as comprehensive genetic maps of our major species become available, as for example in wheat (Gale *et al.*, 1989). When phenotypic selection alone is practised it may be necessary to delay selection until later generations or to use progeny testing, particularly if there is a significant environmental component of response.

Intrinsically, breeding for herbicide resistance when there are readily available sources of resistance presents no more problems, and, indeed probably less than other agronomic traits. However, breeding for herbicide resistance may be low on the plant breeders' list of priorities and, generally, will only be undertaken to increase the attractiveness of a variety to the farmers.

Breeding for herbicide susceptibility

The occurrence of herbicide susceptibility in crop species is generally viewed as a disadvantage. However, the ability to breed both resistant and susceptible varieties could create new opportunities for weed control. For example, if a farmer wishes to grow successive wheat crops, the first could be susceptible to an effective and widely used chemical, such as chlorotoluron and the second crop, resistant. Thus, spraying the second crop will eliminate volunteer plants of the preceding year. This strategy also translates across species. For example, in the UK one of the biggest problems of weed control in wheat is volunteer barley and *vice versa*. Varieties of both crops exhibit differential responses to the phenylureas, chlorotoluron and metoxuron. Thus, a susceptible variety of barley could be rotated with a resistant variety of wheat and *via versa*. Breeding for susceptibility is, however, more complicated than breeding for resistance since direct phenotype selection is, of course, not possible. Thus, progeny testing is required for phenotypic selection, and the use of marker mediated selection could greatly enhance the efficiency of gene transfer in this case.

Clearly there are opportunities for plant breeders to manipulate both "susceptible" and "resistance" alleles to improve methods of weed control. Indeed they could create "cassettes" of different alleles for differential responses to different chemicals to maximise the possibilities for weed control. This is analogous to the strategies of "mix and match" of disease resistance genes to obtain an integrated control over seasons and environments.

Cross-resistance

It has been demonstrated that weed species can show cross-resistance to different herbicides once resistance to one has been developed (see for example, Gressel, 1988). The identification by genetic analysis of individual genes for resistance to specific herbicides within crop species allows us to examine if such genes also show these properties. To study this possibility in wheat we have examined the responses of alternative alleles mediating resistance and susceptibility to chlorotoluron, for response to other phenylureas. Table 5, for example, shows the responses of single chromosome substitution lines developed by transferring chromosomes 6A, 6B and 6D from a variety susceptible to chlorotoluron, Poros, into the resistant variety Cappelle-Desprez.

It will be recalled from earlier discussion that Cappelle-Desprez has an allele for chlorotoluron resistance at the Su1 locus on chromosome 6B. Replacement of the 6B chromosome of Cappelle-Desprez by its homologue

Table 5. Responses to three phenylurea herbicides of single chromosome substitution lines from wheat variety Poros into the variety Cappelle-Desprez.

Genotype	Chlorotoluron	Metoxuron	Isoproturon
Cappelle-Desprez (CD)	Resistant[a]	3.25[b] (R)[c]	7.00[b] (I)[c]
Poros	Susceptible[a]	9.00 (S)***[d]	7.75 (I)
CD (Poros 6A)	Resistant	3.50 (R)	5.75 (R)
CD (Poros 6B)	Susceptible	8.85 (S)***[d]	7.25 (I)
CD (Poros 6D)	Resistant	3.25 (R)	5.50 (R)

a Qualitative classification based on previous experimental data.
b Quantitative measure on scale 1-9 with 1 resistant, 9 susceptible.
c Qualitative classification relative to controls: R = Resistant,
  I = Intermediate, S = Susceptible.
d Significantly different from Cappelle-Desprez, $P<0.001$.

from Poros results in susceptibility to chlorotoluron, further evidence of the importance of this chromosome. However, clearly this gene also mediates the response to metoxuron and, although not so clear cut, has an effect on isoproturon response. This property will simplify the plant breeders' task in breeding resistance to this group of herbicides into new varieties. It will be interesting and important to see if other single gene loci, such as the difenzoquat resistance, Dfq1 gene, have similar properties.

Gene isolation

The primary motivation for establishing the genetical control of a differential response is to develop an efficient strategy for breeding for resistance and to understand the mode of action of the differences. However, these studies can also be preliminaries to the wider use of identified genes for resistance if they can be isolated and cloned at the molecular level and then used in plant transformation, since barriers for using isolated herbicide resistance genes to produce resistance in unrelated species are being progressively broken down as genetic engineering technology develops (Oxtoby and Hughes, 1989). Thus, plant genes for resistance which are identified via a differential response could be used in similar fashion to bacteria and yeast genes.

Generally to isolate a herbicide resistance gene at the molecular level requires knowledge of its mode of action and the specific target proteins affected (Mazur and Falco, 1989). In this way, for example, yeast and bacterial genes for the enzyme, acetolactate synthase, have been isolated and used to develop systems for resistance to sulfonylurea herbicides. If the mode of action of genes mediating differential responses within species can be identified then, of course, these can also be isolated by these technologies. Often however, particularly for "older" herbicides, the mode of action is unknown and these approaches will not suffice. Nevertheless newer technologies are now becoming available to isolate genes even if the mode of action is unknown. In particular methods of long range mapping and chromosome walking are being developed which allow genes to be isolated once their chromosomal position has been pinpointed (Tanksley *et al.*, 1989) as are methods of transposon tagging (Wienand and Saedler, 1987). Thus the location of genes by genetic analysis should soon lead directly to their isolation at the molecular level. Thus, for example, there is the expectation that genes for herbicide resistance in wheat could be used in, say, potato.

CONCLUSIONS

The few studies of the genetical control of differential responses of crop species to widely used herbicides, such as those described here, indicate major qualitative differences in response to single major genes. Nevertheless there are, in addition, quantitative effects attributable to so-called, "modifier" genes. The major genes can give cross-protection to related herbicides but it remains to be discovered if these also give protection to unrelated chemicals. Presumably this will be determined by the particular modes of actions of the herbicides being considered. It is interesting to note that wild species related to the cultivated forms also exhibit differential responses suggesting that the genes controlling these responses have pleiotropic effects on other characters which are being subjected to natural selection.

Overall, our knowledge of the genetical control of responses to widely used herbicides is still permeated by large gaps. However, the identification of specific genes for herbicide responses can greatly increase our understanding of the evolution and mode of actions of such genes as well as providing new strategies for the genetic manipulation of herbicide resistance in plant breeding.

REFERENCES

Bowran, D.G. and Blacklow, W.M. (1987). Sensitivities of spring wheat cultivars to chlorsulfuron measured as inhibitions of leaf elongation rates and there were genotype x environment interactions. Australian Journal of Agricultural Research **38**, 253-262.

Gale, M.D., Sharp, P.J., Chao, S. and Law, C.N. (1989). Applications of genetic markers in cytogenetic manipulation of the wheat genomes. Genome **31**, 137-142.

Gressel, J. (1988). Multiple resistances to wheat selective herbicides: New challenges to molecular biology. Oxford Surveys of Plant Molecular & Cell Biology **5**, 195-203.

Law, C.N., Snape, J.W. and Worland, A.J. (1987). Aneuploidy in wheat and its uses in genetic analysis. In "Wheat breeding. Its Scientific Basis" (F.G.H. Lupton, ed.), pp. 71-108. Chapman & Hall, London.

Leckie, D. (1989). "The genetical control of the response of wheat to the wild oat herbicide difenzoquat". Ph.D. thesis, University of Cambridge.

Mazur, B.J. and Falco, S.C. (1989). The development of herbicide resistant crops. Annual Review of Plant Physiology and Molecular Biology **40**, 441-470.

Nevo, E. and Beiles, A. (1989). Genetic diversity of wild emmer wheat in Israel and Turkey. Structure, evolution and application in breeding. Theoretical and Applied Genetics **77**, 421–455.

Oxtoby, E. and Hughes, M.A. (1989). Breeding for herbicide resistance using molecular and cellular techniques. Euphytica **40**, 173–180.

Snape, J.W. and Parker, B.B. (1988). Chemical response polymorphisms: an additional source of genetic markers in wheat. Proceedings Seventh International Wheat Genetics Symposium. (T.E. Miller and R.D. Koebner, eds), pp. 651–656. Institute of Plant Science Research, Cambridge.

Snape, J.W., Angus, W.J., Parker, B.B. and Leckie, D. (1987). The chromosomal locations in wheat of genes conferring differential responses to the wild oat herbicide, difenzoquat. Journal of Agricultural Science (Cambridge) **108**, 543–548.

Snape, J.W., Parker, B.B., Leckie, D., Rosati-Colarieti, G. and Bozorgipour, R. (1990). Differential responses to herbicides in wheat: uses as genetic markers and target genes for genetic manipulation. In "Biotechnology for the Plant Breeder" (J.Jensen and C.N. Law, eds), (in press).

Tanksley. S.D., Young, N.D., Paterson, A.H. and Bonierdale, M.W. (1989). RFLP mapping in plant breeding – new tools for an old science. Biotechnology **7**, 257–264.

Tottman, D.R., Lupton, F.G.H. and Oliver, R.H. (1984). The tolerance of difenzoquat and diclofop-methyl by winter wheat varieties at different growth stages. Annals of Applied Biology **104**, 151–159.

Tottman, D.R., Lupton, F.G.H., Oliver, R.H. and Preston, R. (1982). Tolerance of several wild oat herbicides by a range of winter wheat varieties. Annals of Applied Biology **100**, 365–373.

Wienand, U. and Saedler, H. (1987). Plant transposable elements, unique structures for gene tagging and cloning. In "Plant Gene Research: Plant Infectious Agents" (T.H. Hohn and J. Schell, eds), pp. 205–227. Springer-Verlag, Vienna.

HERBICIDE RESISTANCE IN THE GRAMINACEAE - A PLANT BREEDER'S VIEW

D.T. Johnston[1] and J.S. Faulkner[2]

[1]Northern Ireland Horticultural and Plant Breeding Station Loughgall, Armagh, UK [2]Conservation Service, Department of Environment (NI), 23 Castle Place, Belfast

Breeding for herbicide tolerance in grasses began at Loughgall in the late 1960s. One of the main factors which stimulated the programme was the realisation that there was genetic variation between and within plant species in the effects of herbicides upon them. This was followed by demonstrations that particular herbicides were more active against some grass species than against others. It could therefore be deduced that selective breeding should produce grasses with enhanced herbicide tolerance thereby creating a means of manipulating the composition of grass swards.

Early work with perennial ryegrass (*Lolium perenne*) confirmed that it was feasible to select for paraquat tolerance. This was soon followed by equally successful selection programmes using the herbicides dalapon, aminotriazole and glyphosate in both amenity and agricultural grasses.

While the practical value of the herbicide tolerance is a unique selling point, the ultimate success of the varieties also depends heavily on their other attributes, such as seed yield, forage yield in agricultural grasses and turf quality in amenity grasses. Cultivars were developed out of several of the herbicide tolerance selection programmes and have met with varying degrees of success in commerce.

INTRODUCTION

The main aim of breeding forage grasses is the improvement of yield and forage quality. Ultimately, swards tend to be replaced by both grass and dicotyledonous weeds, leading to a decline in productivity. Ingress by unsown species will occur even in swards of the most persistent cultivars.

In amenity turf, weed grasses, particularly annual meadow grass (*Poa annua*), are undesirable for aesthetic reasons because they reduce the playing quality of the turf. They may also be more susceptible to drought or disease.

Few herbicides control grass weeds selectively in sown swards because most grasses tend to be similar in physiology and morphology. Rather than searching for the elusive compound that will kill grass weeds and not sown species, a cheaper and simpler alternative is to breed grass cultivars which are tolerant of lethal doses of grass-killing herbicides. Some of the initial research on herbicides at Loughgall in the mid-1960s compared the differential response of perennial ryegrass genotypes to certain herbicides, and led to the conclusion that this approach to the problem of grass weeds had practical possibilities.

Subsequently, breeding programmes were undertaken to select for tolerance to paraquat, dalapon, aminotriazole and glyphosate in both amenity and agricultural grass species.

VARIATION IN HERBICIDE TOLERANCE

Inter-specific

Differential tolerance to herbicides between grass species has been widely reported. Some of the initial work on testing for inter-specific variation in response to grass-killing herbicides was reported by King and Davies (1963) who showed that common grasses differed in their response to dalapon: perennial ryegrass and fescue (*Festuca* spp.) were relatively resistant whereas mat grass (*Nardus stricta*) and purple moor grass (*Molinia coerulea*) were quite susceptible. Interspecific variation in response to dalapon was found not only in temperate grasses but also in tropical species (Wallis, 1962).

In a comparison of the responses of 12 grass species to a range of grass-killing herbicides, Fisher and Faulkner (1975) showed that there was wide variation between grass species in their response to the herbicides. Browntop bent (*Agrostis tenuis*) was relatively susceptible to most herbicides whereas red fescue (*Festuca rubra*) and cocksfoot (*Dactylis glomerata*) were relatively tolerant to several herbicides.

Many further examples of interspecific differences in tolerance to herbicides have been reported to a wide range of perennial and annual species and to herbicides with differing modes of action. Comes *et al.* (1981) observed differential tolerance to dalapon, aminotriazole and glyphosate and concluded that red fescue was relatively tolerant to glyphosate whereas red-top bent (*Agrostis alba*) was quite tolerant of aminotriazole.

Intra-specific

There are fewer published examples of intraspecific variation of grasses in response to herbicides. Wright (1968) reported significant differences between cultivars of perennial ryegrass in their response to paraquat in the field. Subsequently, Faulkner (1974a) demonstrated paraquat tolerance among plants in an experimental population of perennial ryegrass. Although the difference between the most and least tolerant plants was not particularly great, tolerance nevertheless had a distinct genetic component, with narrow sense heritability estimated to be in the range 0.51 to 0.72.

Variation in dalapon tolerance among 35 perennial ryegrass cultivars in field conditions was demonstrated by Faulkner (1974b). The most tolerant cultivar yielded twice as much herbage as the least tolerant during the 12 week period after spraying with a sub-lethal dose. Tetraploid cultivars were more tolerant than diploid ones. Management treatments such as cutting and spraying dates did not materially affect the relative tolerances of the cultivars.

Perennial ryegrass and Italian ryegrass (*Lolium multiflorum*) cultivars have also been shown to vary in their response to ethofumesate (Faulkner, 1984). Seventy cultivars were screened as seedlings in the glasshouse and marked differences found between them. Generally, tetraploids and Italian ryegrass cultivars were the more tolerant, but there were nevertheless statistically significant differences between cultivars of the same species and ploidy levels. Subsequent experiments in the field confirmed these differences.

In red fescue, cultivars vary in their tolerance of glyphosate (Johnston *et al.*, 1989). Seedling plants of four cultivars of slender creeping red fescue (*Festuca rubra* ssp. *trichophylla*) and ten cultivars of Chewing's fescue (*Festuca rubra* ssp. *commutata*) were sprayed with a range of doses at the 2-leaf stage. The amount of herbicide required to kill 50% of seedlings was determined (ED_{50}). The four slender creeping red fescue varieties had above average tolerance to glyphosate (mean ED_{50} : 0.55 kg ai ha^{-1}) whereas the ten Chewing's fescue varieties had generally a lower tolerance (mean ED_{50} : 0.36 kg ai ha^{-1}).

Differences between cultivars in their response to herbicides seem to be common - indeed we suspect it is probably almost universal in perennial grasses. Nevertheless these differences are usually too small to facilitate the selective removal of unsown grasses in swards of the more tolerant cultivars.

BREEDING FOR HERBICIDE TOLERANCE

Paraquat tolerance

Following the preliminary investigations with herbicides at Loughgall, Faulkner (1974a) demonstrated the heritability of paraquat tolerance in perennial ryegrass and concluded that it should be possible to increase tolerance by selective breeding. Subsequently a paraquat-tolerant line, PRP II, was derived by selection and propagation from a wild collection of perennial ryegrass that contained paraquat tolerance genes. The procedure used for selection purposes was to apply critical doses of paraquat to segregating populations of 2-leaf seedlings. Surviving individuals were retained, rescreened and grown to maturity. These were then allowed to inter-pollinate in collective isolation yielding "selection 1" seed. The progeny produced were subjected to recurrent mass selection with an increased herbicide dosage, thus raising the level of selection pressure.

Preliminary experiments in seed trays demonstrated that it was possible to eliminate indigenous grasses such as rough stalked meadow grass (*Poa trivialis*) and Yorkshire fog (*Holcus lanatus*) as well as other standard ryegrass cultivars from PRP II by spraying with paraquat. Further paraquat tolerant lines were produced by crossing the original paraquat tolerant

321

material with existing cultivars of both agricultural and amenity type, and with breeders' stocks and wild collections.

The ED_{50} values of paraquat tolerant lines, control cultivars and two weed grass species were determined and are shown in Table 1. Causeway, one of the first paraquat tolerant cultivars to be produced, was approximately 8 times more tolerant of paraquat than control cultivars and the more recently bred cultivars, such as PAR 5, are much more tolerant than Causeway.

Dalapon tolerance

Dalapon has a mildly selective effect in favour of perennial ryegrass and against certain other grass species. Faulkner (1974b) found differences in degree of dalapon tolerance among 35 cultivars of perennial ryegrass, but these were not large enough to make a substantial difference to the selectivity of the herbicide between perennial ryegrass and other grasses. It was therefore considered worthwhile to try to breed cultivars with a higher level of tolerance.

Table 1. <u>ED_{50} values of 7 paraquat tolerant and 2 control cultivars and 2 weed grasses</u>: paraquat was applied on glasshouse grown seedlings at the 2-leaf stage (unpublished data).

Cultivar	Parentage	ED_{50} (kg ai ha^{-1} paraquat)
Paraquat tolerant	cultivars	
Causeway	Par.tol.collection (PRP)	0.103
PRP48	PRP* x Cropper	0.119
PRP50	PRP* x Melle	0.195
PRP54	PRP* x Perma	0.167
PAR3	Barry x PRP28	0.313
PAR4	HE177 x PRP3	0.206
PAR5	HE145 x PRP45	0.344
Control cultivars		
Barlenna		0.007
Manhattan		0.013
Weed grasses		
Poa trivialis		0.023
Poa annua		0.013
S.E.		0.011

* Selected paraquat tolerant individual plants derived from wild collection.

A recurrent selection programme was adopted which was similar to but less intensive than that used in the paraquat tolerance programme and a range of cultivars with elevated dalapon tolerance was produced.

The results of a two part dalapon tolerance test are shown in Table 2. In the first part of the test seeds were sown in soil treated with dalapon at 0.075, 0.150 and 0.300 g a.e. kg^{-1}. The data given (Table 2) are percentages of germinated seeds that survived, averaged over the three dalapon levels. Differential dalapon tolerance was exhibited to some degree in the ability of the seedlings to survive in dalapon treated soil.

In the second part of the test, plants with several tillers were sprayed with dalapon at 12, 18 and 27 kg a.e. ha^{-1}, and then cut back 2½ weeks later. Regrowth was scored on a scale 0–5, two weeks after cutting and averaged across dalapon levels. The control cultivars were largely killed, but the selected cultivars showed various degrees of tolerance paralleling their tolerance as expressed in dalapon treated soil (Table 2).

Table 2. <u>Comparison of the response of seedlings and small plants of 7 perennial ryegrass cultivars to dalapon</u> (unpublished data).

Cultivar	% Seedling survival in treated soil	Regrowth score after foliar spraying (0–5)
Controls:		
Barlenna	13.4	0.11
Melle	21.0	0.25
Cultivars	selected for dalapon	tolerance
Antrim	51.9	2.09
Rathlin	28.2	1.35
Tyrone	47.7	2.17
Portstewart	61.8	2.68
Carrick	59.9	3.08
S.E.	2.25	0.15

Aminotriazole tolerance

Fisher and Faulkner (1975) showed that both brown-top bent and red fescue were relatively tolerant to aminotriazole. As bents and fescues are commonly sown in mixtures, parallel programmes for selecting for tolerance to aminotriazole were undertaken (Lee and Wright, 1981). A selection technique was developed whereby seeds were sown into a sterilised soil based compost. Prior to sowing, aminotriazole was mixed with the compost at a rate calculated to give a 95% – 99% mortality. The compost was put into 0.5 m x 1.0 m wooden frames which were placed on carpet underlay on a level capillary bed. Surviving seedlings were removed from the compost when they had 4 or 5 leaves and were repotted. A second stage of selection was carried out by applying a foliar dose of aminotriazole to the seedlings when they had several tillers. Surviving seedlings, approximately 100 per variety, were grown to maturity and allowed to inter-pollinate yielding Selection 1 seed. This was subjected to a further cycle of selection by increasing the amino-triazole dose in the compost and at the second stage foliar application.

To assess the increase in tolerance which had occurred with each cycle of selection and to compare the relative tolerance of the final selection with the unselected parent an ED_{50} experiment was carried out. There had been an approximate doubling of resistance with each cycle of selection and the S3 population had an ED_{50} approximately 19 times, and 5 times greater than the unselected parent for bent and fescue respectively. (Fig. 1a and b). Two cultivars, Countess Chewing's fescue and Duchess brown-top bent have already emanated from this programme and several further cultivars are at an advanced stage of development.

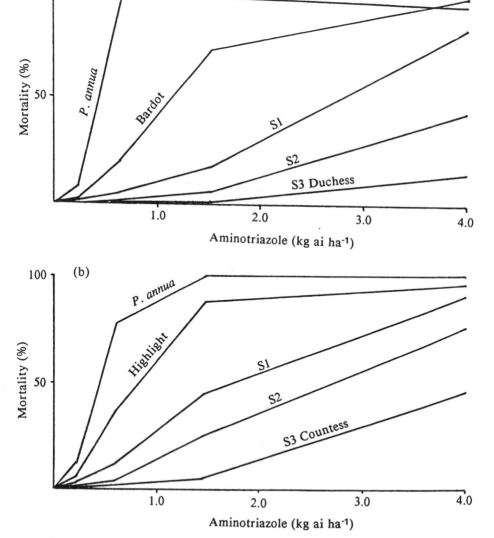

Fig. 1. Percentage mortality of *Poa annua* and of two amenity grass species each represented by three selections (1, 2 and 2) and unselected control cultivar (a). *Agrostis tenuis*; (b) *Festuca rubra* spp. *commutata.*

Glyphosate tolerance

A programme at Loughgall is aimed at increasing levels of tolerance to the herbicide glyphosate in red fescue and perennial ryegrass. Glyphosate was chosen because it is extremely effective against a wide range of perennial and annual species and it is perceived as environmentally benign and has low mammalian toxicity.

Johnston *et al.* (1989) showed that selection for glyphosate tolerance was particularly successful in fescue cultivars: For example, four cycles of selection in the hard fescue cultivar Waldina produced a population (HF2) distinctly more tolerant than the parent cultivar, which in turn is more tolerant than annual meadow grass (Fig 2).

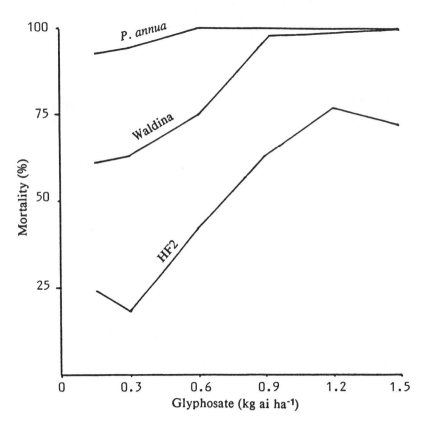

Fig. 2. Percentage mortality of 2-leaf seedlings of *Poa annua,* a control cultivar of hard fescue, and a glyphosate-tolerant selection, HF2.

In perennial ryegrass the increase in tolerance to glyphosate with successive cycles of selection has been much slower than in fescue. Our most advanced line has undergone 11 cycles of recurrent selection and is now approximately five times more tolerant of glyphosate than other control cultivars.

HERBICIDE TOLERANCE IN PRACTICE

Agricultural swards for forage production

In some of the initial work on herbicide tolerance in perennial ryegrass, Faulkner (1975) found that it was possible to use paraquat selectively to remove Yorkshire fog and *Poa* spp. from stands of a paraquat tolerant line PRP II. In a further trial (Faulkner, 1976), based on field plots, the proportion of ryegrass in a sward of paraquat tolerant variety Causeway was increased from 45% to 96% by applying paraquat at 0.3 kg ai ha^{-1}.

Further research (unpublished) demonstrated the feasibility of controlling couch grass (*Agropyron repens*) in swards of paraquat tolerant cultivars. Low doses of paraquat applied twice annually for three successive years eliminated couch and maintained a sward comprised of 99% perennial ryegrass. Plots that were not sprayed had 64.8% of weed grasses, including 13.7% of couch. The dry matter yield over three years, however, was significantly lower in the plots which were recurrently sprayed.

Similar field experiments were carried out with the dalapon- tolerant cultivars. Faulkner (1978) showed that dalapon could be used effectively to increase the proportion of ryegrass in swards of a dalapon tolerant cultivar.

In another trial (unpublished), a single dose of dalapon reduced the amounts of bent, Yorkshire fog and couch. It made little long-term difference whether the dalapon was applied in spring, summer or autumn. Dry matter yields in the years after spraying were not significantly affected by dalapon treatment.

No further experiments on the use of dalapon as a basis for maintaining pure swards of perennial ryegrass are being carried out at Loughgall because the herbicide is no longer widely available.

Only preliminary experiments have been carried out on the use of glyphosate on field plots of glyphosate tolerant cultivars. It has been shown that it is possible to eliminate indigenous grasses by spraying with glyphosate at 0.3 kg ai ha^{-1}. More extensive trials will be carried out when lines with a higher level of glyphosate tolerance have been produced.

Amenity Turf

The use of aminotriazole-tolerant cultivars, Countess and Duchess, will help to facilitate the selective control of weed grasses in fine turf. Johnston and Fisher (1985) showed that it was possible to eliminate annual meadow grass, perennial ryegrass, cocksfoot and Yorkshire fog from turfs of Countess and Duchess by spraying with aminotriazole at a rate equivalent to 1.5 kg ai ha^{-1}, without causing any damage to the Countess and Duchess. Further trials (unpublished) have shown that Countess and Duchess can tolerate aminotriazole rates of up to 8.0 kg ai ha^{-1}.

These cultivars are now being used for the production of fine turf for bowling and golf greens. Aminotriazole is applied in the turf nursery during the time of seedling establishment and a pure turf is produced which is completely free of weed grasses.

As a Distinctness Characteristic

In a wide range of countries including the United Kingdom it is not permissible to market seeds of grass cultivars which are not inscribed on the National List. One of the requirements for obtaining National List status is that new cultivars must be shown to be distinct from all other existing cultivars on the list. Usually it is possible to distinguish cultivars by comparing morphological or phenological characteristics such as height of stem or date of ear emergence, but as the number of cultivars of a species increases this has become more difficult. A lack of distinctness in such characters has been encountered on several occasions with new cultivars bred at Loughgall and in these cases herbicide tolerance has proved to be a useful distinctness character. A typical example of this was using the dalapon tolerance characteristic of the perennial ryegrass cultivar Antrim to distinguish it from the morphologically similar cultivar Perma.

In Seed Production

Herbicide tolerance is a useful characteristic in the production of herbage seed. In order to achieve certification standards seed must be free from contamination with rogues of the same species and weed grasses.

In the UK and the USA, aminotriazole is used routinely at a rate of 1.5 kg ai ha^{-1} on seed crops of Countess Chewing's fescue and Duchess brown-top bent. Paraquat has been used effectively for maintenance of weed-free pre-basic seed crops of paraquat tolerant varieties at Loughgall.

INCIDENTAL EFFECTS OF HERBICIDE TOLERANCE

Herbage Yield

Apart from a few exceptions, the herbage yield of dalapon tolerant cultivars has been consistently better than that of paraquat tolerant cultivars. It has been suggested (Faulkner, 1983) that some or all of the genes controlling paraquat tolerance have an adverse affect on performance, whereas those for dalapon tolerance do not.

To date two paraquat tolerant perennial ryegrass cultivars, Causeway and Portrush, have been inscribed on the National List in the UK although neither of them achieved a sufficiently high level of herbage production to be recommended for use by farmers.

Conversely, the dalapon tolerant cultivars have been extremely successful in official trials and at present four late-maturing cultivars, Antrim, Tyrone, Carrick and Portstewart are on the Recommended List for England and Wales, and a fifth, Rathlin, has been widely used in France for several years.

Table 3. Herbage yield of 4 dalapon tolerant and 4 paraquat tolerant cultivars at Loughgall in 1986, 1987 and 1988: expressed as a percentage of Melle.

	Herbage DM yield : % of Melle		
Cultivar	1986	1987	1988
Dalapon tol.			
DRP 18	109	106	100
DRP 19	106	108	110
DRP 20	104	106	107
DRP 21	109	106	106
Paraquat tol.			
PRP 49	98	100	104
PRP 50	97	92	94
PRP 51	105	101	107
PRP 52	109	101	108

Turf Quality

Glyphosate-tolerant fine-leaved *Festuca* cultivars which were bred at Loughgall are currently undergoing turf trials in Holland. The glyphosate-tolerant cultivars tend to have a darker green colour than the control cultivars and this is considered to be a favourable characteristic in many turfgrass markets.

CONCLUSION

Why were grasses selected as the subject of the herbicide tolerance work at Loughgall? In reality, they weren't. It was more a question of selecting herbicide tolerance as a feature of the grass breeding programme to which the Station was committed.

The work began on forage grasses, in which it appeared that the ability to control invasive species might offer a major advance in prolonging the productive life of the sward. The underlying theory was that competition, particularly from indigenous grasses, was a major cause of sward deterioration.

It has certainly proved technically feasible to breed cultivars with genetically based tolerance to herbicides that kill other grasses. These herbicides can then be used to maintain virtually pure swards of the tolerant cultivars. It has not, however, followed that these swards are necessarily more productive. While competition may be responsible for a decline in the proportion of the sown species present, the yield deficit may be made good in part or in whole by the contribution to overall sward productivity from the competitors. Indeed, in some cases, where the sown grass has lacked persistence, killing invasive species simply prevented them from making their contribution to productivity and yield was consequently reduced.

Because herbicide tolerance has not been perceived as likely to increase output, it has not been a major selling point of forage grasses at farmer level. Nevertheless, we would maintain that herbicide tolerance has an important potential in the long term, in conjunction with forage quality characters. Making full use of breeding advances in such things as nutritive value, mineral content and palatability will progressively depend on a capacity to maintain pure swards, free from invasive species of inferior quality. The greater the gains made by breeders in other characters, the greater will be the incentive to build in a mechanism for keeping the sward free from "weeds".

Although herbicide tolerance in forage grasses has so far had relatively little impact on the farm, it has nevertheless shown itself of value in establishing cultivar distinctness, and in enabling the chemical roguing of seed crops. It has been in the amenity grasses, however, that herbicide tolerance has proved most beneficial. This can be attributed in part to the much greater importance of sward uniformity, and the irrelevance of yield. Perhaps even more important, though, is the fact that it is a technically difficult and unreliable operation to create pure swards of fine leaved turf grasses, and any technique that makes it easier or more reliable is good news to the turf grass manager.

Quite apart from the question of weed control in forage and amenity grasslands, we conclude that herbicide tolerance confers two other important benefits.

1. Stands of tolerant cultivars for seed multiplication can be sprayed to remove weed grasses and rogues of the same species thereby ensuring virtually clean seed.

2. Herbicide tolerance is a unique distinctness characteristic and facilitates the resolution of disputes about cultivar distinctness or authenticity of seed.

ACKNOWLEDGEMENTS

We would like to thank our colleagues at the Northern Ireland Horticultural and Plant Breeding Station who have contributed to this work.

REFERENCES

Comes, R., Marquis, L.Y. and Kelley, A. D. (1981). Response of seedlings of three perennial grasses to dalapon, amitrole and glyphosate. Weed Science **29**, 619-621.

Faulkner, J.S. (1974a). Heritability of paraquat tolerance in *Lolium perenne* Euphytica **23**, 281-288.

Faulkner, J.S. (1974b). The effect of dalapon on thirty five cultivars of *Lolium perenne*. Weed Research **14**, 405-413.

Faulkner, J.S. (1975). A paraquat-tolerant line in *Lolium perenne*. Proceedings of the 1st European Weed Research Society Symposium, Paris, 349–359.

Faulkner, J.S. (1976). A paraquat resistant variety of *Lolium perenne* under field conditions. British Crop Protection Conference – Weeds 485–490.

Faulkner, J.S. (1978). Dalapon tolerant varieties – a possible basis for pure swards of *Lolium perenne* L. British Crop Protection Conference – Weeds 341–348.

Faulkner, J.S. (1983). A plant breeder's view of herbicide tolerance. Aspects of Applied Biology **3**, 47–62.

Faulkner, J.S. (1984). Variation in ethofumesate tolerance in *Lolium* species and cultivars. Weed Research **24**, 153–161.

Fisher, R. and Faulkner, J.S. (1975). The tolerance of twelve grass species to a range of foliar absorbed and root absorbed grass-killing herbicides. Proceedings European Weed Research Society Symposium, 204–215.

Johnston, D.T. and Fisher, R. (1985). Grass weed control in lawns of aminotriazole tolerant cultivars of *Festuca nigrescens_* and *Agrostis tenuis*. Proceedings of the Fifth International Turfgrass Research Conference, 727–734.

Johnston, D.T., Van Wijk, A.J.P. and Kilpatrick, D. (1989). Selection for tolerance to glyphosate in fine-leaved *Festuca_* species. Proceedings of the Sixth International Turfgrass Research Conference, 103–105.

King, J and Davies, G.E. (1963). The effect of dalapon on the species of hill grassland. Journal of the British Grassland Society 18, 52–55.

Lee, H and Wright, C.E. (1981). Effective selection for aminotriazole tolerance in *Festuca* and *Agrostis* turf grasses. Proceedings of the Fourth International Turfgrass Research Conference, 41–46.

Wallis, J.A.N. (1962). Observations on the differential response of 40 grass species and strains to low doses of dalapon. Pesticide Abstracts **8**, 200–202.

Wright, C.E. (1968). A preliminary examination of the differential reaction of perennial ryegrass cultivars to grass-killing herbicides. Proceedings of the 9th British Weed Control Conference, 477 – 483.

IN VITRO TECHNIQUES FOR THE SELECTION OF HERBICIDE RESISTANCE

George Marshall

West of Scotland College, Department of Plant Sciences, Auchincruive, Ayr, KA6 5HW, UK

The ever-increasing range of sophisticated cell and tissue culture techniques in plant biotechnology now offers researchers a wide range of applications. The focus of this review is to consider the application of *in vitro* techniques for the selection of herbicide resistance in crop plants. Accordingly, it is important to outline the general philosophy behind the decision to choose an *in vitro* selection system as compared to an alternative strategy such as classical breeding or genetic engineering. In several situations, *in vitro* techniques play a central role in allowing herbicide tolerance to be transferred in an inter-specific or an intra-specific manner. The diversity of *in vitro* systems will be described including callus, suspension, protoplast, anther, microspore and embryo culture. Each system has been used to different degrees by researchers over the past 10 years and based upon this wealth of published work it is now possible to draw some conclusions on both the value and limitations of each system in relation to a range of target herbicides.

INTRODUCTION

In vitro techniques for the culture of cells and tissues currently play a major role in plant biotechnology research. One of the principal applications of plant cell and tissue culture systems has been for their use in crop improvement (Duncan and Widholm, 1986). While cell selection techniques were pioneered in microbiological situations, plant scientists are now determining the value of *in vitro* selection systems for specific purposes in the genetic improvement of crops. Thus, research programmes are currently well established for the selection of resistance to disease, drought and salt stress and herbicides. It is evident from the number and the comprehensive nature of preceeding reviews (Meredith and Carlson, 1982; Crocomo and Ochoa-Alejo, 1983; Hughes, 1983; Chaleff, 1986a, 1986b; 1988; Gressel, 1987; Maliga *et al.*, 1987) that the *in vitro* selection for herbicide resistance is a topic of considerable interest, research innovation and a degree of controversy. This review will attempt to summarise the theoretical and practical roles which *in vitro* selecion techniques may play in conferring herbicide resistance on plants.

PLANT CELL AND TISSUE CULTURE: THE SYSTEMS

Callus and suspension cultures

It is the totipotency of plant cells which encourages scientists to develop a range of *in vitro* systems for their manipulation. Frequently, the starting point for a system is an explant of differentiated tissue which is then encouraged to proliferate in an undifferentiated mass of cells called a callus. Such culture systems are normally based upon a semi-solid agar base and depend for their success on a growth medium which contains mineral salts, vitamins, plant growth substances and a carbon source (usually sucrose). Callus tissue can subsequently be transferred to a liquid growth medium (without agar) thus providing a suspension of single cells or aggregates each containing several cells. Both calli and suspensions can be maintained for many weeks in this undifferentiated state by sub-culturing techniques.

Protoplast cultures

Unlike callus and suspension cultures where non-photosynthetically active, undifferentiated cells are multiplied, protoplasts cultures are typically derived from green leaf tissues. Protoplast isolation, incubation and culturing procedures are described in detail by Evans and Bravo (1983). The principal feature of protoplast systems is that they can provide genetically homogeneous cultures of single photosynthetic cells. These are more likely than callus and suspension cultured cells to yield mutants that are expressed in differentiated tissues at the whole plant level (Hickok, 1987). In addition, protoplast fusion techniques can now be used to obtain crop plants which carry herbicide resistance genes in the chloroplast genome (Maliga *et al.*, 1987). In traditional or classical breeding programmes the introduction of such a maternally-inherited trait can only be made after several generations of backcrossing as described by Beversdorf *et al.* (1980). Although this is a time-consuming process, it has the incidental benefit that maternally-inherited herbicide resistance in crops cannot spread to weed species by cross-pollination (Souza Machado, 1982). New protoplast fusion techniques and the necessary *in vitro* regeneration systems can now be used to hasten the transfer of cytoplasmically-inherited herbicide resistance, especially using interspecific hybrids.

Anther and microspore cultures

Traditionally, *in vitro* culture systems have been devised to induce haploid embryo production from excised anthers or isolated immature pollen (microspores) in a wide range of crops (Bajaj, 1983). The production of haploid plants via *in vitro* techniques is of immense practical importance in crop improvement for the rapid production of superior homozygous lines. Recently, Hickok (1987) reported a novel application for *in vitro* haploid production and herbicide selection; the use of the gametophytic phase of the fern *Ceratopteris*. This approach utilises whole plants which are photosynthetically active. Following the identification and characterisation of genes conferring herbicide resistance it should be possible to introduce this trait to crop species via recombinant DNA technology. Haploid cells, irrespective of their origin have the advantage over systems where higher levels of ploidy are employed that mutations of either a dominant or recessive type will be expressed in the phenotype, thus allowing a clear genetical analysis of the new herbicide resistance.

SELECTION SYSTEMS

Traditional approaches

History has shown that, as the level of our competence to culture plant cells increases, attempts are made to select for herbicide tolerance in cells in increasingly complex ways. Simple systems have been used in callus and suspension cultures where the plant cells are grown in the presence of a range of herbicide concentrations in an attempt to identify the degree of selection pressure which will allow herbicide resistant variants to be isolated, subcultured and ultimately regenerated. The practical requirements of such systems are reviewed by Hughes (1983). Callus and suspension culture systems have been used successfully to isolate cell lines resistant to a range of herbicides including aminotriazole, 2,4-D, picloram, paraquat, chlorsulfuron, metsulfuron-methyl and imazaquin (Meredith and Carlson, 1982; Chaleff, 1986a, 1986b). The major limitation of this *in vitro* selection procedure is that the degree of herbicide tolerance or resistance which results may be unstable and non-inheritable. Indeed, Meredith and Carlson (1982) are careful to classify a range of four different types of herbicide resistance obtained from tissue culture selections and further to discuss the physiological basis of this variation. Fortunately there are some basic guidelines which we can follow to maximise the opportunity to select stable, inheritable forms of herbicide resistance.

The choice of herbicide is critical for success. If we are culturing undifferentiated cells and intend to select for herbicide resistance then it is logical to use a herbicide which does not rely upon differentiated tissues to express its primary mechanism of action. By contrast, where herbicides require differentiated photosynthesising tissues to effect their toxicity more complex selection systems need to be designed e.g. the use of complementary *in vivo* and *in vitro* techniques as described in tobacco for bentazon and phenmedipham by Radin and Carlson (1978). The classical success stories of *in vitro* selection for herbicide resistance relate to the use of herbicides of the sulfonylurea and imidazolinone families.

Members of the sulfonylurea and the imidazolinone herbicide families inhibit protein synthesis in sensitive species via the enzyme acetolactase synthase (ALS) (Chaleff, 1986b). In the case of the sulfonylurea resistance, Chaleff and Ray (1984) adopted haploid callus cultures derived from tobacco leaves, some of which had previously been chemically mutated in an attempt to increase genetic variation prior to herbicide selection. The calli were subject to at least two cycles on the selective medium. This precaution is taken simply to reduce the possibility that groups of callus cells were apparently herbicide-resistant by virtue of their physical location. In addition, repeated exposure to the herbicide *in vitro* may enhance the ultimate level of resistance achieved. A similar type of cyclic selection protocol was used by Shaner and Anderson (1985). The system used maize callus initiated from isolated embryos and it was found that the level of herbicide resistance in the callus tissues could be usefully increased with increasing selection cycles (up to seven) on relatively low (0.03 mg l^{-1}) concentrations of imazaquin.

Both the studies to select for sulfonylurea and imidazolinone resistance were completed by conducting experiments to determine the mechanism of inheritance and also the biochemical basis for the new trait. Accordingly, chlorsulfuron resistance in tobacco was conferred by a single dominant or semidominant nuclear mutation and an altered form of the target enzyme (ALS), less sensitive to the two sulfonylurea herbicides (Chaleff and Ray, 1984). In maize, imidazolinone resistance was also believed to be controlled by a single nuclear gene resulting in at least a 1000-fold less sensitive target enzyme, acetohydroxyacid synthase (AHAS synonymous with ALS) (Shaner and Anderson, 1985).

Microspore culture

The most recent descriptions of novel selection systems refer to the use of microspores from rapeseed (*Brassica napus* L.) and chlorsulfuron (a sulfonylurea) as the target herbicide. The first reported use of screening microspore cultures for chlorsulfuron resistance was by Keynon *et al.* (1987). In these preliminary experiments, embryos which had begun to produce root and shoot growth were plated on an agar growth medium which contained a range of chlorsulfuron concentrations (selective media). After one month, although the herbicide had a significant effect on reducing embryo regeneration, several putative chlorsulfuron-resistant haploid plants were isolated. Swanson *et al.* (1988) prepared mutagenized and non-mutagenized microspores of rapeseed which were cultured in a microspore medium containing chlorsulfuron. Surviving embryos were then briefly subcultured in a non-selective medium, transferred for a second passage to a selective medium and finally the surviving embryos were regenerated. Ultimately, two microspore-derived chlorsulfuron resistant plants were selected, multiplied and characterised genetically and biochemically. Chlorsulfuron resistance appeared to be inherited in a semi-dominant nuclear manner with an altered target site enzyme activity (AHAS). Levels of chlorsulfuron resistance 10-100 times that of the susceptible parent were generated using this new selection technique.

Protoplast culture

Protoplast cultures of *Nicotiana plumbaginifolia* have been used to select for triazine resistance. Since triazine herbicides are not active on heterotrophic cellular growth, Cseplo *et al.*, (1985) designed a system to recover triazine-resistant mutants using special photomixotrophic cultures of *N. plumbaginifolia* such that they were weaned down from 3.0% to 0.3% sucrose. Subsequent selections on terbutryne, a triazine herbicide used the differential bleaching (sensitive) and non-bleaching (resistant/green) effect to identify mutant cell lines which were triazine resistant. This trait is known in other species to be maternally inherited and is conferred by a mutation of a chloroplast gene encoding a thylakoid membrane protein (Hirschberg and McIntosh, 1983; Hirschberg *et al.*, 1984).

Recently, Swanson *et al.*, (1988) isolated protoplasts from microspore-derived embryos of *B. napus* which had been subcultured on a chlorsulfuron selective medium, then a non-selective medium (previously described). As the protoplasts were isolated and diluted prior to their regeneration, the selection pressure of chlorsulfuron was increased from 0.75 to 2.5 ppb. Finally, one chlorsulfuron-resistant, protoplast-derived plant was selected which showed an especially high level of resistance compared to microspore derived mutants. Unfortunately the mutant plant was aneuploid and consequently was of low fertility. In addition, the biochemical basis for the chlorsulfuron tolerance could not be fully explained but tolerance was not simply due to an altered form of the enzyme AHAS. The authors concluded that the chloroplast selection system and the microspore selection system may have conferred chlorsulfuron resistance through two different mutations even accounting for the different ploidy levels present in the chloroplast and microspore-derived plants. One feature of the system which was common to both selection approaches was that only mutagenised microspores have produced chlorsulfuron resistant plants.

IN VITRO TECHNIQUES AS ADJUNCTS TO SELECTION

Introduction

While the emphasis of this chapter is on *in vitro* selection systems for developing herbicide resistance, it should not be forgotten that *in vitro* culture techniques may be used to improve the value of an existing form of herbicide resistance by intraspecific or even interspecific hybridisation. Two techniques have recently been described in the literature involving protoplast fusion and embryo rescue.

Protoplast fusion

The cytoplasmically inherited gene conferring triazine tolerance has been used commercially in Canada for some time in rapeseed to permit selective control of cruciferous weeds (Marshall, 1987). A novel application for this trait has been proposed originally by Beversdorf *et al.* (1985) and demonstrated by Barsby *et al.* (1987) and Chuong *et al.* (1988). These authors considered that a triazine-resistant, male sterile rape plant could be the maternal parent in a hybrid seed production field containing male fertile pollinators carrying restorer genes. After fertilization, atrazine applications would be used to eliminate the pollinators and the hybrid seed could then be harvested from the field.

Barsby *et al.* (1987) in a series of complex steps isolated protoplasts from a rapeseed genotype carrying the gene for cytoplasmic male sterility (c.m.s.) and another genotype carrying triazine resistance. Following protoplast fusion in polyethylene glycol, the fusion mixtures were cultured and plants regenerated. Several polyploid plants (hybrids) with limited fertility were produced carrying both triazine resistance and c.m.s. traits. After several crosses with maintainers, diploid triazine-resistant/c.m.s. plants with normal fertility were obtained. The introduction of the new cytoplasmic genomes did not appear to induce any undesirable agronomic changes in the progeny.

The logical conclusion of these studies was to attempt the somatic transfer of the cytoplasmic traits using a haploid source of protoplasts for both donor traits (c.m.s. and triazine resistance). Thus, Chuong *et al.* (1988) chose microspore-derived haploid protoplasts of *B. napus* and successfully fused c.m.s. and triazine-resistant genotypes. The resultant fusion products were cultured to produce calli and finally regenerated. This protoplast fusion technique was remarkably successful with 43% of the regenerant plants being diploid. Included in this diploid group was one fully fertile hybrid complete with triazine resistance and c.m.s. This evolution in methodology had clear benefits over the earlier diploid/diploid system (Barsby *et al.*, 1987) in experimental simplicity since no chemical or radiation treatments of chloroplasts were required but also the fertility of the desired hybrids was greatly improved. One of the most important points in governing the success of the above mentioned research on protoplast fusion is that skill and methods must be available for all steps in the *in vitro* culture of microspores, the maintenance of protoplasts, the induction of callus growth from fusion products and plant regeneration.

Embryo rescue

As the title suggests this technique is designed to save the products of fertilization from an incompatible cross. Thus, interspecific hybrids have been rescued using this technique. Applications and techniques of embryo culture are described by Hu and Wang (1986). Although the technique is not new, its application for the transfer of triazine resistance from *B. napus* to *B. oleracea* L. has been recently reported by Ayotte *et al.* (1987). Triazine-resistant rapeseed (*B. napus* ssp. *oleifera*) and a triazine-resistant rutabaga *B.napus* ssp. *rapifera* were self-pollinated and cross-pollinated with *B. oleracea* ssp. *italica*, ssp. *botrystis*, ssp. *capitata* and ssp. *fimbriata*. Although ovule development in the resultant crosses proceeded, degeneration quickly followed and therefore at one day intervals from 11 to 17 days following pollination, embryo rescue was attempted. Embryos were initially cultured in a liquid medium for 2-3 weeks and transferred through three successive regeneration media until finally rooted. The authors concluded that the technique was of real practical value for breeding purposes since interspecific crosses were successfully rescued (71.6% of 67 embryos). Further evaluations of these hybrids by Ayotte *et al.* (1988) confirmed that all the interspecific F_1's were resistant to triazine herbicide (1.0×10^{-4} mol l^{-1} atrazine).

IN VITRO SELECTION: VALUE AND LIMITATIONS

Introduction

It is perhaps fair to suggest that the *in vitro* selection of herbicide resistance has an image of providing a technically simple but somewhat imprecise method for obtaining new crop genotypes. Certainly, this observation might be underlined when we try to identify the number of projects which have produced a commercially viable cultivar. Might the wealth of expertise and technology in plant and cell culture be more effectively exploited? Some observations to address this question are outlined below.

The importance of variation

The *in vitro* selection for herbicide resistance is certainly a technique which can initially be investigated in a wide range of laboratories where modest equipment is available. The culture of plant cells is efficient in terms of its requirement for space and with the adoption of a herbicide selection programme one can attempt to isolate some new genetic material. Although the *in vitro* culture of plant cells via callus and suspension systems may impart heritable variation, described as somaclonal variation (Larkin and Scowcroft, 1981) it is obvious from the previous literature reviews that researchers find this variation is not always sufficient to isolate the required mutants and therefore chemical or radiation types of mutagens are often used as adjuncts. Not all genetic changes are useful. Epigenetic changes although expressed through successive mitotic generations of cells are not transmitted to regenerated plants (Chaleff, 1980). Crops regenerated via tissue culture producing somaclonal variation (without the intervention of herbicides) may possess novel sources of herbicide resistance. So far it would appear that this approach has not been examined.

Variation in the level of herbicide resistance can be influenced by selection methodologies. Cell lines produced from callus culture may exhibit variation in their degree of herbicide resistance. This variation will only be useful if it is inheritable and to establish this the regeneration system must be effective and followed by outcrossing for several generations. Chimeric callus with herbicide-susceptible cells contained within a group of herbicide-resistant types can lead to a loss of apparent herbicide resistance (Hughes, 1983). It is evident that, to avoid this type of non-inherited variation, selection might be transferred to a cell suspension, a protoplast or a microspore system. The choice of system is most likely to be governed by the feasibility to regenerate from the cell source.

Similarly, variation in the degree of resistance or tolerance has been found to be influenced not only by the *in vitro* system but also by the number of times that cells become exposed to the herbicide. It would appear that, assuming the regeneration capacity of selected cells is not completely inhibited by repeated exposure to increasing herbicide concentrations, then this practice can improve the efficiency of selection. The choice of herbicide and its matching to the selection system is also a point of strategic importance. Therefore, herbicides which are toxic in heterotrophic systems such as the sulfonylureas and imidazolinones are eminently suitable for herbicide selection in all *in vitro* systems, from callus culture through to microspores. Conferring resistance to herbicides whose action is expressed only in differentiated structures will require specialised selection systems such as those previously described by Radin and Carlson (1978) and Cseplo *et al.* (1985). Lastly, variation in the level of herbicide resistance produced by different selection techniques e.g. protoplast vs microspore (Chuong *et al.*, 1988) may help to reveal previously unknown genetical changes which confer resistance. The degree of herbicide resistance and the ease of transferring the required gene or genes into new cultivars are important commercial aspects of the exploitation of the fundamental research.

Future exploitation of *in vitro* culture

It is certainly apparent that the range of *in vitro* techniques which are available is now quite comprehensive and can involve a high degree of sophistication. The exploitation of these techniques for the production of herbicide-resistant cultivars cannot really be examined completely in isolation from the alternative strategies of classical plant breeding or genetic engineering. Indeed, *in vitro* techniques have a central role to play in both systems. The reasons are as follows.

In vitro selection systems are not a replacement for classical breeding. If a valuable new source of herbicide resistance is isolated for a particular crop it is quite unlikely that this new line will have all the agronomic characteristics required. Therefore after the cell culturing, selection and regeneration must follow the classical breeding phase. Clearly, *in vitro* selection is unlikely to offer a rapid or inexpensive method of generating new cultivars with a special value but the use of a haploid-based selection system such as the microspore type could produce homozygous elite lines in a rapid and cost-effective manner. It follows that to introduce the herbicide resistance trait to just one cultivar of a crop would be neither attractive to farmers and growers or to the breeding company. Instead it may be more logical and economic to use the herbicide-resistant regenerant, isolate, clone, sequence and transform existing well adapted cultivars as appropriate. Similarly, the genetic engineering approach could be used where only herbicide-resistant cell lines can be isolated because regeneration systems are unreliable or unavailable. In the case of cytoplasmically inherited traits such as triazine resistance it is now possible to use protoplast fusion as an *in vitro* technique which confers an efficient and economical approach to a traditional breeding programme.

Exploitation of *in vitro* selection techniques will also be dependant on the herbicides which are available to researchers, farmers and growers. It has already been established that the sulfonylureas and imidazolinones are very valuable tools in a range of selection systems. There can be no doubt that our understanding of the biochemical and genetical basis of herbicide resistance has been advanced greatly since the introduction of these compounds and the impetus which they have given to research initiatives. However, we have to face the prospect that herbicide market domination by these two herbicide families which possess a similar mechanism of action could indeed hasten the development of resistant weeds and thereby lessen the real value of the herbicides. Perhaps this is the time to reflect on the feasibility of applying *in vitro* selection techniques to a wider range of herbicides in an attempt to maintain diversity in the herbicides available for crop management.

ACKNOWLEDGEMENTS

The author acknowledges the financial support of the Commission of the European Communities (DG XII) for studies related to this review (Contracts: ST2J-0396-C and SC1-0015-C). Special thanks are due to Drs S. Yarrow and E. Swanson, Allelix Crop Technologies for the loan of slide material for this presentation.

REFERENCES

Ayotte, R., Harney, P.M. and Souza Machado, V. (1987). The transfer of triazine resistance from *Brassica napus* L. to *B. oleracea* L. I. Production of F₁ hybrids through embryo rescue. <u>Euphytica</u> **36**, 615-624.

Ayotte, R., Harney, P. M. and Souza Machado, V. (1988). The transfer of triazine resistance from *Brassica napus* L. to *B. oleracea* L. III. First backcross to parental species. <u>Euphytica</u> **38**, 137-142.

Bajaj, Y.P.S. (1983). *In vitro* production of haploids. In "Handbook of Plant Cell Culture", Volume 1. (D.A. Evans, W.R. Sharp, P.V. Ammirato and Y. Yamada, eds), pp. 228-287. Macmillan Publishing Company, New York.

Barsby, T.L., Chuong, P.V. Yarrow, S.A., Wu, S., Coumans, C., Kemble, M., Powell, A.D., Beversdorf, W.D. and Pauls, K.P. (1987). The combination of Polima CMS and cytoplasmic triazine resistance in *Brassica napus*. Theoretical and Applied Genetics 73, 809-814.

Beversdorf, W.D., Erickson, L.R. and Grant, I. (1985). Hybridization process utilising a combination of cytoplasmic male sterility and herbicide tolerance. U.S. Patent and Trademark Office No. 493, 511.

Beversdorf, W.D., Weiss-Lerman, J., Erickson, L.R. and Souza Machado, V. (1980). Transfer of cytoplasmically-inherited triazine resistance from bird's rape to cultivated rapeseed (*Brassica campestris* L. and *B. napus* L.). Canadian Journal of Genetics and Cytology 22, 167-172.

Chaleff, R.S. (1980). Further characterisation of picloram-tolerant mutants of *Nicotiana tabacum*. Theoretical and Applied Genetics 58, 91-95.

Chaleff, R.S. (1986a). Selection for herbicide-resistant mutants. In "Handbook of Plant Cell Culture". Volume 4. (D.A. Evans, W.R. Sharp and P.V. Ammirato, eds), pp. 133-148. Macmillan Publishing Company, New York.

Chaleff, R.S. (1986b). Herbicide resistance. In "BCPC Monograph No. 34, Biotechnology and Crop Improvement", pp. 111-121.

Chaleff, R.S. (1988). Herbicide-resistant plants from cultured cells. In "Applications of Plant Cell and Tissue Culture". Ciba Foundation Symposium 137, pp. 3-20. John Wiley and Sons, Chichester, Sussex.

Chaleff, R.S. and Ray, T.B. (1984). Herbicide-resistant mutants from tobacco cell cultures. Science 223, 1148-1151.

Chuong, P.V., Beversdorf, W.D., Powell, A.D. and Pauls, K.P. (1988). Somatic transfer of cytoplasmic traits in *Brassica napus* L. by haploid protoplast fusion. Molecular and General Genetics 221, 197-201.

Crocomo, O.J. and Ochoa-Alejo, N. (1983). Herbicide tolerance in regenerated plants. In "Handbook of Plant Cell Culture", Volume 1. (D.A. Evans, W.R. Sharp, P.V. Ammirato and Y. Yamada, eds), pp. 770-781. Macmillan Publishing Company, New York.

Cseplo, A., Medgyesy, P., Hideg, E., Demeter, S., Marton, L. and Maliga, P. (1985). Triazine-resistant *Nicotiana* mutants from photomixotrophic cell cultures. Molecular and General Genetics 200, 508-510.

Duncan, D.R. and Widholm, J.M. (1986). Cell selection for crop improvement. Plant Breeding Reviews 4, 153-173.

Evans, D.A. and Bravo, J.E. (1983). Protoplast isolation and culture. In "Handbook of Plant Cell Culture". Volume I. (D.A. Evans, W.R. Sharp, P.V. Ammirato and Y. Yamada, eds), pp. 124-176. Macmillan Publishing Company, New York.

Gressel, J. (1989). Conferring herbicide resistance on susceptible crops. In "Herbicides and Plant Metabolism". (A.D. Dodge, ed.), (In press). Cambridge University Press, Cambridge.

Hickok, L.G. (1987). Applications of *in vitro* selection systems: whole-plant selection using the haploid phase of *Ceratopteris*. In: "Biotechnology in Agricultural Chemistry", pp. 53-65. ACS Symposium Series, American Chemical Society, No. 334.

Hirschberg, J. and McIntosh, L. (1983). Molecular basis of herbicide resistance in *Amaranthus hybridus*. Science **222**, 1346-1348.

Hirschberg, J., Bleecker, A., Kyle, D.J., McIntosh, L. and Arntzen, C.J. (1984). The molecular basis of triazine-herbicide resistance in higher-plant chloroplasts. Zeitschrift für Naturforschung **39c**, 412-420.

Hughes, K. (1983). Selection for herbicide resistance. In "Handbook of Plant Cell Culture", Volume I. (D.A. Evans, W.R. Sharp, P.V. Ammirato and Y. Yamada, eds), pp. 442-460. Macmillan Publishing Company, New York.

Hu, C. and Wang, P. (1986). Embryo culture: technique and applications. In "Handbook of Plant Cell Culture". Volume 4. (D.A. Evans, W.R. Sharp and P.V. Ammirato, eds), pp. 43-96. Macmillan Publishing Company, New York.

Kenyon, P.D., Marshall, G. and Morrison, I.N. (1987). Selection for sulfonylurea herbicide tolerance in oilseed rape (*Brassica napus*) using microspore culture. British Crop Protection Conference - Weeds 871-877.

Larkin, P.J. and Scowcroft, W.R. (1981). Somaclonal variation - a novel source of variability from cell cultures for plant improvement. Theoretical and Applied Genetics **60**, 197-214.

Maliga, P., Fejes, E., Steinback, K. and Menczel, L. (1987). Cell culture approaches for obtaining herbicide-resistant chloroplasts in crop plants. In "Biotechnology in Agricultural Chemistry", pp. 115-124. ACS Symposium Series, American Chemical Society, No. 334.

Marshall, G. (1987). Implications of herbicide-tolerant cultivars and herbicide-resistant weeds for weed control management. British Crop Protection Conference - Weeds 489-498.

Meredith, C.P. and Carlson, P.S. (1982). Herbicide resistance in plant cell cultures. In "Herbicide Resistance in Plants" (H.M. LeBaron and J. Gressel, eds), pp. 275-291. John Wiley, New York.

Radin, D.N. and Carlson, P.S. (1978). Herbicide-tolerant tobacco mutant selected *in situ* and recovered via regeneration from cell culture. Genetical Research **32**, 85-89.

Shaner, D.L. and Anderson, P.C. (1985). Mechanism of action of the imidazolinones and cell culture selection of tolerant maize. In: "Biotechnology in Plant Science". (M. Zaitlin, P.R. Day and A. Hollaender, eds), pp. 287-299. Academic Press, Orlando, FL.

Souza Machado, V. 1982). Inheritance and breeding potential of triazine tolerance and resistance in plants. In "Herbicide resistance in plants". (H.M. LeBaron and J. Gressel, eds), pp. 257-273. John Wiley, New York.

Swanson, E.B., Coumans, M.P., Brown, G.L., Patel, J.D. and Beversdorf, W.D. (1988). The characterization of herbicide tolerant plants in *Brassica napus* L. after *in vitro* selection of microspores and protoplasts. Plant Cell Reports 7, 83-87.

MOLECULAR ANALYSIS OF SULFONYLUREA HERBICIDE RESISTANT ALS GENES

M.E. Hartnett, C.-F. Chui, S.C. Falco, S. Knowlton, C.J. Mauvais, and B.J. Mazur

Agricultural Products Department, E.I. du Pont de Nemours & Co., (Inc.), Experimental Station 402, Wilmington, Delaware 19880-0402, U.S.A.

The enzyme acetolactate synthase (ALS) is the target of several classes of herbicides, including the sulfonylureas, the imidazolinones, and the triazolopyrimidines. ALS is required for the biosynthesis of isoleucine, leucine and valine. We have cloned ALS genes from a variety of monocotyledonous and dicotyledonous plants. The number of ALS genes in these species varies from one, as in *Arabidopsis* and sugar beets, to two or more, as in corn and soybeans. The deduced amino acid sequences of the mature plant ALS genes which have been sequenced are closely related. Three domains of the plant ALS proteins are also conserved relative to bacterial and yeast ALS proteins. ALS genes have been isolated and sequenced from sulfonylurea herbicide resistant plants and the resistance trait shown to result from single or double mutations in the conserved domains of the genes. To produce additional genes able to confer sulfonylurea herbicide resistance in transgenic crops, *in vitro*-generated mutations were introduced into the coding regions of cloned plant ALS genes. A number of single and double mutations were shown to be effective in conferring resistance, upon subsequent introduction of the altered genes into plants. The mutations conferred varying degrees of resistance, depending upon the site of the resulting amino acid substitution and upon the amino acid substituted at that site. Replacement of ALS regulatory sequences led to increases in the proportion of herbicide resistant enzyme such that up to 90% of the total ALS activity was herbicide resistant. The altered genes conferred sulfonylurea herbicide resistance in glasshouse trials of a wide variety of transgenic crop species. Additionally, transgenic tobacco, tomato and oilseed rape were shown to be herbicide resistant in field trials.

INTRODUCTION

Although the main thrust of this Symposium is the disadvantageous development of herbicide-resistant weeds, the herbicide resistance trait can also be used to advantage to create crop species able to withstand herbicide treatments. The most cost-effective and environmentally acceptable herbicides, which might otherwise be phytotoxic, can then be applied to such crops for weed control. Herbicide-resistant crops can also increase the margin of crop safety to chemically selective compounds, preventing unexpected phytotoxicity due to reductions in plant metabolism during adverse growing conditions. In addition, herbicide-resistant crops can increase the grower's flexibility in choosing crops for rotations by providing resistance to herbicide residues in the soil. In this chapter, we review our work on developing crops resistant to the sulfonylurea herbicides.

The sulfonylurea herbicides belong to a large class of compounds; over 28,000 have been synthesized and characterized. The herbicides in this class are environmentally favorable because of their very low use rates and very low toxicity to animals. The high biological activity of these compounds, which permit field application rates as low as 2 g ha^{-1}, result from the high specificity that they have for inhibition of a particular plant enzyme. Their low toxicity is due, in part, to the absence of this target enzyme in animals. Soil residual properties of these herbicides vary, with some having half-lives measured in days, and others having extended residual activity that can be used for fallow fields or for vegetation management. A number of the compounds are selectively toxic because they are metabolized to inactive intermediates by some crops but not by weeds. However, such chemically selective compounds have not been discovered for all crop species, so that biological selectivity achieved through genetic engineering of the crop plant would extend the range of utility of sulfonylurea herbicides.

In order to create herbicide-resistant plants we have used three different experimental strategies. The first method exploited plant cell culture. Individual plant cells were selected for herbicide resistance *in vitro* and plants were regenerated from the resistant cells. The second method was mutation breeding. Plant seeds or pollen were treated with a mutagen and mutant lines, selected for the ability to grow in the presence of the herbicide, were isolated. The third method employed plant transformation. A gene coding for a herbicide-resistant target protein was isolated from resistant plants or constructed by *in vitro* site-specific mutation, introduced into plant cells via transformation, and the transformed cells were regenerated into resistant plants.

MODE OF ACTION

Studies on the mode of action of the sulfonylurea herbicide, chlorsulfuron, showed that inhibition of cell division in plant tissue was an early response to treatment (Ray 1982a, 1982b). This, along with the ability to isolate resistant mutants in cultured plant cells, suggested that the herbicide antagonized a single basic cellular function, and encouraged the use of microbial models to investigate herbicide action. Physiological studies in *Salmonella typhimurium* suggested that the target of the sulfonylurea herbicide sulfometuron-methyl was the enzyme acetolactate synthase (ALS, also known as acetohydroxy acid

344

synthase or AHAS), which is required for the synthesis of isoleucine, leucine and valine (LaRossa and Schloss, 1984). *In vitro* analyses of ALS activity from yeast, pea, tobacco and *Chlamydomonas* demonstrated that the eukaryotic enzymes are very sensitive to chlorsulfuron or sulfometuron-methyl (Chaleff and Mauvais, 1984; Falco and Dumas, 1985; Ray, 1984; Hartnett *et al.* 1987).

Proof that the sulfonylurea herbicides act by inhibition of ALS came from a combination of genetic and biochemical studies. Sulfonylurea-resistant mutants of *S. typhimurium*, yeast, tobacco and *Arabidopsis* were isolated. Most of the mutants from each organism were shown to produce ALS activity that was insensitive to the herbicide, and the resistant enzyme activity was shown to cosegregate with cellular resistance in genetic crosses (Chaleff and Mauvais, 1984; Falco and Dumas, 1985; Haughn and Somerville, 1986; LaRossa and Schloss, 1984). Surprisingly, ALS is also the target of two other structurally distinct classes of herbicides, the imidazolinones, (Muhitch *et al.*, 1987; Shaner *et al.*, 1984) and the triazolopyrimidine sulfonamides (Hawkes *et al.* 1988, Kleswick *et al.*, 1984). Thus, ALS may be a particularly efficacious target for herbicides.

ALS GENES

Cloned yeast and bacterial ALS genes were used to investigate the molecular basis for resistance to the sulfonylurea herbicides. Mutations that resulted in the production of sulfonylurea-resistant ALS were isolated in the yeast *ILV2* and *Escherichia coli ilvG* genes by genetic selection (Yadav *et al.*, 1986). DNA sequencing showed that each mutant gene contained a single nucleotide change, resulting in a single amino acid substitution in the ALS protein. The yeast system was used to discover other possible mutations that resulted in sulfonylurea herbicide resistance. The isolation and sequencing of 41 mutations in the yeast gene revealed 24 different amino acid substitutions that occurred at 10 different sites, ranging from the amino to the carboxy ends of the protein (Falco *et al.*, 1989) (Fig. 1).

Efforts to use these microbial genes to generate herbicide-resistant transgenic plants faced a number of difficulties and uncertainties. For example, plant ALS is localized in the chloroplast (Miflin, 1974; Jones *et al.*, 1985) and bacterial ALS is composed of two different subunits. Thus, the generation of herbicide resistant plants using bacterial genes required not only expression of the two protein subunits of the enzyme, but also their translocation and assembly in plant chloroplasts. Since ALS had not been purified from yeast, the subunit structure of the enzyme, as well as the amino terminus of the mature protein present in the mitochondria was unknown, adding considerable uncertainty to any effort to express that enzyme in plants. For these reasons the isolation and use of plant ALS genes appeared to be a more attractive route for engineering sulfonylurea-resistant transgenic plants.

After the cloning of the yeast ALS gene we observed that significant cross hybridization occurred between the yeast and *S. typhimurium* ALS genes (Mazur *et al.*, 1985). This hybridization indicated sequence con-servation between ALS genes from prokaryotic and eukaryotic organisms. Subsequent comparisons of the deduced amino acid sequences from the yeast and bacterial genes revealed three regions of homology (Falco *et al.*, 1985).

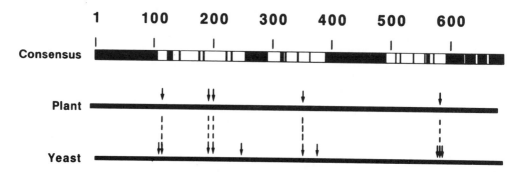

Fig. 1. <u>Comparison of ALS amino acid sequences.</u> The top bar shows a schematic representation of homology between *E. coli* ALS I, II, and III (large subunits), yeast ALS, and ALS from the plants tobacco and *Arabidopsis*. The white boxes show stretches of amino acids that are conserved between all enzymes; the black boxes depict nonconserved regions. The arrows on the horizontal lines show positions in yeast and plant sequences where substitutions can produce a sulfonylurea herbicide resistant enzyme. Dashed vertical lines indicate analogous positions where such substitutions occur in both yeast and plants. Numbers refer to amino acids.

These results suggested that hybridization probes could be designed to isolate ALS genes from other distantly related organisms. A fragment encompassing the majority of the ALS coding region of the yeast gene was used to screen genomic libraries of *A. thaliana* and *Nicotiana tabacum* (tobacco). Genomic clones from both species were isolated using low stringency hybridization conditions (Mazur *et al.*, 1987a).

Complete sequence analyses of the *Arabidopsis* and tobacco ALS genes indicated that they code for proteins of 670 (2013 bp) and 667 (2004 bp) amino acids, respectively. The two plant genes are highly conserved, with 75% sequence identity at the nucleotide level and 85% at the amino acid level (Mazur *et al.*, 1987b). When compared to the deduced amino acid sequences of ALS from yeast and bacteria, both plant ALS enzymes maintain the three conserved regions previously noted in the microbial enzymes. The sequence similarity between the plant genes extends into the non-conserved regions from bacteria and yeast, however. One region of the plant genes, the 5' end of the coding sequence, is not conserved between *Arabidopsis* and tobacco. This portion of each gene codes for a chloroplast transit sequence (Miflin, 1974; Chaleff and Mauvais, 1984; Chaleff and Ray, 1984; Jones *et al.*, 1985).

The tobacco ALS gene was used as a probe to screen Southern blots and genomic DNA libraries from soybean, corn, and sugar beet. Several fragments hybridized to the tobacco probe on the Southern blots of both corn and soybean DNA indicating that these crops contain multiple ALS genes. Genomic clones from all three species were isolated and complete sequence analyses were carried out on two soybean (each with distinct restriction digest patterns), one corn and one sugar beet ALS gene. Comparison of the deduced amino acid sequences from the plant genes indicated that the ALS enzymes are highly conserved with identical amino acids varying between 80-90%. The majority of the divergence between the plant proteins occurs in the presumed transit peptide regions.

The cloned *Arabidopsis* and tobacco ALS genes were used as homologous hybridization probes to isolate genes carrying ALS mutations from herbicide-resistant mutants of *Arabidopsis* and tobacco (Lee *et al.* 1988; Haughn *et al.* 1988). DNA sequence comparisons of the mutant and wild-type genes from each plant line permitted the determination of the amino acid substitutions in the mutant enzymes. One mutant tobacco gene coded for a substitution of glutamine for proline at amino acid position 194. A mutation at the analogous site in the yeast ALS gene had previously been shown to confer resistance (Yadav *et al.*, 1986). Another tobacco gene, from a mutant line isolated by two successive rounds of selection to obtain a highly resistant mutant, coded for an alanine for proline substitution at the analogous position, as well as a leucine for tryptophan substitution at amino acid position 568 (Lee *et al.*, 1988). The two mutations were consistent with the two cycles of selection used to isolate the line (Chaleff *et al.*, 1987). Similarly, a gene that conferred sulfonylurea herbicide resistance was isolated from *Arabidopsis* and sequenced; this gene encoded a serine substitution for the analogous proline in the *Arabidopsis* enzyme (Haughn and Somerville, 1986; Haughn *et al.*, 1988). Both the proline to glutamine mutation in the tobacco gene and the proline to serine mutation in the *Arabidopsis* gene conferred resistance to sulfonylurea herbicides but not to imidazolinone herbicides. The double mutant gene was cross-resistant to both classes of herbicides.

Oligonucleotide site-directed mutagenesis of the tobacco ALS gene was used to identify additional mutations in plant ALS genes that confer herbicide resistance. Among the three mutant genes isolated from the herbicide-resistant tobacco and *Arabidopsis* plants, two mutation sites were identified. Mutations were made in a wild type tobacco gene at these sites and at additional sites predicted to confer resistance from the yeast experiments. The mutations were tested in tobacco cells in culture and in transgenic tobacco plants. Four novel herbicide resistance mutations in the tobacco gene were identified in this way, confirming the utility of the microbial systems for predicting herbicide resistance mutations and supporting the postulate that mutations resulting in herbicide resistance are conserved.

HERBICIDE RESISTANT PLANTS

The first sulfonylurea herbicide resistant plants were obtained using tobacco cell culture (Chaleff and Ray, 1984). Callus cultures initiated from haploid tobacco lines were selected on either chlorsulfuron or sulfometuron-methyl. Diploid plants derived from regenerated resistant calli were crossed and the mutations were found to define two genetic loci. ALS activity in the resistant lines was shown to be relatively insensitive to inhibition by these compounds, and the resistant enzyme activity was further shown to co-segregate with the resistance phenotype (Chaleff and Mauvais, 1984). One mutant line was subjected to a second round of selection at a higher herbicide concentration. A new line was recovered which exhibited higher levels of herbicide resistance. The second mutation in this line was shown to be linked to the original mutation. Plants from both lines were resistant to foliar applications of a sulfonylurea herbicide in field tests. These tobacco mutants served as sources for the isolation of genes coding for herbicide-resistant ALS as described above.

In the case of soybeans, cell culture methods that would permit the selection of herbicide-resistant lines had not been developed. Therefore, soybean seeds were subjected to EMS mutagenesis, and resistant lines were selected using a hydroponic screening procedure. Initial screens of the mutagenized seeds resulted in the isolation of herbicide-tolerant lines. The tolerant soybean lines did not express a herbicide-resistant ALS enzyme and the herbicide tolerance phenotype segregated as a recessive trait. Additional screening resulted in the isolation of a dominant herbicide-resistant mutant line. The resistance was due to an altered ALS enzyme and the resistance phenotype co-segregated with the resistant enzyme. The resistant line has shown excellent tolerance to sulfonylurea herbicides in field tests, and the resistance trait is being crossed into a number of elite soybean cultivars (Sebastian and Chaleff 1987).

The mutant genes isolated from the two mutant lines of tobacco and the *Arabidopsis* mutant gene were re-introduced into tobacco and tested for the ability to confer herbicide resistance (Mazur *et al.*, 1987b; Lee *et al.*, 1988; Haughn *et al.* 1988). These experiments demonstrated that either the single or double mutant tobacco genes as well as the *Arabidopsis* mutant gene could confer herbicide resistance in transgenic tobacco plants. Since the introduction of the double mutant tobacco gene resulted in higher levels of resistance, this gene was transferred to commercial lines of tobacco, and regenerated plants were assayed for levels of sulfonylurea resistance by several methods. Resistance was measured by assaying leaf ALS activity, callus growth, and seed germination and growth in the presence of herbicide, and by monitoring plant damage after foliar spray applications of the herbicide. The results of these tests were consistent, yet indicated the need to monitor resistance by several methods in order to identify those lines most suitable for crop breeding.

To determine the agronomic utility of the transgenic herbicide resistant plants, some of the tobacco lines were evaluated in field tests conducted in North Carolina in collaboration with Northrup King Co. Foliar sprays of the herbicide chlorimuron-ethyl were applied to transplanted seedlings at rates of 0, 8, 16 and 32 g ha^{-1}. Transformed plants showed no damage at the highest application rate tested, which was four times that of a typical field application. Thus, transformation with the double mutant gene provided an effective means of conferring sulfonylurea resistance in crop plants.

The tobacco double mutant gene was also used to transform a number of heterologous species to herbicide resistance at the cellular level, and in some cases, whole plant level. These species included oilseed rape, tomato, sugar beet, alfalfa, lettuce and melon. In one example, transgenic tomato plants were assayed for the level of herbicide-resistant ALS enzyme activity. The expression of the resistant enzyme in the tomato plants was efficient; it ranged between 30 and 65 percent of the total ALS activity in the transformed plants. In others, such as rape, only a low level of resistance was observed. Thus, the effectiveness of heterologous genes must be evaluated on a case-by-case basis. Transgenic tomato lines were tested in the field in Delaware, Florida and France in 1988 (Fig. 2). These lines were resistant to herbicide application rates several fold higher than that needed for weed control and showed no adverse agronomic effects.

Fig. 2. <u>Field trial of herbicide-resistant transgenic tomato plants.</u> Tomato was transformed with the double mutant tobacco ALS gene. The resulting transgenic tomato plants were tested for sulfonylurea resistance in a field trial. Transgenic plants (right and left) were healthy, while nontransformed plants (centre) were severely stunted by the treatment.

In order to improve the level of expression of heterologous genes in transgenic plants, the Cauliflower Mosaic Virus (CaMV) 35S promoter was linked to the tobacco double mutant gene. Expression of the double mutant gene from this promoter was analyzed in both transgenic tobacco and transgenic oilseed rape. In tobacco, the proportion of herbicide-resistant enzyme was increased from 40% of the total ALS activity at 100 ppb chlorsulfuron with the native tobacco ALS promoter to 90% with the 35S CaMV promoter. The expression of the mutant gene in oilseed rape increased from less than 10% resistant enzyme with the tobacco native promoter to 40% with the 35S promoter. A field test of two transgenic Canola lines in Canada in 1989 demonstrated resistance to field use rates of two sulfonylurea herbicides.

HERBICIDE RESISTANT WEEDS

The relative ease of isolation of resistant mutants of tobacco and

Arabidopsis and the fitness of the mutant plants were early indications that herbicide resistant weeds might arise. It is now clear that many of the characteristics which combine to make ALS an excellent target for engineering beneficial herbicide resistance in crop plants may also lead to the proliferation of herbicide-resistant weeds. These characteristics include the following: sulfonylurea herbicide resistance is a semi-dominant trait that is carried on a nuclear gene(s); ALS is the single primary site of action; there are multiple positions in ALS that can be mutated to confer herbicide resistance; mutant ALS enzymes can possess full catalytic activity.

The properties of the herbicides that target ALS can also contribute to the emergence of resistant weeds. There are several classes of compounds that target ALS including the sulfonylureas, the imidazolinones and the triazolopyrimidines. Within these classes are a number of herbicides that are used at rates that kill a high proportion of the weeds, thus increasing the likelihood that resistant biotypes will rapidly dominate the weed population. In addition, many of the compounds have extended residual activity in soil, which creates continuous selection pressure for the emergence of herbicide-resistant weeds. Finally, crop management practices, such as the tendency to use these herbicides continually in mono-culture cropping systems, can accelerate the proliferation of herbicide-resistant weeds.

Recently, resistant biotypes of *L. serriola, S. iberica, S. media* and *Kochia,* which exhibit herbicide-resistant ALS activity, have been identified at a number of locations, following prolonged use of long-residual sulfonylurea herbicides (Saari *et al.*, 1990). In order to avoid further increases in the resistant biotypes, a number of control measures have been implemented. These include changes in the recommended herbicide use patterns, such as reduction in use rates as well as number of applications of the same herbicide in a growing season, and elimination of the use of the same herbicide in successive seasons. The replacement of long soil-residual herbicides with short soil-residual herbicides has also been recommended. Finally, the use of tank mixes of herbicides with differing modes of action has been advised. These changes in management practices should delay the proliferation of herbicide-resistant weeds and allow these environmentally favourable compounds to continue to be used for weed control. The engineering of herbicide-resistant crops will provide additional flexibility, for example by allowing the use of a short-residual herbicide on a crop where no such selective herbicide was available. In addition, the engineering of crops resistant to multiple herbicides will expand the opportunities for the use of tank mixes of herbicides with differing modes of action.

ACKNOWLEDGEMENTS

We would like to acknowledge our many colleagues who have participated in these studies. Our co-workers at DuPont include Anthony Guida, Sharon J. Keeler, Raymond E. McDevitt, Julie K. Smith and R. Timothy Ward on molecular analyses of the ALS genes, Todd Houser, Christina Kostow, Craig Sanders and Carole Beaman on plant transformations and Gary Fader on glasshouse spray tests. Joan Odell and Perry Caimi have also contributed to this project. Finally, we have enjoyed collaboration with John Bedbrook, Jeff Townsend, Kathy Lee and Pamela Dunsmuir at Advanced Genetic Sciences, Inc.

REFERENCES

Chaleff, R.S., and Mauvais, C.J. (1984). Acetolactate synthase is the site of action of two sulfonylurea herbicides in higher plants. Science **224**, 1443–1445.

Chaleff, R.S. and Ray, T.B. (1984). Herbicide-resistant mutants from tobacco cell cultures. Science **223**, 1148–1151.

Chaleff, R.S., Sebastian, S.A., Creason, G.L., Mazur, B.J., Falco, S.C., Ray, T.B., Mauvais, C.J. and Yadav, N.S. (1987). Developing plant varieties resistant to sulfonylurea herbicides. In "Molecular Strategies for Crop Protection" (C.J. Arntzen, and C. Ryan, eds). A.R. Liss, Inc. NY.

Falco, S.C. and Dumas, K.D. (1985). Genetic analysis of mutants of *Saccharomyces cerevisiae* resistant to the herbicide sulfometuron methyl. Genetics **109**, 21–35.

Falco, S.C., Dumas, K.D. and Livak, K.J. (1985). Nucleotide sequence of the yeast *ILV2* gene which encodes acetolactate synthase. Nucleic Acids Research **13**, 4011–4027.

Falco, S.C., McDevitt, R.E., Chui, C.-F., Hartnett, M.E., Knowlton, S., Mauvais, C.J., Smith, J.K. and Mazur, B.J. (1989). Engineering herbicide-resistant acetolactate synthase. In "Developments in Industrial Microbiology" (J. Cooney and O. Sebek, eds), Journal of Industrial Microbiology, Supplement **4**, 187–194.

Hartnett, M.E., Newcomb, J.R. and Hodson, R.C. (1987). Mutations in *Chlamydomonas reinhardtii* conferring resistance to the herbicide sulfometuron methyl. Plant Physiology **85**, 898–901.

Haughn, G.W., Smith, J., Mazur, B.J. and Somerville, C. (1988). Transformation with a mutant *Arabidopsis* acetolactate synthase gene renders tobacco resistant to sulfonylurea herbicides. Molecular and General Genetics **211**, 266–271.

Haughn, G.W. and Somerville, C.R. (1986). Sulfonylurea-resistant mutants of *Arabidopsis thaliana*. Molecular and General Genetics **204**, 430–434.

Hawkes, R., Howard, J.L., and Pontin, S.E. (1988). Herbicidal inhibition of branched amino acid biosynthesis. In "Herbicides and Plant Metabolism" (A.D. Dodge, ed.), pp.. Cambridge Academic Press, Cambridge.

Jones, A.V., Young, R.M. and Leto, K.J. (1985). Subcellular localization and properties for acetolactate synthase, target site of the sulfonylurea herbicides. Plant Physiology **77**, 5293–5297.

Kleswick, W.A., Ehr, R.J., Gerwick, B.C., Monte, W.T., Pearson, N.R., Costales, M.J. and Meikle, R.W. (1984). New 2-arylamino-sulphonyl-1,2,4-triazolo-(1,5-a)-pyrimidine(s) useful as selective herbicides and to suppress nitrification in soil. European Patent Application 0142152.

LaRossa, R.A. and Schloss, J.V. (1984). The sulfonylurea herbicide sulfometuron methyl is an extremely potent and selective inhibitor of acetolactate synthase in *Salmonella typhimurium*. Journal of Biological Chemistry 259, 8753-8757.

Lee, K.Y., Townsend, J., Tepperman, J., Black, M., Chui, C.-F., Mazur, B., Dunsmuir, P. and Bedbrook, J. (1988). The molecular basis of sulfonylurea herbicide resistance in higher plants. EMBO Journal 7, 1241-1248.

Mazur, B.J., Chui, C.-F., Falco, S.C., Mauvais, C.J. and Chaleff, R.S. (1985). Cloning herbicide resistance genes into and out of plants. In "The World Biotech Report" Online International, New York.

Mazur, B.J., Chui, C.-F. and Smith, J.K. (1987a). Isolation and characterization of plant genes coding for acetolactate synthase, the target enzyme for two classes of herbicides. Plant Physiology 85, 1110-1117.

Mazur, B.J., Falco, S.C., Knowlton, S. and Smith, J.K. (1987b). Acetolactate synthase, the target enzyme of the sulfonylurea herbicides. In "Plant Molecular Biology" (D. von Wettstein and N.-H. Chua, eds) Plenum Press, New York, Vol. 140.

Miflin, B.J. (1974). The location of nitrite reductase and other enzymes related to amino acid biosynthesis in the plastids of roots and leaves. Plant Physiology 54, 550-555.

Muhitch, M.J., Shaner, D.L. and Stidham, M.A. (1987). Imidazolinones and acetohydroxyacid synthase from higher plants. Plant Physiology 83, 451-456.

Ray, T.B. (1982a). The mode of action of chlorsulfuron: a new herbicide for cereals. Pesticide Biochemistry and Physiology 17, 10-17.

Ray, T.B. (1982b). The mode of action of chlorsulfuron: the lack of direct inhibition of plant DNA synthesis. Pesticide Biochemistry and Physiology 18, 262-266.

Ray, T.B. (1984). Site of action of chlorsulfuron. Plant Physiology 75, 827-831.

Saari, L.L., Cotterman, J.C. and Primiani, M.M. (1990). Mechanism of sulfonylurea herbicide resistance in the broadleaf weed *Kochia scoparia*. Plant Physiology 93, 55-61.

Sebastian, S.A. and Chaleff, R.S. (1987). Soybean mutants with increased tolerance for sulfonylurea herbicides. Crop Science, 27, 948-952.

Shaner, D.L., Anderson, P.C. and Stidham, M.A. (1984). Imidazolinones (potent inhibitors of acetohydroxyacid synthase). Plant Physiology **76**, 545-546.

Yadav, N., McDevitt, R.E., Benard, S. and Falco, S.C. (1986). Single amino acid substitutions in the enzyme acetolactate synthase confer resistance to the herbicide sulfometuron methyl. Proceedings of the National Academy of Science USA, **83**, 4418-4422.

ENGINEERING OF GLUFOSINATE RESISTANCE AND EVALUATION UNDER FIELD CONDITIONS

Johan Botterman, Kathleen D'Halluin, Marc De Block, Willy De Greef and Jan Leemans

Plant Genetic Systems N.V. , J. Plateaustraat 22, 9000 Gent, Belgium

Glufosinate and bialaphos are non-selective herbicides which act by inhibiting glutamine synthetase. A gene which confers resistance to bialaphos (bar) was isolated from *Streptomyces hygroscopicus*, the organism which produces bialaphos. The gene was shown to encode a phosphinothricin acetyl transferase. Using *Agrobacterium*-mediated Ti plasmid transformation methodology, the bar gene has been introduced into and expressed in several plant species. Transgenic tobacco, tomato, potato, poplar, alfalfa, oilseed rape and sugar beet plants show a complete resistance towards the commercial preparations of glufosinate and bialaphos. The strategy thus provides a successful approach to obtain herbicide-resistant plants by introducing a pathway for detoxification of the herbicide. At present, these crops are being evaluated under open field conditions. Results thus far demonstrate a complete resistance to field dose applications, no yield penalties, and confirmed that glufosinate can be applied as a selective post-emergence herbicide on engineered crops.

INTRODUCTION

In the last decade, remarkably fast progress in plant molecular biology has taken place. The transformation and regeneration of more than 20 different plant species including several important field crops has been achieved. At the same time, dramatic progress has been made in the identification and improvement of genes encoding valuable agronomic traits (for review see Gasser and Fraley, 1989).

Engineering crops resistant to herbicides has been one of the first issues targeted and in recent years, alternative ways to obtain herbicide selectivity have been established (for review, see Botterman and Leemans, 1988). Indeed, the engineering of plants resistant to herbicides would allow the selective use of these chemicals for post-emergence applications in more effective and flexible weed control programmes. Research has largely concentrated on those herbicides with properties such as high unit activity, low toxicity, low soil mobility and rapid biodegradation and with a broad spectrum against various weeds. Success with resistance engineered towards a number of commercially used herbicides has been reported (for review, see Botterman and Leemans, 1988). Several of these achievements have already entered a next phase of development. At present, several crops engineered for herbicide tolerance are being evaluated under open field conditions.

Conceptually, two approaches have been followed in engineering herbicide tolerance. The first consists of altering the level and sensitivity of the target enzyme for the herbicide, while in the second a gene that detoxifies the herbicide is incorporated in the plant genome. In practice, these approaches are based on the selection for resistance in tissue culture or seedlings or on genetic engineering methodology. We described the engineering of plants resistant to the non-selective herbicides glufosinate or phosphinothricin (PPT) and bialaphos by transferring and expressing a herbicide inactivating enzyme in transgenic plants.

THE HERBICIDE RESISTANCE TRAIT

Bialaphos (Kondo *et al.*, 1973) and phosphinothricin (PPT) (Bayer *et al.*, 1972) have recently been developed as new broad-spectrum contact herbicides that are highly effective, safe and rapidly biodegraded. Bialaphos is a tripeptide antibiotic produced by *Streptomyces hygroscopicus*. It consists of PPT and two L-alanine residues. PPT itself is an inhibitor of glutamine synthetase (GS). GS plays a central role in the assimilation of ammonia and in the regulation of nitrogen metabolism in plants. It is the only enzyme in plants that can detoxify ammonia released by nitrate reduction, amino acid degradation and photorespiration. Inhibition of GS by PPT causes rapid accumulation of ammonia which leads to death of the plant cell (Tachibana *et al.*, 1986). The herbicides are available as the chemical synthesized PPT, glufosinate-ammonium (Basta®, 200 g a.i. l^{-1}, Hoechst AR, FRG), while bialaphos is produced by fermentation of *S. hygroscopicus* ('Herbiace®', 330 g a.i. l^{-1}, Meji Seika, Japan).

One approach to engineer tolerance was based on the selection of mutants. A PPT-tolerant alfalfa suspension culture line has been selected; gene amplification resulted in enhanced GS expression and the gene copy number correlated with the level of resistance (Donn *et al.*, 1984). The creation of mutants with tolerant GS enzymes is very unlikely in view of the complex structure of the holoenzyme. Moreover, mixed complexes of mutant and wild-type subunits might yield an inconsistent phenotype.

Possible sources of detoxification systems are plants or microorganisms. Detoxification pathways exist in plant species that are naturally tolerant to specific herbicides and these systems have been exploited in agriculture for selective use on crops. However, plant detoxification systems are in general poorly characterized. Different classes of microorganisms which degrade glufosinate in the soil have been isolated (Tebbe and Reber, 1988; Bartsch and Tebbe, 1989), but single genes responsible for detoxifying the herbicide have not been isolated yet. On the other hand, it has been observed that *Streptomyces* strains which produce antibiotics have evolved mechanisms to avoid the toxicity of their own products. As observed with other clusters of antibiotic production genes, a gene coding for antibiotic resistance is physically linked to the biosynthesis pathway for bialaphos (Murakami *et al.*, 1986). It has been shown that bialaphos is synthesized from three carbon precursors in a series of at least 13 conversions. Many of the genes possible for these conversions have been defined by blocked mutants. One of these intermediate steps (step 10) contained an acetyl-coenzyme A-dependent

356

activity which can modify either an intermediate in the pathway, or PPT itself. The gene encoding this enzyme, which confers resistance to bialaphos (bar) has been isolated from *S. hygroscopicus* and characterized (Thompson *et al.*, 1987). It has been shown that the bar gene encodes a protein involved in both antibiotic resistance and biosynthesis since it plays a role in both self-defense and bialaphos biosynthesis (Kumada *et al.*, 1988). The gene was shown to comprise a polypeptide of 22 kDa with acetyl transferase activity (phosphinothricin acetyl transferase – PAT). It converts PPT into a non-herbicidal acetylated form, by transfering the acetyl group of acetyl-coenzyme A on the free amino group of PPT. Analysis of the substrate specificity of the acetyl transferase by determining the K_m and kcat value for PPT and chemically related compounds, showed that the enzyme has a very specific substrate requirement and preferentially acetylates PPT (Thompson *et al.*, 1987). Optimal activity of the enzyme was found at 35°C and neutral pH. These enzymatic characteristics looked *a priori* suitable for the application of this gene as a herbicide resistance trait in transgenic plants.

EXPRESSION OF THE RESISTANCE GENE IN TRANSGENIC PLANTS

To express the gene in plants, transcriptional and translational regulatory signals were provided. The coding sequence with a GTG initiation codon was adapted by introducing an ATG codon, and chimeric gene constructs with the bar coding region under control of different promoters such as the CaMV 35S promoter, the octopine T-DNA TR promoter and a Rubisco small subunit promoter from *Arabidopsis thaliana*, and provided with a 3' untranslated sequence containing signals for 3' end formation and polyadenylation have been designed. Using *Agrobacterium* mediated Ti plasmid transformation methodology, these chimeric genes have been transferred in tobacco, tomato, potato, oilseed rape, poplar, alfalfa or sugar beet (De Block *et al.*, 1987; De Block, 1988; De Almeida *et al.*, 1989; D'Halluin *et al.*, submitted).

The bar gene proved to be a convenient dominant selectable marker in tissue culture for plant transformation in several of these crops (De Block *et al.*, 1987; 1988; D'Halluin *et al.*, submitted). Transgenic tobacco, tomato and potato plants expressing the bar gene are resistant to applications of glufosinate under glasshouse conditions (De Block *et al.*, 1987). Glufosinate at 400 g a.i. ha^{-1} killed control plants in ten days, whereas transgenic plants were resistant to herbicide treatments of up to 4000 g a.i. ha^{-1}. Doses recommended by the supplier for agriculture vary from 800 to 2000 g a.i. ha^{-1}. We also observed that additional applications of the herbicide did not affect growth of the plants. In general the growth of transgenic plants in the glasshouse was indistinguishable from non-transformed control plants. Biochemical analysis also showed that an active PAT enzyme was produced in the plant cells. Treated resistant plants did not show any increase in NH_4^+ content and showed a complete protection of the plant glutamine synthetase from the action of the herbicide. Treated tobacco and tomato plants also flowered and produced normal amounts of seed in the glasshouse. The trait was also shown to be inherited as a dominant Mendelian trait. Since bialaphos has an activity similar to glufosinate-ammonium, these plants also proved resistant to applications of bialaphos, as expected.

Similar results obtained with transgenic sugar beet, oilseed rape, poplar and alfalfa proved that a complete resistance towards high doses of glufosinate was obtained independent of the plant species used. We also noted that resistance was observed with all of the chimeric genes used, although the promoters are known to show tissue specific expression patterns.

Beyond its practical use, the <u>bar</u> gene also revealed a convenient selectable and scoreable marker gene in plant transformation and expression technology. The enzymatic activity can be detected chromatographically and a spectrophotometric acetyl transferase assay allows quantification by measuring enzyme kinetics (De Block *et al.*, 1987; Denecke *et al.* 1989; De Almeida *et al.*, 1989).

FIELD TRIALS WITH TRANSGENIC CROPS

Although the results with transgenic plants in the glasshouse looked very promising, different important questions were still to be addressed. Observations in the laboratory and the glasshouse, which are needed to monitor the expression of the introduced gene and the behaviour of the genetically modified crop have to be followed by open field trials to prove a statistically significant performance and to provide yield data. They have to answer two recurring questions : (1) how stable is the expression of the introduced gene under the highly variable conditions in the field, opposed to the tightly controlled conditions in the glasshouse, and (2) do plants suffer any undesired effects as a result of the genetic modification? Moreover, these field trials with engineered crops have to teach us more about the qualitative and quantitative characteristics of engineered crops relative to competitive existing products.

Important considerations are also the potential risks coupled with the cultivation of genetically engineered plants. The following risks are to be considered : (1) the offspring of the transgenic plant may become a pest; (2) the newly introduced information may be transmitted to a related wild species; (3) the transgenic crop may have an ecological impact. The most significant risk with genetically modified plants is the transfer of newly-acquired genes to wild and weedy relatives. Several precautions can be taken with small-scale field trials to eliminate the risk of spread of the recombinant genes : absence of relatives in test areas, prevention of pollen transfer (by early harvesting, buffer zones or deflowering), daily plant monitoring, access restricted to authorized personnel, destruction of the removed plant and seed material, control of volunteer weeds and follow-up of the field in subsequent sessions.

Since 1986, several small-scale field trials have been conducted in different countries. In 1988, about half of the field tests performed were for herbicide tolerance in tobacco and tomato. The remaining half were nearly all for insect and disease resistance. During 1989, field experiments involve a greater range of crops, including potato, corn, soybean, oilseed rape, cotton and sugar beet. At present, it is difficult to estimate the number of field trials with transgenic plants and one can expect that requests for new trials will increase in the coming years. Presently, field trials (1) with glyphosate-resistant tomato, oilseed rape, tobacco, cotton and soybean (Fraley, pers. comm.); (2) with sulfonylurea-resistant tobacco and tomato (Falco, pers. comm.); (3) with bromoxynil-resistant tobacco, tomato and cotton (Stalker, pers. comm.) and (4) with glufosinate-resistant tobacco, potato, tomato, poplar, sugar beet and alfalfa have been or are being conducted. In most cases a herbicide resistant or tolerant phenotype was observed and no yield penalties were observed. However, data of in depth analysis on qualitative and quantitative chracteristics are not published yet.

GLUFOSINATE-RESISTANT CROPS UNDER FIELD CONDITIONS

Since 1987, PGS has conducted several field tests with glufosinate-resistant crops. The strategy followed was focused on a co-operation with government agencies and independent research institutes. For example, trials have been conducted at France's Institut du Tabac and the Centrum ter Bevordering van de Aardappelteelt in Belgium. This type of co-operation has benefited both parties, since for public institutions, it means access to new technology and advance experience with potentially new commercial varieties. Moreover, it allowed work with skilled agronomists who are alert to economically important traits and crop production methods.

In a first trial in 1987, two transgenic tobacco and four potato lines were evaluated under field conditions (De Greef *et al.*, 1989). Transgenic plants of the tobacco dwarf variety SR1 as well as untransformed control plants were planted in the field according to a factorial design to compare different genotypes and four levels of herbicide application. Untransformed plants were used as control instead of plants obtained from the same regeneration protocol in order to evaluate the agronomic performance of transgenic crops. In the weeks following the herbicide applications, no visible effects or damage were observed on the resistant plants. The tobacco trial was assessed by measuring the length of the largest leaf before and after herbicide application. The analysis of variance demonstrated that there was no significant difference between the treatments. Since flowers had to be removed from the plants according to the field test protocol, only a subset of plants were qualitatively analysed for flowering and were bagged for selfing. The flowering and seed set were not influenced by the herbicide treatments and confirmed previous glasshouse tests. This first field test proved the complete resistance to field dose applications of glufosinate in both lines tested, even although the expression of the resistance gene in these lines varied by two orders of magnitude.

Transgenic lines obtained from three commercial potato cultivars were chosen for field tests : Bintje, Berolina and Desirée. In the weeks following the herbicide treatment, there was no difference between unsprayed controls and transformed plants sprayed with the herbicide, even at the highest dose. At harvest, the number of surviving plants per plot and the tuber weight were recorded. In general, there was no difference between the unsprayed controls and the sprayed transformed lines in percentage of surviving plants. The production of tubers varied strongly depending on the source of planting material but these differences were observed in the control material as well as in the transgenic lines, and are most likely a consequence of the *in vitro* propagation conditions of the planting material. For the statistical analysis, the fresh tuber weight per surviving plant was determined for each plot. The analysis of variance indicated no significant difference between the control and any of the sprayed herbicide resistant lines.

The growth of the transgenic tobacco and potato lines was

indistinguishable in all treatments from the non-transformed non-treated control lines over the whole crop cycle and they showed the same agronomic performance (De Greef *et al.*, 1989).

In 1988, field trials have been performed with transgenic tobacco plants of the commercial line PDB6. The results showed that the presence of the bar gene conferred total resistance to the herbicide. There were no phenotypic differences observed between the transgenic lines and the untransformed control lines. Moreover, weed control with the commercial preparation of glufosinate allowed optimal weed control during the whole growth season.

During the same season, trials of transgenic lines of potato, alfalfa, tomato and oilseed rape were performed. Analysis of several independent transgenic lines containing two different chimeric genes with the octopine T-DNA TR2' promoter and the 35S promoter illustrated the importance of the analysis of several lines. Whereas tobacco lines containing these chimeric genes generally were fully resistant, it was observed that alfalfa lines gave different results. This analysis suggested the importance of combining field test data with molecular analysis for several individual transformants (D'Halluin *et al.*, submitted).

In 1989, field trials are planned with herbicide-resistant tobacco, tomato, oilseed rape, poplar and sugar beet. At the end of this year, PGS will have conducted about 20 field trials with glufosinate-resistant lines of seven crops (tobacco, several cultivars of potato, poplar, alfalfa, winter and spring varieties of oilseed rape, tomato and sugar beet) in five different countries. Statistical analysis of field data should allow us to evaluate in depth the behaviour of crops engineered resistant to glufosinate in normal conditions of agricultural practice.

PERSPECTIVES FOR ENGINEERED HERBICIDE RESISTANT PLANTS

The present results and experience with glufosinate-resistant plants have really demonstrated the technical feasibility of the new trait in plants. Similarly, as with other developments in the engineering of herbicide tolerance, these products are approaching the phase of commercialization. The commercial strategy in engineering herbicide tolerance is basically focused on the gain of a market share through a shift in the application of herbicides and not to increase the overall use of herbicides, as is frequently believed. Herbicide-resistant plants will have the positive impact of reducing overall herbicide use through substitution of more effective and environmentally acceptable products. It is also important to notice that a herbicide resistant trait is also applicable as a selectable marker in both basic research and plant breeding. For basic research, the resistance gene can serve as an analogue of the widely used antibiotic resistance genes in selecting transformants. For example, they have recently been used with success in rice and corn. For plant breeding, the resistance genes can be physically linked to other genes conferring agronomically useful traits which are not easy to follow during transformation and regeneration procedures. They are introduced in plants via transformation using selection for resistance to the herbicide. Subsequently, in a plant breeding program the trait can be transferred to various cultivars through standard genetic crosses by following the easily assayable herbicide resistance phenotype associated with the linked marker gene. Moreover, the spraying of the seedlings can be used to follow segregation and offers a great advantage for selection of progeny grown in soil, this in contrast with many other marker genes. For example, the breeding of hybrids will be facilitated by using herbicide resistant male tester lines.

The strategy followed here to engineer plants resistant against a broad spectrum herbicide clearly illustrate the advantage for using bacterial detoxifying enzymes. This system is independent from the plant species, is highly effective and has no effect on crop performance. However, since the herbicide is detoxified *in planta*, the fate and toxicology of the metabolized herbicide also require careful examination. The biological activities of these metabolites, if existing, will have to be analysed according to existing chemical residue regulations.

Before transgenic plants can be commercialized, important aspects will have to be considered. Factors such as herbicide performance, crop and chemical registration costs, potential for out-crossing to weed species, proprietary right issues and competing herbicides must all be considered before final decisions on commercialization of specific herbicide-tolerant crops can be made (Fraley *et al.*, 1987). Other issues that will affect the introduction of genetically engineered plants include regulatory approval and public perception. Although plant breeding products have always been freely distributed, transgenic plants require regulatory approval before even small-scale field testing can be performed. In 1989, the regulatory situation in Europe varies from one country to another. Since the potential problems, if existing, associated with genetically modified plants will not be confined to national territories, a common guideline for deliberate release of genetically engineered organisms in the environment is desired. It is of extreme importance that the process for evaluating field tests of genetically modified crops responds quickly to the need for testing plants at multiple locations and under normal agronomic practices including completion of the crop reproduction cycle in normal areas of production. Such a regulatory process should satisfy the concerns regarding environmental impact and health and the need to let research and development proceed in a rational and efficient way. Those regulatory structures should focus not on how a particular crop is made but on what new traits it has and how it will be used. Also, it is necessary that the regulation of the commercialization of these crops be formulated and harmonized in a way which does not discriminate the particular process used to improve the varieties.

ACKNOWLEDGMENTS

We thank our collaborators D. Rahier at PGS and C. Thompson at Institut Pasteur, Paris.

REFERENCES

Bayer, E., Gugel, K., Hagele, K., Hagenmaier, H., Jessipow, S., Koning, W. and Zahner, H. (1972). Stoffwechselprodukte von Mikroorganismen. Phosphinothricin und Phosphinothricyl-alanyl-alanin. <u>Helvetica Chimica Acta</u> **55**, 224-239.

Botterman, J. and Leemans, J. (1988). Engineering herbicide resistance in plants. _Trends in Genetics_ **4**, 219-222.

Bartsch, K. and Tebbe, C. (1989). Initial steps in the degradation of phosphinothricin (glufosinate) by soil bacteria. _Applied and Environmental Microbiology_ **55**, 711-716.

De Almeida, E., Gossele, V., Muller, C., Dockx, J., Reynaerts, A., Botterman, J., Krebbers, E. and Timko, M. (1989). Transgenic expression of two marker genes under the control of an _Arabidopsis_ rbcS promoter : sequences encoding the Rubisco transit peptide increase expression levels. _Molecular and General Genetics_ **218**, 78-86.

De Block, M., Botterman, J., Vandewiele, M., Dockx, J., Thoen, C., Gossele, V., Movva, N.R., Thompson, C., Van Montagu, M. and Leemans, J. (1987). Engineering herbicide resistance in plants by expression of a detoxifying enzyme. _The EMBO Journal_ **6**, 2513-2518.

De Block, M. (1988). Genotype-independent leaf disc transformation of potato (_Solanum tuberosum_) using _Agrobacterium tumefaciens_. _Theoretical and Applied Genetics_ **76**, 767-774.

De Greef, W., Delon, R., De Block, M., Leemans, J. and Botterman, J. (1989). Evaluation of herbicide resistance in transgenic crops under field conditions. _Bio/Technology_ **7**, 61-64.

Denecke, J., Gossele, V., Botterman, J. and Cornelissen, M. (1989). Quantitative analysis of transiently expressed genes in plant cells. _Methods in Molecular and Cellular Biology_ **1**, 19-27.

Donn, G. , Tischer, E., Smith, J. and Goodman, H. (1984). Herbicide-resistant alfalfa cells : an example of gene amplification in plants. _Journal of Molecular and Applied Genetics_ **2**, 621-635.

Fraley, R., Kishore, G., Gasser, C., Padgette, S., Horsch, R., Rogers, S., Della-Cioppa, G. and Shah, D. (1987). Genetically-engineered herbicide-tolerance - technical and commercial considerations. _British Crop Protection Conference - Weeds_ 463-470.

Gasser, C. and Fraley, R. (1989). Genetically engineering plants for crop improvement. _Science_ **244**, 1293-1299.

Kondo, Y., Shomura, T., Ogawa, Y., Tsuruoka, T., Watanabe, K, Totukawa, T., Suzuki, T., Moriyama, C., Yoshida, J., Inouye, S. and Niida, T. (1973). Isolation and physicochemical and biological characterization of SF-1293 substances. _Scientific Reports of Meiji Seika Kaisha_ **13**, 34-43.

Kumada, Y., Anzai, H., Takano, E., Murakami, T., Hara, O., Itoh, R., Imai, S., Satoh, A. and Nagaoka, K. (1988). The bialaphos resistance gene (_bar_) plays a role in both self-defense and bialaphos in _Streptomyces hygroscopicus_. _The Journal of Antibiotics_ **16**, 1839-1845.

Murakami, T., Anzai, H., Imai, S., Satoh, A., Nagaoka, K. and Thompson, C. (1986). The bialaphos biosynthetic genes of _Streptomyces hygroscopicus_ : Molecular cloning and characterization of the gene cluster. _Molecular and General Genetics_ **205**, 42-50.

Tachibana, K., Watanabe, T., Sekizawa, Y. and Takematsu, T. (1986). Inhibition of glutamine synthetase and quantitative changes of free amino acids in shoots of bialaphos-treated Japanese barnyard millet. <u>Journal of Pesticide Science</u> **11**, 27–31.

Tebbe, C. and Reber, H. (1988). Utilization of the herbicide phosphinothricin as a nitrogen source by soil bacteria. <u>Applied Microbiology and Biotechnology</u> **29**, 103–105.

Thompson, C.J., Movva, N.R., Tizard, R., Crameri, R., Davies, J.E., Lauwereys, M. and Botterman, J. (1987). Characterization of the herbicide-resistance gene <u>bar</u> from *Streptomyces hygroscopicus*. <u>The EMBO Journal</u> **6**, 2519–2523.

OPPORTUNITIES FOR INTRODUCING HERBICIDE-RESISTANT CROPS

S W J Bright

Plant Biotechnology Section, ICI Seeds, Jealott's Hill Research Station, Bracknell, Berks RG12 6EY, UK.

Inventions in plant biotechnology will be brought to the market as crop varieties for sale to farmers. A few examples of such crops are now in development. This technology is likely to be widely applicable, giving the opportunity to choose from a range of herbicides when designing an integrated weed management programme. The value of resistant varieties will be determined by their ability to deliver better or cheaper weed control. The yield of the crops will need to be maintained, a point about which data is still scarce. Problems will need to be overcome in the areas of patents and regulatory requirements. Environmental questions on the effect of resistant crops on herbicide usage, weed resistance and the spread of resistance from crops to wild species must also be addressed. These problems are soluble. The resultant varieties will give beneficial choices allowing the development of integrated weed management practices to combat weed resistance and reduce chemical loads in the environment.

INTRODUCTION

Selective herbicides kill weeds and leave crop plants unharmed. There are now genetic as well as chemical approaches to providing new selectivities. There has been much publicity about new, biotechnological, routes to herbicide-resistant crops. This has ranged from learned journals through to the popular press. Nevertheless, the product in the market place, a variety resistant to a herbicide will be bought by farmers only when they can see a clear benefit. This paper examines the state of the technology, what it can deliver as products, and some of the key factors which could affect choices between resistant varieties. Finally, some of the unknown quantities which could have a big impact on the pace at which herbicide resistant crops are introduced to the seed market are addressed.

TECHNICAL ACHIEVEMENTS

The possibility of deliberately producing herbicide resistant crops has been talked about for over 20 years. Real progress has been made in recent years towards release of varieties which are resistant to particular herbicides (reviewed in Mazur and Falco, 1989). There are a number of overlapping stages on the road to a variety. The process starts with a clear difference between sensitive and resistant experimental lines, either in cell culture or as seedlings in the laboratory. Many experiments have got to this stage. A harder task is to demonstrate in glasshouse and field tests that resistance is genetically stable and sufficient to allow use of field rates of herbicides. A safety margin of at least three times the standard field use rate is probably required. The final stage is to determine efficacy in commercial use. Only one specialised example, canola resistant to atrazine, has made this final step (Beversdorf *et al.*, 1980; Beversdorf and Hume, 1984). Some of the examples where resistance has been demonstrated to approximately field use rates are given in Table 1. Levels of resistance at or above standard field rates have now been found with eight structurally unrelated classes of herbicides. There is still much to do particularly in terms of backcrossing and yield trials before these examples are translated to varieties on sale to the farmer. This process can take 6-8 years (Newhouse *et al.*, 1989). However, technical success can be anticipated on the basis of one of three general mechanisms of resistance; namely (i) altered herbicide target enzyme (ii) over-production of target enzyme and (iii) metabolism or degradation of the herbicide.

The three general mechanisms have been found to work in a variety of cases involving herbicides of widely differing structure and modes of action. Indeed, where metabolism or degradation is involved, it is not necessary to know anything about the mode of action of the herbicides in question.

Other mechanisms can be considered, such as exclusion (lack of uptake or transport to the target site), a common mechanism for resistance to antimetabolites. Such mutants have been used to study transport and metabolic processes (Bright *et al.*, 1983a), but have not provided a useful source of herbicide resistance. Lack of a required metabolising enzyme can also give resistance, as with the herbicide chlorate. Selected mutants are resistant to this herbicide, but are not viable as they lack a functional nitrate reductase (Bright *et al.*, 1983b).

Resistance to a herbicide is, therefore, a technically achievable goal. This is important for a seed company, because it leads on to the conclusion that just as there are many disease resistance genes which need to be handled in breeding programmes, there are likely to be many herbicide resistance genes. Choosing the most advantageous ones and mixing them in the germplasm will be an important activity in years to come. The choices are likely to be determined not by technical effects but by market requirements.

MARKET REQUIREMENTS FOR HERBICIDE RESISTANCE

Herbicides are big business, and farmers are offered a wide selection of herbicides to control their weed problems. For instance in the UK in 1987 there were 131 products (containing 37 different active ingredients) which were recommended for use on winter wheat (Flint, 1987). The utility of a crop variety resistant to any particular herbicide will depend on it providing a better solution to weed

Table 1. Herbicide resistance at field rates.

Herbicide	Crop	Herbicide rate (kg ha^{-1})			Ref.[1]
		Field use	Resistant plants		
Bromoxynil	Tobacco	0.2–0.5	4.5	(G)[2]	a
Chlorsulfuron	Tobacco	0.017	0.016	(G)	b
Glyphosate	Petunia	0.5–1.5	0.9	(G)	c
Imazethapyr	Maize	0.1	2	(F)	d
Phosphinothricin	Tobacco	1	4	(F)	e
	Potato	1	1–4	(F)	e
Sethoxydim	Maize	0.35	0.44	(F)	f
2,4–D	Tobacco	0.3–1.3	1.1	(G)	g

[1]: References, a: Stalker *et al.*, 1988, b: Falco *et al.*, 1987, c: Shah *et al.*, 1986, d: Shaner *et al.*, 1985, e: deGreef *et al.*, 1989, f: Wyse *et al.*, these Proceedings; g: Streber & Willmitzer, 1989.
[2]: G, tests in glasshouse; F, tests in field.

problems. One way to examine this is to look at current herbicide usage in different crops (Table 2). The total value of herbicide expenditure is greatest in the crops with the greatest area of land, but the average spent per hectare does not follow the same pattern. Thus, sugar beet and rice are the most expensive crops for weed control for various specialised reasons.

These world average figures are simplified estimates that conceal great fluctuations between regions and countries. For instance, winter wheat crops are treated very differently in extensive agricultural systems such as in Australia ($6 ha^{-1}) and the USA ($21 ha^{-1}) compared with intensive systems in the UK ($42 ha^{-1}) or Japan ($52 ha^{-1}). A first place to look for potential value from a resistant crop is to achieve the same level of weed control using cheaper herbicides. The figures of expenditure can be used to estimate the magnitude of savings which potentially could be shared between the farmer and the seed company. The equation would have to take into account the costs of applying the substituting herbicide and then be used to work out some split of the benefit between the herbicide manufacturer, seed company and farmer.

The higher the current cost of a herbicide programme the more potential there is for making significant improvements in weed control costs. Thus it seems unlikely that Australian wheat varieties with herbicide resistance would be commercially successful on this basis

Table 2.   <u>Herbicide usage in different crops worldwide in 1986.</u>

Crop	Herbicide expenditure	
	Total ($M)	Average ($ ha^{-1})
Maize	1500	32
Soybean	1300	38
Wheat	800	17
Rice	614	95
Barley	400	18
Sugar Beet	300	99
Oilseed Rape	107	39
Sunflower	50	20

Source: Internal ICI estimates

alone, but would need to bring some other benefit.

One reason for high herbicide costs might be the presence of problem weeds that are poorly controlled. In Europe, atrazine is a cheap compound to control broadleaved weeds in maize. Resistant weeds require treatments that may be many times as expensive, and still give incomplete control. A significant additional opportunity exists therefore to provide improved weed control. The value of this would depend on the benefit to the farmer of the improvement. This is a more sizeable market sector, and one where both Agrochemical and Seed companies have a chance to provide new products.

Weeds closest in genetic terms to crops are generally the hardest to control. Selectivity is more difficult to find in this circumstance. Wild relatives of canola are extremely troublesome (Beversdorf *et al.*, 1988) and, similarly, *Avena* spp. are major weeds in small grain cereal crops worldwide (Holm *et al.*, 1977). In these cases, breeding genetic herbicide resistance into the crop provides an attractive way of providing, or improving, herbicide selectivity.

The weed control situation is not static. Losses from particular weeds can vary significantly from year to year even in the same area (Dexter *et al.*, 1981). The weed populations adapt, as in the case of atrazine-resistant weeds, and soils can adapt to change the activity of a herbicide, such as EPTC which is used to control grass weeds in corn (Kaufman *et al.*, 1985). Commercially also there are changes as companies compete for markets, and compounds come off patent. The costs and benefits of introducing herbicide-resistant varieties are therefore uncertain estimates at present. These uncertainties are compounded by other factors affecting the production, registration and use of resistant varieties.

PROBLEMS IN COMMERCIALISATION OF HERBICIDE-RESISTANT CROPS

Problems in the areas of the Patents, Regulation, Efficacy and the Environment will need to be solved before herbicide-resistant crops are commonplace and unremarkable. Each area has uncertainties at present but does not present insuperable difficulties.

Patents

In common with many areas of biotechnology there is considerable uncertainty as to the scope of patents in this area with major differences apparent between Europe and the USA (Curry, 1987). Partly as a consequence of the newness of the area, patents can take many years to be reviewed and granted. Legal battles have so far been largely confined to the pharmaceutical arena where products from biotechnology are already in the marketplace. The legitimate role and scope of plant patents, plant variety protection and other forms of intellectual property protection are actively being debated in many countries. The outcome of these debates of principle and practice will be important in determining much about how herbicide resistant crops appear in the market. For instance it will affect how broadly or narrowly the individual sources of resistance are available within the seed industry, their rate of introduction and the prices charged.

Regulation

Many routes to herbicide resistance involve production of transgenic plants. These plants will be regulated, and it is in the interest of all parties to have a clear and effective system of regulation for the release of transgenic plants based on scientific principles. This will apply from field testing through to commercial production. There have now been widespread small scale field trials of transgenic plants in the USA, 48 permits being granted by August 1989 for trials in 19 different states. Field tests have also been undertaken in various European countries, including the UK.

In the UK the Advisory Committee for Genetic manipulation (ACGM) is the guiding body, on a case by case basis responsible to the Health and Safety Executive. It includes representatives of relevant Government departments, Local Authorities, employers and employees as well as scientific expertise. Each European country has a national system, and the CEC is considering the overall framework for regulation. This CEC system needs to be one of setting standards for the national systems and checking that these are met rather than having another level of regulatory approval in Brussels. At the time of writing there is a concern that there may be a burdensome triple review in prospect within the CEC. There is a further threat of duplication between the regulatory sectors involved in environmental release and the product itself.

Once the crop is grown it will be used for food, feed or industrial purposes. There are already well established regulations governing the extension of herbicide uses to new crops. These require studies of the metabolism and residues of the herbicide in the new crop. In addition, there will be specific consideration of transgenic crops as sources of food. The regulations in this area are not yet formulated and could be a limiting factor in determining the timescales for the first introductions of transgenic, herbicide resistant crops.

A sound regulatory framework is welcomed by industry, but it is axiomatic that regulation costs time and money. The eventual shape of the regulatory framework will therefore have a big effect on the costs of getting a resistant variety to the market and hence on the economics of any herbicide programme involving resistant varieties.

Efficacy

In an earlier section I argued that resistance to any herbicide is, in the longer term, technically feasible. That is not to say that it is a simple or routine exercise. A key factor for success will be the extent to which the resistant variety will perform as well as a normal one in terms of yield and quality of harvested produce. Equivalent yield in the absence of weed competition is the goal to be aimed at, but in the final analysis it will be the yield with a particular herbicide programme compared with conventional variety in a normal weed control programme. This will translate to return to the farmer per hectare of land. It is unlikely that any significant yield penalty will be allowed and this is an area where data is at present very scarce. Atrazine-resistant canola has an economic benefit because of the improved seed quality gained by the absence of wild mustard seed in the harvested crop. Even in that case a yield penalty of 20-25% (Beversdorf *et al.*, 1988) is too much. Many countries have state-run trials under standard conditions which give rise to National Lists. It may be decided to be necessary for a variety to be assessed for herbicide resistance as it is for disease resistance currently. Comparisons based on different herbicide programmes would certainly be more complex than those that are currently performed. The requirements for farmer support and training from the agricultural industry are also likely to increase initially as new resistant varieties are introduced.

Environmental issues

The issues around the principles of release of transgenic crop plants in general are being addressed in developing the regulatory framework. Very strongly and sincerely held views are currently being debated. Three specific issues have been raised with respect to herbicide resistance. These are their impact on total herbicide usage, the development of herbicide resistant weeds and the movement of crop plants or their genes into natural populations. Each of these issues will be best addressed on a crop by crop basis. Much relevant information is already known.

EFFECT ON HERBICIDE USAGE

It has been argued that herbicide resistant crops must lead to greater herbicide usage. This is not necessary. Indeed, one of the greatest benefits of biotechnology in this area is that it gives meaningful choices. One choice might be to reduce specifically the amount of a particular chemical for one year, or as part of a rotation if, for instance, there were problems with ground water. This end could be achieved without total disruption to farmers if an alternative was available in conjunction with herbicide resistance. Another choice might be to maximise the use of new, safe, low rate herbicides by enhancing the range of crops which can be treated. In effect this would relax the requirements for new herbicides to meet stringent crop selectivity requirements allowing more effort to go towards designing molecules with benign environmental profiles or utilising those which have these characteristics already. Although there are a plethora of herbicides available to farmers, in any particular area it is likely that a core herbicide programme would be common to most farmers. Thus the same few herbicides are likely to be widely used to provide the basic broadleaf and grass weed control. In the USA some 95% of the maize acreage is treated with one of two triazine herbicides and over 50% with alachlor or metolachlor. The advent of herbicide resistant crops should increase choices allowing, for instance, the chance of a cost effective replacement for atrazine in areas where resistant weeds are a problem. The choice could be left to the farmer or be imposed by regulation. In either case it must be beneficial to have choices available.

WEED RESISTANCE

The problems of weed resistance are increased by greater exposure of weeds to selective pressure. Increasing the use of individual herbicides by introducing resistant varieties has the potential therefore to accelerate the development of weeds resistant to that compound. The well documented increases in populations resistant to triazines and sulfonylureas (see these Proceedings) have occurred in areas with high selection pressure. Mathematical modelling to understand and simulate the effects of different herbicide programmes is now being undertaken (Gressel, 1986). It seems therefore that the problems are soluble if the control of weeds is considered in the same way as the control of other plant pests i.e in an integrated way. Integrated weed management should utilise the choices provided by herbicide resistance to give the farmer sustainable and cost effective weed control by mixing herbicides and varieties. This will be a growth area for specialist research and advice as the resistant varieties come to the market place.

HERBICIDE-RESISTANT CROPS AS WEEDS

If a crop plant remains in a following crop because of the establishment of a stray seed or propagule, then it becomes a weed (being a plant in the wrong place at the wrong time). If it is resistant to a particular herbicide then a different one will be required for control. This is a specific problem for an integrated weed management programme and should be considered along with control of other weeds. It only becomes a significant problem in a single herbicide scenario where all crops are sprayed with only one herbicide. I have already argued that this is unrealistic because it would lead to the development of weed resistance.

A more general concern is that the genes conferring resistance will transfer by sexual crossing into wild relatives. This would render the resistance genes less useful and therefore this gene flow must be minimised. Furthermore there is a the possibility that such weeds may become super weeds by virtue of these new genes. The extent of this form of gene transfer will differ from crop to crop. Much is already known about the extent to which crosses can be successful for the purposes of introgression of genes from wild species. Thus in the genus *Hordeum* 650/880 possible combinations of thirty species have been tried and 45% of these produced plants (Jacobsen and Von Bothmer, 1981). This is a different figure to that which could occur in nature and this general question is being addressed as part of studies on the release of transgenic plants (e.g. Drozd, 1989). There are few studies of what specific genes make weeds competitive. However, it is unlikely that single gene effects are involved. The only data on fitness of herbicide resistant weeds is in relation to atrazine resistance where the fitness is reduced (Gressel, 1986). Thus, while there is scope for more research it seems highly unlikely that a single herbicide-resistance gene can confer superior fitness (except in the presence of the herbicide). Similarly, the volunteer plants from the previous crop are unlikely to become persistent weeds solely by virtue of their herbicide resistance although there are cases, such as Proso millet in Canada, where "normal" volunteers have become a significant weed control problem (Bough and Cavers, 1989)

CONCLUSIONS

A few examples of herbicide resistant crops are now in development. This technology is likely to be widely applicable, giving the opportunity for farmers to choose from a range of herbicides when designing an integrated weed management programme. The price of weed control programmes is likely to be a key factor as this technology matures. The rate that it is introduced will be determined at least in part by the evolution of appropriate regulatory regimes. Concerns over the inappropriate application of resistance technology can be overcome if it is seen as providing beneficial choices to manage chemical usage and weed resistance.

ACKNOWLEDGEMENTS

I am grateful for stimulating discussions with Ian Evans, Ian Bryan, and Tim Hawkes.

REFERENCES

Beversdorf, W.D., Weiss-Lerman, J., Erickson, L.R. and Souza Machado, V. (1980). Transfer of cytoplasmically-inherited triazine resistance from bird's rape to cultivated *Brassica campestris* and *Brassica napus*. Canadian Journal of Genetics and Cytology **22**, 167-172.

Beversdorf, W.D. and Hume, D.J. (1984). OAC Triton spring rapeseed. Canadian Journal of Plant Science **64**, 1007-1009.

Beversdorf, W.D. Hume. D.J. and Donelly-Vanderloo, M.J. (1988). Agronomic performance of triazine-resistant and susceptible reciprocal spring canola hybrids. Crop Science **22**, 932-934.

Bough, M.A and Cavers, P.B. (1989). Proso millet, a crop gone wild. Canadian Journal of Plant Science **69**, 265.

Bright, S.W.J., Kueh, J.S.H. and Rognes, S.E. (1983a) Lysine transport in two barley mutants deficient in a basic amino acid uptake system in the root. Plant Physiology **72**, 821-824.

Bright, S.W.J., Norbury, P.B., Franklin, J., Kirk D. and Wray, J.S. (1983b). A conditional lethal cnx-type nitrate reductase deficient barley mutant. Molecular and General Genetics **189**, 240-244.

Curry, J.R. (1987) "The patentability of genetically engineered plants and animals in the US and Europe". Intellectual Property Publishing Limited, London. 41pp.

De Greef, W., Delon, R., DeBlock, M., Leemans J. and Bottermans, J. (1989). Evaluation of herbicide resistance in transgenic crops under field conditions. Bio/technology **7**, 61-64.

Dexter, A.G., Nalewaja, J.D.,Rasmusson, D.D. and Buchli, J. (1981). Survey of wild oats and other weeds in North Dakota 1978 and 1979. North Dakota Reports. **79**.

Drozd, J. (1989). PROSAMO - a DTI/Industry/AFRC initiative on the ecological aspects of agricultural biotechnology. In "Biotech 89", pp. 85-89. Blenheim On-line, London.

Falco, S.C., Knowlton, S., Larossa, R.A., Smith, J.K. and Mazur, B.J. (1987). Herbicides that inhibit amino acid biosynthesis; the sulfonylureas - a case study. British Crop Protection Conference - Weeds 149-158.

Flint, C. (1987) "Crops Guide to Herbicides". Reed Business Publishing, Wallington, UK. 731 pp.

Gressel, J. (1986) Modes and genetics of herbicide resistance in plants. In "Pesticide Resistance: strategies and tactics for management" (Board of Agriculture, eds), pp. 54-73. National Academy Press, Washington D.C.

Holm, L.G., Plucknett, D.L., Pancho, J.V. and Herberger, J.P. (1977). "Worlds worst weeds, distribution and biology". University Press of Hawaii, Honolulu. 609 pp.

Jacobson, N. and Von Bothmer, R. (1981) Interspecific hybridization in the genus *Hordeum*. In "Barley Genetics IV" (R.N.H. Whitehouse ed.) pp. 710-715. Edinburgh University Press, UK.

Kaufman, D.D., Katan, Y., Edwards, D.F. and Jordan, E.G 1985. Microbial adaption and metabolism of pesticides. In "Agricultural Chemicals of the Future", BARC Symposium number 8, (J.L. Hilton, ed.), pp. 437-451. Rowman and Attenheld.

Mazur, B.J. and Falco, S.C. (1989). The development of herbicide resistant crops. Annual Review of Plant Physiology and Plant Molecular Biology **40**, 441-470.

Newhouse, K.E., Shaner, D.L., Wang, T. and Fincher, R. (1989). Genetic modification of crop responses to imidazolinone herbicides. In "Managing Resistance to Agrochemicals; from Fundamental Research to Practical Strategies". (M.B. Green, H. Le Baron and W.K. Moberg, eds) pp. 474-481. American Chemical Society, Washington D.C.

Shah, D.M., Horsch, R.B., Klee, H.J., Kishore, G.M., Winter, J.A., Turner, N.E., Hironaka. C.M., Sanders, P.R., Gasser, C.S., Aykent, S., Siegel, N.R., Rogers, S.G. and Fraley, R.T. (1986). Engineering herbicide tolerance in transgenic plants. Science, **233**, 478-481.

Shaner, D., Malefyt, T. and Anderson, P. (1985). Herbicide resistant maize through cell culture selection. In "Biotechnology and its application to Agriculture", Monograph No. 32, pp. 45-50. BCPC Publications, Thornton Heath, UK.

Stalker, D.M., McBurke, K.E. and Malyi, L.D. (1988). Herbicide resistance in transgenic plants expressing a bacterial detoxification gene. Science **242**, 419-423.

Streber, W.R. and Willmitzer, L. (1989). Transgenic tobacco plants expressing a bacterial detoxifying enzyme are resistant to 2,4-D. Bio/technology **7**, 811-816.

ASSESSMENT OF THE HAZARDS FROM GENETICALLY-ENGINEERED PLANTS : THE WORK OF THE ADVISORY COMMITTEE ON GENETIC MANIPULATION INTENTIONAL INTRODUCTION SUB-COMMITTEE

Mark Williamson

Department of Biology, University of York, York YO1 5DD, England

The development of herbicide-resistant crop plants may give rise to some hazards that are not easily quantified. These include various pleiotropic effects on the plant itself, particularly toxicity and weediness. The engineered genes may spread, particularly by pollen, to other species, which could produce new weeds, or plants undesirable for other reasons. The genes may affect the pathogens and other organisms associated with the crop. It is the job of ACGM/IISC to assess and advise on the hazard of introducing into the environment genetically-engineered organisms of all kinds. The sub-committee is moving towards a structured method of assessment, a process that may be accelerated by the report of the Royal Commission on Environmental Pollution. It is not the sub-committee's job to assess the hazards, *per se*, of increased spraying of herbicides, or the hazard from the evolution of unrelated herbicide-resistant weeds that may result. Genetically-engineered organisms are being developed in many countries. There are problems of public perception, and public participation in the discussion of hazards. Attempts are being made by CEC and by OECD to harmonise national approaches, and progress here is mentioned.

Since the Long Ashton Symposium there have been some important developments in the British regulatory system. Part VI of the Environmental Protection Act, 1990, which became law at the end of that year, covers the release of genetically modified organisms to the environment. A new joint HSE and Department of the Environment committee, the Advisory Committee on Releases to the Environment (ACRE) has replaced ACGM/IISC but the description and comments on the working of ACGM/IISC in the paper remain broadly true of ACRE. Meanwhile, a European Communities Directive (90/220/EEC) on the deliberate release of genetically modified organisms has been adopted, and the twelve Member States should have legislation in place to implement it by 23 October 1991.

INTRODUCTION

This paper is concerned with the risk assessments of the introduction into agriculture of herbicide-resistant crops, and, in particular, with the work of the Advisory Committee on Genetic Manipulations Intentional Introduction Sub-committee (ACGM/IISC), and the related committees outside the UK. ACGM/IISC, which comes under the Health & Safety Executive has to start with a "presumption of guilt". Before saying that it has no objection to a proposal, it has to be satisfied that the particular proposal is, on such evidence as may reasonably be made available, reasonably certainly safe. The public has collectively a great variety of views on this subject, indeed most commonly no view at all. However, there is an appreciable proportion of the public that feels that there may well be hazards, though they are uncertain of what they are. While some think that all genetic engineering is too risky to be allowed, most of those concerned with environmental issues wish to be reassured that there is some system in place to protect the environment against possible hazards.

HAZARDS

Introducing a herbicide resistant crop which has a genetic basis means the introduction into the agricultural environment of a genotype and phenotype that has not been seen before. It is often argued that agriculture has been doing work of this sort for millenia, although much more rapidly in this century, and that no problems have ever arisen. For instance, a recent French directive on this topic says "In fact, since the start of agriculture, no exponential or anarchic spread of cultivated plants has been observed" (Afnor, 1988). Unfortunately, this is not true. There have indeed been problems produced by agricultural crops. The frequency with which these problems arise is low, but their consequences can be serious.

Problems from introduced plants

One analogy with the introduction of genetically-engineered crops is the introduction of foreign species into a habitat. There is an extensive literature on the ecology of biological invasions, including a recent SCOPE programme, which has culminated in a global synthesis (Drake *et al.*, 1989). A new species introduced is, of course, genetically much more distinct from those present than a new variant of a crop. Nevertheless, the problems produced by introduced species are a useful indication of the sort of problems that need to be looked for when examining the hazards and assessing the risks of genetically-engineered crops. The frequency with which introduced species have become pests has been estimated as being about 1% in the United Kingdom (Williamson and Brown, 1986) and about 2.5% in Australia (Groves, 1986). Genetically- engineered crops ought to be appreciably safer, but that is likely only to mean an order of magnitude or so, so one could expect problems from between 1 in 100 and 1 in 1000 new genetically-engineered crops. This is a low frequency, but still enough to cause quite serious environmental problems if steps are not taken to make it even lower. All these figures are very rough estimates, but from the point of view of the sub-committee, the frequency is not very relevant. What it is concerned with is a discussion of particular proposals, with assessing the possible risks to people and the environment of a proposal to introduce one particular genetically-engineered crop at a time.

There are four hazards from the introduction of plants to new habitats that we might be concerned with. These are weediness, toxicity, the possibility of carrying pathogens, and the possibility of hybridization. The first three would all be pleiotropic effects of the herbicide-resistant gene. Hybridization is likely to be independent of the gene, and the consequences to be thought of are what happens to the new gene when it occurs in a different genome.

Weediness
The evolution of weeds from crops is quite a well-known phenomenon. Weeds and crops are not well-distinguished categories. "Those who are not acquainted with the distribution across the world of our worst weeds often seem confused that a plant which is an important weed in one area may be a valuable crop in another place" (Holm *et al.*, 1977). That is, one genotype might be simultaneously both a crop and a weed. In other cases the weeds are merely related to the crop. "There are weed barleys and weed wheats. There are weed ryes, weed rices, weed sorghums, weed sugarcanes, weed maize and weed oats...there are weed carrots, weed beets, weed radishes, weed lettuce, weed peppers, weed potatoes, weed tomatoes, weed sunflowers, weed safflower, weed hemp, weed watermelon and many, many more. Weed races are not confined to annuals nor to small herbs. *Citrus* spp., *Manihot, Psidium, Carica, Punica, Mangifera, Passiflora* and *Prunus*, have weed races" (Harlan, 1985). That quotation does not distinguish between those weeds which have been derived from a crop, and those which have been derived from close relatives to the crop. In quite a few cases it is uncertain which has happened.

In the case of genetically-engineered herbicide-resistant crops, the hazard we have to consider is whether the pleiotropic effect of the herbicide resistance confers characters that might make the crop weedy, either where it is going to be grown, or, and this is a much more difficult question, in other places where it might grow. Succinctly, those assessing the hazards have to look at reproduction, survival, and growth of a new form, and assess whether these variables are sufficiently different from the original crop to constitute a potential hazard. In other words, the committee will want comparative data on the crop before modification and on the modified crop.

Toxicity
Again, there could be a pleiotropic effect of the gene for herbicide resistance that induces toxicity. Two types of toxicity may be distinguished, from the point of view of risk assessment. The first is direct toxicity in the crop product, whether for man or farm animals. Details of the biochemistry of the pathway by which herbicide resistance is produced may well indicate whether or not this is a hazard that needs particular tests, or can be ignored. It is, however, a problem for another committee, not the ACGM/IISC.

What would be a problem for ACGM/IISC, though one which it has not yet considered in detail, is whether the herbicide-resistant plant will affect the phenotype of the plant in such a way that it becomes toxic to other organisms. It is possible to imagine that the crop does not become a weed because it is attacked by certain insects; if these insects themselves are killed by a side effect of the gene, then the crop might be more likely to become a weed or an invasive species in natural habitats. Again, one could imagine that the nectar in the flowers might become poisonous to bees. So far, considerations of this sort have not been worth pursuing, because the plants that have come before the committee, notably potatoes, are most unlikely to produce hazards of this sort. This is perhaps the point to say that the committee is not able, and would not be willing, to pursue every conceivable minute hazard. Its job is to make a reasonable assessment in a reasonable time. One consequence that has to be faced is that there is a finite possibility of it overlooking something unexpected. This is a real, though not a deep, concern to me; it may explain the real and deep concern felt by certain environmental activists.

Carrying pathogens

Some of the nastiest consequences from introduced plants are the pathogens they bear (von Broembsen, 1989). Perhaps the best known example is that of chestnut blight in the United States. In this case an important and conspicuous member of the natural forest, namely the American Chestnut, *Castanea dentata*, was devastated by a fungus, *Chryphonectria parasitica* (formerly *Endothia parasitica*) which was introduced on a related Asiatic species. The European Chestnut, *Castanea sativa*, is also attacked in parts of its range. Another example is the Pine blister rust, *Cronartium ribicola*, probably introduced on a two-needled pine, which prevents the cultivation of five-needled pines in Britain (Williamson, 1988). This is the sort of hazard for which it is difficult to suggest appropriate experimental protocols, and so it is unlikely to be looked for unless some pathogen is already known to be a potential hazard either to the crop itself, or to some relative.

Hybridization

One of the major pathways in plant evolution seems to have been hybridization between related species leading, usually through polyploidy, to a new species with ecological characteristics different from either parent. There are plenty of examples in the European flora, but the best known one involving an introduced species is the Cord Grass, *Spartina anglica*. This has arisen from hybridization between *S. maritima*, a European species, and *S. alterniflora*, an American one. With herbicide-resistant crops produced by single gene differences, this is a hazard that could probably be ignored. However, much plant improvement is being done by cell-cell hybridization, and it is conceivable that herbicide resistance might be introduced into a distant related species by such a technique. This is then the equivalent of the hybridization type seen in *Spartina*, and so the ecological characteristics of the hybrid are of considerable interest to anyone assessing risks. The more distant the two hybrid parents, the more likely one is to get an ecologically novel product. It is interesting to note that there is no record in the published literature of anyone managing to duplicate the *Spartina* hybridization in the laboratory (Gray, 1986). With the techniques that are at present being used to produce herbicide-resistant crops, committees are not likely to be concerned with this hazard. Hybridization between species that cannot be crossed by classical methods is covered by the Genetic Manipulation Regulations.

A more immediate hazard, and one more amenable to regulation than the creation of new hybrid species, is the possibility of transferring genes by pollen to related species (Ellstrand, 1988). With herbicide-resistant plants, there would be particular concern about the possible transfer of pollen to weedy relatives of the crop plant. As was noted above, many crop plants do indeed have weedy relatives. The insertion of herbicide resistance into a plant which is not weedy might nevertheless turn it weedy. The reason is that a weed is not merely a plant growing in the wrong place, which is the traditional definition, but more particularly a plant that grows successfully and vigorously in a habitat of interest to someone. So fields intensively treated with herbicides are a likely habitat in which all sorts of species that were not weeds before might become weeds. Committees will therefore be interested in all species that can be pollinated successfully by a genetically manipulated plant.

Management changes

The effect of producing a herbicide-resistant crop is to change the management of the field in which it will be grown. This is rather obvious: the amount of herbicide used in the field will be greatly increased. It would be naive to suppose that the amount of herbicide would not also be increased in neighbouring habitats. Consequently, the introduction of a herbicide-resistant crop produces, in one particular sense, a range of new habitats. Again, as with the hybridization hazard, it is difficult to suggest simple experiments that would give useful information on the size and nature of such a risk. One particular hazard has of course been brought out strongly in this Symposium. That is the selection pressure that will produce herbicide-resistant strains in weed species which are taxonomically unrelated to the crop. But as is noted above, there may also be weed species which are quite close taxonomic relatives to the crop.

If the progress of agricultural science leads to a wide variety of different herbicide-resistant crops, then we have to consider the possibility of multiple herbicide resistance, particularly in weeds. The questions asked by committees always include questions about the effect of the genetically-engineered organism on ecosystems, and sometimes on other species. Nevertheless, it is difficult to see how in practice a committee can reasonably take action which will prevent the evolution of herbicide-resistant weeds. The only effective way of doing this is to ban the use of a herbicide-resistant crop, and hence of the extensive use of the herbicide. Such a ban could only be justified if it could be shown that using the herbicide will result in such a weed problem that the net benefits are negative. I find it hard to imagine that anybody could produce a convincing case for this, but nevertheless quite easy to imagine that such cases may arise.

The most likely hazard from the widespread introduction of herbicide-resistant plants is, then, the production of herbicide-resistant races of other plants, which will flourish as weeds. At the same time, this is the most difficult hazard to guard against by any regulatory process. What regulators can do fairly readily are three things. They can require tests on the characteristics, the phenotype, of the herbicide-resistant crop, with particular attention to the possibility that its ecological characteristics could turn it into an invasive species, or a species that would result in pathogen invasion. Second, they can look at the dispersal of pollen, and whether the gene can be transferred to other species. Longer term effects, caused by management practices resulting from the availability of herbicide-resistant crops, including the evolution of new strains of plants or of herbivores, and the evolution of new species by allopolyploidy, will need to be controlled by management practices rather than by committee regulation. The importance of good management in this area has been stressed by several other speakers.

PUBLIC PERCEPTION

Work with genetically-modified organisms, including plants, is regulated because of earlier concern about the laboratory safety of genetically-engineered organisms. In the event, safety in the laboratory has been ensured by adherence to fairly standard medical microbiological containment practices. However, "it is possible that the failure to detect hazards during this period is a consequence of successful containment, rather than the intrinsic safety of genetically-modified organisms constructed by recombinant DNA procedures" (Simonsen and Levin, 1988). With the release of genetically-engineered organisms we are, by definition, passing beyond containment. It is scarcely surprising, then, that there has been much public concern about the release of genetically-engineered organisms. The concerns arise from a number of sources. Lack of containment, and the ability of organisms to reproduce and disperse are fully rational reasons for being concerned about this process. Molecular biology is a young subject, and we are still learning about the way genes interact. With the powerful techniques we have for forming new genetical combinations, it is still possible to be rationally concerned that dangerous combinations might appear by accident. Beyond that, there are worries widely expressed by the public, which can neither be said to be rational or irrational. For example, there is concern about our perceived dependence on technology, and about the loss of naturalness and natural habitats. There is a general dislike of the unknown.

Faced with this variety of responses to genetic engineering, what should be the response of the agricultural scientist? It is perhaps natural that some, wishing to avoid trouble, or hoping to avoid getting involved in public arguments, or just not wishing to go to the trouble of explaining themselves, would like to keep their processes and products confidential. This indeed does happen in some European countries. Civil servants sometimes talk about "the need to know". This approach may avoid trouble in the short run, but it may create more trouble in the long run. Keeping the public informed from the earliest moment possible is the simplest way of keeping the public reassured.

Practice in Britain has been to require proposers, once they have been told there is no objection, to make an announcement in the local press, and maybe to hold meetings locally. Certainly they are well-advised to keep local politicians fully informed, and it is now the expectation that the local Environmental Health Officer would be a member of the relevant local Safety Committee. To many this is insufficient. Particularly in Germany, there is pressure to have public participation as well as public information. That is, there is the view that members of the public should be on committees advising on the release of genetically-engineered organisms. Amongst the EC member states, Germany is the only one to have pressed this point of view. A difficulty normally seen with the idea of public participation is that it is very difficult to decide who should represent the public. As scientists, we have a natural preference for people who are well-informed, and who can understand the issue easily. That would suggest that for us, the public should be represented by people who understand agriculture, genetics, ecology and related topics. To some extent this is the approach that has been adopted in the United Kingdom. Difficulties can arise when parts of the public, and particularly vocal and politically active parts of the public, perceive the risks as greater than these are perceived by regulatory authorities. This problem can be seen in the United States. It seems that all the release proposals in the United States so far have met with some resistance at local level, that is to say the county or municipal level. The consequence has been that quite a number of states, including Illinois, Minnesota, New Jersey, New York, North Carolina, Washington and Wisconsin have produced or are considering producing state legislation for the approval of such tests (Ezzell, 1989). In South Carolina (Sun, 1989) the State Health Authority has objected to a proposal that had been approved at Federal level. It appears that regulatory bodies that take insufficient account of grassroot opinion are only laying up trouble for themselves.

COMMITTEE PROCEDURES

Committee structure

The Advisory Committee on Genetic Manipulation Intentional Introduction Sub-committee, originally known as the Planned Release Sub-committee, has been assessing proposals for the introduction of genetically-engineered organisms into the general environment since 1986. This is under a voluntary system of notification operated by the Health and Safety Executive. Statutory powers for notification are included in the revised Health and Safety (Genetic Manipulation) Regulations 1989.

The Intentional Introduction Sub-committee includes members with a range of specialities. It is a sub-committee of ACGM, which is a tripartite committee, containing members appointed by industry, and by trade unions, as well as scientific experts. Their affiliations, as analysed in Table 1, fail to indicate members' fields of interest, and, more particularly, the attitudes they are likely to take. The variety of interests and experience means that there is no simple classification of the committee, and it shows in practice. The committee does not divide into factions, it works in a remarkably coherent way.

Table 1. Composition of ACGM/II.

Members :			
Government service	5	Local government	1
Universities	6	Industry	1
Assessors :			
MAFF	3	DAFS	1
DOE	2	DTI	1
DH	1	HSE	1

ACGM and ACGM/IISC are both serviced by the Health and Safety Executive, one of whose primary concerns is the operation of the Health and Safety at Work Act etc. 1974 (HSWA). The Health and Safety Executive is the operating arm of the Health & Safety Commission, established under the HSWA and appointed by the Secretary of State for Employment.

ACGM/IISC has been working satisfactorily for three years now, but in a fast changing and politically sensitive field like this it is inevitable that changes should be proposed. These should be considered in an international context.

The international scene

The main international body is the Organisation for Economic Cooperation and Development, the OECD. One of the main concerns of OECD is to harmonize approaches to regulations in order to facilitate international trade. OECD represents all the member states of the EC, the other Scandinavian states, the United States and Canada on the American continent, Japan, Australia and New Zealand and some others. It is an association of 'western', developed, states. In 1986 a committee of experts prepared a most influential report (OECD, 1986) which contained specimen guidelines for the release of genetically-engineered organisms. These guidelines have been taken into account in many countries, including the UK, and the proposed EC directive on the release of genetically-engineered organisms (Commission of the European Communities, 1988) are modelled closely on them.

The Council of Ministers has recently agreed a directive on the contained use of genetically-engineered organisms. The related directive on release has not yet been agreed, but this is likely within the next year. If it is agreed, then all member states of the EC will have to establish a competent authority to implement the directive, and these national authorities will have to exchange proposals amongst themselves. The present British system would fit quite nicely with this aspect of the proposed directive; some member states might have more difficulties.

Proposed changes in British legislation

The Royal Commission on Environmental Pollution (1989) has issued a comprehensive and interesting review of this field and made proposals that there should be new legislation. They have also proposed that the Department of the Environment should take a more prominent role. So, not surprisingly, the Department of the Environment (1989) has issued a consultative paper about possible legislation. The situation is somewhat confusing in that the Department of the Environment is also responsible for the Wildlife and Countryside Act (1981) which controls introductions of non-indigenous kinds of animals into this country. At the same time, other ministries, such as MAFF, have responsibilities in relation to certain types of introductions. In MAFF's case these include any living herbicides or pesticides. The situation is a complicated one, with much potential for overlap of interests, and I would not wish to predict what the likely outcome would be in legislation in the next parliamentary session. Meanwhile, all the various Government departments concerned have got together and produced an interdepartmental form to guide proposers through the present maze of legislation. This can be obtained from the Health and Safety Executive. The Genetic Manipulation

Regulations (1989) are likely to be in force in the near future, perhaps by the time this paper is published. Under these regulations, notification to HSE of a proposal to introduce a genetically-engineered organism is required.

Role of ACGM/IISC

Since 1986, this has been the British committee concerned with proposals for the intentional introduction, or planned release, of genetically-engineered organisms. It is advisory, as the name of its parent committee indicates, formally advising HSC, HSE and Ministers who have an interest in this area. Nevertheless, as so often in the British system, a large company or Government-funded organisation is unlikely to think it sensible to go against the sub-committee's advice. The role of the sub-committee is neither to protect industry from public criticism, nor to protect the public against unreasonable proposals by industry, though the effect of the committee is to do both those to some extent. The primary role of the sub-committee, as indicated by its coming under the Health and Safety Commission, is to ensure safety, particularly safety of workers and the general public who may be affected by work activities, under the HSWA. It also aims to advise on safety of the environment.

The procedure at the moment is that someone with a proposal to release a genetically-engineered organism into any environment is well-advised to seek the advice of the Health and Safety Executive. If the proposal needs to come before ACGM/IISC, or any of the other committees that might be concerned with aspects of this matter, the proposer fills in the interdepartmental proposal form, and answers the questions indicated in guidelines. The latest published guidelines are of 1986 (Advisory Committee on Genetic Manipulation, 1986), revised ones are in preparation (Advisory Committee on Genetic Manipulation, in prep). The revisions are primarily of detail, but one interesting innovation is that in addition to the straight factual information, such as the name of proposer, there is also a risk assessment section set out in the form of questions.

The Royal Commission (the Royal Commission on Environmental Pollution, 1989) has made some trials with systematic methods of identifying hazards. They found that biological knowledge is insufficiently precise for the HAZOP type of assessment used in the chemical industry to be practicable. Nevertheless, they have made some progress in producing a system they call GENHAZ. If this system can be developed into a working system it may be adopted in guidance by ACGM/IISC, or whatever committee is concerned at that stage. Meanwhile, proposers are asked to follow the guidelines of ACGM/HSE/NOTE 3, and make a risk assessment.

The British system of control is built on the general duties of HSWA, specific regulations, published guidance and on site inspection by HSE. Under the revised regulations a local Safety Committee must be established with the appropriate expertise to make a preliminary judgement of the proposal before it comes to ACGM/IISC. In general, the local committee will include the local Environmental Health Officer.

So far there have not been a great number of proposals and some of these are repeat proposals. A recent count is set out in Table 2.

Table 2. Proposals to ACGM/II in its first 3 years.

Release proposals	Done	Not yet done	Not regulated
Micro-organisms	5	1	2
Crop plants	9	2	1
Invertebrates	0	1	0
Total	14	4	3

Many people expected that most of the proposals would relate to bacteria, no doubt because bacteria have been so much to the fore in the contained use of genetically-engineered organisms, and in the development of molecular biology in general. In fact, as can be seen, proposals for crop plants have been commoner. The proposals for micro-organisms contain a series of proposals for genetically-engineered viruses. So bacteria are distinctly in the minority. ACGM/IISC has up to now always invited a proposer, at least on the first occasion that he makes a proposal, to present his work to the sub-committee. The intention of this has been to help both the proposer and the committee. In all cases, so far, the sub-committee has asked for modifications to a proposal, but the column labelled 'Not yet done' in Table 2 relates to only one case where the sub-committee asked for an essentially new proposal. The others are cases where, for reasons of timing or whatever, the particular experiment was not in fact done within the time agreed, and therefore a new proposal had to be made.

The OECD report (OECD, 1986) recommended that proposals should be considered on a case by case basis. ACGM/IISC is still in this stage. Nevertheless, it is perfectly clear that follow-up proposals and related proposals can, even on a case by case basis, be considered much more rapidly than an initial new proposal. So we would expect that, as experience accumulates, many proposals would be able to be dealt with rapidly, even routinely.

CONCLUSION

Hazards from the introduction of herbicide-resistant crops are, except for the production of herbicide-resistant weeds, likely to occur infrequently, but to be important when they do occur. ACGM/IISC has the unenviable task of trying to spot these rare hazards in a mass of sometimes incomplete data and assess the risks involved.

I end with a plea that scientists bringing proposals to the sub-committee should, in their own interests, put to it all the relevant information they have openly and freely.

ACKNOWLEDGEMENTS

The two secretaries of ACGM/IISC, Brian Ager and Tina McGowan, have been most helpful. For comments on this paper I thank Tina McGowan and Charlotte Williamson.

REFERENCES

Advisory Committee on Genetic Manipulation (1986). The planned release of genetically manipulated organisms for agricultural and environmental purposes. Guidelines for risk assessment and for the notification of proposals for such work. ACGM/HSE/NOTE 3. London, ACGM.

Advisory Committee on Genetic Manipulation (in press). The intentional introduction of genetically manipulated organisms into the environment. Guidelines for risk assessment and for the notification of proposals for such work. ACGM/HSE/NOTE 3 (revised). London, ACGM.

Afnor (1988). Guide de bonnes pratiques de recherche et d'essai au champ (Developpement en conditions naturelles produites en station experimentale) des plantes transgeniques (de classe 1). Norme Francaise NF X 42-071, 1-14.

Commission of the European Communities (1988). Proposal for a Council Directive on the deliberate release to the environment of genetically modified organisms. Official Journal of the European Communities C **198**, 19-27.

Department of the Environment, Welsh Office and Scottish Office (1989). Environmental protection. Proposals for additional legislation on the intentional release of genetically manipulated organisms. A consultation paper. Department of the Environment, London.

Drake, J.A., Mooney, H.A., di Castri, F., Groves, R.H., Kruger, F.J., Rejmanek, M. and Williamson, M. (1989). "Biological Invasions, a Global Perspective". (SCOPE 37). John Wiley & Sons, Chichester.

Ellstrand, N.C. (1988). Pollen as a vehicle for the escape of engineered genes? TREE **3(4)** TIBTECH **6(4)**, S30-S32.

Ezzell, C. (1989). North Carolina adopts its own rules. Nature **340**, 497.

Genetic Manipulation Regulations (1989). London, HMSO.

Gray, A.J. (1986). Do invading species have definable genetic characteristics? Philosophical Transactions of the Royal Society B, **314**, 655-674.

Groves, R.H. (1986) Plant invasions of Australia : an overview. In " Ecology of Biological Invasions : An Australian Perspective". (R.H. Groves and J.J. Burdon, eds), pp. 137-149. Australian Academy of Science and Cambridge University Press.

Harlan, J.R. (1985). The possible role of weed races in the evolution of cultivated plants. Euphytica **14**, 173-176.

Holm, L.G., Plucknett, D.L., Pancho, J.V. and Herberger, J.P. (1977). "The world's Worst Weeds". University Press of Hawaii, Honolulu.

OECD (1986). Recombinant DNA Safety Considerations. OECD, Paris.

Royal Commission on Environmental Pollution (1989). Thirteenth Report. The release of genetically engineered organisms to the environment. Cm 720. London, HMSO.

Simonsen, L. and Levin, B.R. (1988). Evaluating the risk of releasing genetically engineered organisms. TREE **3(4)** TIBTECH **6(4)**, S27-S30.

Sun, M. (1989). South Carolina blocks test of Rabies vaccine. Science **244**, 1535.

von Broembsen, S.L. (1989). Invasions of natural ecosystems by plant pathogens. In "Biological invasions, a global perspective" (J.A. Drake, H.A. Mooney, F. di Castri, R.H. Groves, F.J. Kruger, M. Rejmanek and M. Williamson, eds), pp. 77-83.

Williamson, M. (1988). Potential effects of recombinant DNA organisms on ecosystems and their components. TREE **3(4)** TIBTECH **6(4)**, S32-S35.

Williamson, M.H. and Brown, K.C. (1986). The analysis and modelling of British invasions. Philosophical Transactions of the Royal Society B, **314**, 505-522.

HERBICIDE RESISTANCE IN WEEDS AND CROPS, PROGRESS AND PROSPECTS

Baruch Rubin

Department of Field and Vegetable Crops, Faculty of Agriculture, The Hebrew University of Jerusalem, Rehovot 76100, Israel

Herbicide resistance has appeared in numerous plant species and to several herbicide classes throughout the world, particularly when monoculture, monoherbicide and minimum tillage are widely practised. Much research has been directed at the elucidation of the resistance mechanisms, their molecular bases and their implications for agriculture and the environment. These studies have significantly contributed to our basic knowledge on the physiological, biochemical and molecular mode of action of herbicides and their selectivity mechanisms. Altered herbicide target site, enhanced detoxification and sequestration of the herbicide away from its target site are the major reported resistance mechanisms. Multidisciplinary models have been developed to better understand the population dynamics processes involved in the evolution and spread of herbicide resistance, and to evaluate management strategies to overcome and delay its appearance. Reduction of selection pressure, through rational use of lower rates of herbicides, crop and herbicide rotations, mixtures of herbicides, safeners, and synergists are suggested to delay and combat resistance. Classical breeding, *in vitro* and *in vivo* selection and genetic engineering techniques have been employed to improve crop selectivity and develop herbicide-resistant crops. Proper utilization of resistant crops and awareness of the potential risks involved in their introduction may improve our ability to develop cost-effective and flexible weed management programmes.

INTRODUCTION

The large increase in reported cases of weeds resistant to different classes of herbicides, and from different parts of the world (LeBaron, these Proceedings; LeBaron and McFarland, 1990), may indicate that the problem is with us to stay. The rapid spread of herbicide resistance endangers the usefulness of valuable old and new herbicide classes, on which adequate crop production relies. It also increases the cost of weed control and poses a real threat to the environment.

387

In most cases, herbicide resistance is associated with high selection pressure imposed by an intensive agriculture where monoculture, mono-herbicide and minimum tillage practices are widely exercised. The repetitive use of high rates of the same residual herbicide (mostly triazines), along roadside, railways and fences, where cultivation is impractical or impossible, also led to the appearance of herbicide-resistant populations (Rubin *et al.*, 1985; Burnet *et al.*, these Proceedings).

Herbicide resistance has appeared not only when residual herbicides were heavily used, but also in cases where selection pressure is inflicted by the repeated use of herbicides like paraquat and diquat which lack soil activity (Fuerst and Vaughn, 1990; Powles and Howat, 1990; Dodge, these Proceedings; Matsunaka and Ito, these Proceedings).

Herbicide resistance – a worldwide problem

The number and distribution of herbicide-resistant weeds has increased dramatically in the last ten years (Holt and LeBaron, 1990; LeBaron and McFarland, 1990; LeBaron, these Proceedings). In their comprehensive survey, LeBaron and McFarland (1990) documented the worldwide distribution of the phenomenon and its implications for modern agriculture. There are more than 100 weed biotypes resistant to at least 15 classes of herbicides, in virtually all continents. It is difficult to state the real number of herbicide-resistant weed populations, as it is highly dependent on the intensity of the search (van Oorschot, these Proceedings). Furthermore, there is a clear decline in new reported cases of triazine resistance, presumably because of lack in public interest or greater awareness of the farmers, who have switched to other alternative herbicides (van Oorschot, these Proceedings). Nevertheless, new cases of weeds resistant to various herbicides have already appeared but not yet been reported (e.g. paraquat-resistant watergrass (*Commelina elegans* or *C. diffusa*) a dominant perennial weed in banana plantations in the Caribbean Islands which withstand paraquat at 15 kg ha^{-1} (M. Parham, pers. comm., 1990).

Nomenclature

There is much confusion in the terminology used in the literature, regarding the use of terms such as "tolerance", "resistance", "cross-resistance" and "multiple-resistance". Based on the definition suggested by Holt and LeBaron (1990), "tolerance" will be used in cases where weeds and crops have a low level of resistance which is dose dependent. "Resistance" will be used where weed populations have evolved mechanism(s) to withstand herbicide applied at field dosage or above.

Cross-resistance will be used to describe cases in which a weed population is inheritably resistant to two or more different chemicals which either act at the same target site (e.g. triazines, triazinones etc.), or are degraded by the same enzyme system (e.g. cytochrome P_{450} mixed function oxidases, glutathione-S-transferases etc.).

The term "multiple-resistance" will be used to describe cases where a weed population such as *Kochia scoparia* has evolved resistance due to two or more different mechanisms. In this species there is evidence of two altered targets, 32 KD binding protein in the chloroplasts for triazines and an insensitive ALS (acetolactate synthase) for sulfonylureas (Thill *et al.*, these Proceedings).

The author is aware that, in many cases of herbicide-resistance, the exact mechanism involved is not yet fully understood. This may require future changes in this proposed nomenclature.

MECHANISMS OF HERBICIDE RESISTANCE

Altered site of action

The occurrence of herbicide resistance in many weeds is often attributed to an inherited modification of the herbicide site of action. This is evidently true for most cases of resistance to the triazines, sulfonylureas and dinitroanilines.

Triazine resistance is generally associated with a point mutation of the chloroplast gene psbA, encoding for the PSII protein Q_B – the herbicide-binding site (see Trebst, these Proceedings, for details). In all higher plants that have been studied so far, this mutation involves a substitution of one amino acid at position 264 of the 32 KD protein, resulting in loss of the herbicide-protein affinity. This mutation leads also to an alteration of the electron transport mechanism, which is expressed as reduced quantum yield for whole chain electron transport (Arntzen *et al.*, 1982). Triazine-resistant weeds often exhibit cross-resistance to other herbicides having different chemical structures, such as the triazinones, pyridazinones, uracils and some cyanoacrylates, that bind to the same site. Conversely, these plants appear to remain sensitive to the urea herbicide diuron which is also a PSII inhibitor (Fuerst *et al.*, 1986; Yaacoby *et al.*, 1986; van Oorschot, these Proceedings).

Most weeds which have evolved resistance to the ALS inhibitors were controlled with the sulfonylurea herbicide, chlorsulfuron (Mallory-Smith *et al.*, 1990; Primiani *et al.*, 1990; Saari *et al.*, 1990; Thill *et al.*, these Proceedings). One exception is *Ixophorus unisetus*, which evolved resistance following repeated application for five years of the imidazolinone herbicide, imazapyr, in Costa Rica (B.E. Valverde, 1990, pers. comm.).

The mechanism of resistance appears to be due to an alteration of the gene(s) encoding for ALS (acetolactate synthase) enzyme. This enzyme, ALS or AHAS (acetohydroxy acid synthase), is the target for inhibition by five different chemical groups, namely the sulfonylureas, imidazolinones, triazolopyrimidines, sulfonylcarboximides and N-phthalyl-L-valine anilides (Shaner, these Proceedings). Hartnett *et al.* (1990 and these Proceedings), using comparative analyses of the ALS gene sequences from various resistant and wild-type organisms, have identified several mutations at different loci which involved substitution of various amino acids. These mutations lead to varying degrees of insensitivity of the ALS enzyme to sulfonylureas and a different pattern of cross-resistance to imidazolinone herbicides (see also Saxena and King, 1988). A similar diversity in the degree of resistance among the sulfonylureas and cross-resistance to imidazolinones has also been demonstrated at the whole plant level in transgenic crops as well as crops and weeds selected for resistance to ALS inhibitors (Gabard *et al.*, 1989; Mallory-Smith *et al.*, 1990; Newhouse *et al.*, 1990; Primiani *et al.*, 1990; Saari *et al.*, 1990; Shaner, these Proceedings; Thill *et al.*, these Proceedings). The diversity in response of plant populations to different ALS inhibitors is hard to predict and indicates that some of the imidazolinones may bind to a slightly different site on the ALS enzyme (Saxena and King, 1988; Shaner, these Proceedings).

An altered target site was also found in a biotype of *Eleusine indica* highly resistant to dinitroaniline herbicides. It was shown that an altered tubulin, the major constituent of the microtubule, is insensitive to all dinitroanilines and to amiprophosmethyl, as determined by lack of increase in mitotic index or disruption of spindle and cortical microtubules (Vaughn *et al.*, 1987). Vaughn and Vaughan (1990) have recently identified a novel form of ß-tubulin present in the resistant (R) biotype of *E. indica*. Based on the differential response of sensitive (S) and R cells to taxol, an agent which hyperstabilizes microtubules, Vaughn and Vaughan (1990, and these Proceedings) concluded that in R *Eleusine*, hyperstability caused by the novel ß-tubulin form, plays a major role in dinitroaniline resistance. It may also explain the observed increased sensitivity of R plants to the phenylcarbamate herbicides such as IPC and CIPC. The resistance mechanism of other reported dinitroaniline-resistant weeds including an intermediate biotype of *E. indica* (Vaughn *et al.*, 1990), *Amaranthus palmeri* (Gossett *et al.*, 1990), *Setaria viridis* (Morrison and Beckie, these Proceedings), has not yet been elucidated.

Another case of resistance to herbicide which might be (among other factors) due to an altered binding site involves the resistance of *Stellaria media* to mecoprop – an auxin-type herbicide (Lutman and Heath, 1990). Based on an explant stem elongation test (Barnwell and Cobb, 1989), and the mecoprop-induced proton efflux kinetics (Cobb *et al.*, these Proceedings), it was concluded that the R biotype is far less receptive to mecoprop than the S biotype. Although not ruling out the possible involvement of differential metabolism among the biotypes, the authors suggest that the differences in mecoprop binding may form the basis for mecoprop-resistance in this weed population. It will be of interest to examine if a similar mechanism also exists in other reported cases of weeds resistant to other auxin-type herbicides (Bourdot *et al.*, 1989; Popay *et al.*, these Proceedings).

Acetyl coenzyme A carboxylase (ACCase) inhibition was recently discovered to be the major mechanism by which aryloxyphenoxypropionate and cyclo-hexanedione herbicides affect grasses (see Rendina *et al.*, 1989 and Owen, these Proceedings for details). In spite of being applied several times a year in many field situations (e.g. for perennial grass control in broadleaf crops), the author is not aware of any reported cases of weed resistance evolving due to an altered ACCase (see also Powles *et al.*, 1990). In maize cell culture however, several resistant lines were selected in the presence of either sethoxydim or haloxyfop. In line – S2, the ACCase activity was less sensitive to inhibition by sethoxydim and haloxyfop, compared to unselected lines. In the other resistant line – H2, ACCase activity was five-fold higher than in the unselected line, but the enzyme was equally sensitive to inhibition by the herbicides (Gronwald *et al.*, 1989; Wyse *et al.*, these Proceedings). The nature of the change in the ACCase of line S2 is not yet known, but it was stable and the resistance was also exhibited in regenerated plants from the S2 line. One may wonder if the ease by which these resistant lines were selected in the laboratory is indicative of what may occur in the field in the future.

Similarly, there are no reports on the evolution of weeds resistant to norflurazon, a wide-spectrum bleaching herbicide which has a long residual activity in the soil. However, mutants of the cyanobacterium *Synechococcus*, resistant to norflurazon, were generated and selected in the laboratory (Linden *et al.*, 1990), which exhibited less sensitive target site - phytoene desaturase. Some mutants have also exhibited cross-resistance to other bleaching herbicides. Although it is hard to conclude from a chemically-mutated cyanobacterium to higher plants, it may indicate the vulnerability of this residual herbicide to evolve resistance in the field.

It is interesting to note that unlike in transgenic plants, there are no reports to date on over-expression of the target site as a resistance mechanism in any of the field evolved resistant weeds.

Enhanced detoxification

As mentioned above, differential metabolism is a major mechanism of plant selectivity to herbicides. Several weed biotypes have recently been reported to evolve resistance to herbicides due to a capacity to rapidly degrade and/or conjugate the toxic compound, hence forming less or non-toxic products, as occurs in many resistant crops.

Gronwald *et al.* (1989) have shown that *Abutilon theophrasti* from Maryland is resistant to atrazine due to an enhanced capacity to detoxify the herbicide via glutathione conjugation, as occurs in maize (see Lamoroureux, these Proceedings). Although no details were given, enhanced metabolism was also postulated for a triazine-resistant biotype of *Echinochloa crus-galli* from France (LeBaron and McFarland, 1990).

In both multiple-resistant black-grass and annual ryegrass, no alteration of the target site nor differential uptake and translocation were observed between R and S biotypes (Kemp, pers. comm.; Powles *et al.*, 1990). Kemp and his co-workers (1990; and these Proceedings) have clearly shown that chlorotoluron is more rapidly metabolised (oxidation followed by rapid conjugation) in the Peldon (resistant) black-grass, as compared to the sensitive biotype (Rothamsted). The possible involvement of microsomal cytochrome P_{450} mono-oxygenases in various oxidative reactions leading to detoxification of several herbicides in plants was recently reviewed (see Jones and Caseley, 1989; Jones, these Proceedings). These studies suggest that multiple forms of cytochrome P_{450} are present or can be induced in higher plants. Based on the reaction product(s), requirement of NADPH and O_2, inhibition by CO, use of specific inhibitors and partial purification of the microsomal fraction, it was concluded that in tolerant plants, cytochrome P_{450} mono-oxygenases are catalyzing the initial detoxification reactions of dimethylureas (Frear *et al.*, 1969; Cabanne *et al.*, 1987; Mougin *et al.*, these Proceedings) and diclofop (McFadden *et al.*, 1989; Zimmerlin and Durst, these Proceedings).

Similarly, using known P_{450} inhibitors such as ABT (1-aminobenzotriazole) and PB (piperonyl butoxide), in combination with the respective herbicides, resulted in a significant reduction in R plant growth (Kemp *et al.*, 1988; Kemp *et al.*, 1990; Powles *et al.*, 1990), and also inhibited chlortoluron degradation in R blackgrass (Kemp *et al.*, 1990). Furthermore, Jones and Caseley (1989) have found, in a cell culture derived from the resistant Peldon blackgrass, increased levels of P_{450} present in the microsomes isolated from the R plants, which associated with increased rate of degradation of chlorotoluron.

Although only very few studies directly associate enhanced activity of monooxygenases with resistance, when considering the pattern of cross-resistance in both *A. myosuroides* and *L. rigidum* to other herbicides having different chemistry and mode of action, this degradative mechanism is considered to be sufficient to account for the observed multiple-resistance.

Sequestration and compartmentation

The possible involvement of sequestration and compartmentation of herbicides or their phytotoxic metabolites as mechanisms of resistance is discussed in detail by Coupland (these Proceedings). Unfortunately, little information is available which relates herbicide resistance in weeds solely to one of these processes.

It was shown (Coupland *et al.*, 1990) that mecoprop-resistant *Stellaria media* metabolises the herbicide to conjugates significantly more than the susceptible plant. As ester conjugates predominate, this implies that vacuolar compartmentation may be responsible for the recovery of the resistant plants (Coupland, these Proceedings).

Sequestration of paraquat, thus excluding it from the site of action in the chloroplast, either by its binding to unidentified cellular components, or by its storage in the vacuole, was strongly advocated by several workers as the major mechanism for resistance (for details see Fuerst and Vaughn, 1990; Coupland, these Proceedings; Dodge, these Proceedings). This theory is based on the limited mobility of paraquat observed in resistant biotypes of *Conyza bonariensis* (Fuerst *et al.*, 1985), *Erigeron philadelphicus* and *E. canadensis* (Tanaka *et al.*, 1986) and *Hordeum glaucum* (Bishop *et al.*, 1987). It is also based on the rapid recovery of chloroplast functions of the R plants, such as CO_2 fixation and chlorophyll fluorescence quenching, indicating removal of the herbicide from its site of action (Fuerst and Vaughn, 1990).

Shaaltiel and Gressel (1986, 1987) however suggested that the resistance mechanism is associated with the chloroplast itself. They found in "class A" intact chloroplasts isolated from a paraquat-resistant biotype of *Conyza bonariensis*, enhanced activities of enzymes which are able to detoxify the oxygen radicals generated in the chloroplast by the herbicide. The activities of these enzymes, namely, superoxide dismutase, ascorbate peroxidase and glutathione reductase (referred to as "Halliwell-Asada system"), were elevated in the R chloroplasts by 1.6, 2.5 and 2.9 times, respectively, compared with the activities found in the S chloroplasts. Similar increases in these detoxifying enzymes were found in paraquat-resistant *Erigeron canadensis* (Matsunaka and Ito, these Proceedings).

Unfortunately until now, only a small number of paraquat-resistant weeds have been thoroughly studied, and much more research is needed to elucidate the relative importance of either the compartmentation or the detoxification theory to explain the resistance to paraquat.

Although much progress has been made during the last decade in understanding the physiological and biochemical bases of herbicide action, selectivity and resistance further research is required, as there is a relatively large number of resistant weeds whose mechanisms of resistance have not yet been elucidated or even studied. From the data accumulated so far we know that the same species can evolve resistance to different herbicides in different locations, and also that resistance to the same herbicide can be based on different mechanisms in different species.

MODELLING HERBICIDE RESISTANCE

Purpose of modelling

Several attempts have been made to develop multi-disciplinary models, which integrate the genetic, ecological and physiological processes involved in the evolution of a herbicide resistant population (cf. Maxwell *et al.*, 1990; Gressel, these Proceedings; Radosevich *et al.*, these Proceedings). These models, based on detailed analyses of case histories, highlight the relative importance of factors controlling the development and spread of herbicide resistance. They may indicate how to modify our current practices in order to overcome the already existing resistance problems, and may serve as a predictive tool for evaluation of new management strategies to prevent, or at least delay, the appearance of others. These models also provide invaluable information on the detrimental effects imposed by monoculture and monoherbicide regimes, and lack of cultivation on weed populations.

In addition, the models are helpful in indicating the type and importance of missing data for future research. There is a general agreement that we are lacking in our knowledge and interpretation of weed/crop population ecology and its interaction with the current control strategies as will be outlined below.

Gressel's model (these Proceedings) estimates the increase in the proportion of resistant plants in a population over time (years of treatment) in a population, based on the initial frequency of the resistance genotype (R) in the population, its relative ecological fitness (when the herbicide is absent), the selection pressure imposed by the herbicide (effective kill, persistence in soil) and the longevity of the weed's seed bank in soil. Another model (Maxwell *et al.*, 1990, Radosevich *et al.*, these Proceedings), stresses the importance of two major processes that determine the dynamics of herbicide resistance. First, the processes such as survivorship, fecundity of pollen and seeds, that influence the fitness of R biotypes relative to that of the S biotype and crop species under competition in a mixed population. Second, the processes that contribute to gene flow in space and time, that alter the frequencies of R and S alleles in a population, such as immigration of pollen and seeds, seed dormancy, and type of breeding.

In spite of the differences between models in emphasising the relative importance of specific biological processes, their predictions are similar and conclude that selection pressure must be reduced by using lower rates of herbicide. However, they differ considerably in their strategy tactics and management approaches, as will be discussed later.

Although the present models provide useful and important information essential for a rational and successful approach to combat or reduce the problem, there are several discrepancies which call for more research in order to refine them.

Fitness and resistance traits

It was stated that reduced vigour and ecological fitness is an intrinsic feature of the herbicide resistance trait, and may be considered as a "pleiotropic effect" as the "cost" or "penalty" for resistance (Radosevich and Holt, 1982; Gressel, 1985). The relative ecological fitness of the R biotypes is highest when the herbicide (or a mixture of herbicides with the same mode of action), with high "effective kill" of the S plants, is present. When the selector (e.g. the herbicide) is removed however, by rotating to another crop or herbicide, the demography of the weed population will inevitably be different, the inferior R type will be gradually replaced by the more fit S plants (see Gressel, these Proceedings; Radosevich *et al.*, these Proceedings). Indeed, in the case of triazine-resistant plants, due to the alteration in the herbicide binding site, a less efficient electron transport is considered to be the reason for the lower photosynthetic potential, and thus, reduced vigour and lower overall ecological fitness (Conard and Radosevich, 1979; Radosevich and Holt, 1982; Ahrens and Stoller, 1983; Ort *et al.*, 1983; Stowe and Holt, 1988; Holt, 1990; Benyamini *et al.*, these Proceedings). Moreover, LeBaron (these Proceedings) concluded that "the lack of fitness in most triazine-resistant weeds is a very important reason why they have been fairly easily controlled, and why there have been few problems of cross-resistance or multiple-resistance where both triazine and other types of herbicides have been used repeatedly together".

Lower ecological and physiological fitness was also reported in paraquat-resistant biotypes of *Erigeron philadelphicus* and *E. canadensis* from Japan (Matsunaka and Ito, these Proceedings), and *Hordeum glaucum* from Australia (Powles and Howat, 1990). Johnston and Faulkner (these Proceedings) suggested that in perennial ryegrass, some or all of the genes controlling paraquat tolerance have an adverse effect on herbage yield performance, whereas those for dalapon tolerance do not.

The generalization of this phenomenon must be questioned, as several studies (Rubin *et al.*, 1985; Jansen *et al.*, 1986; Schonfeld *et al.*, 1987; Chauvel and Gasquez, these Proceedings) have shown that photosynthetic potential and growth parameters of triazine-resistant populations are at least comparable to those found in S populations, both under competitive and noncompetitive conditions. Similarly, Gasquez (these Proceedings) has found very little difference in fitness among three isogenic biotypes of triazine-resistant *Chenopodium album*.

There are indications, however, that some nuclear control of the fitness also exists in triazine-resistant weeds (Stowe and Holt, 1988; Gasquez, these Proceedings), which means that fitness is difficult to predict and has to be studied separately in each case after the evolution of the resistance trait. Furthermore, no apparent reduction in growth was observed in chlorotoluron-resistant *Alopecurus myosuroides* (Chauvel and Gasquez, these Proceedings; Moss and Cussans, these Proceedings), among biotypes of *Eleusine indica* resistant to dinitroaniline herbicides (Valverde *et al.*, 1988), and in mecoprop-resistant and -susceptible *Stellaria media* (P.J.W. Lutman, 1990, pers. comm.). In addition, no adverse agronomic effects have been observed in various selected or transgenic crop plants engineered for herbicide resistance (Freyssinet *et al.*, 1989; Hartnett *et al.*, 1990; these Proceedings; Botterman *et al.*, these Proceedings; Shaner, these Proceedings).

More research is needed to elucidate the relationship between the resistance trait and ecological fitness, particularly studies conducted under real field conditions, e.g. competition of the R biotypes with the S plants, other weeds and crops.

Selection pressure

As mentioned above, application of one herbicide with long lasting soil activity (e.g. atrazine) every season, or repetitive application of a post-emergence herbicide (e.g. paraquat) several times a year, may result in an increase in selection pressure, thus imposing or facilitating the appearance of resistance. Careful examination of the literature, however, shows that it is quite difficult to predict or to quantify the level of selection pressure which eventually leads to resistance.

In both reported cases of multiple-resistance that have appeared in arable crops in the UK (Moss and Cussans, these Proceedings) and in Australia (Heap and Knight, 1982), no apparent strong selection pressure was imposed. In the case of black-grass (*A. myosuroides*), the challenging herbicides were not usually effective enough to create a high level of selection pressure. In Australia, resistant annual ryegrass (*L. rigidum*) has appeared within only four years of use of diclofop-methyl which has practically no soil activity, and is usually applied only once a year. In both cases it was impossible to link the occurrence of resistance solely with the intensity and amount of herbicide used, indicating that other factors may be involved in the development of resistance. There are some similarities between black-grass and ryegrass, which may increase their vulnerability to acquire resistance to various herbicides more than other weed species. For example, both species evolved resistance to tri- azines in Israel (Rubin *et al.*, 1985), while *A. myosuroides* populations developed resistance to urea-type wheat herbicides in Germany (Niemann and Pestemer, 1984), and *L. rigidum* resistance to amitrole, triazines and urea herbicides was found in Western Australia (Burnet *et al.*, these Proceedings). Both weeds are out-crossing, have high reproductive capacity and relatively short-lived seeds.

It was assumed that either there may be no genes in weeds for 2,4-D resistance, or due to its low selection pressure and rapid degradation in soil there is little prospect for such a resistance to evolve, in spite of its continuous use for many years (Gressel, these Proceedings). Recently however, phenoxy-herbicide resistance was reported in populations of *Ranunculus acris* and *Carduus nutans* from New Zealand, following 15 to 30 annual applications of either MCPA or 2,4-D (Bourdot *et al.*, 1989; Popay *et al.*, these Proceedings). Similarly, no apparent selection pressure was imposed in the evolution of mecoprop-resistant populations of *Stellaria media*, as the herbicide usage was not intensive, and application of mecoprop alone was not frequent (Lutman and Heath, 1990).

One may question also the intensity of the selection pressure involved in the development of paraquat-resistant populations of *Epilobium ciliatum* in hop gardens in England, as paraquat was applied only once or twice a year (Clay, 1989).

A better understanding of the complex interactions between the herbicide mode of action, weed ecology and genetics and resistance mechanism are crucial for more precise prediction and quantification of selection pressure.

Initial frequency of the resistance trait

There is ample evidence that due to polymorphism in plant populations, resistant genotypes are present at varying frequencies before any exposure to herbicides. Even in the more uniform crop plants, genetic variation exists as evident by differential response of crop varieties to herbicides (cf. Snape *et al.*, these Proceedings). It is very difficult, however, to predict the initial frequency of the genotype resistant to a given herbicide in a weed population.

The initial frequency of resistant genotypes in a population varies with plant species and the type of resistance. This frequency might also be increased by the effect of nuclear gene(s) coding for a "plastom mutator" as suggested by Arntzen and Duesing (1983) and Gasquez (these Proceedings). The initial frequency of triazine resistance (which comprises the majority of resistance "cases") has been estimated to be 10^{-10} to 10^{-20}, compared with the higher initial frequency (10^{-6}) of ALS-type resistance. These differences accounted for the relatively "delayed" appearance of triazine resistance (10 years of continuous use), and the rapid evolution of ALS-type resistant (3-4 years) (Gressel, these Proceedings).

Gasquez (these Proceedings) and Darmency and Gasquez (1990) in their studies of the population dynamics of *Chenopodium album* in France have shown that certain natural populations (Sp) contain a high frequency (10^{-4} to 10^{-3}) of an intermediate biotype (I). This intermediate biotype, although possessing a mutated *psbA* gene similar to that of R plant, can survive only very low doses of atrazine, indicating a nuclear control of the expression of the chloroplastic gene. After treatment with a sublethal dose of atrazine or other systemic pesticides, the progeny of the I biotype are completely resistant (R), and can survive very high doses of atrazine. The high frequency of I plants could account for the rapid spread of R *C. album* in meadows recently turned to maize fields in France. It is of great importance to know whether this phenomenon is a universal one or whether it is unique to these French populations.

Supporting this theory are the reports on various levels of a plastidic resistance found in populations of *Solanum nigrum* and *Amaranthus cruentus* from Italy, *Poa annua* from France, and *C. album* from Hungary (see Gasquez, these Proceedings).

Thus, the initial frequency of resistant individuals in a population to a given herbicide, is difficult to predict and might be estimated for each herbicide separately before its launch. This is highly unrealistic. An interesting and relatively simple approach to determine the proportion of chlorsulfuron-resistant individuals in alfalfa was conducted by Stannard and Fay (1987), who found 15 out of 20 million seedlings that survived the herbicide, of which several have shown high tolerance. This frequency agrees with previous estimations conducted in cell culture.

HOW CAN HERBICIDE-RESISTANCE BE MANAGED?

The fact that herbicide-resistance is predominantly associated with monocultural crops, intensive use of herbicides and reduced cultivation or even no-tillage practices, strongly indicates that in any attempt to delay or prevent resistance, measures must be taken to modify these practices. Indeed, both models (Gressel, these Poceedings; Radosevich *et al.*, these Proceedings), emphasise the importance of reducing the selection pressure imposed by the current practices. Several approaches have been proposed and will be discussed in the following sections.

Rational use of herbicides

a) Alternative herbicides. The simplest response of a farmer, upon detection of a reduced performance or failure, is to raise the herbicide dose and/or to switch to alternative products which are usually available, while maintaining the normal practices. Such an approach might solve the problem temporarily as suggested by Stephenson *et al.* (1990), but it is bound to fail in the long run. This was the case with insecticide-resistance, leading to the "pest treadmill" phenomenon (Ruscoe, 1987). It seems that many weed scientists, and to some extent the industry, tend to fall into this trap (see van Oorschot, these Proceedings). This is best demonstrated by Lehoczki *et al.* (these Proceedings) who found that attempts to control plastid-type triazine-resistant *Amaranthus retroflexus* and *C. canadensis* in vineyards with diuron, resulted within 3 to 4 years in a new type of non-plastidic resistance to diuron. Moreover, switching to another herbicide which controls the resistant biotype, even for several years, does not mean that the resistant population will dissipate (Thill *et al.*, these Proceedings).

b) Herbicide mixtures. Mixing herbicides having different modes of action as a primary preventive measure both to avoid (or delay) and to combat herbicide-resistance has been suggested in numerous studies (Stephenson *et al.*, 1990; Gressel, these Proceedings; Hartnett *et al.*, these Proceedings; LeBaron, these Proceedings; Thill *et al.*, these Proceedings). Using such mixtures will obviously lower the initial frequency of the R type in a compounding rate and broaden the weed spectrum controlled, but it may also result in elevated selection pressure (see Radosevich *et al.*, these Proceedings). Apart from being less cost-effective for the farmer, such mixtures may not always be available. The feasibility of this approach is further diminished by cross- and multiple-resistance, a major problem which complicates the battle against herbicide resistance. Weeds may evolve resistance not only to the challenging chemical but also to herbicides from other chemical groups and modes of action to which they were not exposed before. Generally, plants which are resistant to a particular herbicide due to an altered binding site, are also resistant to chemicals which act at the same site. For example, triazine-resistant weeds are cross-resistant to triazinones, pyridazinones and uracils, but remain sensitive to the urea herbicide diuron (Fuerst *et al.*, 1986; Yaacoby *et al.*, 1986; van Oorschot, these Proceedings). Similarly, weeds and crop plants which evolved resistance to sulfonylurea herbicides in the field or selected in the laboratory, are cross-resistant in varying degrees to some, but not to all ALS inhibitor herbicides belonging to the imidazolinone, sulfonanilide and triazolopyrimidine groups (Saxena and King, 1988; Gabard *et al.*, 1989; Saari *et al.*, 1990; Hartnett *et al.*, 1990; these Proceedings; Shaner, these Proceedings; Thill *et al.*, these Proceedings).

Triazine-resistant grass weeds however, were found to be resistant also at the chloroplast level to urea type herbicide - methabenzthiazuron, cyanoacrylates and bis-phenylcarbamates as well as having increased tolerance at the whole plant level to diclofop-methyl (Rubin *et al.*, 1985; Yaacoby *et al.*, 1986; Yaacoby, 1989). Similarly, triazine-resistant *Conyza canadensis* exhibited cross-resistance to chlorbromuron and metobromuron (Lehoczki *et al.*, these Proceedings).

Several weed populations treated for many years with tank-mixes of herbicides differing in their mode of action, mechanism of degradation or residual activity, evolved resistance to both chemicals. Populations of *C. canadensis* (Pölös *et al.*, 1988), *Epilobium ciliatum* (Clay, 1989) and *Poa annua* (Clay and Hadleigh, these Proceedings), became resistant to both triazines and paraquat (and to some extent also to diquat). Similarly, annual treatment with an atrazine-amitrole mixture for more than 10 years, resulted in selection of a ryegrass population which is resistant not only to both herbicides, but also to triazinones and phenyl-ureas (Powles and Howat, 1990).

In addition, different populations of the same weed species, may evolve several types of resistance, as well as different patterns of cross-resistance. Clay (1989) found populations of *E. ciliatum* in hop gardens in England which are either triazine-resistant, paraquat-resistant (but triazine sensitive), resistant to both herbicides, or sensitive to both classes. Similarly, Thill *et al.* (these Proceedings) reports that there are biotypes of *Kochia scoparia* which are either resistant to triazines only, sulfonylurea only, triazine and sulfonylurea, or sensitive to both herbicide classes.

The most complex cases which may render ineffectual any attempt to combat resistance by herbicide mixtures are the "multiple-resistant" weeds (Moss, 1987, 1990; Heap, these Proceedings; Holtum *et al.*, these Proceedings; Powles *et al.*, these Proceedings). In these weeds, the relative level of resistance and the spectrum of herbicides to which resistance is exhibited may vary among populations, and not always associated with the previous herbicide exposure. This substantial diversity in response of weeds to herbicides must be considered when developing management practices. Thus, before any recommendations of herbicide mixtures are made, the user must be acquainted with the weed populations and their response to a wide range of herbicides. This information must be generated by the scientific community and effectively transferred to the user.

c) Use of herbicide-synergists and safeners. Effective usage of synergists and safeners in weed management is based upon a successful manipulation of the physiological and biochemical response of plants, weed and crop, to a given herbicide (see Hatzios and Hoagland, 1989; Hatzios, these Proceedings; Kemp and Caseley, these Proceedings).

By selectively protecting the crop, safeners permit the use of otherwise nonselective or less selective chemicals, hence extending the spectrum of weeds controlled. The recent introduction of the combination of several graminicides with safeners such as fenoxaprop-ethyl with the safener fenchlorazole for grass weed control in cereals (Bieringer *et al.*, 1989), not only increases the spectrum of herbicides available, but also provides control of triazine-resistant grass weeds including *Phalaris paradoxa* biotypes with cross-tolerance to diclofop (Rubin, unpublished data). Unfortunately, several populations of *A. myosuroides* resistant to chlorotoluron were also found to be less sensitive to fenoxaprop-ethyl (Moss and Rubin, unpublished data).

Synergistic effects may be attributed to modification of either the uptake, translocation or metabolism of the herbicide (Kemp and Caseley, these Proceedings). Successful implementation of synergy in resistance management requires better understanding of the mechanism of herbicide action and resistance. In cases where resistance is based on enzymatic enhanced detoxification of the herbicide, identification and characterization of the enzyme(s) involved, in terms of specific and alternative substrates, is critical.

The possible use of synergistic combinations as a tool for managing herbicide-resistance is described in detail elsewhere (Kemp *et al.*, 1988; Varsano and Rubin, 1988; Caseley *et al.*, these Proceedings; Kemp and Caseley, these Proceedings).

Application of known cytochrome P_{450} inhibitors such as piperonyl butoxide (PB) and 1-aminobenzotriazole (ABT) in combination with several herbicides improved their efficacy in controlling resistant weeds in a synergistic manner (Varsano, 1987; Varsano and Rubin, 1988; Kemp *et al.*, 1988, 1990; Powles *et al.*, 1990).

Varsano and Rubin (1988) suggested that the observed synergy of piperonyl butoxide with atrazine, terbutryne and methabenz-thiazuron in triazine-resistant ryegrass, in whole plant and at the PSII electron transport levels, might be due to a rapid and light-dependent damage caused by the combination to membrane integrity.

Tridiphane is at present the only commercially available herbicide-synergist, which selectively enhances the control of several grass weeds by herbicides such as triazines (Boydston and Slife, 1987), alachlor and EPTC (Ezra *et al.*, 1985) and propanil (Street and Snipe, 1989). In grass weeds, tridiphane alters atrazine uptake and metabolism by inhibiting the glutathione S-transferase (GST) mediated conjugation of the herbicide with glutathione (Boydston and Slife, 1986; Lamoureux and Rusness, 1986). The control of a chlorotoluron-resistant biotype of black-grass was synergistically enhanced by a combination of tridiphane with chlortoluron and isoproturon (Caseley *et al.*, these Proceedings). These rather unexpected effects might be attributed to the inhibitory effect of tridiphane on cytochrome P_{450} isozymes as found in mouse hepatic microsomes (Moreland *et al.*, 1989). Tridiphane also synergises atrazine phytotoxicity in some broadleaved weeds such as *Abutilon theophrasti* (Ehr and Burroughs, 1986), but contributed little to the control of triazine-resistant *C. album* (Myers and Harvey, 1986).

Although the practical implementation of synergists in weed management programs is very limited at present, it may allow a reduction in herbicide rate which is an economical and environmental advantage. Using synergists and safeners, however, is expected to maintain or even improve the effective kill of weeds, and therefore may not reduce the selection pressure.

e) Using herbicides at the threshold level. Based on their simulation models Radosevich and his co-workers (Maxwell *et al.*, 1990; Roush *et al.*, 1990; Radosevich *et al.*, these Proceedings) have suggested a different approach. The authors advocate the importance of optimizing herbicide input to the economical threshold level (see Cussans *et al.*, 1987), which will be more cost-effective and will also lower the selection pressure. Reduction of the herbicide rate and deliberately leaving some weeds within the field and along its margins, may allow the S plants to compete with the R plants, further diluting their proportion of the weed population through fitness and gene flow of seeds and pollen.

This approach requires much more interdisciplinary and long-term experimental work, conducted under various field condition, to be validated. Moreover, even if proved correct, this approach will require a great deal of educational effort amongst weed scientists, extension workers and farmers before being fully implemented.

Crop and herbicide rotations and cultivation

Various aspects and implications of crop rotation to avoid monoherbicide culture systems and delay resistance are discussed in depth elsewhere (see Gressel, these Proceedings). Rational crop rotation is valuable in several ways. Many serious weeds are strongly associated with specific crops, because of seedbed requirement, patterns of growth etc. Crop rotation reduces the intrinsic success of such weeds and also imposes the use of alternative herbicides. Crop rotation may also result in reduction of the relative fitness of the R biotype by using more competitive crops and different patterns of cultivation. Unfortunately, most farmers are reluctant to change their cropping system, as in many cases alternative crops and cultivation are not feasible due to man power and equipment specialisation, edaphic, climatic and economical reasons (Parochetti *et al.*, 1982).

One should remember also, that the required number of "years-off" of a given crop and/or herbicide, is highly dependent on the size, longevity and flux of the R and S weed seed-banks and the interaction with cultivation method (Watson *et al.*, 1987). For instance, in spite of the relatively short persistence of *A. myosuroides* seeds (see Moss and Cussans, these Proceedings), long term stability of its distribution in a cereal field was observed over ten years, despite the use of herbicides and grass breaks in the rotation (Wilson and Brain, 1990). Moreover, in minimum-tillage practice, rotation of winter wheat with oil-seed rape (where good control of the weed was achieved) once every fourth year, was not enough to reduce the infestation level in the successive years. Similarly, sulfonylurea-resistant *Lactuca serriola* was still present in the majority of fields in which it was originally detected, in spite of three years' treatment with non-AHAS herbicides (Thill *et al.*, these Proceedings).

Minimum- or no-tillage systems are now widely practised throughout the world for soil, water and energy conservation. In many crops, cultivation, particularly ploughing, has been reduced dramatically, which has resulted in less weed seed burial, thus the farmer has to rely more on chemical means. In order to maintain static weed populations, the prevalent higher infestation will require higher herbicide doses to achieve effective kill (Cussans *et al.*, 1987), which in turn will increase the selection pressure. Moreover, lack of ploughing maximises the proportion of the weed population derived from seeds shed in the previous crop (after being further selected for resistance), and minimises the probability of back-crossing with earlier unselected generations (Moss and Cussans, these Proceedings). Minimum tillage in small grain monoculture in the UK has been associated with straw burning, which may result in development of an adsorptive layer on the soil surface. This adsorptive layer reduces the activity of soil-acting herbicides, which encourages the farmer to apply higher rates of herbicides (Moss and Cussans, these Proceedings).

Rotating cereal crops with row crops, may allow inter-row cultivation which is a cost-effective means of controlling emerging weed seedlings. Lack of interrow cultivation in maize fields in south-east Ontario was associated with high occurrence of triazine-resistant weeds (Stephenson *et al.*, 1990). Unfortunately, farmers are reluctant to re-adopt mechanical control methods.

Management programmes

We have to divide our strategies against herbicide resistance into two different categories, both of them require awareness by the public of the worldwide magnitude of the problem and a willingness to adopt the suggested measures.

a) Preventing and delaying resistance. This goal is attainable, but it requires careful analysis of the current agronomic practices, and readiness to modify them when necessary. This examination may prove fruitful not only in preventing evolution of herbicide-resistant weed biotypes, but also to reduce the possible detrimental impact of the chemicals on the soil and environment. We must be prepared to develop and adopt biological control methods, and to incorporate them into the weed management programmes. Based on the currently available information regarding case histories, the predictive models, and cooperation with the industry (Anon., 1990), we are now well equipped to employ preventive measures through education and specific label recommendations which must reach the end user. Practices which emphasize the importance of herbicide usage at the economical threshold and at the proper timing, rotating crops and herbicides (even mixtures thereof), introducing more mechanical control, and preventing weed seed dispersal, are advised. Programmes should be designed on a local basis, perhaps even "tailor made" for a specific farm. In the long run, most of these measures will prove to be more cost-effective and inexpensive compared with the cost of combating resistance after its evolution and spread. Obviously, it requires an integrated educational effort among all parties involved - scientists, extension agronomists, industry - and above all, farmers.

b) Controlling herbicide-resistant weeds. There is no general way to combat herbicide resistance which will fit all possible circumstances.

Nevertheless, early identification, containment and eradication of resistant individual plants are of vital importance before they spread further. To achieve these goals, rapid and reliable methods for identifying resistant plants must be developed, as was done in the case of triazine-resistant weeds (see van Oorschot, these Proceedings). Recruiting the already existing services of the Pest Diagnostic and Advisory Clinics for this purpose is now routine in Ontario (Stephenson *et al.*, 1990).

A thorough investigation and analysis of all aspects involved in the evolution of each particular case should be performed. It is best illustrated by the evolution of herbicide-resistant weeds in France (Gasquez, these Proceedings), Ontario, Canada (Stephenson *et al.*, 1990), and Australia (Powles and Howat, 1990). Such analyses are imperative before any specific management guidelines can be introduced.

Detailed integrated management systems either to minimize or to overcome confirmed herbicide resistance should be proposed by a joint effort of the agro-chemical industry, weed scientists, regulation authorities and extension personnel, who have a good knowledge of the prevailing conditions *in situ*. These programmes must be, above all, feasible and practical to be adopted by farmers. Integrated programmes were proposed in Australia (Powles and Howat, 1990), which include practical advice on crop alternation, rational choice of herbicides and cultural methods. Although it is too early to assess the long-term success of this approach, similar attempts are necessary in each country.

INTRODUCTION OF HERBICIDE-RESISTANT CROPS: BENEFITS AND LIMITATIONS

The impressive progress in introducing herbicide-resistance traits into crops using breeding, biotechnological and genetic-engineering techniques was the subject of several recent reviews (Botterman, 1989; Mazur and Falco, 1989; Padgette *et al.*, 1989; Hartnett *et al.*, 1990; these Proceedings; Bright, these Proceedings; Shaner, these Proceedings). These achievements were greatly aided by the progress made in understanding the genetic, biochemical and physiological bases of the herbicide's mode of action and selectivity mechanisms, and the proficiency developed in regenerating plants from cultured cells and tissues.

The economical value of herbicides in the market (for details see LeBaron and McFarland, 1990; Bright, these Proceedings) and the cost and complications in registration of new compounds, have encouraged both academic and private institutes to establish research and development programmes in this area (Mazur and Falco, 1989; Bright, these Proceedings).

Various methods of incorporation of herbicide-resistance traits into beneficial crops have been employed with varying degrees of success.

Selection and breeding

Genetical analyses of the differential response to herbicides of grass crop and weed species have shown that genes for herbicide-tolerance may have evolved before the domestication of the cultivated crops (Snape *et al.*, these Proceedings). Indeed, numerous studies have shown that inter- and intra-specific variation in response to herbicides at the whole plant and cellular levels exist at low frequency. Exposure of seeds and cell cultures to mutagens was also successfully used in order to increase the frequency (Pinthus *et al.*, 1972; Haughn and Somerville, 1986; Mazur and Falco, 1989). Polymorphism in response to herbicides has been exploited for selection of those individual plants exhibiting phenotypic and/or genotypic tolerance for further breeding of herbicide-resistant crops. At present, several cultivars of amenity and agricultural grasses with resistance to paraquat, dalapon and aminotriazole are registered, and glyphosate-resistant types are under development (Johnston and Faulkner, these Proceedings), as well as sulfonylurea-resistant soybean cultivars with an altered ALS (Mazur and Falco, 1989).

The possible uses of cell cultures for selecting herbicide-resistant strains have been reviewed elsewhere (e.g. Meredith and Carlson, 1982). Marshall (these Proceedings) reviewed the currently available *in vitro* methods for selection for herbicide resistance at the tissue, cell, protoplast, anther and microspore levels, as well as the advantages and drawbacks that may be involved in exploiting such variation for herbicide-resistance. The potential of these methods is well demonstrated in the recent selections of maize mutants resistant to cyclohexanedione and aryloxyphenoxy propionate and imidazolinone herbicides in tissue culture, with the resistance expressed in regenerated plants and their progeny (Gronwald *et al.*, 1989; Newhouse *et al.*, 1990; Wyse *et al.*, these Proceedings). In addition, the significant progress accomplished with *in vitro* techniques, particularly in protoplast fusion, microinjection and regeneration of plants from culture, played an important role in making genetic engineering methods feasible.

Traditional sexual crosses have been employed to transfer the maternally inherited atrazine-resistance trait from *Brassica campestris* to valuable *Brassica* crops such as canola, rutabaga, oilseed rape and broccoli (Beversdorf *et al.*, 1980; Souza Machado, 1982; Souza Machado and Hume, 1987). In spite of the significant reduction in yield and quality, the triazine-resistant canola cultivars are grown in Canada in locations where weeds (particularly cruciferous weeds) are difficult or expensive to control by other means. A similar approach was used to transfer triazine-resistance from green foxtail (*Setaria viridis*) to the related crop foxtail millet (*S. italica*) (Darmency and Pernes, 1985).

Gene transfer

A more thorough understanding of the biochemical and genetical bases of herbicide mode of action in microorganisms and higher plants at the molecular level has led to an impressive advancement in transferring resistant traits to valuable crops. Three approaches have been employed for this purpose.

a) Introduction of less sensitive target sites. Mutant ALS genes encoding for less-sensitive ALS enzyme were identified and introduced by genetic-engineering techniques to crop species such as tobacco, tomato, sugar beet, oilseed rape, alfalfa, lettuce and melon (Mazur and Falco, 1989; Hartnett *et al.*, 1990; these Proceedings). These transgenic plants showed varying degrees of resistance to sulfonylurea herbicides and cross-resistance to other ALS inhibiting herbicides, depending upon the site and type of the mutation. Several transgenic plants were shown to be resistant in glasshouse and field trials.

Similarly, glyphosate-resistant plants were generated by transferring mutant *aroA* genes encoding for a less sensitive EPSPS enzyme to valuable crops (see Padgette *et al.*, 1989 for recent review). Successful targeting of the mutated EPSPS enzyme to the chloroplast, where the entire shikimate cycle is located was found to increase the resistance level as compared to cytosol-targeted mutant enzyme.

b) Overexpression of target site. Glyphosate-resistant transgenic plants were produced by transferring and expressing *aroA* genes from bacteria and plants, which were shown to overproduce EPSPS by approximately 20-fold due to gene amplification. Some of these regenerated transgenic plants tolerated above use levels of glyphosate (Padgette *et al.*, 1989).

Over-production of ACCase was suggested as the basis for the relatively small increase in tolerance of H2 maize line to haloxyfop and sethoxydim (Gronwald *et al.*, 1989).

A ten-fold amplification of glutamine synthetase, the target enzyme for phosphinotricin (glufosinate) and bialaphos inhibition, was found in a resistant alfalfa cell line (Donn *et al.*, 1984; Botterman *et al.*, these Proceedings). This alfalfa gene was cloned and transferred to other plants, which showed only a low level of tolerance to phosphinotricin (Mazur and Falco, 1989). The prospect of this approach in developing plants resistant to these herbicides appears to be rather small (Botterman *et al.*, these Proceedings).

c) Introduction of detoxifying enzymes. This strategy is based on the concept of improving the desired crop selectivity to a herbicide, by introduction of genes encoding for detoxifying enzymes originated from other organisms. The successful examples of this approach used bacteria with high capacity to detoxify the phytotoxic compound to non-phytotoxic products. This method was exploited to engineer crop plants resistant to phosphinotricin (glufosinate) and bialaphos (Botterman *et al.*, these Proceedings), bromoxynil (Stalker *et al.*, 1988; Freyssinet *et al.*, 1989) and 2,4-D (Streber and Willmitzer, 1989). The major advantage of this strategy is that the transgenic plants produced so far do not exhibit any reduction in growth, seed setting or yield capacities. Further detailed toxicological studies are still needed to ascertain that the terminal detoxification products are safe to animals, humans and the environment.

A gene (bar), which confers resistance to bialaphos was isolated from *Streptomyces hygroscopicus* strains that synthesize bialaphos. The bar gene serves in *S. hygroscopicus* as a bialaphos biosynthetic gene and as a self-defence mechanism to avoid toxicity of its own products. This gene encodes for an enzyme – acetyl transferase, which specifically catalyses the conversion of the herbicide to a non-active acetylated form. The bar gene was transferred and expressed in several crops which exhibited an inherited resistance to high doses of phosphinotricin and bialaphos, both under glasshouse and field conditions. (Botterman, 1989; Botterman *et al.*, these Proceedings).

The gene bxn encoding for a specific nitrilase enzyme that hydrolyses bromoxynil to its non-phytotoxic metabolite 3,5-dibromo-4-hydroxy benzoic acid, was isolated and cloned from the soil bacterium *Klebsiella pneumonia* subsp. *ozanea* which uses bromoxynil as a sole nitrogen source. The gene was expressed in tobacco, mainly photosynthetic tissue, under the control of a light-inducible promoter of the sunflower Rubisco small subunit gene. The transgenic tobacco plants rapidly metabolise bromoxynil to a free and conjugated metabolite as in *K. pneumonia* and were resistant to high doses of the herbicide, in both glasshouse and field trials (Stalker *et al.*, 1988; Freyssinet *et al.*, 1989).

Similarly, Streber and Willmitzer (1989) describe the introduction and expression of a 2,4-D degradation mechanism from the soil bacterium *Alcaligenes eutrophus* into tobacco plants. The bacterial gene *tfdA*, encodes a monoxygenase (DAPM) enzyme catalyzing the first step in bacterial conversion of 2,4-D to its much less phytotoxic metabolite, 2,4,-dichlorophenol. Glasshouse experiments have shown that this trait was expressed in several transgenic tobacco lines with varying degrees of resistance.

Other attempts to engineer herbicide-resistant crops by introducing genes encoding detoxifying enzymes from other organisms have been made but have proven difficult or not suitable for further development (for recent reviews see Mazur and Falco, 1989; Padgette *et al.*, 1989).

It is believed, particularly amongst those who are directly involved, that herbicide-resistance is a valuable trait. It has the potential to enhance significantly the use of environmentally safer and broad-spectrum herbicides, to lower herbicide use rates, provide the grower more flexibility in choosing crops for rotation, and to design cost-effective weed management programmes. Furthermore, Snape *et al.* (these Proceedings) suggested exploiting these techniques to breed not only resistant crop varieties, but also sensitive ones to allow elimination of volunteer resistant plants of the preceding year. However, Williamson (these Proceedings) presenting the views of the UK Advisory Committee on Genetic Manipulation (ACGM) concluded that among other hazards, introduction of herbicide-resistant crops will increase the use of herbicide not only in the field where these plants are grown, but also in the neighbouring habitats. This in turn, will increase both the selection pressure (which may further enhance the evolution of resistant weeds) as well as the hazard to the environment.

Amongst other hazards which may be increased with the introduction of herbicide-resistant crops, one of major concern is hybridization, namely the possible spread of the resistance trait from transgenic plants to related wild species by gene flow through pollen (see also Darmency *et al.*, these Proceedings). This threat of forming "super weeds" is further amplified by the fact that in most cases the resistance trait is inherited as a dominant or semi-dominant single nuclear gene, with no apparent reduction in ecological fitness.

Another hazard considered by the ACGM is "weediness", or the possible escape of a resistant crop from cultivation. This hazard is illustrated in the case of proso millet (*Panicum miliaceum*) which has been cultivated as a crop in Canada since the early European settlement, but became a troublesome weed due to its vigorous growth, huge seed production and resistance to many commonly-used herbicides (Bough and Cavers, 1989).

Bright (these Proceedings) and Marshall (1987), reviewed the potential benefits as well as the problems regarding patents, regulation, efficacy and environmental hazards that should be solved prior to the introduction of herbicide-resistant crops. Bright concluded that the value of resistant crops will be determined by their ability to provide cheaper or better weed control, without reducing crop yield and quality.

At present, one of the biggest problems in the area of weed control in minor crops (e.g. vegetable and ornamental crops), is the small number of selective herbicides available. This stems mainly from the lack of interest of the agrochemical industry in developing appropriate selective herbicides, due to the limited acreage and market. Thus, minor crop may benefit most from the introduction of herbicide-resistant cultivars.

Parasite weeds such as *Striga* spp. (Lane, 1989), and *Orobanche* spp. (Pieterse, 1979) are major weeds in many countries throughout the warm climate zones of the world. Most of the available weed control measures failed to reduce the vast yield losses caused by these weeds (Ramaiah, 1987). Introduction of crops resistant to a phloem-mobile broad spectrum herbicide such as glyphosate, which will selectively control the weeds, may prove to be a highly beneficial solution to many growers.

It is the author's belief that the future introduction and commercialization of transgenic plants conferring herbicide resistance is an inevitable process. It will necessitate joint efforts from all parties involved to assess carefully and minimize all possible risks to protect the ecosystem.

CONCLUDING REMARKS

Herbicide resistance is now a widespread phenomenon throughout the world to many classes of herbicides, with the frightening potential of increasing yield losses and subsequent reduction in farming profitability. There is a crucial need for governments, industry, the weed science community, and farmers to set up new priorities in future research. We have to examine the present agronomic practices and find out where mistakes have been made that have resulted in the development of resistance. Alternative and rational integrated weed management programmes must be developed, through research, in order to meet the ever changing situations in the field with different environmental and cropping systems. Above all, farmers should realize that if they want to continue to rely on cost-effective and selective herbicides, they must be prepared to adopt changes and modify their farming and weed control practices continuously.

ACKNOWLEDGEMENTS

I would like to thank B.J. Gossett,M.S. Kemp, P.J.W. Lutman, S.R Moss, M. Parham, L.L. Saari, and B.E. Valverde for sharing with me their unpublished data. I am also grateful to my colleagues J. Caseley, D. Coupland, G.W. Cussans, M.S. Kemp and S.R. Moss for their co-operation and stimulating discussions. This study was supported in part by the Fund for Basic Research administered by The Israel Academy of Sciences and Humanities. The author acknowledges the scholarships provided by the Israel Academy of Sciences and Humanities, The Royal Society and The Harry and Abe Sherman Fund.

REFERENCES

Ahrens, W.H. and Stoller, E.W. (1983). Competition, growth and rate of CO_2 fixation in triazine susceptible and resistant smooth pigweed (*Amaranthus hybridus*). Weed Science **31**, 438–444.

Anon. (1990). Herbicide resistance – a call for industry action. Weed Technology **4**, 215–219.

Arntzen, C.J. and Duesing, J.H. (1983). Chloroplast-encoded herbicide resistance. In "Advances in Gene Technology" (F. Ahamand, K. Downey, J. Schultz and R.W. Voellmy, eds), pp. 273–299. Academic Press, New York.

Arntzen, C.J., Pfister, K. and Steinback, K.E. (1982). The mechanism of chloroplast triazine resistance: alteration of the site of herbicide action. In "Herbicide Resistance in Plants" (H. M. LeBaron and J. Gressel, eds), pp. 185–214. John Wiley and Sons, New York.

Barnwell, P. and Cobb, A.H. (1989). Physiological studies of mecoprop-resistance in chickweed (*Stellaria media* L.). Weed Research **29**, 135–140.

Beversdorf, W.D., Weiss-Lerman, J., Erickson, L.R. and Souza Machado, V. (1980). Transfer of cytoplasmically-inherited triazine resistance from bird's rape to cultivated oilseed rape *Brassica campestris* and *B. napus*. Canadian Journal of Genetics and Cytology **22**, 167–172.

Bieringer, H., Bauer, K., Hacker, E., Heubach, G., Leist, K.H. and Ebert, E. (1989). HOE-70542 – a new molecule for use in combination with fenoxaprop-ethyl allowing selective post-emergence grass weed control in wheat. Brighton Crop Protection Conference – Weeds 77–82.

Bishop, T., Powles, S.B. and Cornic, G. (1987). Mechanism of paraquat resistance in *Hordeum glaucum*. II. Paraquat uptake and translocation. Australian Journal of Plant Physiology **14**, 539–547.

Botterman, J. (1989). Advances in engineering herbicide resistance in plants. Brighton Crop Protection Conference – Weeds 979–985.

Bough, M.A. and Cavers, P.B. (1989). Proso millet, a crop gone wild. Canadian Journal of Plant Science **69**, 265.

Bourdot, G.W., Harrington, K.C. and Popay, A.I. (1989). The appearance of phenoxy-herbicide resistance in New Zealand. Brighton Crop Protection Conference – Weeds 309–316.

Boydston, R.A. and Slife F.W. (1986). Alteration of atrazine uptake and metabolism by tridiphane in giant foxtail (*Setaria faberi*) and corn (*Zea mays*). Weed Science **34**, 850–858.

Boydston, R.A. and Slife F.W. (1987). Post-emergence control of giant foxtail (*Setaria faberi*) in corn (*Zea mays*) with tridiphane and triazine combinations. Weed Science **35**, 103–108.

Cabanne, F., Huby, D., Gaillardon, P., Scalla, R. and Durst, F. (1987). Effect of the cytochrome P_{450} inactivator 1-aminobenzotriazole on the metabolism of chlorotoluron and isoproturon in wheat. Pesticide Biochemistry and Physiology **28**, 371–380.

Clay, D. (1989). New developments in triazine and paraquat resistance and co-resistance in weed species in England. Brighton Crop Protection Conference - Weeds 317-324.

Conard, S.G. and Radosevich, S.R. (1979). Ecological fitness of *Senecio vulgaris* and *Amaranthus retroflexus* biotypes susceptible or resistant to triazines. Journal of Applied Ecology 16, 171-177.

Coupland, D., Lutman, P.J.W. and Heath C. (1990). Uptake, translocation, and metabolism of mecoprop in a sensitive and a resistant biotype of *Stellaria media*. Pesticide Biochemistry and Physiology 36, 61-67.

Cussans, G.W., Cousens, R.D. and Wilson, B.J. (1987). Progress towards rational weed control strategies. In "Rational Pesticide Use"(K.J. Brent and R.K. Atkin, eds), pp. 301-314. Cambridge University Press, Cambridge.

Darmency, H. and Gasquez, J. (1990). Appearance and spread of triazine resistance in common lambsquarters (*Chenopodium album*). Weed Technology 4, 173-177.

Darmency, H. and Pernes, J. (1985). Use of wild *Setaria viridis* (L.) Beauv. to improve triazine resistance in cultivated *S. italica* (L.) by hybridization. Weed Research 25, 175-179.

Donn, G., Tischer, E., Smith, J.A. and Goodman, H.M. (1984). Herbicide-resistant alfalfa cells: an example of gene amplification in plants. Journal of Molecular and Applied Genetics 2, 621-635.

Ehr, R.J. and Burroughs, F.G. (1986). A study of the interaction between tridiphane and atrazine on broadleaf weeds. Proceedings of the north Central Weed Control Conference 41, 78-79.

Ezra, G., Dekker, J.H. and Stephenson, G.R. (1985). Tridiphane as a synergist for herbicides in corn (*Zea mays*) and proso millet (*Panicum milliaceum*). Weed Science 33, 287-290.

Frear, D.S., Swanson, H.R. and Tanaka, F.S. (1969). N-demethylation of substituted 3-(phenyl)-1-dimethylureas: Isolation and characterization of a microsomal mixed function oxidase from cotton. Phytochemistry 8, 2157-2169.

Freyssinet, G., Leroux, B., Lebrun, M., Pelissier, B., Sailland, A. and Pallett, K.E. (1989). Transfer of bromoxynil resistance into crops. Brighton Crop Protection Conference - Weeds 1225-1234.

Fuerst, E.P., Nakatani, H.Y., Dodge, A.D., Penner, D. and Arntzen, C.J. (1985). Paraquat resistance in *Conyza*. Plant Physiology 77, 984-989.

Fuerst, E.P., Arntzen, C.J. and Penner, D. (1986). Herbicide cross-resistance in triazine-resistant biotypes of four species. Weed Science 34, 344-353.

Fuerst, E.P. and Vaughn, K.C. (1990). Mechanism of paraquat resistance. Weed Technology 4, 150-156.

Gabard, J.M., Charest, P.J., Iyer, V.N. and Miki, B.L. (1989). Cross-resistance to short residual sulfonylurea herbicides in transgenic tobacco plants. Plant Physiology **91**, 574-580.

Gossett, B.J., Murduck, E.C., Toler, J.E. and Harris, J.R. (1990). Palmer amaranth resistance to the dinitroaniline herbicides. Proceedings of Southern Weed Science Society **43**, 327.

Gressel, J. (1985). Herbicide tolerance and resistance: alteration of site of activity. In "Weed Physiology" (S.O. Duke, ed.), pp. 159-189. CRC Press, Boca Raton, Florida.

Gronwald, J.W., Andersen, R.N. and Yee, C. (1989). Atrazine resistance in velvetleaf (*Abutilon theophrasti*) due to enhanced atrazine detoxification. Pesticide Biochemistry and Physiology **34**, 149-163.

Gronwald, J.W., Parker, W.B., Somers, D.A., Wyse, D.L. and Genenbach, B.G. (1989). Selection for tolerance to graminicide herbicides in maize tissue culture. Brighton Crop Protection Conference - Weeds 1217-1224.

Hartnett, M.E., Chui, C.F, Mauvais, C.J., McDevitt, R.E., Knowlton, S., Smith, J.K. Falco, S.C. and Mazur, B.J. (1990). Herbicide-resistant plants carrying mutated acetolactate synthase genes. In "Managing Resistance to Agrochemicals, From Fundamental Research to Practical Strategies" (M. B. Green, H. M. LeBaron and W. M. Moberg, eds), pp. 459-473. ACS Symposium Series No. 421, Washington D.C..

Hatzios, K.K. and Hoagland, R.E. (1989). "Crop Safeners for Herbicides: Development, Uses and Mechanisms of Action". Academic Press, San Diego.

Haughn, G.W. and Somerville, C.R. (1986). Sulfonylurea-resistant mutants of *Arabidopsis thaliana*. Molecular and General Genetics **204**, 430-434.

Heap, I. and Knight, R. (1982). A population of ryegrass tolerant to the herbicide diclofop-methyl. The Journal of the Australian Institute of Agricultural Science **48**, 156-157.

Holt, J.S. (1990). Fitness and ecological adaptability of herbicide-resistant biotypes. In "Managing Resistance to Agrochemicals, From Fundamental Research to Practical Strategies" (M. B. Green, H. M. LeBaron and W. M. Moberg, eds), pp. 419-429. ACS Symposium Series No. 421, Washington D.C..

Holt, J.S. and LeBaron, H.M. (1990). Significance and distribution of herbicide resistance. Weed Technology **4**, 141-149.

Jansen, M.A.K., Hobe, J.H., Wesselius, J.C. and van Rensen, J.J.S. (1986). Comparison of photosynthetic activity and growth performance in triazine-resistant and -susceptible biotypes of *Chenopodium album*. Physiologie Vegetale **24**, 475-484.

Jones, O.T.G. and Caseley, J.C. (1989). Role of cytochrome P_{450} in herbicide metabolism. Brighton Crop Protection Conference - Weeds 1175-1184.

Kemp, M.S., Newton, L.V. and Caseley, J.C. (1988). Synergistic effects of some P_{450} oxidase inhibitors on the phytotoxicity of chlortoluron in a resistant population of black-grass (*Alopecurus myosuroides*). Proceedings of the European Weed Research Society Symposium, "Factors Affecting Herbicidal Activity and Selectivity" 121-126.

Kemp, M.S., Moss, S.R. and Thomas, T.H. (1990). Herbicide resistance in *Alopecurus myosuroides*. In "Managing Resistance to Agrochemicals, From Fundamental Research to Practical Strategies" (M. B. Green, H. M. LeBaron and W. M. Moberg, eds), pp. 376-393. ACS Symposium Series No. 421, Washington D.C..

Lamoureux, G.L. and Rusness, D.G. (1986). Tridiphane [2-(3,5-dichlorophenyl)-2-(2,2,2-trichloroethyl)oxirane] an atrazine synergist: enzymatic conversion to a potent glutathione S-transferase inhibitor. Pesticide Biochemistry and Physiology **26**, 323-342.

Lane, A. (1989). Prospect for the control of *Striga*, a noxious parasite weed of tropical crops. Rural Development in Practice **1**, 9-10.

LeBaron, H.M. and McFarland, J. (1990). Overview and prognosis of herbicide resistance in weeds and crops. In "Managing Resistance to Agrochemicals, From Fundamental Research to Practical Strategies" (M. B. Green, H. M. LeBaron and W. M. Moberg, eds), pp. 336-352. ACS Symposium Series No. 421, Washington D.C..

Linden, H., Sandmann, G., Chamovitz, D., Hirschberg, J. and Boger, P. (1990). Biological characterization of *Synechococcus* mutants selected against the bleaching herbicide norflurazon. Pesticide Biochemistry and Physiology **36**, 46-51.

Lutman, P.J.W. and Heath, C.R. (1990). Variation in the resistance of *Stellaria media* to mecoprop due to biotype, application method and 1-aminobenzotriazole. Weed Research **30**, 129-137.

Mallory-Smith, C.A., Thill, D.C. and Dial, M.J. (1990). Identification of sulfonylurea herbicide-resistant prickly lettuce (*Lactuca serriola*). Weed Technology **4**, 163-168.

Marshall, G. (1987). Implications of herbicide-tolerant cultivars and herbicide-resistant weeds for weed control management. British Crop Protection Conference – Weeds 489-498.

Maxwell, B.D., Roush, M.L. and Radosevich, S.R. (1990). Predicting the evolution and dynamics of herbicide resistance in weed populations. Weed Technology **4**, 2-13.

Mazur, B.J. and Falco, S.C. (1989). The development of herbicide resistant crops. Annual Review of Plant Physiology and Plant Molecular Biology **40**, 441-470.

McFadden, J.J., Frear, D.S. and Mansager, E.R. (1989). Aryl hydroxylation of diclofop by cytochrome P_{450} dependent monooxygenase from wheat. Pesticide Biochemistry and Physiology **34**, 92-100.

Meredith, C.P. and Carlson, P.S. (1982). Herbicide resistance in plant cell cultures. In "Herbicide Resistance in Plants" (H. M. LeBaron and J. Gressel, eds), pp. 275-291. John Wiley, and Sons, New York.

Moreland, D.E., Novitzky, W.P. and Levi, P.E. (1989). Selective inhibition of cytochrome P_{450} isozymes by the herbicide synergist tridiphane. Pesticide Biochemistry and Physiology 35, 42-49.

Moss, S.R. (1987). Herbicide resistance in black-grass (*Alopecurus myosuroides*). British Crop Protection Conference - Weeds 879-886.

Moss, S.R. (1990). Herbicide cross-resistance in slender foxtail (*Alopecurus myosuroides*). Weed Science (in press).

Myers, M.G. and Harvey, R.G. (1986). Control of triazine resistant common lambsquarters in corn. Proceedings of the North Central Weed Control Conference 41, 78.

Newhouse, K.E., Shaner, D.L., Wang, T. and Fincher, R. (1990). Genetic modification of crop responses to imidazolinone herbicides. In "Managing Resistance to Agrochemicals, From Fundamental Research to Practical Strategies" (M.B. Green, H.M. LeBaron and W.M. Moberg, eds), pp. 474-481. ACS Symposium Series No. 421, Washington D.C.

Niemann, V.P. and Pestemer, W. (1984). Resistenz verschiedener Herkunfte von Acker-Fuchsschwanz (*Alopecurus myosuroides*) gegenuber Herbizidbehandlungen. Nachrichtenblatt des Deutschen Pflanzenschutz-dienstes 36, 113-118.

Ort, D.R., Ahrens, W.H., Martin, B. and Stoller, E.W. (1983). Comparison of photosynthetic performance in triazine resistant and susceptible biotypes of *Amaranthus hybridus*. Plant Physiology 72, 925-930.

Padgette, S.R., della-Cioppa, G., Shah, D.P., Fraley, R.T. and Kishore, G.M. (1989). Selective herbicide tolerance through protein engineering. In "Cell Culture and Somatic Cell Genetics of Plants" (J. Schell and I. Vasil, eds), Vol. 6, pp. 441-476. Academic Press, Inc., New York.

Parochetti, J.V., Schnappinger, M.G.,Ryan, G.F. and Collins, H.A. (1982). Practical significance and means of control of herbicide-resistant weeds. In "Herbicide Resistance in Plants" (H. M. LeBaron and J. Gressel, eds), pp. 309-323. John Wiley and Sons, New York.

Pieterse, A.A. (1979). The broomrapes (*Orobanceae*) - a review. Abstracts on Tropical Agriculture 5, 9-35.

Pinthus, M.J., Eshel, Y. and Shchori, Y. (1972). Field and vegetable crops mutants with increased resistance to herbicides. Science 177, 715-716.

Pölös, E., Mikulas, J., Szigeti, Z., Matkovics, B., Hai, D.Q., Parducz, A. and Lehoczki, E. (1988). Paraquat and atrazine co-resistance in *Conyza canadensis* (L.) Cronq. Pesticide Biochemistry and Physiology 30, 142-154.

Powles, S.B., Holtum, J.A.M., Matthews, J.M. and Liljegren, D.R. (1990). Herbicide cross-resistance in annual ryegrass (*Lolium rigidum* Gaud): The search for a mechanism. In "Managing Resistance to Agrochemicals, From Fundamental Research to Practical Strategies" (M.B. Green, H.M. LeBaron and W.M. Moberg, eds), pp. 394–406. ACS Symposium Series No. 421, Washington D.C.

Powles, S.B. and Howat, P.D. (1990). A review of weeds in Australia resistant to herbicides. Weed Technology 4, 178–185.

Primiani, M.M., Cotterman, J.C. and Saari, L.L. (1990). Resistance of kochia (*Kochia scoparia*) to sulfonylurea and imidazolinone herbicides. Weed Technology 4, 169–172.

Radosevich, S.R. and Holt, J.S. (1982). Physiological responses and fitness of susceptible and resistant weed biotypes to triazine herbicides. In "Herbicide Resistance in Plants" (H. M. LeBaron and J. Gressel, eds), pp. 163–183. John Wiley and Sons, New York.

Ramaiah, K.V. (1987). Control of *Striga* and *Orobanche* species – a review. In "Parasitic Flowering Plants" (H.C. Weber and W. Forstreuter, eds), pp. 637–664. Maburg, F.R.G..

Rendina, A.R., Beaudoin, A.C., Craig-Kennard, and Breen, M.K. (1989). Kinetics of inhibition of acetyl-coenzyme A carboxylase by the aryloxyphenoxypropionate and cyclohexanedione graminicides. Brighton Crop Protection Conference – Weeds 163–172.

Roush, M.L., Radosevich, S.R. and Maxwell, B.D. (1990). Future outlook for herbicide-resistance research. Weed Technology 4, 208–214.

Rubin, B., Yaacoby, T. and Schonfeld, M. (1985). Triazine resistant grass weeds: cross resistance with wheat herbicide, a possible threat to cereal crops. British Crop Protection Conference – Weeds 1171–1178.

Ruscoe, C.N.E. (1987). Pesticide resistance: strategies and co-operation in the agrochemical industry. In "Rational Pesticide Use" (K.J. Brent and R.K. Atkin, eds), pp. 197–208. Cambridge University Press, Cambridge.

Saari, L.L., Cotterman, J.C. and Primiani, M.M. (1990). Mechanism of sulfonylurea herbicide resistance in broadleaf weed, *Kochia scoparia*. Plant Physiology 93, 55–61.

Saxena, P.K. and King, J. (1988). Herbicide resistance in *Datura innoxia*, cross-resistance of sulfonylurea-resistant cell lines to imidazolinones. Plant Physiology 86, 863–867.

Schonfeld, M., Yaacoby, T., Michael, O., and Rubin, B. (1987). Triazine resistance without reduced vigor in *Phalaris paradoxa*. Plant Physiology 83, 329–333.

Shaaltiel, Y. and Gressel J. (1986). Multienzyme oxygen radical detoxifying system correlated with paraquat resistance in *Conyza bonariensis*. Pesticide Biochemistry and Physiology 26, 22–28.

Shaaltiel, Y. and Gressel J. (1987). Kinetic analysis of resistance to paraquat in *Conyza*. Plant Physiology **85**, 869–871.

Souza Machado, V. (1982). Inheritance and breeding potential of triazine tolerance and resistance in plants. In "Herbicide Resistance in Plants" (H. M. LeBaron and J. Gressel, eds), pp. 257–273. John Wiley and Sons, New York.

Souza Machado, V. and Hume, D.J. (1987). Breeding herbicide-tolerant cultivars – a Canadian experience. British Crop Protection Conference – Weeds 473–477.

Stalker, D.M., McBride, K.E. and Malyj, L.D. (1988). Herbicide resistance in transgenic plants expressing a bacterial detoxification gene. Science **242**, 419–423.

Stannard, M.E. and Fay, P.K. (1987). Selection of alfalfa seedlings for tolerance to chlorsulfuron. Weed Science Society of America Abstracts **27**, 61.

Stephenson, G.R., Dykstra, M.D., McLaren, R.D. and Hamill, A.S. (1990). Agronomic practices influencing triazine-resistant weed distribution in Ontario. Weed Technology **4**, 199–207.

Stowe, A.E. and Holt, J.S. (1988). Comparison of triazine-resistant and -susceptible biotypes of *Senecio vulgaris* and their F_1 hybrids. Plant Physiology **87**, 183–189.

Streber, W.R. and Willmitzer, L. (1989). Transgenic tobacco plants expressing a bacterial detoxifying enzyme are resistant to 2,4–D. Bio/Technology **7**, 811–816.

Street, J.E. and Snipe, C.E. (1989). Propanil plus tridiphane for barnyardgrass (*Echinochloa crus-galli*) control in rice (*Oryza sativa*). Weed Technology **3**, 632–635.

Tanaka, Y., Chisaka, H. and Saka, H. (1986). Movement of paraquat in resistant and susceptible biotypes of *Erigeron philadelphicus* and *E. canadensis*. Physiologia Plantarum **66**, 605–608.

Valverde, B.E., Radosevich, S.R. and Appleby, A.P. (1988). Growth and competitive ability of dinitroaniline-herbicide resistant and susceptible goosegrass (*Eleusine indica*) biotypes. Proceedings of the Western Society of Weed Science **41**, 81.

Varsano, R. (1987). "Enhancement of Herbicide Activity by Piperonyl Butoxide", pp. 118. Ph.D. Thesis, The Hebrew University of Jerusalem, Israel.

Varsano, R. and Rubin, B. (1988). Synergistic effect of piperonyl butoxide with triazine herbicides in plants. Proceedings of the European Weed Research Society Symposium, "Factors Affecting Herbicidal Activity and Selectivity" 127–132.

Vaughn, K.C., Marks, M.D. and Weeks, D.P. (1987). A dinitroaniline-resistant mutant of *Eleusine indica* exhibits cross-resistance and supersensitivity to antimicrotubule herbicides and drugs. Plant Physiology **83**, 956-964.

Vaughn, K.C. and Vaughan, M.A. (1990). Structural and biochemical characterization of dinitroaniline-resistant *Eleusine*. In "Managing Resistance to Agrochemicals, From Fundamental Research to Practical Strategies" (M.B. Green, H.M. LeBaron and W.M. Moberg, eds), pp. 364-375. ACS Symposium Series No. 421, Washington D.C.

Vaughn, K.C., Vaughan, M.A. and Gossett, B.J. (1990). A biotype of goosegrass (*Eleusine indica*) with an intermediate level of dinitroaniline herbicide resistance. Weed Technology **4**, 157-162.

Watson, D., Mortimer, A.M. and Putwain, P.D. (1987). The seed bank dynamics of triazine resistant and susceptible biotypes of *Senecio vulgaris* - implication for control strategies. British Crop Protection Conference - Weeds 917-924.

Wilson, B.J. and Brain, P. (1990). Weed Monitoring on a whole farm - patchiness and the stability of distribution of *Alopecurus myosuroides* over a ten year period. Proceedings of the European Weed Research Society Symposium, "Integrated Weed Management in Cereals" 45-52.

Yaacoby, T. (1989). "Characterization of Triazine Resistance in Grass Weeds: *Phalaris paradoxa*, *Lolium rigidum* and *Alopecurus myosuroides*" pp. 121. Ph.D. Thesis, The Hebrew University of Jerusalem, Israel.

Yaacoby, T., Schonfeld, M. and Rubin, B. (1986). Characteristics of atrazine resistant biotypes of three grass weeds. Weed Science **34**, 181-184.

ABSTRACTS OF

POSTERS

SYNERGISED MYCOHERBICIDES – POTENTIAL AGENTS FOR CONTROLLING HERBICIDE-RESISTANT WEEDS

Z. Amsellem, A. Sharon and J. Gressel

Dept of Plant Genetics, Weizmann Institute of Science, Rehovot, Israel

Host-specific mycoherbicides have the potential of specifically controlling problem weeds, especially when the weeds are closely related to the crop, precluding the use of selective herbicides. One of their widely discussed disadvantages is that each mycoherbicidal organism usually controls only one weed species. This disadvantage is of less importance when herbicide resistance breaks out, as usually only one species evolves resistance in a given field. These resistant weeds could easily be controlled by species specific mycoherbicides. The herbicide-resistant weed may be even better controlled than the wild type due to its lack of fitness or due to negative cross resistance.

As mycoherbicides are "self-replicating" herbicides, very low inocula should be needed. Instead over a thousand-fold more inoculum is usually needed to infect weeds than seems necessary, i.e. usually much more than 1000 spores per square centimetre of leaf surface for leaf pathogenic mycoherbicides (Walker and Riley, 1982; Boyette and Turfitt, 1988) or 10,000 spores per gram of soil for applied organisms. We have found the accepted requirements for high threshold levels of inoculum unnecessary if micro-environmental conditions are improved; basically the inoculum threshold could be abolished (Amsellem *et al.*, 1990).

We used an invert-emulsion developed by Quimby *et al.* (1988); this contains water on the inside and a mixture of vegetable and mineral oils and paraffin wax on the outside. With the emulsion, one spore per infection site was sufficient to infect *Cassia obtusifolia* with *Alternaria cassiae* and *Datura stramonium* with *A. crassa*. Even at low humidity, *C. obtusifolia* plants were infected by one spore in a droplet of the invert emulsion whereas 1000 spores per droplet of water only infected a fifth of the sites (Fig. 1).

The rate and intensity of infection could be enhanced, even with the invert-emulsion, when antimetabolites were added to suppress the host plants' "immunological" system. *Cassia* contains phytoalexins from the phenyl-propanoid pathway. Glyphosate (an inhibitor of EPSP synthase) at a sublethal 50 µM and *O*-benzylhydroxyamine (an inhibitor of phenylalanine-ammonia-lyase) at 10 µM both enhanced infection in the above pathogen/weed system, even in the presence of the invert-emulsion but their effect was greater with emulsion. The rates of antimetabolites used were too low to cause phytotoxicity by themselves. This was probably due to the improvement of flavanoid phytoalexin biosynthesis, in response to the infection.

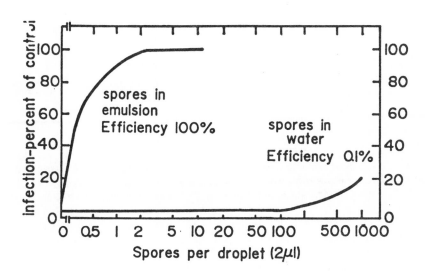

Fig. 1. <u>Abolition of inoculum threshold at low humidity</u>. *Cassia obtusifolia* plants were grown in a controlled temperature, high humidity glasshouse (60-80% RH) at 24°C day (14 h) and 22°C high (10 h). The first leaf was treated on 12-day-old plants bearing one leaf. One 2 µl droplet was applied to each opposite lobe of each leaf with 20 replicates per data point. Plants were put on open shelves at low humidity (50% RH) immediately after infection. (Modified from Amsellem *et al.*, 1990.)

The use of synergising adjuvants that provide an optimal micro-environment for fungal germination and growth, and the use of synergising antimetabolites that suppress the plants' defences, provide the possibility of using mycoherbicidal organisms cost effectively for weed control, including when resistance has evolved to chemical herbicides.

REFERENCES

Amsellem, Z., Sharon, A., Gressel, J. and Quimby, P.C. (1990). Complete abolition of inoculum threshold of two mycoherbicides (*Alternaria cassiae* and *A. crassa*) when applied in invert emulsion. <u>Phytopathology</u> (In press).

Boyette, C.D. and Turfitt, L.B. (1988). Factors influencing biocontrol of Jimson weed (*Datura stramonium* L.) with the leaf-spotting fungus, *Alternaria crassa*. <u>Plant Science</u> **56**, 261-264.

Quimby, P.C. Jr., Fulgham, F.E., Boyette, C.D. and Conick, W.J. Jr. (1988). An emulsion replaces dew in biocontrol of sicklepod – a preliminary study. <u>In</u>: "Pesticide Formulation and Application Systems". Vol. 8. (D.A. Hoved and G.A. Beestman, eds). American Society for Testing and Materials, Philadephia.

Walker, H.L. and Riley, J.A. (1982). Evaluation of *Alternaria cassiae* for the biocontrol of sicklepod (*Cassia obtusifolia*). <u>Weed Science</u> **30**, 651-654.

HERBICIDE-RESISTANT WEEDS AND ALTERNATIVE CONTROL MEASURES IN YUGOSLAVIA

M. Arsenovic[1], Z. Ostojic[2] and Z. Klokocar Smit[1]

[1] Plant Protection Institute, Agriculture Faculty, Novi Sad, Yugoslavia
[2] Plant Protection Institute, Agriculture Faculty, Zagreb, Yugoslavia

Regular yearly application of herbicides with the same or similar site of action has induced the appearance and propagation of resistant weed biotypes. The phenomenon of resistance have been detected in different parts of the world in some 100 species (LeBaron and Gressel, 1982).

The appearance of resistance to triazine in Yugoslavia was first noticed in 1981 in maize grown in monoculture in S.R. Hrvatska where atrazine and simazine had been used repeatedly. Tests revealed the high tolerance of the *Chenopodium album* to atrazine, simazine and prometryne (Bulcke *et al.*, 1985).

For the total weed eradication on the non-cropped areas, high dosages of triazine herbicides have been used regularly. The resistant biotypes of *Amaranthus retroflexus* and *Erigeron canadensis* (Conyza) were detected on such sites as canal slopes and railway tracks.

The biotest in the chernozem soil was done with different concentrations of atrazine under controlled conditions. Seeds of *A. retroflexus* collected on the sites treated continuously with atrazine and from untreated plots were germinated. The decrease in biomass net weight of young plants derived from the seed of the plants suspected of resistance correlated with the increase of the atrazine content in the soil. At the same time, the seed from the susceptible biotype did not grow in soil with atrazine treatments as low as 0.5 kg ha^{-1}.

Chemical and biological measures were applied to prevent further development of the resistant biotypes. The weeds on non-cropped land were controlled by applying herbicides different from triazines.

To avoid the problems of appearance of atrazine-resistant biotypes of *Chenopodium album* in corn, reduced doses of triazine herbicide were combined with pyridate, bentazone, bromoxynil and ICI-0051. The efficacy index ranged from 56.5 to 100%. We have shown that linuron at 0.75 kg a.s. added to metalachlor or alachlor reduced the expression of resistance. Late treatment with pyridate, bentazone or bromoxynil was also effective.
The efficiency of those combinations was confirmed using atrazine plus lower dosages of linuron. The level of resistance of weeds in maize plots, treated in this way, remained the same or slightly increased.

For the last three years, due to the existence of resistant biotypes, we have attempted to find and develop alternative biological control methods by surveying the health of the more common weed species. We have observed the following diseases and micro-organisms:

- downy mildew (*Peronospora farinosa*) was found on *C. album* and *C. hybridum*, while an *Alternaria* sp. caused brown leaf spot and gradual desiccation.

- white rust (*Albugo bliti*) often appeared in epiphytotic association as reported by Gilreath and Jones (1985). An *Alternaria* sp. isolate was obtained from spotted leaves of *Amaranthus retroflexus* and its pathogenicity was confirmed in glasshouse experiments.

- One isolate of *Drechslera* sp. and one of *Alternaria* sp. has been obtained from the seeds of *Erigeron canadensis*. Both were pathogenic to young seedlings. *Alternaria* sp. and *Drechslera* sp. were considered to be promising candidates for further investigation and are being tested as biocontrol agents.

REFERENCES

Bulcke, R., De Vleeschauwer, J., Vercruysse, J. and Stryckers, J. (1985). Comparison between triazine-resistant and -suceptible biotypes of *Chenopodium album* L. and *Solanum nigrum* (L.)(X). <u>Mededelingen van de Faculteit Landbouwetenschappen Rijksuniversiteit Gent</u>, 211-220.

Gilreath, J.P., and Jones, J.P. (1985). White rust epiphytotic on *Amaranthus hybridus* in South Florida infected by *Albugo bliti*. <u>Plant Disease</u> **6**, 542.

Klokocar-Smit, Z. and Arsenovic, M. (1989). Mikroflora emerznih biljnih vrsta u kanalima Dunav-Tisa-Dunav. <u>Fragmenta herbologica Yugoslavica</u>, 145-150.

LeBaron, H.M. and Gressel, J. (1982). Herbicide Resistance in Plants. Wiley-InterScience, New York.

RELATIONSHIP BETWEEN LIGHT INTENSITY AND GROWTH PARAMETERS IN TRIAZINE-RESISTANT *AMARANTHUS HYBRIDUS* AND *A. BLITOIDES*

Y. Benyamini, H. Schickler, M. Schonfeld and B. Rubin

The Hebrew University of Jerusalem, Faculty of Agriculture, Rehovot 76100, Israel

Several studies have indicated that triazine-resistant (R) biotypes of broad-leaf plants are inferior in growth as compared to the sensitive (S) wild types. In contrast, R and S biotypes of several grass weeds found in Israel, exhibited similar growth both under competitive and non-competitive conditions (Schonfeld *et al.*, 1987). Differences in light intensities prevailing during plant growth in different parts of the world, may account for the variable results reported. High light intensities, however, are frequently associated with high temperatures, and it was shown that triazine resistance is accompanied by increased sensitivity to high temperatures. heat sensitivity was suggested to be involved in determining the geographical distribution of resistant weeds (Ducruet and Lemoine, 1985). Although light intensities and temperatures during the Israeli summer are above the average in Europe, triazine resistance was recently found in two summer weeds *Amaranthus hybridus* and *A. blitoides*. The purpose of this study was to investigate the relationship between light intensities during growth, the relative growth, and the response to high temperatures in R and S biotypes of these species.

Plants were grown in a growth chamber (28/20°C day/night, 14 h light) at three light intensities: 75, 250, 500 μE m^{-2} s^{-1}, and in a phytotron at two light intensities: 650, 1400 μE m^{-2} s^{-1}. Carbon dioxide fixation was measured in intact leaves with an infra-red gas analyser. Chlorophyll fluorescence was monitored in leaf discs with a PAM fluorometer.

The rate of CO_2 fixation was determined as a function of light intensity in intact leaves of R and S plants of *A. hybridus* and *A. blitoides* grown at a high light intensity. Carbon dioxide fixation in the S biotypes was higher than in the R biotypes at all light intensities tested (50 to 1100 μE m^{-2} s^{-1}). However, while at low light intensities the advantage of S plants was about 30%, it was only a few percent at high light intensities.

When the two biotypes of both species were grown separately at a low light intensity (75 μE m^{-2} s^{-1}), biomass production of the S plants was approx-imately five times higher than that of R plants. The difference between biotypes decreased with increase in light intensity, and tended to disappear for plants grown at high light intensities. A significant advantage of the S biotype was observed even at high light intensities, under competitive conditions, where light within the plants' canopy was a limiting factor. These results clearly indicate, for both *Amaranthus* species, that the differential growth of R and S biotypes is highly dependent on light intensity.

Mature leaves taken from plants of both species grown under 75, 250 and 500 μE m^{-2} s^{-1} were exposed to high temperatures (38 to 47°C) for up to 10 min. Heat sensitivity was determined by measuring the initial level of chlorophyll fluorescence (Fo) of dark-adapted leaf discs before and after the heat treatment. R plants grown at all light intensities, were significantly more sensitive to heat treatment than S plants. Heat sensitivity of both biotypes was similarly increased with the light intensity used during plant growth.

An increase in light intensity evidently affects the relative fitness of R plants in two opposing ways. On the one hand it facilitates higher photosynthesis rates and on the other it increases the heat sensitivity of these plants (which is higher than in S plants, even at low light intensities). R plants of both *Amaranthus* species were found to persist during the Israeli summer, characterized by both high temperatures and high light intensities. This seems to indicate that the beneficial effects of high light intensities on the growth vigour of R plants overcomes the detrimental effects of high temperatures.

ACKNOWLEDGEMENT

This research was supported in part by the Fund for Basic Research administered by the Israel Academy of Science and Humanities.

REFERENCES

Ducruet. J.V. and Lemoine, Y. (1985). Increased heat sensitivity of the photosynthetic apparatus in triazine-resistant biotypes from different plant species. Plant and Cell Physiology **26**, 419-429.

Schonfeld, M., Yaacoby, T, Michael, O and Rubin, B. (1987). Triazine resistance without recorded vigour in *Phalaris paradoxa*. Plant Physiology **83**, 329-333.

IN VITRO SELECTION OF HERBICIDE-TOLERANT VARIANTS OF WHEAT

R. Bozorgipour and J.W. Snape

AFRC Institute of Plant Science Research, Cambridge Laboratory, Trumpington, Cambridge, CB2 2LQ UK

In vitro culture methods may be useful for the generation and selection of resistant analogues of susceptible cultivars through exploiting somaclonal variation. To investigate this possibility in wheat, protocols have been developed and tested using model systems for responses to the herbicides difenzoquat, chlorotoluron and atrazine. Two genetically defined varieties, Chinese Spring and Sicco, were tested. These were chosen because Chinese Spring is resistant to difenzoquat and susceptible to chlorotoluron, whilst Sicco is susceptible to difenzoquat but resistant to chlorotoluron. Like all other bread wheat varieties they are both sensitive to atrazine.

Callus was initiated from immature embryos and then subcultured on media containing different concentrations, 5, 10 and 50 μM, of the active ingredients. Response was measured by calculating callus growth rate, rate of green spot formation, regeneration rate and regeneration frequency.

Both genotypes showed a linear decrease in callus growth rate in response to increasing amounts of difenzoquat. However, the response of Sicco was more marked than that of Chinese Spring. Similar results were obtained for response to atrazine although, no difference between varieties was observed. With chlorotoluron no consistent patterns of response were observed. Indeed, the susceptible variety Chinese Spring had a higher growth rate at 50 μM than at 10 μM. Similarly, Sicco had a higher growth rate at 10 μM than 5 μM. Regeneration ability was adversely affected by all three herbicides and fewer plants were regenerated as the concentration of chemical increased. Nevertheless, some plants were regenerated from all concentrations, apart from the highest concentration of atrazine.

The regenerated plants (R_o generation), the presumptive mutants, were grown to maturity and self-pollinated. Their R_1 progenies were then tested as whole plants for their responses to spray applications of the herbicides under controlled conditions. For difenzoquat, all progenies of the regenerated plants of Chinese Spring were, as expected, resistant. For Sicco, variation in response was observed from extreme susceptibility to resistance. Three plants as tolerant as the Chinese Spring controls were observed and selected. With chlorotoluron, two tolerant plants of Chinese Spring were isolated. No progenies tolerant to atrazine were found and all R_1 plants were killed by application of the chemical.

Selected tolerant R_1 plants were selfed and also test crossed to their respective parents to examine the genetical control of any induced resistance. Overall, no clear evidence of single gene mutations were obtained with either Chinese Spring lines resistant to chlorotoluron or

Sicco lines resistant to difenzoquat. Nevertheless, some R_2-resistant progenies were obtained. These will be examined further to eludicate the genetical control of presumptive induced resistance.

REFERENCES

Chaleff, R.S. and Parson, M.F. (1978). Direct selection *in vitro* for herbicide-resistant mutants of *Nicotiana tabacum*. Proceedings of the National Academy of Sciences USA, **75**, 5104-5107.

Mathias, R.J. and Simpson, E.S. (1986). The interaction of genotype and culture medium on the tissue culture response of wheat (*Triticum aestivum* L. em. thell) callus. Plant Cell, Tissue and Organ Culture **7**, 31-37.

Murashige, T. and Skoog, F. (1962). A revised medium for rapid growth and bioassays with tobacco tissue culture. Physiologia Plantarum **15**, 473-497.

Snape, J.W. and Parker, B.B. (1988). Chemical response polymorphism: an additional source of genetic markers in wheat. In "Proceedings of Seventh International Wheat Genetics Symposium" (T.E. Miller and R.M.D. Koebner, eds), 651-656. AFRC Institute of Plant Science Research, Cambridge.

Snape, J.W., Angus, W.J., Parker, B.B. and Leckie, D. (1986). The chromosomal location in wheat of genes conferring differential response to the wild oat herbicide, difenzoquat. Journal of Agricultural Sciences, Cambridge **108**, 543-548.

INITIAL EFFECTS OF THE HERBICIDE MECOPROP ON PHOTOSYNTHESIS AND TRANSPIRATION BY A MECOPROP-RESISTANT BIOTYPE OF *STELLARIA MEDIA*

V.G. Breeze, D. Coupland, P.J.W. Lutman and A. Hutchings

Department of Agricultural Sciences, University of Bristol, AFRC Institute of Arable Crops Research, Long Ashton Research Station, Bristol, BS18 9AF, UK

Herbicide-resistant weeds present both a problem of control to the farmer and agrochemical industry and a challenge to the physiologist to explain the mechanisms of resistance. Populations of *Stellaria media* resistant to mecoprop have been identified (Lutman and Snow, 1987) in which a possible mechanism of resistance is the de-activation of the herbicide by conjugate formation (Coupland *et al.*, 1989). Surprisingly, the initial response of mecoprop-resistant and -susceptible populations, such as leaf epinasty, is the same, although the resistant plants later recover. We have therefore compared the time courses of carbon dioxide and water vapour exchange by whole plants following application of mecoprop, for resistant and sensitive biotypes.

Plants were grown in 7.5 cm pots in soil or sand under glasshouse conditions for about 6 weeks. Plants were sprayed using a laboratory pot sprayer, at herbicide rates of 3.6 kg a.e. ha^{-1}. Immediately after spraying, plants were sealed into chambers for gas-exchange measurements using the system described by Breeze and West (1987). Essentially, this consists of 30 l transparent polycarbonate chambers with fans to circulate the air. The light intensity on the leaves was approximately 200 W m^2 of photosynthetically-active radiation and the inlet air temperature was controlled. Photoperiod was 12 h. The air circulation around the shoot was isolated from that around the root. Carbon dioxide and water vapour concentrations were measured by infra-red gas analysers connected to an automatic sampler. Depending on the size of the plant, air flow rates of 35 - 50 l min^{-1} were used.

Typically, sprayed plants of both biotypes showed a decline in both CO_2 and water vapour exchange following spraying, whereas the gas exchange of the untreated plants was approximately steady over the experimental period. The rate of the sprayed plants reached a minimum after about 6 h or by the end of the first light periods, by which time the rate of gas exchange was about 50% of the initial value. Thus during this phase, the pattern for both CO_2 and water vapour was similar.

Visible symptoms of epinasty and stem bending were noted on the sprayed plants during the first light period. These usually appeared after the decline in gas exchange, or about half-way through the first light period.

During the second light period, CO_2 exchange and transpiration by the sprayed, mecoprop-susceptible biotype remained at its previous, low rate. In contrast, the gas exchange of the sprayed, resistant biotype began to recover, so that by the end of the light period, it had regained its former value. This remained the case in the third light period, so that the two controls and the sprayed, resistant biotypes had CO_2 exchange and transpiration rates approximately twice that of the sprayed, susceptible biotype. Further measurements for up to 5 light periods did not indicate any recovery of the susceptible biotype.

Another experiment investigated the gas exchange of plants in continuous light, for up to 26 h after spraying. In these conditions, the sprayed, resistant plants did not show a pronounced recovery by either photosynthesis or transpiration. It would therefore appear that a dark period is necessary for the resistance to mecoprop to be manifest.

It seems that the initial response by *S. media* to mecoprop is the same for both resistant and susceptible biotypes. However, long-term irreversible effects occur in the susceptible plants whereas the resistant biotype is not affected to the same extent. Further investigations are in progress to clarify these responses, but it may be that stomatal closure is the first response and that subsequent effects are due to the direct action of the herbicide on the process of CO_2 fixation.

REFERENCES

Breeze, V.G. and West, C.J. (1987).   Long- and short-term effects of vapour of the herbicide 2,4-D butyl on the growth of tomato plants. Weed Research **27**, 13-21.

Coupland, D., Lutman, P.J.W. and Heath, C. (1989). Uptake, translocation and metabolism of mecoprop in a sensitive and resistant biotype of *Stellaria media*. Pesticide Biochemistry and Physiology **36**, 61-67.

Lutman, P.J.W. and Snow, H.S. (1987). Further investigations into the resistance of chickweed (*Stellaria media*) to mecoprop. Proceedings 1987 British Crop Protection Conference - Weeds, 901-908.

STRESS PROTEINS INDUCED BY ATRAZINE AND HEAT SHOCK IN *RHODOBACTER SPHAEROIDES*

A.E. Brown, C.T. Highfill and V.P. Adkison

Department of Botany and Microbiology, Auburn University, Auburn, Alabama 36849-4201, USA

The induction of specific proteins in response to heat stress and atrazine treatment was compared in the photosynthetic bacterium, *Rhodobacter sphaeroides*. When cells were subjected to heat shock by shifting cultures from a normal growth temperature of 30°C to elevated temperatures between 39–45°C, pulse labelling studies showed that polypeptides at 93.5, 88, 84.5, 76, 66.3, 20.8 and 18.5 kDa were synthesised in response to the temperature shift. When cultures were treated with 100 μM atrazine, 23 polypeptides were detected by pulse labelling as compared to the non-atrazine culture. Common polypeptides to the heat shock response and atrazine treatment were detected at 93, 88, 84, 66.5, 20.8 and 18.5 kDa. Both the heat shock and atrazine stress proteins were partially characterised in the major cell fractions and they were found to be primarily associated with the crude membrane fraction of the cell. In some cases, polypeptides not previously seen in whole cell protein extracts were detected by pulse labelling in membrane fractions from cells treated with the herbicide or by heat shock. An atrazine-resistant strain of *Rhodobacter sphaeroides* was also pulse labelled during growth in the presence of 100 μM atrazine. Proteins at 63.5, 53.4, 25 and 21.5 kDa were detected in this organism. It appears that atrazine can induce many of the same proteins as heat shock. Furthermore, some polypetides are specific for atrazine stress including two proteins found in an atrazine-resistant strain.

TRIAZOLE, TRIAZINE, TRIAZINONE, AND PHENYLUREA RESISTANCE IN ANNUAL RYEGRASS (*LOLIUM RIGIDUM*)

M.W.M. Burnet, O.B. Hildebrand, J.A.M. Holtum and S.B. Powles

Department of Agronomy, Waite Agricultural Research Institute, University of Adelaide, P.M.B 1 Glen Osmond, S.A. 5064, Australia

In Australia, *Lolium rigidum* Gaud. (annual ryegrass) is a widely dispersed, economically important annual weed. It is possibly the most important weed of cropping in Southern Australia. In Western Australia, a biotype infesting *ca.* 2000 km of railway track has developed resistance to a mixture of the triazine herbicide, atrazine, and the triazole herbicide, amitrole. This multiple resistance has appeared after ten years of annual treatment with a mixture of the two herbicides. Diminishing control was recognised for the last three years of that period. The addition of diuron to the herbicide mixture did not improve control in the field.

Seed collected in 1988 and 1987 was tested during the winter of 1988 and 1989 to determine the extent of cross-resistance. Ryegrass seedlings were grown outdoors in pots during the normal growing season. For post-emergent applications of herbicide, the seedlings were sprayed at the three-leaf stage. A biotype taken from an area with no history of herbicide application was included as a susceptible control.

Resistance to amitrole was most significant with an 11-fold difference in sensitivity between the susceptible and resistant biotypes. The biotype is resistant to atrazine and to the other chloro-s-triazines; simazine, cyanazine and propazine as well as the thiomethyl-s-triazines, prometryne, terbutryne and ametryne. Resistance extends to the hexazinone and the triazinone herbicide, metribuzin. The biotype is also resistant to the phenylurea herbicides; chlorotoluron, isoproturon, metoxuron, diuron, linuron, fluometuron and methabenzthiazuron. With the exception of amitrole, which inhibits carotenoid biosynthesis, all the foregoing are inhibitors of photosystem II (PS II). Resistance to the herbicide methazole was also observed. While methazole itself is not an inhibitor of PS II, its dichlorophenylmethyl urea metabolite is (Suzuki and Casida, 1981). Resistance to methazole is related to the resistance observed to the phenylurea herbicides.

There were no differences in susceptibility to the herbicides ethidimuron, karbutilate, norflurazon, diclofop-methyl, fluazifop-butyl, sethoxydim, tralkoxydim, chlorsulfuron, metsulfuron methyl, sulfometuron methyl, paraquat, glyphosate amd carbetamide. The susceptibility of the Western Australian biotype to ethidimuron and karbutilate was significant given that they are thought to have a similar mode of action to other substituted ureas.

Resistance to the triazine herbicides is widespread in the Northern Hemisphere. For 54 of the 55 triazine resistant weed species reported in 20 countries, resistance is caused by an alteration to the PS II target site, the 32 kDa herbicide binding protein (LeBaron and McFarland,

427

1989). We have isolated chloroplast thylakoids from the biotypes of *L. rigidum* under study. Evolution of oxygen by isolated thylakoids from resistant and susceptible leaves was equally inhibited by atrazine and diuron.

In another experiment, intact plants were treated with a sub-lethal dose of chlorotoluron. Photosynthetic activity was measured as net CO_2 exchange. Initially, resistant and suceptible plants were equally affected but resistant plants recovered photosynthetic activity more rapidly than susceptibles. Similar recoveries in photosynthetic activity have been associated with rapid metabolism in resistant species such as wheat (Ryan and Owen, 1982).

Since the inhibition of electron transfer is the same in both biotypes it is, therefore, unlikely that an insensitive target site is responsible for the resistance phenomenon. Given that resistance is not caused by an alteration in the 32 kDa binding protein it is possible that resistance may be due to detoxification or sequestration. Further experiments to identify the physiological and genetic basis of this cross-resistance phenomenon are being conducted.

REFERENCES

LeBaron, H.M. and McFarland, J. (1989). Overview and prognosis of herbicide resistance in weeds and crops. In "Fundamental and Practical Approaches to Combating Resistance" (W.K. Moberg and H.M. LeBaron, eds). American Chemical Society, Washington (in press).

Ryan, P.J. and Owen, W.J. (1982). The mechanism of selectivity of chlorotoluron between cereals and grass weeds. Proceedings of the 1982 British Crop Protection Conference - Weeds, 317-324.

Suzuki, T. and Casida, J.E. (1981). Metabolites of diuron, linuron and methazole formed by liver microsomal enzymes and spinach plants. Journal of Agricultural and Food Chemistry 29, 1027-1033.

CONTROL OF HERBICIDE RESISTANT BLACK-GRASS WITH HERBICIDE MIXTURES CONTAINING TRIDIPHANE

J.C. Caseley, L. Copping[1] and D. Mason

Department of Agricultural Sciences, University of Bristol, AFRC Institute of Arable Crops Research, Long Ashton Research Station, Bristol, BS18 9AF, UK.
[1] Dow Chemical Company Limited, Letcombe Laboratory, Letcombe Regis, Wantage, Oxon, OX12 9JT, UK.

Intensive use of chlorotoluron and isoproturon has led to the development of populations of black-grass (*Alopecurus myosuroides*) which are resistant to these herbicides. Unlike triazine-resistant weeds, these black-grass populations are cross-resistant to herbicides with different modes of action and chemistry (Moss, 1987). Kemp *et al.* (1989) found enhanced degradation of ^{14}C-labelled chlorotoluron in the resistant 'Peldon' compared to the tolerant 'Rothamsted' biotype suggesting that resistance is due to enhanced degradation. Post-emergence application of tridiphane controls giant foxtail (*Setaria faberi* L.) and crabgrass (*Digitaria* spp.) and it has a synergistic effect when used in combination with atrazine for the control of several grass weeds in corn, without decreasing selectivity in the latter (Ezra *et al.*, 1985). Our aim was to evaluate the activity of tridiphane alone and in combination with selected herbicides used to control black-grass in wheat and barley, with particular emphasis on performance against the resistant biotype.

Imbibed seeds of 'Peldon' resistant and 'Rothamsted' herbicide susceptible biotypes of black-grass were exposed to light for 8 h and were sown 1 cm deep in sandy loam soil in 9 cm diameter plastic pots. For post-emergence herbicide treatments, the pots were watered from above immediately after sowing and thereafter by sub-irrigation. The pre-emergence treatments received an additional top watering one hour after spraying. The herbicides, applied with a laboratory pot sprayer calibrated to deliver 200 l ha^{-1}, were used pre- and post-emergence at the 1-2 and 3-4 leaf growth stage. The plants were grown in conventional or gauze-sided glasshouses at $18/10 \pm 8°C$ (day/night respectively) and humidity was uncontrolled.

Tridiphane (e.c. 500 g a.i. l^{-1}) was applied at 0.25 and 0.50 kg ha^{-1} and the other herbicides at 15, 30 and 60% of the 'recommended' field dose – cyanazine 1.75 kg a.i. ha^{-1}, isoproturon, 2.0 kg a.i. ha^{-1}, terbutryne 2.8 kg a.i. ha^{-1}, and triallate 1.4 kg a.i. ha^{-1}. The combinations were 'tank mixed' and 1% Atplus 411F was added to the spray solution of the post-emergence treatments. Following application of the pre-emergence treatments, c. 5 g of dry soil was applied to the surface to reduce loss of tridiphane and triallate by volatility. Fresh and dry weight of foliage were recorded three and four weeks after spraying for the post- and pre-emergence treatments respectively.

In the pre-emergence study, the shoot weight of the susceptible biotype was substantially reduced by both tridiphane and isoproturon alone. Isoproturon at 1.20 kg ha^{-1} killed the susceptible biotype. Combinations of the lowest doses of both compounds were additive and lethal. Activity of the individual compounds was far less on the resistant biotype, but complete kill was achieved with both doses of tridiphane plus 1.20 kg ha^{-1} isoproturon and with 0.5 kg ha^{-1} tridiphane plus 0.6 kg ha^{-1} isoproturon.

In the post-emergence study, the lower doses of isoproturon alone failed to give good control of susceptible plants with 1-2 leaves, but with 0.5 kg ha^{-1} of tridiphane the 0.6 kg ha^{-1} dose was lethal. Compared to the untreated control plants the foliage weight of the resistant biotype was reduced more by 0.5 kg ha^{-1} of tridiphane than by the high dose of isoproturon, and the top doses of the combined herbicides were lethal and synergistic (Colby 1967). At the 3-4 leaf growth stage, the effect of both compounds alone and in combination, diminished on the resistant, but not the susceptible biotype. Tridiphane alone and its combination with atrazine were also most effective on the early growth stages of giant foxtail (Ezra *et al.* 1985).

These synergistic and additive effects seen with tridiphane/isoproturon combinations were observed when tridiphane was tank mixed with cyanazine, terbutryne and triallate. Tridiphane activity appears to be affected by weather factors as its phytotoxic and synergistic effects varied significantly between experiments conducted at different times of the year. The mechanism of tridiphane synergistic action with atrazine involves inhibition of glutathione-S-transferase (Lamoureux and Rusness, 1986). This seems a less likely explanation for the herbicide used here in view of their known metabolites and we are currently investigating the metabolic profiles of several herbicides in plants and suspension cell cultures with and without tridiphane.

REFERENCES

Colby, S.R. (1967). Calculating synergistic and antagonistic responses of herbicide combinations. Weeds **15**, 20-22.

Ezra G., Dekker, J.H. and Stephenson, G.R. (1985). Tridiphane as a synergist for herbicides in corn (*Zea mays*) and proso millet (*Panicum miliaceum*). Weed Science **33**, 287-290.

Kemp, M.S., Moss, S.R. and Thomas, T.H. (1989). Herbicide resistance in *Alopecurus myosuroides*. In "Fundamental and Practical Approaches to Combating Resistance", ACS Series from Proceedings of the 196th National Meeting of the American Chemical Society, Los Angeles, September, 1988.

Lamoureux, G.L. and Rusness, D.G. (1986). Tridiphane [2-(3,5-Dichloro-phenyl)-2-(2,2,2-trichloroethyl)oxirane] an atrazine synergist: enzymatic conversion to a potent glutathione S-transferase inhibitor. Pesticide Biochemistry & Physiology **26**, 323-342.

Moss, S.R. (1987). Herbicide resistance in black-grass (*Alopecurus myosuroides*). Proc. British Crop Protection Conference - Weeds, **3**, 879-886.

STUDY OF THE GROWTH OF HERBICIDE-RESISTANT BLACK-GRASS POPULATIONS

B. Chauval and J. Gasquez

INRA, Malherbologie, BV 1540, 21034 Dijon, France

During the last ten years, many biotypes of several weed species have become resistant to herbicides (Yaacoby *et al.*, 1986; Moss, 1987). The most widespread are triazine-resistant biotypes, which generally grow slower and are less fitted than susceptible biotypes. However, one can ask whether all herbicide resistances always lead to loss of plant fitness.

To answer this question, we used triazine-resistant and chlorotoluron-resistant biotypes and susceptible populations of *Alopecurus myosuroides* Huds. (black-grass). These were:

- a triazine-resistant population (R_t) and a susceptible population (S_t) collected in the same area of Israel (Yizreel Valley).

- susceptible (S_c) and resistant (R_c) chlorotoluron populations collected in the United Kingdom (Peldon).

Samples of these populations were grown in controlled conditions. Relative growth rate, length and area of leaf, number of tillers and dry weight of roots and leaves were measured after 40 days of growth.

As we observed that only a part (25%) of the chlorotoluron-resistant population was resistant and that only 65% of the progeny of treated plants were resistant, the resistance was confirmed for each plant by differential chlorophyll fluorescence. Leaf pieces were soaked in chlorotoluron (15 ppm) for 6 hours in the light. The fluorescence results showed that all leaves had a blocked photosynthesis. Then the leaves were left in water. In order to test the level of detoxication, the fluorescence was recorded after 24, 30 and 48 hours in darkness. The leaves whose photosynthesis was still blocked after 48 hours were assumed to be susceptible.

Whatever the resistance level, Israeli biotypes grow slowly, produce less tillers but are taller than European biotypes. Therefore, we compared only susceptible and resistant biotypes of the same origin.

Within Israeli biotypes, R_t and S_t have the same growth rate. The resistant biotype produces more tillers, but the susceptible biotype is taller with longer leaves and larger area. Nevertheless, root and leaf dry weight are not significantly different. The growth rate of the triazine-resistant biotype in these conditions is lower but there is no evidence that its fitness is reduced.

Although enzyme polymorphism has been observed (Chauval and Gasquez, 1988), the growth rate of susceptible (S_c) and resistant (R_c) biotypes is very similar, the only difference being a lower dry weight of roots for the resistant biotype. But this difference which was observed in other studies whatever the conditions does not seem to be correlated with the resistance mechanism.

We have also crossed (R_t) and (R_c) biotypes. In the progenies of the (R_t) plants, some plants are both triazine- and chlorotoluron-resistant, the others are only triazine-resistant. In controlled conditions, there is no significant difference between the two groups of hybrid plants and the mother plants. Thus within the progeny, the growth of young plants is not affected by the chlorotoluron resistance.

In conclusion, we can suggest that in contrast to triazine resistance, where generally the lower fitness of the resistant plants can be observed from the first stages of development, the biological cost of chlorotoluron resistance seems to be null for black-grass. But a more complete study of flowering time, seed production etc. is necessary to estimate the overall fitness.

REFERENCES

Chauval, B. and Gasquez, J. (1988). Polymorphisme enzymatique de populations sensibles et résistantes au chlortoluron chez *Alopecurus myosuroides* Huds. VIII ème Colloque sur la Biologie, l'écologie et la systématique des mauvaises herbes, 237-246.

Moss, S.R. (1987). Herbicide resistance in black-grass (*Alopecurus myosuroides*). Proceedings 1987 British Crop Protection Conference - Weeds, 879-886.

Yaacoby, T., Schonfeld, M. and Rubin, B. (1986). Characteristics of atrazine-resistant biotypes of three grass weeds. Weed Science 34, 181-184.

CROSS-RESISTANCE OF *POA ANNUA* BIOTYPES TO PARAQUAT AND TRIAZINES

D.V. Clay and R. Hadleigh

Department of Agricultural Sciences, University of Bristol, AFRC Institute of Arable Crops Research, Long Ashton Research Station, Bristol, BS18 9AF.

Triazine-resistant biotypes of *Poa annua* were first reported from France in 1978 (Ducruet and Gasquez, 1978) and from the UK in 1982 (Putwain, 1982). The mechanism of resistance appeared to be a modification of the chloroplast thylakoid membrane binding site, since large doses of triazines did not kill the plants. Paraquat resistance of certain biotypes of *P. annua* was reported in 1982 (LeBaron and Gressel, 1982), resistant plants requiring a four-fold increase in dose for control. For some years, there have been reports of failure to control *P. annua* with paraquat and simazine in some hop gardens in Kent; the areas had been treated annually with recommended doses of these herbicides for 25 years. In the experiments reported here, the response of these biotypes to paraquat and triazines has been compared with that of triazine-resistant and -susceptible types.

Plants were taken from two hop gardens near Maidstone, Kent (M1 and M2), grown in the glasshouse and seed collected. Seed was also used from a biotype from Berkshire (B) with no previous history of herbicide treatment and a known triazine-resistant biotype from a forest nursery near Chard, Somerset (C), where simazine had been used annually for 10 years. For tests of response to paraquat, seeds of each biotype were germinated in trays of peat/sand compost, five seedlings transplanted to 7.5 cm diam. pots of soil and grown on capillary netting. For tests of herbicide response to simazine, 12 seeds were sown in 7.5 cm diam. pots in sandy loam soil, covered with a 2.5 mm soil layer and the soil wetted from overhead. Paraquat as Gramoxone 100, and simazine as Gesatop 500FW were applied at the doses shown in Tables 1-2 using a laboratory track sprayer at a spray volume of 370 l ha^{-1}. After spraying, pots were set out in a heated glasshouse with supplementary illumination in two to four randomised blocks. Simazine-treated pots were watered lightly from overhead as required. Foliage fresh weight was recorded at the end of the experiments and dose response curves fitted using MLP (Ross, 1987).

In a preliminary experiment when large *P. annua* plants were treated with herbicides, biotypes M1 and M2 were relatively resistant to atrazine and slightly resistant to paraquat compared with biotype B (Clay, 1989). Biotype C was resistant to all doses of atrazine but killed by all doses of paraquat. In the paraquat dose-response experiment (Table 1), the M1 and M2 biotypes were more resistant than the B or C biotypes; the difference in dose required to give equivalent damage was two- to three-fold. In the test of pre-emergence activity of simazine (Table), the difference in dose giving equivalent damage to the B compared with the M1 and M2 biotypes was around two-fold, suggesting a small degree of resistance in these biotypes; this contrasted with the C biotype which was not damaged by any dose. The high degree of resistance to triazines of biotype C corresponds with that found previously with *P. annua* (Ducruet and Gasquez, 1978; Putwain, 1982). The M1 and M2 biotypes, however, showed a relatively small increase in resistance to pre-emergence simazine but greater resistance to atrazine applied to large plants. This low level of resistance has not been reported before with *P. annua*. Similarly, the resistance to paraquat is not as extreme as has been reported for *Erigeron* spp.(LeBaron and Gressel, 1982) but comparable to that observed with *Epilobium ciliatum* in the U.K. (Clay, 1989). These results suggest that the resistance mechanism of *Poa annua* biotypes from hop gardens may be different from that previously reported for triazines and be due to an enhanced capacity to inactivate the herbicides before reaching their sites of action.

Table 1. <u>The effect of paraquat applied to *Poa annua* at nine doses from 0.01 to 0.40 kg ha^{-1} on 13 January, 1989.</u>

Biotype	C	B	M1	M2
ED$_{50}$ (kg a.i./ha^{-1})	0.063[a]	0.081	0.279	0.179
ED$_{50}$ (Ln dose)	-2.765	-2.512	-1.278	-1.720
SE$\pm$	0.109	0.081	0.063	0.070

[a] detransformed value, based on foliage fresh wt recorded 23 February 1989

Table 2. <u>The effect of simazine applied pre-emergence to *Poa annua* at five doses from 0.025 to 0.40 kg ha^{-1} on 8 June, 1989.</u>

Biotype	C	B	M1	M2
ED$_{50}$ (kg a.n. ha^{-1})	0.40	0.011[a]	0.023	0.027
ED$_{50}$ (Ln dose)	–	-4.550	-3.768	-3.596
SE$\pm$	–	0.636	0.105	0.062

REFERENCES

Clay, D.V. (1989). New developments in triazine and paraquat resistance and co-resistance in weed species. <u>Proceedings 1989 Brighton Crop Protection Conference - Weeds</u>, 317-324.

Ducruet, J.M. and Gasquez, J. (1978). Observations de la fluorescence sur feuille entière et mise en évidence de la résistance chloro-plastique à L'atrazine chez *Chenopodium album* L. et *Poa annua* L. <u>Chemosphere</u> **6**, 691-696.

LeBaron, H.M. and Gressel, J. (1982). "Herbicide Resistance in Plants" pp. 401. J. Wiley, New York.

Putwain, P.D. (1982). Herbicide resistance in weeds - an inevitable consequence of herbicide usage. <u>Proceedings of 1982 British Crop Protection. Conference - Weeds</u>, 719-728.

Ross, G.J.S. (1987). Maximum Likelihood Programme 3.08 NAG Ltd, Oxford.

IS MECOPROP-RESISTANCE IN CHICKWEED DUE TO ALTERED AUXIN SENSITIVITY?

A.H. Cobb, C. Early and P. Barnwell

Herbicide Research Group, Department of Life Sciences, Faculty of Science, Nottingham Polytechnic, Clifton Lane, Nottingham, NG11 8NS

Although auxin-type herbicides have been extensively used for over forty years, they are foliar-applied and generally non-persistent. Consequently, few cases of resistance to these herbicides have been reported. However, Lutman and colleagues (1985 and 1987) have identified resistance to mecoprop in *Stellaria media* L. (chickweed), and these observations have been confirmed in this laboratory (Barnwell and Cobb, 1989). In this paper, we propose that the basis of this resistance could be due to reduced mecoprop binding to an active site, although the involvement of uptake, movement and metabolism could have similar importance. Here, we report some findings of a further study of mecoprop-induced proton efflux in mecoprop-resistant and -susceptible *S. media* seedlings.

An investigation of the kinetics of proton efflux in etiolated *S. media* shoots was performed as in previous studies with *Avena sativa* L. coleoptiles (Fitzsimons *et al.*, 1988). Dose-response analysis of mecoprop-induced proton efflux enabled the relative affinities of mecoprop to receptors in both resistant (Bath 0) and susceptible (WR0) biotypes of *S. media* to be computed using a curve fitting programme (Fitzsimons, 1989).

Dose-response analysis of mecoprop-induced proton efflux for mecoprop-resistant (Bath 0) and mecoprop-susceptible (WR0) biotypes is presented in Table 1. The H_{50} values (i.e. the mecoprop concentration giving half-maximal response, which may be taken as a measure of receptor affinity), showed that the resistant biotype (Bath 0) was over 10^5 times less receptive to mecoprop than the susceptible biotype (WR0). These observations indicate that differential auxin-sensitivity may form the basis of mecoprop resistance in this species.

Table 1. H_{50} values for mecoprop-induced proton efflux.

Biotype	H_{50} values (mM)
Bath 0	2.49×10^{-3}
WR0	1.46×10^{-8}

ACKNOWLEDGEMENTS

We thank Dr P.J.W. Lutman (IACR, Rothamsted) for providing seed of both *S. media* biotypes, the National Advisory Body (UK) for financial support and Sharon Morrison for secretarial skills.

REFERENCES

Barnwell, P. and Cobb, A.H. (1989). Physiological studies of mecoprop resistance in chickweed (*Stellaria media* L.). Weed Research **29**, 135-140.

Fitzsimons, P.J. (1989). The determination of sensitivity parameters for auxin-determined H^+-efflux from *Avena* coleoptile segments. Plant, Cell and Environment **12**, 595-614.

Fitzsimons, P.J., Barnwell, P. and Cobb, A.H. (1988). A study of auxin-type herbicide action based on a dose-response analysis of H^+-efflux. In "Factors Affecting Herbicidal Activity and Selectivity" Proceedings of the European Weed Research Society Symposium 63-68.

Lutman, P.J.W. and Lovegrove, A.W. (1985). Variations in the tolerance of *Galium aparine* (cleavers) and *Stellaria media* (chickweed) to mecoprop. Proceedings British Crop Protection Conference - Weeds **2**, 411-418.

Lutman, P.J.W. and Snow, H.S. (1987). Further investigations into the resistance of chickweed (*Stellaria media*) to mecoprop. Proceedings British Crop Protection Conference - Weeds **3**, 901-908.

ESCAPE OF HERBICIDE RESISTANCE GENE FROM TRANSGENIC CROPS

H. Darmency[1], I. Till [2] and X. Reboud[2]

[1] INRA, Malherbologie, BV 1540, 21034 Dijon, France
[2] Université de Paris 11, Evolution et Systématique des Végétaux, Bat 362, 91405 Orsay, France

Prior to the release of transgenic crops for large scale use in agriculture, we need to estimate the risks associated with gene dispersal by interspecific crossing. Indeed, gene flow between crops and wild plants may occur through pollen. Many crops are affected by this phenomenon, as they belong to botanical genera or families which contain closely related wild plants e.g. rice, millet, wheat, sorghum, oats, rapeseed, sugar beet, sunflower, potato, alfalfa, pea and many other legumes.

These risks may be estimated at two different levels:

> (1) what is the chance of getting viable hybrid seeds? It may be very low but even a minute proportion of the millions of seeds produced each year by crops and wild plants could represent a significant risk.

> (2) what is the survival of such hybrid plants? It may be very variable according to different environments. This second step depends mainly on the nature of the gene that has been introduced by genetic engineering into the crop genome. If it confers a high adaptive value, it is likely that plants expressing this gene will be at advantage with respect to other plants.

Genes used for crop improvement could be useful for wild plants in arable fields since weed populations grow in the same ecological niche as crops. Herbicide resistance genes are a good example of useful genes both for crops and weeds in sprayed fields. Foxtail millet (*Setaria italica*) is one of the crops where a breeding programme has led to a herbicide-resistant cultivar. This cultivar is resistant to the triazine herbicides (maternal inheritance) and has not been obtained through genetic engineering but by classical breeding using a resistant wild genitor. It could be used as a model of a crop into which a foreign gene has been introduced in the cytoplasm in order to prevent gene dispersal through pollen. In this case, experimentation can proceed without the restrictions due to government regulations on transgenic organisms but would simulate future genetic engineering on the cytoplasmic genomes.

We present the different steps that could lead to the release of resistant wild *Setaria* seeds in the field. Evidence of field hybridization of cultivated plants by pollen from wild *Setaria* is shown by the finding of "off type" plants whose progeny segregates for several characters involved in the domestication of the foxtail millet: seed shedding lemna and pericarp pigmentation, and presence of polyphenol oxidases in seeds (Darmency and Pernès, 1987). Hybrids and their descendants are intermediary growing forms that represent more troublesome weeds than their parents and are already known from botanists as *S. viridis* var. *major* (Gaud) Posp. (Darmency *et al.*, 1987).

Experiments are being done to estimate the hybridization rate using dominant red colour marker on the coleoptile of young seedlings (monogeni nuclear inheritance). From 0.02 to 0.5% of resistant hybrids are foun among the progeny of mother cultivated plants according to experimenta designs and weed genotypes. In contrast, hybrid production on wild mothe plants is higher ($\geq 2\%$), and this is probably due to the great amount o cultivated type pollen in the field. Therefore, the strategy of havin transgenic genes in organelle DNA may reduce by 100-fold the risk of gen dispersal.

Mathematical models are studied and computer simulations used to determin the relative and absolute importance of each biological step in the lif span of resistant hybrid material. Preliminary estimates of each paramete are made using simple experimental designs: seed survival and germination hybrid seedling survival and plant growth, fertility and reproduction i presence of herbicide or not, etc.

Other studies are carried out on different crops and herbicide resistances For instance, we are now studying the gene flow between transgenic rap seeds engineered by Plant Genetic Systems to show resistance to glufosinat and wild *Sinapsis*. These experiments must be performed in confine environments to prevent the actual spread of recombined genes.

REFERENCES

Darmency, H. and Pernès, J. (1987). An inheritance study of domesticatio in foxtail millet using an interspecific cross. Plant Breeding **99**, 30-33

Darmency, H., Zangre, G.R. and Pernès, J. (1987). The wild-weed-cro complex in *Setaria*: a hybridization study. Genetica **75**, 103-107

LUMINESCENCE AND FLUORESCENCE STUDY OF PHOTOSYSTEM II ELECTRON TRANSFER IN TRIAZINE RESISTANT MUTANTS OF WEED PLANTS. COMPARISON WITH HERBICIDE RESISTANT MUTANTS FROM CYANOBACTERIA

J.M. Ducruet[1], G.Ajlani[2], C. Astier[2], S. Creuzet[1], A.L. Etienne[2] and C. Vernotte[2],
[1] CEA/INRA Saclay-SBPH/Division de Biologie, 91191 Gif-sur-Yvette, France
[2] Laboratoire de Photosynthèse-CNRS, 91190 Gif-sur-Yvette, France

The influence of herbicide resistance mutations of the D1 protein of photosystem II on the photosynthetic electron flow (Q_A----$>Q_B^-$ electron and Q_B^- charge stabilisation) was studied in higher plants and cyanobacteria.

In *Chenopodium album* and quasi-isogenic lines of *Solanum nigrum* (Ser 264 changed to Gly), the rate of fluorescence decay after one flash (Bowes *et al.*, 1980) was the same in R and S leaves in the first part of the kinetics (<200 µs), but became slower for R biotype at longer times. In thermoluminescence measurements on leaf fragments and on chloroplasts, the emission peak due to the $Q_B^-S_2$ recombination was shifted to lower temperatures in R, as reported for *Erigeron canadensis* (Demeter *et al.*, 1985). Luminescence emission after a saturating flash, at different temperatures from 5°C to 40°C, showed two exponential decay phases. In R leaves or chloroplasts, the $Q_B^-S_2$ recombination (slow phase) was faster and the ratio of $Q_A^-S_2/Q_B^-S_2$ recombination was increased more rapidly with temperature elevation.

For three herbicide-resistant mutants from a cyanobacteria, *Synechocystis* PCC6714, with single or double substitutions of amino-acids within the Q_B pocket (Ajlani *et al.*, 1989a,b), results were similar to those obtained on higher plants. In contrast, an ioxynil-resistant mutant (IoxI; Asn 266---> Thr (6)) showed a slightly increased stability of Q_B^- (Etienne *et al.*, 1990).

These results confirm a higher dark concentration of Q_A^- and a displaced Q_A<--->Q_B equilibrium in R. This can be explained by a higher proportion of empty Q_B sites (no plastoquinone bound) in the R-mutated D1 protein.

The influence of the R-impaired PS II electron transfer on the overall photosynthetic activity depends on external conditions, mainly light intensity (Ort *et al.*, 1983) and temperature. Although no temperature dependence could be found in the light-saturated Hill reaction of R and S chloroplasts below 35°C, previous studies (Ducruet and Ort, 1988; Ducruet, 1988) showed that a sudden temperature increase above 38°C induced greater damage in R plants. This could result from:

- a thermally-induced distortion of the D1 protein leading to an irreversible alteration. The results presented above suggest that D1 with an empty Q_B pocket could be more sensitive to thermal denaturation.
- the "shade type" modifications of R plant chloroplasts could cause a higher thermal sensitivity. This agrees with the absence

of such a shade adaptation in resistant plants growing under elevated temperatures (Schonfeld *et al.*, 1988).

This high-temperature susceptibility (Havaux, 1989) of R weed or crop plants is likely to cause them a competitive disadvantage in warm areas, able to slow down, although not to preclude the emergence of resistant weeds under the continuous pressure of herbicide treatment.

REFERENCES

Ajlani, G., Kirilovsky, D., Picaud, M. and Astier, C. (1989b). Molecular analysis of psb A mutations responsible for various herbicide resistance phenotypes in *Synechocystis* 6714. <u>Plant Molecular Biology</u> **13**, 469-479.

Ajlani, G., Meyer, I., Vernotte, C. and Astier, C. (1989b). Mutation in phenol-type herbicide resistance maps with the psb A gene in *Synechocystis* 6714. <u>FEBS</u> **246**, 207-210.

Bowes, J., Crofts, A.R. and Arntzen, C.J. (1980). Redox reactions on the reducing side of photosystem II in chloroplasts with altered herbicide binding properties. <u>Archives of Biochemistry and Biophysics</u> **200**, 303-308.

Demeter, S., Vass, I., Hideg, E. and Sallai, A. (1985). Comparative thermoluminescence study of triazine-resistant and -susceptible biotypes of *Erigeron canadensis* L. <u>Biochimica et Biophysica Acta</u> **806**, 16-24.

Ducruet, J.M. (1988). Increased heat-susceptibility of a triazine-resistant line of *Solanum nigrum* L.: effects of heat shocks on whole plants. <u>Proceedings EWRS Symposium Wageningen</u>, 37-44.

Ducruet, J.M. and Ort, D.B. (1988). Enhanced susceptibility of photosynthesis to high leaf temperature in triazine-resistant *Solanum nigrum* L. Evidence for Photosystem IID, protein site of action. <u>Plant Science</u> **56**, 39-48.

Etienne, A.L. *et al.* (1990) (submitted).

Havaux, M. (1989) Comparison of atrazine-resistant and -susceptible biotypes of *Senecio vulgaris* L. : effects of high and low temperatures in the *in vivo* photosynthetic electron transfer in intact leaves. <u>Journal of Experimental Botany</u> **40**, 217, 849-854.

Ort, D.B., Ahrens, W.H., Martin, B. and Stoller, E.W. (1983). Comparison of photosynthetic performance in triazine-resistant and susceptible biotypes of *Amaranthus hybridus*. <u>Plant Physiology</u> **72**, 925-930.

Schonfield, M., Yaacoby, T., Benyamini, Y., Michael, O. and Rubin, B. (1988). Photosynthetic capacity of triazine-resistant weeds. <u>Proceedings EWRS Symposium Wageningen</u>, 361-366.

COMPARATIVE METABOLISM OF CINNAMIC ACID AND BENTAZON BY SOYBEAN, RICE AND VELVETLEAF (*ABUTILON THEOPHRASTI*) CULTURED CELLS

A.E. Haack and N.E. Balke

Department of Agronomy, University of Wisconsin, Madison 53706, USA

Conversion of cinnamic acid to p-coumaric acid is catalyzed by cinnamic acid 4-hydroxylase (CA4H), a cytochrome P_{450}-mediated mono-oxygenase (Reichhart *et al.*, 1980). This aromatic ring hydroxylation is important for the synthesis of many simple phenolic compounds and the complex polymer (i.e. lignin) in plants. Detoxification of many herbicides also involves aromatic ring hydroxylation, which is often followed by conjugation with sugars. We have previously demonstrated that bentazon detoxification in suspension-cultured soybean and rice cells is via ring hydroxylation and subsequent glucosylation (Sterling and Balke, 1988). Soybean cells metabolise bentazon to both 6- and 8-0-glucoside bentazon, whereas rice cells metabolise bentazon to only 6-0-glucoside bentazon. Velvetleaf cells, which are sensitive to bentazon, do not metabolise bentazon. Because the benzene ring of both cinnamic acid and bentazon are being ring hydroxylated by plant cells, it is possible that the naturally-occurring CA4H catalyses the hydroxylation of bentazon, as well as cinnamic acid. The objective of our present research is to compare the metabolism of cinnamic acid and bentazon by soybean, rice, and velvetleaf suspension cells.

Cultured cells of each species were incubated at 1 µM ^{14}C-cinnamic acid or ^{14}C-bentazon at 3.0 µCi µmol^{-1}. After incubation for times up to 4 h, the cells were homogenised and extracted with 100% methanol. The crude extracts were centrifuged, the supernatants concentrated, and the resulting radiolabelled chemicals were separated by HPLC (C_{18} column) and quantitated by liquid scintillation spectrometry to determine cinnamic acid metabolism.

Preliminary results indicate that uptake of cinnamic acid (nmol g fr. wt^{-1}) is 1.5-fold greater by rice than by either soybean or velvetleaf cells. The identity of the radiolabelled chemicals in the incubation medium at termination of uptake (40% of the original radioactivity) is unknown.

All three species metabolised cinnamic acid to more polar compounds. Resolution of the unknown metabolites extracted from cells was poor using an isocratic solvent system of 45% MeOH and 55% water (containing 0.02% acetic acid, pH 3.6). β-Glucosidase (E.C. 3.2.1.21) treatment (24 h) of the unknown metabolites extracted from soybean cells yielded p-coumaric acid, although not all of the unknowns were cleaved. This result indicated that a portion of the cinnamic acid metabolites was the glucoside of p-coumaric acid. As compared to soybean cells, a lower percentage of the cinnamic acid metabolites extracted from rice was hydrolysed by β-glucosidase, although this treatment did yield a small amount of p-coumaric acid. β-Glucosidase had no effect on the metabolites of cinnamic acid extracted from velvetleaf cells.

A fraction of the unknown metabolites from each species that remained after β–glucosidase treatment was collected and submitted to acid (HCl) hydrolysis. Acid treatment of the pooled soybean unknowns resulted in hydrolysis of only a portion of them to a new unknown compound which eluted from the HPLC column after *p*-coumaric acid and before cinnamic acid. There was no effect of acid hydrolysis on the unknowns pooled from either rice or velvetleaf.

These preliminary results indicate that all three plants metabolise cinnamic acid, but the metabolism is different for each species. Until the identities of the unknown metabolites are determined, we will be unable to compare quantitatively the relative metabolism of cinnamic acid and bentazon by the three species. Currently we are developing methods to provide better resolution and identification of the unknown metabolites of cinnamic acid.

REFERENCES

Reichhart, D., Salaun, J., Benveniste, I. and Durst, F. (1980). Time course of induction of cytochrome P-450, NADPH-cytochrome c reductase, and cinnamic acid hydroxylase by phenobarbitol, ethanol, herbicides and manganese in higher plant microsomes. Plant Physiology **66**, 600-604.

Sterling, T. and Balke, N. (1988). Use of soybean (*Glycine max*) and velvet-leaf (*Abutilon theophrasti*) suspension cultured cells to study bentazon metabolism. Weed Science **36**, 558-565.

INTRA-SPECIFIC VARIATION OF *GALIUM APARINE* TO FLUROXYPYR FROM A RANGE OF SITES IN EUROPE

A.L. Hill and A.D. Courtney

Department of Agricultural Botany, Queen's University, Belfast BT9 5PX

There have been a number of reports of variation in *Galium aparine* sensitivity to mecoprop. Cussans (1984) and Lutman and Lovegrove (1985) have suggested that the degree of tolerance was not linked to previous history of mecoprop use. In an initial series of experiments (Courtney and Hill, 1988), significant differences in response to fluroxypyr were observed in nine populations of *Galium aparine* from a range of European countries. The most tolerant population was one from Italy (Bologna).

In this investigation, 50 populations of *G. aparine* from 12 countries have been compared in a set of seedling tolerance tests.

The populations were sown in seed trays filled with peat-based compost mixed with sand and gravel in a ratio of 2:1:1 and 15 plants of each population were sprayed with five rates of herbicide (120, 100, 80, 60 and 40 g ai ha^{-1}) and replicated three times. Five weeks after spraying, the numbers of live and dead plants were recorded. Two experiments were carried out. The first with 24 populations, was sown in October 1988 and sprayed on 21 December 1988, and maintained in a glasshouse at a mean temperature of 13.1°C. The second, with 26 populations, was sown in February 1989 and sprayed on 14 March 1989, and maintained in a glasshouse at a mean temperature of 15.2°C. From these results, the LD$_{50}$ doses were calculated for each population (Table 1).

Highly significant differences in response to fluroypyr by *G. aparine* were observed, resulting in a 3-fold range of responses. This was similar to the range observed by Lutman and Lovegrove (1985) and by Courtney and Hill (1988) when the reduction in fresh weight was used to calculate the results.

With the more comprehensive range of populations than previously tested (Courtney and Hill, 1988), it can be seen that a range of response is evident in all of the countries featured, with large differences from adjacent sites and little evidence that previous herbicide usage influences weed tolerance.

Table 1. Fluroxypyr LD_{50} doses for populations of *Galium aparine* from Europe.

1st Experiment		LD_{50}	2nd Experiment		LD_{50}
Population			Population		
Bologna	Ita	117.75	Bologna	Ita	97.37
Soulezbach	Fra	102.25	Vatry	Fra	86.93
Gottingen	WG	95.40	Wemeldingen	Hol	85.00
Oberkirchen	WG	92.85	Royer	Fra	82.27
Riddinghausen	WG	84.43	Bologna 87	Ita	82.17
Waringstown	NI	83.93	Long Ashton	Eng	77.77
Mageragall	NI	82.20	Burgundy 88	Fra	73.60
Wendhausen	WG	80.78	Bzouidan	Fra	72.93
Stamfordham	Eng	78.85	Berlin	EG	72.70
Wellesbourne	Eng	73.73	Burgundy 87	Fra	71.27
Kilpatricks	NI	65.53	Landskrona	Swe	69.73
Muhlenbrook	WG	64.63	Semur	Fra	68.00
Darmstadt	WG	59.60	Bridgets	Eng	66.73
Hotzum	WG	59.33	Enstone	Eng	64.60
Sitzborne	WG	58.48	Burgundy 85	Fra	64.03
Wolfach	WG	56.83	Salzdahlum	WG	63.76
Billingsbear	Eng	54.95	Waringstown	NI	63.47
Bonn	WG	53.93	Som	Hun	62.37
Murlough	NI	52.45	Magharafelt	NI	60.67
Hordorf	WG	52.23	Gleadthorpe	Eng	60.60
Prudhoe	Eng	44.03	Drayton	Eng	59.73
Wihr au Val	Fra	40.48	Nyerfesfau	Hun	58.50
Kobiercyze	Pol	40.15	Forli	Ita	57.13
Schobull	WG	37.85	Basel	Swi	49.47
			Zaragoza	Sp	46.67
			Wihr au Val	Fra	31.67

$F = 2.8^{***}$, $CV = 37.4\%$,
 LSD (5%) = 50.88

$F = 3.0^{***}$, $CV = 20.2\%$,
 LSD (5%) = 31.44

REFERENCES

Courtney A.D. and Hill, A. (1988). A preliminary study of variation in response to fluroxypyr in *Galium aparine* from a range of sites in Europe. VIIIème Colloque International sur la Biologie, L'Ecologie et la Systematique des Mauvaises Herbes, Dijon, September 1988.

Cussans, G.W. (1984). Is *Galium aparine* becoming resistant to mecoprop? In "Understanding Cleavers and their Control in Cereals and Oilseed Rape", p. 12. Abstract, Association of Applied Biology Conference.

Lutman, P.J.W. and Lovegrove, A.W. (1985). Variations in the tolerance of *Galium aparine* (cleavers) and *Stellaria media* (chickweed) to mecoprop. Proceedings 1985 British Crop Protection Conference - Weeds 411-418.

ON THE MECHANISMS OF RESISTANCE TO ARYLOXYPHENOXYPROPIONATE, CYCLOHEXANEDIONE AND SULFONYLUREA HERBICIDES IN ANNUAL RYEGRASS (*LOLIUM RIGIDUM*)

J.A.M. Holtum, J.M. Matthews, D.R. Liljegren and S.R. Powles

Waite Agricultural Research Institute, University of Adelaide,
P.M.B. 1, Glen Osmond, Adelaide 5064, South Australia, Australia

Heap and Knight (see these Proceedings) have described biotypes of the grass weed *Lolium rigidum* (annual ryegrass) with resistance to the selective, post-emergent graminicide diclofop-methyl. In all biotypes, which have been identified in all major cereal and grain- legume cropping regions of mainland Australia, resistance has developed following exposure to diclofop-methyl. The resistant biotypes exhibit resistance to a range of other herbicides including the aryloxyphenoxy-propionate (APP) graminicides haloxyfop acid, quizalofop acid and fluazifop acid, the cyclohexanedione (CHD) graminicides sethoxydim, alloxydim and tralkoxydim, the sulfonylureas chlorsulfuron, triasulfuron and metsulfuron-methyl, and to the dinitroaniline herbicide trifluralin. The spectrum of multiple- and cross-resistance is biotype-dependent and, with the exception of resistance to diclofop and trifluralin, is not necessarily associated with any history of exposure to the herbicides against which there is resistance. Between them the biotypes of *L. rigidum* exhibit resistance to every selective post-emergent graminicide registered for use in Australia and are resistant to several compounds which have not been released.

The biochemical mechanism, or mechanisms, responsible for cross-resistance in *L. rigidum* are not known. At least two general, not necessarily mutually exclusive mechanisms could account for the phenomena. These are (i) the altered sensitivity of one or more herbicide target sites, and (ii) a reduction in the concentration of herbicides at their respective target sites. The latter phenomenon could be influenced by factors including a reduction in the rates and amounts of herbicides which enter the resistant plants, reduced conversion of herbicides to their active forms, a reduction in the rates or amounts of herbicides translocated to their target sites, changes in the inter- or intracellular sequestration of herbicides or an increased capacity to detoxify the herbicides. Any mechanism, or mechanisms, must be general enough to account for resistance to a range of structurally distinct herbicides yet specific enough to account for the herbicide specificity which is still observed.

The APP and CHD herbicides inhibit acetyl CoA carboxylase (ACC). ACC extracted from susceptible and resistant *L. rigidum* does not differ in sensitivity to inhibition by APP or CHD herbicides by chlorsulfuron or trifluralin. The extractable activities and acetyl CoA affinities of ACC from the susceptible and resistant biotypes are similar between 5 d post-germination until the onset of tillering. Exposure of plants to diclofop-methyl does not induce changes in either the expression or herbicide sensitivity of the enzyme. It is concluded that differences in expression or inhibition kinetics of ACC are not responsible for the resistance to the APP graminicides and trifluralin, nor are they

445

responsible for cross-resistance to the CHD and sulfonylurea herbicides.

Acetolactate synthase (ALS), a target enzyme of the sulfonylurea herbicides, was extracted from susceptible and resistant biotypes. The sensitivity of the enzyme to inhibition by chlorsulfuron was the same for ALS from both biotypes irrespective of whether the source plants had been exposed to chlorsulfuron or not. The extractable activity of ALS from untreated plants did not suffer. ALS activity from susceptible plants exposed to 100 nM chlorsulfuron was less than that from resistant plants, however, the activity from susceptible plants recovered if the exposure to chlorsulfuron was transient. It is concluded that cross-resistance to chlorsulfuron is not due to differences in either the expression or the chlorsulfuron inhibition kinetics of ALS. An oxidation-based detoxification mechanism is proposed.

^{14}C-Diclofop-methyl was fed in 1 µl droplets to the leaf axils of two-leaved susceptible and resistant plants to determine whether resistance to this herbicide is due to differences in the rate of uptake, transport or metabolism. The rates of uptake and the root/meristem/leaf distribution were similar for both biotypes. Diclofop-methyl was rapidly taken up by both biotypes but, during the 24 h experiment, less than 15% of the radioactivity was transported to the leaves or roots. Both biotypes rapidly demethylated diclofop-methyl to form the biocidal diclofop acid. Radioactivity was subsequently detected in non-biocidal ester glucosides and aryl-O-glucosides of diclofop acid. Susceptible plants accumulated about 15% more radioactivity in diclofop acid than did resistant plants. Resistant plants had a slightly greater capacity to form glucosides. Differences in the glucoside composition of the biotypes appeared to be minor. The small differences in the pool sizes of the active and inactive metabolites, although indicative of different capacities to detoxify the herbicide, are by themselves unlikely to account for a 30-fold difference in sensitivity to the herbicide at the whole plant level. A model, which incorporates observations by us and others, is presented to account for the selectivity and mode of action of diclofop-methyl upon susceptible and resistant biotypes of annual ryegrass.

NON-PLASTID RESISTANCE TO DIURON IN TRIAZINE-RESISTANT WEED BIOTYPES

E. Lehoczki[1], P. Solymosi[2], G. Laskay[1] and E. Pölös[2]

[1] Research Group of the Hungarian Academy of Sciences, Department of Botany, József Atilla University, H-6722 Szeged
[2] Plant Protection Institute, Hungarian Academy of Sciences, H-1525 Budapest
[3] Research Institute for Viticulture and Enology, H-6000 Kecskemét, Hungary

Weed biotypes with plastid-level resistance to atrazine are also resistant to all s-triazines and some asymmetric triazinones. Such resistance is due to a mutation altering the atrazine binding site in the thylakoids (Arntzen *et al.*, 1982). Some triazine-resistant weed biotypes exhibited a different level and pattern of resistance to various phenylurea, pyridazinone and uracil herbicides (Oettmeier *et al.*, 1982; Fuerst *et al.*, 1986; Pölös *et al.*, 1987; Solymosi and Lehoczki, 1988), suggesting that photosystem II herbicides do not have fully identical binding sites and/or the binding sites are altered by different mutations. Since the triazine-resistant biotype always retained its sensitivity to diuron, it was suggested previously that weeds could not develop resistance to diuron. Indeed, cross-resistance between atrazine and diuron in weeds can be precluded on a molecular basis at a plastid level (Gressel, 1986). This study verifies the non-plastid level resistance to diuron in triazine-resistant *Amaranthus retroflexus* and *Conyza canadensis* populations.

Seeds were collected from Hungarian vineyards. A diuron-resistant population of *Amaranthus retroflexus* was found in vineyards where diuron had been applied for 3-4 years to an atrazine-resistant weed population. The atrazine-resistant *Conyza* biotype exhibited cross-resistance at a plastid level to chlorbromuron and metobromuron, and tolerance to linuron (Pölös *et al.*, 1987). Growth conditions, herbicide treatments and fluorescence induction measurements are described by Pölös *et al.* (1987) and Solymosi and Lehoczki (1988).

Atrazine-susceptible plants died within 3 weeks of pre-emergence treatment with 0.5 kg ha^{-1} diuron, whereas atrazine-resistant plants showed no visual symptoms of phytotoxicity even at 2-3 kg ha^{-1} diuron.

Surviving plants of resistant *Amaranthus* and *Conyza* biotypes were also affected by the electron transport-blocking action of diuron, as revealed by fluorescence induction kinetics in intact leaves. This inhibition disappeared within 30-60 days in *Amaranthus* and 3-4 months in *Conyza*.

If previously untreated 2-3 month-old plants were treated post-emergent with 2 kg ha^{-1} diruon, the fluorescence induction curves demonstrated the uptake and translocation of diuron to its active site in both susceptible- and -resistant biotypes. Susceptible plants died within three weeks, but resistant plants gradually recovered their photosynthetic competence, faster in *Amaranthus* than in *Conyza*.

When leaf-discs were floated on diuron solutions, both biotypes of both species showed a similar sensitivity to diuron after 24 h of treatment.

We conclude that the atrazine-resistant populations of *Amaranthus* and *Conyza* contain sub-populations with resistance to diuron. This resistance is probably unrelated to alteration in the binding site of diuron.

REFERENCES

Arntzen, C.J., Pfister, K. and Steinback, K.E. (1982). The mechanism of chloroplast triazine resistance: Alterations in the site of herbicide action. In "Herbicide Resistance in Plants" (H.M. LeBaron and J. Gressel, eds), pp. 158-214. John Wiley and Sons, New York.

Fuerst, E.P., Arntzen, C.J., Pfister. K. and Penner, D. (1986). Herbicide cross-resistance in triazine-resistant biotypes of four species. Weed Science 34, 344-353.

Gressel, J. (1986). Modes and genetics of herbicide resistance in plants. In "Pesticide Resistance: Strategies and Tactics for Management" 54-73. National Academic Press, Washington.

Oettmeier, W., Masson, K., Fedtke, V., Konze, J. and Schmidt, R.R. (1982). Effect of photosystem II inhibitors on chloroplasts isolated from species either susceptible or resistant towards triazine herbicides. Pesticide Biochemistry and Physiology, 18, 347-367.

Pölös, E., Laskay, G., Szigeti, Z., Pataki, Sz. and Lehoczki, E. (1987). Photosynthetic properties and cross-resistance to some urea herbicides of triazine-resistant *Conyza canadensis* Cronq (L.). Zeitschrift für Naturforschung 24c, 783-793.

Solymosi, P. and Lehoczki, E. (1988). Co-resistance of atrazine-resistant *Chenopodium* and *Amaranthus* biotypes to other photosystem II inhibiting herbicides. Zeitschrift für Naturforschung 44c, 119-127.

THE OCCURRENCE OF HERBICIDE-RESISTANT WEEDS IN TREATED STRIPS IN ORCHARDS

J. Lipecki[1] and R. Stanek[2]

[1] Department of Pomology, Agricultural University, Lublin, Poland
[2] Department of Plant Physiology, Agricultural University, Lublin, Poland

Reports of some weed species resistant to triazines have increased in many countries in recent years (LeBaron, see these Proceedings). We have tried to find out whether the specific conditions occurring in herbicide strips in orchards (low pH, high nitrogen level, shadow under trees) could stimulate the growth and development of such weeds. Pot experiments in 1984–1989 examined the effect of simazine and/or nitrogen on resistant biotypes of *Erigeron canadensis* L., *Amaranthus retroflexus* L. and recently *Capsella bursa-pastoris* (L.) Med. The results are given in Tables 1 and 2.

Table 1. **Effect of nitrogen and/or simazine on** *Capsella bursa-pastoris*[*].

Treatment	Length of shoots/ plant (cm)	Fresh weight per plant (g)	No. of lateral shoots/ plant	No. of siliques /plant
Control	73	2.46	2.25	88.9
100 kg N ha^{-1} as ammonium nitrate (N)	140	4.71	4.53	144.1
Simazine,3 kg^{-1} (S)	87	2.61	2.12	107.9
N + S as above	122	4.08	4.50	131.6

[*] Treated 3 June, measured 21 June, 1989.

These results show that nitrogen fertilisation stimulates the biomass production of resistant weeds; it is also possible that these plants will produce more seeds than without nitrogen – see for example the number of laterals and siliques in *Capsella bursa-pastoris* (L.) Med. (Table 1) and the number of laterals in *Erigeron canadensis* L. (Table 2). This would explain, at least partially, the high density of these weeds in herbicide strips in orchards. Simazine alone did not show such a clear effect in most cases, as reported by Weaver and Thompson (1982) in relation to atrazine, and also the combined use of nitrogen + simazine was slightly less than that of nitrogen alone. Further experiments including other weed species are under way.

Table 2. **Effect of nitrogen and/or simazine on the growth of** *Erigeron canadensis* **L. and** *Amaranthus retroflexus* **L. (1988).**

Treatment	*Erigeron canadensis* L.		*Amaranthus retroflexus* L.	
	Height of plant (cm)	Fresh weight /plant (g)	No. of laterals /plant	Height of plant (cm)
Control	18.0	1.29	13.8	10.3
Nitrogen*	18.1	2.27	17.4	20.3
Simazine	18.9	1.21	15.0	11.3
N + S	19.2	1.88	18.8	20.3

* Treatments as in Table 1. Experiments done in 1988.

REFERENCE

Weaver, S.I. and Thompson, B.K. (1982). Comparative growth and atrazine response of resistant and susceptible populations of *Amaranthus* from Southern Ontario. <u>Journal of Applied Ecology</u> **19**, 611–620.

EFFECT OF HEAT AND HERBICIDE STRESS IN CULTURED SOYBEAN CELLS

M. McElwee, R.H. Burdon, R.C. Kirkwood and P.A. Boyd

Division of Biology, Department of Bioscience and Biotechnology, Todd Centre, University of Strathclyde, Glasgow G4 0NR, Scotland

The heat shock response is a fundamental homeostatic phenomenon whereby brief exposure of animal, plant or bacterial cells to sublethal temperatures subsequently confers tolerance to normally lethal temperatures. Because prior heat stress is also known to increase resistance of mammalian cells to certain cytotoxic anti-cancer drugs, the effect of prior heat stress on the susceptibility of cultured plant cells (soybean) to certain herbicides was initially investigated.

Preliminary studies using Northern blotting and specific cloned gene sequences showed that brief heat stress of cultured soybean cells (40°C, 30 min) was sufficient to induce elevated levels of mRNAs for HSP70 as well as for ubiquitin and catalase. In subsequent experiments the soybean cells were heat stressed (40°C, 3 h) prior to exposure to the herbicide chlorsulfuron (which inhibits the enzyme acetolactate synthase). Such heat stress leads to a reduced susceptibility to the toxic effects of the herbicide (at 1 ppm) as judged by assay of cellular respiratory function.

Experiments were then carried out to examine the effect of herbicide stress alone. When soybean cells were exposed to chlorsulfuron on a continuous basis, the induction of novel proteins was detectable after 3 days even by direct silver staining of SDS-PAGE gels. The level of these particular herbicide-stress proteins however began to decline by 7 days of herbicide exposure. Heat stress prior to herbicide exposure had no effect on the induction of these herbicide-stress proteins.

If the exposure of the soybean cells to herbicide is extended to 10 weeks and the spectrum of proteins in the resultant herbicide-tolerant cells is examined a different set of herbicide-stress proteins appear to have been induced, particularly in the 90 kDa region.

If herbicide is withdrawn from the culture medium of such cells there was a reduction in the cellular level of these novel stress proteins. Nevertheless they reappeared if the chlorsulfuron was reintroduced into the medium.

Further investigation of these herbicide-stress proteins may provide a route to novel methods for resistance breeding.

INHERITANCE OF SULFONYLUREA HERBICIDE RESISTANCE IN PRICKLY LETTUCE (*LACTUCA SERRIOLA*) AND DOMESTIC LETTUCE (*LACTUCA SATIVA*)

C. Mallory-Smith, D.C. Thill, M.J. Dial and R.S. Zemetra

Department of Plant, Soil and Entomological Sciences, University of Idaho, Moscow, Idaho 83843, USA

A sulfonylurea herbicide-resistant (R) biotype of *Lactuca serriola* L. (prickly lettuce), was identified in a no-till winter wheat field in northern Idaho, USA (Mallory *et al.*, 1989, 1990). The field had been treated annually for five years with sulfonylurea herbicides. In field and glasshouse studies, the R biotype was always more resistant than was the susceptible (S) biotype to eight sulfonylurea herbicides. In order to determine the genetic control of resistance, the R biotype was crossed with the S biotype and with Bibb (B) domestic lettuce (*L. sativa*) (Thill *et al.*, 1989).

The S biotype and B domestic lettuce were used as pollen acceptor flowers, and open flowers from the R biotype were used as pollen donors.

A foliar and soil drench of 500 ppb a.i. w/v metsulfuron was applied to the parents, the F_1 and F_2 generations of the S x R cross. The F_2 generation of the B x R cross and the F_3 generations of both crosses were treated with 13 g a.i. ha^{-1} of metsulfuron applied through a glasshouse spray chamber. A non-ionic surfactant (0.5% v/v) was added to all herbicide treatments.

F_1 plants that survived the herbicide treatment were allowed to self to produce F_2 seed. The F_2 seedlings were treated with metsulfuron. The plants were rated as resistant, intermediate and susceptible. The resistant and intermediate plants were transplanted and grown to maturity for F_3 seed production. The F_3 generations were treated with metsulfuron and evaluated as before for segregation of the resistance trait.

Chi Square analysis was used to determine the genetic control of sulfonylurea resistance. The best fit for Chi Square analysis for the generation was a 1:2:1 ratio of resistant, intermediate and susceptible seedlings (S x R χ^2 = 2.21, 0.25 < $\underline{P}$ < 0.50; B x R χ^2 = 0.1084, 0.90 < $\underline{P}$ < 0.95) indicating the trait is controlled by a single nuclear gene with incomplete dominance.

In the F_3 generation, all the seedlings from the resistant plants were resistant while the seedlings from the intermediate plants again segregated (S x R 3:1, χ^2 = 3.26, 0.10 < $\underline{P}$ < 0.25; B x R 1:2:1, χ^2 = 0.04, 0.975 < $\underline{P}$ < 0.99). These results confirm the analysis for the F_2 generation that the trait is controlled by a single gene with incomplete dominance.

REFERENCES

Mallory, C.A., Dial, M.J. and Thill, D.C. (1989). Sulfonylurea resistant prickly lettuce (*Lactuca serriola*). Weed Science Society of American Abstracts **29**, 175.

Mallory-Smith, C.A., Thill, D.C. and Dial, M.J. (1990). Identification of sulfonylurea herbicide resistant prickly lettuce (*Lactuca serriola*). Weed Technology **4**, 163-168.

Thill, D.C., Mallory, S.A., Saari, L.L., Cotterman, J.C. and Primiana, M.M. (1989). Sulfonylurea resistance – mechanism of resistance and cross-resistance. Weed Science Society of American Abstracts **29**, 297.

USE OF CHLOROPHYLL FLUORESCENCE TO STUDY THE DISTRIBUTION OF PHOTOSYSTEM II- INHIBITING HERBICIDES IN LEAVES

S. Mona, J-M. Ducruet[1], P. Ravanel and M. Tissut

Laboratoire de Physiologie Cellulaire Végétale, Université J. Fourier, BP 53X, 38041 Grenoble
[1] CEA/INRA Saclay, SBPH Division de Biologie, 91191, Gif-sur-Yvette, France

The tolerance of crops or weed species to herbicides is, in most cases, ascribable to differential foliar or root absorption, translocation, detoxification and distribution of active compounds in different compartments of plant cells. These factors of selectivity are generally investigated by chemical analysis or by using radiolabelled compounds. However, the proportion of the active compounds which is effectively bound to the cellular target remains unknown.

In the particular case of photosystem II-inhibiting herbicides, advantage can be taken from their unique property of inducing a high chlorophyll fluorescence yield by binding to their active site, in order to estimate their root absorption and detoxification (Cadahia *et al.*, 1982; Ducruet *et al.*, 1984) or their foliar absorption (Voss *et al.*, 1984; Habash *et al.*, 1985). This fluorescence emission reflects the blocking of the electron flow between the primary and the secondary acceptors Q_A and Q_B of PS II. As described earlier (Ducruet *et al.*, 1984), the fluorescence signal is digitized by an A/D conversion card plugged into a microcomputer (Apple II or PC), which provides a large memory to store the signal and computes ratios of O, I and P levels and other parameters. In the case of isolated chloroplast suspensions, identical dose response curves are obtained by using either increases of fluorescence I level or uncoupled Hill reaction rates (Ahrens *et al.*, 1981). In contrast, leaf fragments infiltrated by PS II herbicides exhibit a higher level of inhibition by fluorescence kinetics, which directly reflect the proportion of the blocked centres, than by oxygen evolution, due to the non-limiting role of PS II in the overall photosynthetic electron flow in leaves (Mona *et al.*, 1989). Therefore, taking into account the amount of compound introduced in leaf fragments and the dose/response curves in isolated chloroplasts, this provides a measuring tool for the "active" fraction of the herbicide (actually bound to the inhibitory site).

The binding to the site itself represents a major cause of depletion of a herbicide in the leaf and the site concentration, in the micromolar range, can be estimated from the chlorophyll concentration per gram fresh leaves. Atrazine-resistant (R) plants can be used, in comparison with atrazine-susceptible (S) plants, to assess the influence of the site-induced depletion. The I_{50} values measured on isolated *Chenopodium album* chloroplasts at low chlorophyll concentrations are 500-fold higher in R compared to S chloroplasts. In excised leaves dipped in atrazine, the time-dependent development of inhibition for R leaves in 100 µM atrazine was approximately half that observed for S leaves in 10 µM atrazine. From these leaf measurements, the ratio between the capacities of inhibition by atrazine in S versus R plants could be estimated in the 20- to 50-fold range, Hence, the actual level of resistance in leaves is lower than estimated from Hill reaction I_{50} on isolated chloroplasts, mainly due to depletion by binding on S sites.

REFERENCES

Ahrens, W.H., Arntzen, C.J. and Stoller, E.W. (1981).   Chlorophyll fluorescence assay for the determination of triazine resistance.   Weed Science **29**, 316-322.

Cadahia, E., Ducruet, J-M. and Gaillardon, P. (1982).   Whole leaf fluorescence as a quantitative probe of detoxification of the herbicide chlortoluron in wheat.   Chemosphere **11**, 445-450.

Habash, D., Percival, M.P. and Baker, N.R. (1985).   Rapid chlorophyll fluorescence technique for the study of penetration of photosynthetically active herbicides in leaf tissue.   Weed Research **25**, 389-395.

Mona, S., Ducruet, J-M., Nurit, F., Ravanel, P. and Tissut, M. (1989). Distribution of phenmedipham and other bicarbamates inside leaf fragments of *Chenopodium album* and *Spiracea oleracea* plants.   Chemosphere **18**, 2077-2082.

Voss, M., Renger, G., Gräber, P. and Kotter, C. (1983).   Measurements of penetration and detoxification of PS II herbicides in whole leaves by a fluorometric method.   Zeitschrift Naturforschung **39c**, 359-361.

DIFFERENTIAL RESPONSES OF WHEATS AND RYE TO A SULFONYLUREA HERBICIDE

A. Monteira, I. Moreira and W.S. Viegas,

Departmento de Botânica, Instituto Superior de Agronomia, Tapada da Ajuda, 1300 Lisboa, Portugal

Some of the herbicides introduced in recent years behave mainly as inhibitors of specific enzymes responsible for the production of essential amino acids. Sulfonylureas, for instance, block the production of the essential amino acids valine and isoleucine by inhibiting the enzyme acetolactate synthase (Chaleff and Mauvais, 1984; Chaleff and Ray, 1984; La Rosa and Schloss, 1984; Lee *et al.*, 1988). There is a well documented report of the existence of diverse differential responses of wheat varieties to several herbicides (Tottman *et al.*, 1983). Therefore it was decided to analyse responses of different wheat cultivars to a sulfonylurea herbicide.

We compared the tolerance of four cultivars of *Triticum durum* L. (cvs 'Celta', 'Timpasnas', 'Castiça" and 'Faia'). and of *T. aestivum* cv. 'Chinese Spring', *Secalis cereale* cv. 'Centeio do Alto' and Triticosecale (6n) cv. 'Borba' to a methyl sulfonylurea herbicide (DPX-L5300).

In all experiments, seeds were germinated in different aqueous solutions of the herbicide at pH 6.75, and the results were compared with water only. For each herbicide concentration 75 seeds were used, distributed in three Petri dishes, kept during three days in the dark at $20\pm1°C$. After this period the coleoptile length of each seedling was measured and the mitotic index was determined in root tips from untreated and treated seeds, which were directly fixed in acetic acid: alcohol (3:1).

The most striking differences observed in responses to the herbicide were between hexaploid and tetraploid wheats. Although the susceptible behaviour of hexaploid wheat cv. Chinese Spring could be accounted for in this particular cultivar, it is important to note that besides the marked differences in the degree of susceptibility, even at the lower herbicide concentration (10 ppb) between both species (an average reduction in coleoptile length of 62% in the hexaploid compared with 35% in the tetraploids), there were also differences between their patterns of response to increasing concentrations (cf. Fig. 1). The hexaploid wheat showed, after a drastic initial reduction in coleoptile length, a continuous decrease in the development of seedlings germinated in increasing herbicide concentrations. In contrast, the pattern in tetraploid wheats showed no significant difference in development between treatments with 10 to 50 ppb of the herbicide. Similar patterns were also observed in triticosecale and in rye.

Only a small difference was observed on mitotic indexes between root tips from seeds germinated in water (4.4%) or in the herbicide (3.6%). However, in the herbicide-treated tips, we found a high number of cytologically disturbed cells, mainly with errors in chromosome orientation and segregation in mitosis; this might account for the low rates of development observed in treated seedlings.

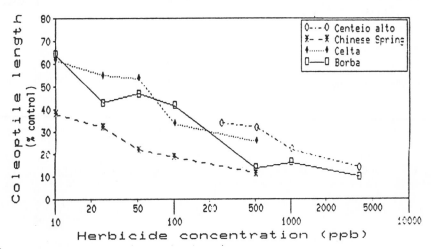

Fig. 1.

Further experiments are in progress to determine whether the differential responses observed between hexaploid and tetraploid wheats are due to differences in absorption rates, in metabolism or in sensitivity of the acetolactate synthase to the herbicide (4). The possible role of genes on D genome chromosomes in herbicide resistance is also of interest.

ACKNOWLEDGEMENTS

We thank Mrs Augusta Barão for her precious technical assistance. The work was supported by a research grant from the Instituto Nacional de Investigacão Scientifica.

REFERENCES

Chaleff, R.S. and Mauvais, C.J. (1984). Science **224**, 1443-1445.
Chaleff, R.S. and Ray, T.B. (1984). Science **223**, 1148-115
La Rosa, R.A. and Schloss, J.V. (1984). Journal Biological Chemistry **259**, 8753-8757.
Lee, K.Y., Townsend, J., Tepperman, J., Black, M., Chui, C.J., Mazur, B., Dansmuir, P. and Bedbrook, J. (1988). The EMBO Journal **7**(5), 1241-1248.
Tottman, D.R., Lupton, F. and Oliver, R.H. (1983). Annals of Applied Biology **104**, 151-159.

OCCURRENCE OF CYTOCHROME P_{450} MONO-OXYGENASES IN THE METABOLISM OF CHLOROTOLURON BY WHEAT MICROSOMES

C. Mougin, R. Scalla and F. Cabanne

Laboratoire des Herbicides (INRA), BV 1540, 20134 Dijon Cedex, France

Wheat cell suspension cultures metabolise the herbicide chlorotoluron by ring-methyl hydroxylation, N-demethylation and sugar conjugation (Cole and Owen, 1987; Canivenc *et al.*, 1989). Cell suspension cultures were used to study oxidative transformation of chlorotoluron *in vitro*. Achlorophyllous cells of wheat cv. Koga II were cultivated as previously described (Canivenc *et al.*, 1989). Microsomes were isolated by differential centrifugation from cells treated with cyometrinil.

Aerobic incubations of microsomes with chlorotoluron and NADPH led to the formation of ring-methyl hydroxylated and N-monodemethylated chlorotoluron. Metabolites were analysed in subcellular fractions after differential centrifugation. They were mainly found in the microsomal fraction in mixture with NADPH-cytochrome P_{450} (cytochrome c) reductase activity. Microsomes metabolised the herbicide at rates ranging from 600 to 650 and 200 to 250 pmoles mg^{-1} 30 min^{-1} for the ring-methyl hydroxylase and N-demethylase, respectively. Microsomal pellets contained substantial levels of cytochrome P_{450} (300 pmol mg $protein^{-1}$).

In cells grown in the presence of cyometrinil, chlorotoluron ring-methyl hydroxylase, N-demethylase, and lauric acid hydroxylase activities were stimulated respectively 1.6-, 3.3- and 1.4-fold compared to controls. The cytochrome P_{450} content of microsomes was increased 1.8-fold by cyometrinil. Ring-methyl hydroxylase and N-demethylase activities exhibited their highest rates with NADPH. Other reductants such as NADH, dithionite or ascorbate were ineffective. Molecular oxygen was also necessary for the NADPH-dependent activities. Herbicide hydroxylase and N-demethylase were drastically reduced by inhibitors of the P_{450} system, namely ABT, *p*-CMB, menadione and cytochrome c.

Carbon monoxide (CO) inhibited the hydroxylase activity with partial reversion by light. In contrast, N-demethylase activity was apparently insensitive to CO. Only the monodemethylated metabolite could also be formed in the presence of cumene hydroperoxide as oxygen donor.

These data allow us to postulate that chlorotoluron ring-methyl hydroxylase and N-demethylase activities probably belong to the family of cytochrome P_{450} mono-oxygenases. The fact that the N-demethylation can consume cumene hydroperoxide leads to a possible role of another type of mono-oxygenase, for example a peroxygenase-type enzyme, in addition to cytochrome P_{450}.

REFERENCES

Canivenc, M.C., Cagnac B., Cabanne, F. and Scalla, R. (1989). Induced changes in chlorotoluron metabolism in wheat cell suspension cultures. Plant Physiology and Biochemistry **27**, 193-201.

Cole, D.J. and Owen, W.J. (1987). Influence of mono-oxygenase inhibitors on the metabolism of the herbicides chlortoluron and metolachlor in cell suspension cultures. Plant Science **50**, 13-20.

ARTHROBACTER SP. NI 86/21 AGAINST PHYTOTOXICITY OF THIOCARBAMATE HERBICIDES

I. Nagy and J. Nagy

North Hungarian Chemical Works, Sajóbábony, H-3792, Hungary

Non-spore forming irregular rod shape bacteria have been reported to affect thiocarbamate decomposition (Tam *et al.*, 1987; Nagy *et al.*, 1987; Miwa *et al.*, 1988). *Pseudomonas aureofaciens* 37 was reported (Baliczka *et al.*, 1985) to metabolise cycloate thereby decreasing the phytotoxicity of this compound on wheat.

We have investigated the ability of *Arthrobacter* sp. NI 86/21 (HE2) to decompose thiocarbamate in soil and basal salt medium as well as its safening effect in glasshouse experiments. This isolate of *Arthrobacter* was obtained from the rhizosphere of corn which was sown in soil previously twice exposed to EPTC.

Arthrobacter sp. NI 86/21 was able to grow on EPTC, butylate and vernolate (100 mg l^{-1}) as sole carbon and energy source. The metabolism of cycloate by this strain seemed to be inhibited by cycloate or its metabolite(s) as was established by mathematical analysis of decomposition curves of thiocarbamates. The strain was not able to grow on molinate. Our experiments revealed that *Arthrobacter* sp. NI 86/21 contains certain materials with safening properties against the phytotoxic effect of thiocarbamates in corn. Application of *Arthrobacter* sp. NI 86/21 together with phytotoxic doses of EPTC, butylate and vernolate on corn decreased or diminished phytotoxic damage. Inoculating the soil with this strain at ≥ 3.3-9.3×10^4 cells g^{-1} protected corn from the phytotoxic effect of EPTC and butylate. The strain was ineffective against cycloate and molinate injury (Tables 1 and 2).

Table 1. <u>Safening effect of *Arthrobacter* sp. NI 86/21 against EPTC, butylate, vernolate, cycloate and molinate, as indicated by shoot length or corn. (Data as % of untreated control.)</u>

Thiocarbamate (1 ha^{-1})	*Arthrobacter* sp. NI 86/21 9.3 x 10 cell g^{-1} soil^{-1}					
	0	10^2	10^3	10^4	10^5	10^6
EPTC (6)	40	41	40	88	104	103
Butylate (12)	73	85	78	100	106	102
Vernolate (6)	39	40	46	55	54	111
Cycloate (6)	31	31	35	33	32	33
Molinate (6)	23	31	27	33	29	19

LSD = 14.

Table 2.   Safening effect of *Arthrobacter* sp. NI 86/21 against EPTC, butylate, vernolate, cycloate and molinate, as indicated by fresh weight of corn. (Data as % of untreated control).

| Thiocarbamate | | *Arthrobacter* sp. NI 86/21 9.3×10 cell g^{-1} soil^{-1} | | | | |
(1 ha^{-1})	0	10^2	10^3	10^4	10^5	10^6
EPTC (6)	77	76	76	98	107	102
Butylate (12)	90	103	103	111	114	103
Vernolate (6)	75	72	88	92	90	121
Cycloate (6)	43	42	42	46	45	43
Molinate (6)	63	72	62	76	65	63

LSD = 17.

REFERENCES

Balicka N., Kosinkiewicz, B. and Wegryzn, T. (1985). Effect of *Pseudomonas aureofaciens* 37 metabolites on the activity of Roneet. Les Colloques de INRA No. 31. Versailles 3-4 June, 1984.

Miwa, N., Takeda, Y. and Kuwatsuka, S. (1988). Plasmid in the degrader of the herbicide thiobencarb isolated from soil. A possible mechanism for enrichment of pesticide degraders in soil. Journal of Pesticide Science **13** (2), 291-294.

Nagy, I., Nagy, J., Mátyás, J. and Kecskés, M. (1987). Decomposition of EPTC by soil microbes in two soils. British Crop Protection Council Conference - Weeds, Brighton **1/4**, 525-530.

Tam, C.A., Behki, R.M. and Khan, S.U. (1987). Isolation and characterization of an S-ethyl-N,N-dipropylthiocarbamate-degrading *Arthrobacter* strain and evidence for plasmid-associated S-ethyl-N,N-dipropylthiocarbamate degradation. Applied and Environmental Microbiology **53**, 1088-1093.

J. Nagy, I. Nagy and E. Lorik

North Hungarian Chemical Works, Sajóbábony, H-3792 Hungary

Pre-sowing treatment of corn with 6 kg ha^{-1} EPTC causes severe damage to the plants. We found however, in laboratory experiments that after EPTC treatment harvesting the 5-10 day-old seedlings and repeating the experiment in the same soil, the second crop showed increasing tolerance to EPTC. We assumed there is a relationship between the increased tolerance of corn and the presence of the roots of the previously harvested plants left in the soil.

To test this hypothesis we did further laboratory experiments assaying the safening activity of extracts from EPTC-treated corn roots and shoots.

Pi-3707 corn hybrid was grown without seed dressing in soil in pots. The pots were treated with 6 kg ai ha^{-1} EPTC at the time of sowing and half of the pots received a second EPTC treatment on the 11th day. Shoots and roots were harvested on the 13th day and extracted twice, first with water followed by acetone. The extracts were then tested for biological activity. Corn was also grown in soil in pots after treating pre-emergence with the shoot-extracts or root-extracts of untreated or EPTC-treated corn.

In control plants, shoot length was 19 cm and general appearance was scored as 100; in EPTC-treated plants the corresponding measurements were 9 cm and 59 (Tables 1 and 2). General appearance of the seedlings was evaluated using the scale below: 100% normal growth without distortion; 99-87% slight distortion of shoots; 86-75% medium distortion and/or twist of the shoots; 74-51% strong distortion; 50% death.

Table 1.　Shoot length and general appearance of corn seedlings treated with shoot-extracts of untreated or EPTC-treated corn.

Pre-treatment of extracted corn	Solvent	Shoot length (cm)		Appearance (%)	
		Control	EPTC	Control	EPTC
Untreated	water	21	9	100	50
EPTC (once)	water	20	16	100	100
EPTC (twice)	water	2	3	50	50
Untreated	acetone	19	16	100	100
EPTC (once)	acetone	18	18	99	100
EPTC (twice)	acetone	19	13	100	74

Table 2. Shoot length and general appearance of corn seedlings treated with root-extracts of untreated or EPTC-treated corn.

Pre-treatment of extracted corn	Solvent	Shoot length (cm)		Appearance %	
		Control	EPTC	Control	EPTC
Untreated	water	6	3	50	50
EPTC (once)	water	7	5	72	50
EPTC (twice)	water	6	7	50	50
Untreated	acetone	18	17	100	100
EPTC (once)	acetone	19	16	100	95
EPTC (twice)	acetone	18	16	100	100

The results show that the extracts of shoots and roots of EPTC-treated corn contain unknown substances exhibiting safening or herbicidal activity depending on previous EPTC treatments of the extracted plant and the solvent used for extraction. Safening or phytotoxic substances are also present in the extracts of untreated corn.

It is suggested that certain substances from higher plants may decrease or eliminate herbicide damage to crop plants. A new antagonistic interaction is indicated.

INVESTIGATION OF SIMAZINE, LINURON AND TRIFLURALIN RESISTANT AND SUSCEPTIBLE WEED SPECIES IN BULGARIA

G.K. Nikolova, K.A. Konstaninov and V.A. Nikolova

Plant Protection Institute, Kostinbrod 2230, Bulgaria

Several plant species have been reported to have developed populations showing resistance to simazine, linuron and trifluralin herbicides (Gressel *et al.*, 1982). In Bulgaria, some weeds species have been found in areas where simazine, linuron and trifluralin had been applied annually for at least 6–8 years (Fetvadjieva *et al.*, 1985).

The purpose of this study was to determine the influence of simazine and linuron doses on growth and seed germination of resistant (R) and susceptible (S) biotypes of *Aramanthus retroflexus* and trifluralin on *Echinochloa crus-galli* respectively.

Seeds of R *A. retroflexus* were collected from plants grown in apple and raspberry plantations, treated with simazine for at least 6–8 years, and R *E. crus-galli* seeds, from vegetable crops treated with trifluralin. The seeds of S plants of *A. retroflexus* and *E. crus-galli* were collected from plants in an area that had never received a herbicide application. Seeds of R and S plants of *A. retroflexus* and *E. crus-galli* were sown in plastic pots

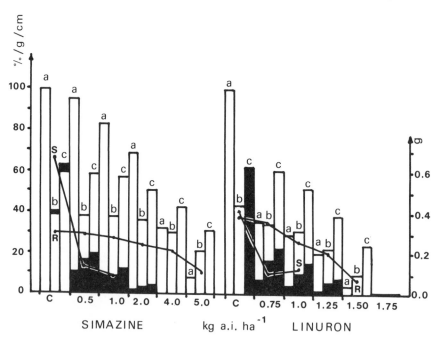

Fig. 1. Effect of simazine and linuron on (a) germination of the seeds (%); (b) height (cm); (c) total above ground fresh weight (g); and seed dry weight (g) of resistant (R, •—•) and susceptible (S, •—•) *Amaranthus retroflexus* compared to the controls.

Table 1. **The influence of trifluralin on % seed germination (G); plant height in cm (PH); total above ground fresh weight (AGF wt.); and seed dry weight in g (SD wt.) of the resistant (R) and susceptible (S)** *E. crus-galli.*

Rate kg ai ha⁻¹	R				S			
	G	PH	AGF wt.	SD wt.	G	PH	AGF wt.	SD wt.
Control	100.0	46.6	58.3	0.76	100.0	44.2	64.4	0.52
0.48	94.4	38.9	48.9	0.53	60.9	18.3	34.8	0.33
0.72	87.7	34.7	40.5	0.39	53.3	21.0	23.9	0.26
0.84	76.5	32.1	35.0	0.30	27.9	14.8	16.5	0.18
0.96	47.9	16.6	24.8	0.19	4.1	9.6	11.3	0.06
1.08	7.1	9.5	10.9	0.13	–	–	–	–

containing sandy loam soil (2.6% o.m; pH 6.2), under glasshouse conditions (daily temp. 22–26°C). Soil in pots was treated with graduated rates of simazine – 0.5, 1.0, 2.0, 4.0, 5.0; linuron – 0.75, 1.00, 1.25, 1.50, 1.75; and trifluralin –0.48, 0.72, 0.84, 0.96, 1.78 kg a.i. ha⁻¹. Seed germination, the height, and fresh weight of the stems and leaves and dry weight of the seed were estimated at maturity. All glasshouse experiments were repeated four times.

Plant height and above ground fresh weight are more for the simazine and linuron R compared to S biotype of *A. retroflexus* (Fig. 1); the R plants also had more seeds. The R biotype of *E. crus-galli* produced more above ground biomass than the S biotype (Table 1).

The results show that simazine R biotypes of *A. retroflexus* occur in horticultural crops in Bulgaria. In vegetable crops, linuron R biotypes of *A. retroflexus* and trifluralin R biotypes of *A. crus-galli* have been found.

REFERENCES

Gressel, J., Ammon, H.I., Fogelfords, H., Gasquez, J., Kay, Q.O.N. and Kees, H. (1982). Discovery and distribution of herbicide resistant weeds outside North America. In "Herbicide Resistance in Plants", (H.M. Le Baron and J. Gressel, eds) 31–35. Wiley Interscience, New York.

Fetvadjieva, N., Nikolova, G. and Konstaninov, K. (1985). Investigation of *Amaranthus retroflexus* and *Echinochloa crus-galli* resistance to herbicides. Symposium of XII Scientific Meeting of CMEA, 37–49.

EFFECTS OF THE HERBICIDE SAFENER NAPHTHALIC ANHYDRIDE ON THE GROWTH OF A *ZEA MAYS* L. CELL SUSPENSION CULTURE: INTERACTION WITH THE HERBICIDE METSULFURON-METHYL AND 2,4-D

N.D. Polge[1], A.D. Dodge[1] and J.C. Caseley[2]

[1]School of Biological Sciences, University of Bath, Bath, UK
[2]Department of Agricultural Sciences, University of Bristol, AFRC Institute of Arable Crops Research, Long Ashton Research Station, Long Ashton, Bristol, BS18 9AF, UK

Naphthalic anhydride (NA) enhances the tolerance of maize and some other cereals to several "growth inhibitor-type" herbicides such as thiocarbamates and acetanilides (Hatzios, 1983). It has also been reported to provide partial protection against the sulfonylurea, metsulfuron-methyl (Richardson and West, 1984).

The mechanism(s) of NA action are not clearly understood (Hatzios, 1983). Effective safener rates cause a slight inhibition of early plant growth (Hickey and Krueger, 1974), which may suggest a direct hormonal effect of NA, or an indirect effect on plant hormone levels. Exogenous applications of the plant hormones abscisic acid (ABA) and gibberellic acid (GA_3), and the auxins indoleacetic acid (IAA) and phenylacetic acid (PAA) have been found to provide protection against "growth inhibitor-type" herbicides (Field and Caseley, 1987; Wilkinson, 1989). Naphthalic anhydride has also been reported to produce auxin-like responses in maize (Field *et al.*, 1987). Enhanced metabolism of herbicides via the mixed function oxidases (MFOs), has also been suggested as a possible mode of NA action (Sweetser, 1985).

The effect of NA on the growth of rapidly dividing maize cells was investigated using cell suspension cultures. Possible safening agents metsulfuron-methyl in this system, and interaction with 2,4-D used as an artificial hormone in the growth media were also evaluated.

<u>Zea mays</u> L. var. Black Mexican Sweet suspension cultures were maintained in Murashige and Skoog media containing 2 mg l^{-1} 2,4-D and subcultured at 7 day-old intervals. Ten ml of 7 day cells were added to the 40 ml media in 250 ml conical flasks, and placed on an orbital shaker (120 rpm) in a growth room, at 25°C and 16 h light.

Naphthalic anhydride (97% a.i.) was dissolved in acetone before addition to cultures. Final acetone concentration in all cultures was 0.5% v/v. Technical grade metsulfuron-methyl was dissolved in growth media and filter sterilised before addition to the cultures.

Growth was measured by settled cell volumes of 10 ml aliquots of cultures taken at intervals during the growth period, and/or by culture dry weights at 7 or 9 days' growth.

Naphthalic anhydride treatments (10^{-7}M to 5 x 10^{-5}M) applied at subculture caused a concentration-dependent reduction in culture growth at 10^{-6}M and above. This response was associated with an increase in the generation time of cultures. Fluorescein-diacetate/Evans Blue staining of cells revealed no loss of viability associated with NA treatments, except at 5 x 10^{-5}M.

Metsulfuron-methyl (1 ppm) applied 24 h after subculture resulted in a 70% reduction in culture dry weights after 7 days growth, compared to controls. Treatment of cultures with 5 x 10^{-6}M NA at subculture gave no subsequent protection against metsulfuron-methyl applied at 24 h.

Altering the 2,4-D concentration of the culture media was found to have a significant effect on the growth response of cultures to NA treatments. At 0.2 mg l^{-1} 2,4-D, growth inhibition induced by NA increased, and significant effects occurred at 10^{-7}M NA and above. Increasing 2,4-D to 10 mg/l^{-1} caused some reduction of culture growth. However, the response to NA was reversed, with an increase in culture growth with NA concentration up to 10^{-5}M.

The inhibitory effect of NA on cell growth and antagonistic interaction with 2,4-D, may suggest that a direct hormonal-type mechanism is involved in NA action. However, NA may also enhance MFO directed metabolism of 2,4-D in the cells, thus providing partial protection against toxic levels of 2,4-D, whilst lowering levels below that required for maximal growth at lower 2,4-D concentrations.

REFERENCES

Field, R.J. and Caseley, J.C. (1987). Abscisic acid as a protectant of *Avena fatua* L. against diclofop-methyl activity. Weed Research **27**, 237-244.

Frear, D.S., Swanson, H.R. and Mansager, E.R. (1987). 1,8-naphthalic anhydride/auxin protection against chlorosulfuron inhibition of corn seedling growth. In "Pesticide Science and Biotechnology", (R. Greenhalgh and T.R. Roberts, eds), 499-503. Blackwell, Oxford.

Hatzios, K.K. (1983). Herbicide antidotes: development, chemistry and mode of action. Advances in Agronomy **36**, 265-316.

Hickey, J.S. and Krueger, W.A. (1974). Alachlor and 1,8-naphthalic anhydride effects on corn coleoptiles. Weed Science **22**, 89-90.

Richardson, W.G. and West, T.M. (1984). The activity and pre-emergence selectivity of some recently developed herbicides: imazaquin, isoxaben, metsulfuron-methyl, achlorifen and orbencarb. Weed Research Organization Technical Report **80**, 25-27.

Sweetser, P.B. (1985). Safening of sulfonylurea herbicides to cereal crops: mode of herbicide antidote action. Procedings British Crop Protection Conference - Weeds **3**, 1147-1154.

Wilkinson, R.E. (1989). Terpenoid biosynthesis as a site of action for herbicide safeners. In "Crop Safeners for herbicides: Development, Uses and Mechanisms of Action" (K.K. Hatzios and R.E. Hoagland, eds), 221-240, Academic Press, San Diego.

DIQUAT RESISTANCE IN PARAQUAT/ATRAZINE CORESISTANT *CONYZA CANADENSIS*

E. Pölös[1], Z. Szigeti[2], Gy. Váradi[1] and E. Lehoczki[3]

[1] Research Institute for Viticulture and Enology, Kecskemét, H-6000
[2] Department of Plant Physiology, Eötvös University, Budapest, H-1445
[3] Botanical Research Group of Hungarian Academy of Sciences, József Attila University, Szeged, H-6722, Hungary

An atrazine-resistant *Conyza canadensis* population was found in Hungarian vineyards several years ago (Lehoczki *et al.*, 1984). This biotype also showed a strong resistance against paraquat, as demonstrated by fluorescence induction kinetics and other photosynthetic activity measurements (Pölös *et al.*, 1988). The resistance factor of 450 was calculated from fluorescence induction curves (Szigeti *et al.*, 1988). In other paraquat-resistant *Erigeron* or *Conyza* biotypes, the resistance factor is usually about 50-100 (Watanabe *et al.*, 1982; Fuerst *et al.*, 1985; and Shaaltiel and Gressel, 1986). These *Conyza* populations, in most cases, showed only a moderate (approximately 10-fold) resistance to diquat (Watanabe *et al.*, 1982; Vaughn *et al.*, 1989). In the present study, the responses of the atrazine/paraquat co-resistant and susceptible biotypes of *Conyza canadensis* were investigated to check the presence or absence of resistance to diquat.

Seeds were collected from Hungarian vineyards near Kecskemét. Growth conditions, diquat treatment and photosynthetic activity measurements are described by Pölös *et al.* (1988) and Szigeti *et al.* (1988).

In field experiments, we observed that diquat was ineffective towards the paraquat/atrazine resistant *Conyza* population.

Cross-resistance of paraquat-resistant *Conyza* to diquat was demonstrated by chlorophyll bleaching, *in vivo* CO_2 fixation and fluorescence induction measurements.

The resistance factors determined on the basis of fluorescence induction parameters of susceptible and paraquat/atrazine resistant *Conyza* leaves floated on diquat solutions showed that our paraquat-resistant *Conyza* biotype is approximately 50-fold more resistant to diquat than the susceptible biotype and 5-fold more resistant to diquat than other paraquat-resistant weed biotypes.

We demonstrated earlier (Szigeti *et al.*, 1988) the transient inhibition of photosynthetic activity in paraquat sprayed paraquat-resistant *Conyza* plants. For comparison, we investigated the behaviour of paraquat/atrazine resistant *Conyza* after spraying with diquat. The observed time curves show a similar shape to that of paraquat-treated intact plants.

We conclude that (i) the paraquat/atrazine resistant biotype of *Conyza canadensis* from Hungarian vineyards exhibited 50-fold cross-resistance to diquat; (ii) the higher degree of resistance to diquat can be connected to a higher resistance to paraquat; and (iii) the recoveries of CO_2 fixation capacity and fluorescence induction curves indicate that diquat can penetrate into chloroplasts of paraquat/atrazine resistant plants, and suggest that in resistant plants a mechanism responsible should exist for the gradual elimination of diquat from the main site of action.

REFERENCES

Fuerst, E.P., Nakatani, H.Y., Dodge, A.D., Penner, D. and Arntzen, C.J. (1985). Paraquat-resistance in *Conyza*. Plant Physiology 77, 984-989.

Lehoczki, E., Laskay, G., Pölös, E. and Mikulás, J. (1984). Resistance to triazine herbicides in horseweed (*Conyza canadensis*). Weed Science 32, 669-674.

Pölös, E., Mikulás, J. Szigeti, Z., Mathovics, B., Do Quy Hai, Párducz, A. and Lehoczki, E. (1988). Paraquat and atrazine co-resistance in *Conyza canadensis* (L.) Cronq. Pesticide Biochemistry and Physiology 30, 142-145.

Shaaltiel, A. and Gressel, J. (1986). Multi-enzyme oxygen radical detoxification system correlated with paraquat resistance in *Conyza bonariensis*. Pesticide Biochemistry and Physiology 26, 22-28.

Szigeti, Z., Pölös, E. and Lehoczki, E. (1988). Fluorescence properties of paraquat-resistant *Conyza* leaves. In "Applications of Chlorophyll Fluorescence in Photosynthesis Research, Stress Physiology, Hydrobiology and Remote Sensing" (H.K. Lichtenthaler, ed.) pp. 109-115. Kluwer Academic Publishers, Dordrecht/Boston/London.

Vaughn, K.C., Vaughan, M.A. and Camilleri, P. (1989). Lack of cross-resistance of paraquat-resistant hairy fleabane (*Conyza bonariensis*) to other toxic oxygen generators indicates enzymatic protection is not the resistance mechanism. Weed Science 37, 5-11.

Watanabe, Y., Honma, T., Itoh, K. and Miyahara, M. (1982). Paraquat-resistance in *Erigeron philadelphicus* L. Weed Research (Japan) 27, 49-54.

HERBICIDE RESISTANCE IN WEEDS IN NEW ZEALAND

A.I. Popay, G.W. Bourdot, K.C. Harrington[1] and A. Rahman

MAFTech., New Zealand Ministry of Agriculture and Fisheries, Palmerst
North, New Zealand

[1] Massey University, Palmerston North, New Zealand

Weeds are not always easy to kill with herbicides. The susceptibility
a weed species to a herbicide may depend on many factors, including a
and condition of plants, growth rate, inter- and intra-specific pla
competition, soil moisture and temperature. Sometimes, however, differe
biotypes of a weed species may show differences in susceptibility to
herbicide when grown together under uniform conditions. Four New Zeala
weed species have so far been confirmed as showing biotype variation
herbicide tolerance.

Chenopodium album was the first of these species. Its tolerance to atrazi
was noted in 1979-80.   Limited work suggested that it could tolera
between 35 and 60 kg ha^{-1} of atrazine (Rahman *et al.*, 1983).

Tolerance of *Polygonum persicaria* to atrazine was noted in 1980 in a fie
which had been in successive crops of maize, treated with atrazine, sin
1970. Later tests (Rahman and Patterson, 1978) showed that the resista
biotypes could tolerate up to 20 kg ha^{-1} of atrazine, and that they we
also resistant to 12 other atrazine herbicides.

Harrington and Popay (1987) tested a biotype of *Carduus nutans* for possib
resistance to MCPA and 2,4-D.   Results in the glasshouse show
differences in susceptibility between resistant and tolerant biotypes of
to 30 times.   Under field conditions, a 6-fold increase in MCPA w
required for equivalent control of the resistant biotype (Harrington *et al*
1988).

Bourdot and Hurrell (1988) confirmed that of two populations of *Ranuncu
acris*, suspected of differences in tolerance to MCPA, one was tolerant
4.8 times more MCPA than the other. The susceptible population had had
MCPA treatment for at least 20 years, whilst the tolerant population h
been treated every one, two or three years for over 30 years.

For all four cases, the evidence strongly suggests that enrichment f
resistance has occurred as a result of herbicide selection pressure.

Although herbicide tolerance has been confirmed in these four specie
there are a number of other pastoral weeds which sometimes prove difficu
to control in the field.   It is possible that resistance to pheno
herbicides has also developed in biotypes of *Carduus pycnocephalus*, *C. tenuiflor*
Cirsium arvense, *Rubus fruticosus*, *Senecio jacobaea* and *Silybum marianum*. This has yet
be confirmed.

REFERENCES

Bourdot, G.W. and Hurrell, G.A. (1988). Differential tolerance of MCP among giant buttercup (*Ranunculus acris*) biotypes in Takaka, Golden Bay Proceedings 41st New Zealand Weed and Pest Control Conference 231-234.

Harrington, K.C. and Popay, A.I. (1987). Differences in susceptibility o nodding thistle populations to phenoxy herbicides. Proceedings 8t Australian Weeds Conference 126-129.

Harrington, K.C., Popay, A.I., Robertson, A.G. and McPherson, H.G. (1988) Resistance of nodding thistle to MCPA in Hawkes Bay. Proceedings 41st Ne Zealand Weed and Pest Control Conference 219-222.

Rahman, A., James, T.K. and Mortimer, J. (1983). Control o atrazine-resistant fathen in maize. Proceedings 36th New Zealand Weed an Pest Control Conference 229-232.

Rahman, A. and Patterson, T.M. (1987). *Polygonum persicaria* triazine-resistant biotype. Proceedings 40th New Zealand Weed and Pes Control Conference 186-188.

COMPARISON OF TRIAZINE–RESISTANT AND –SUSCEPTIBLE BIOTYPES OF *SOLANUM NIGRUM*

R. de Prado, C. Dominguez, E. Romera, and M. Tena

Departmento de Bioquímica y Biología Molecular E.T.S. Ingenieros Agrónomos, Universidad de Córdoba, Córdoba, Spain

Since the first report in 1970, triazine-resistant biotypes have been described for more than 30 common annual weed species. Recent studies have shown that the molecular basis for triazine resistance is a point mutation in the quinone/ herbicide binding protein of PS II (Hirschberg *et al.*, 1984). This mutation leads to both a sharp decrease in affinity for s-triazine binding and an impaired rate of electron transport between the primary (Q_A) and the secondary (Q_B) quinone acceptors at the reducing side of PS II (LeBaron and Gressel, 1982). As resistant biotypes only appear after several years of repeated applications of triazine herbicides, it has been suggested that these biotypes have lower ecological fitness than their susceptible, wild-type counter-parts. However, the diminished fitness of triazine-resistant biotypes and the possible connection of this trait with the biochemical alteration which confers resistance has not been yet established(LeBaron and Gressel, 1982).   To gain further insight in this subject, we have compared photosynthetic activity and growth performance between triazine-resistant (R) and –susceptible (S) biotypes of *Solanum nigrum.*

As R and S biotypes can differ in growth rate, we measured their chlorophyll a/b ratio and photosynthetic O_2 production at the following six states of plant development: (a) 4 buds, (b) 8 buds, (c) floral initiation, (d) pre-anthesis, (e) anthesis and (f) plants with fruits. The two measured parameters attained maximum values at the floral initiation state and they were slightly higher in the R than in the S biotype. At this optimal state of plant development, we compared the following parameters of photosynthetic activity in R and S plants: (a) O_2 production by intact leaves at various light intensities, (b) O_2 production by intact leaves at various temperatures, and (c) activity of various parts of the electron chain in isolated chloroplasts. In the first case, a slightly higher photosynthesis rate in R than in S plants was obtained at all the light intensities studied.   In the 10–45°C temperature range, photosynthesis rate was also slightly higher in R than in S plants, the optimum temperature for this process being 40°C in both cases. Finally, for the three parts of the chloroplast electron chain studied (Table 1), the two biotypes differed little in both Hill reaction ($H_2O \rightarrow FeCy$) activity and whole chain ($H_2O \rightarrow MV$) activity, whereas in R chloroplasts the activity of reaction $H_2O \rightarrow$ duroquinone (DQ) was half of that in S chloroplasts.

Studies on growth performance indicated (Table 2) that plants of the S biotype showed significantly higher values than those of the R biotype in the following growth parameters: plant height, leaf number, leaf area and biomass. By contrast, the seed number and the dry weight of 100 seeds were significantly higher in R than in S plants.

Our results indicate that, as for other triazine-resistant biotypes, the R biotype of *S. nigrum* has an impaired rate of electron transport between Q_A and Q_B (although this is not associated to any decrease in photosynthetic activity) and a diminished growth performance(Jansen *et al.*, 1986; Stowe and Holt, 1988).

Table 1. <u>Rate of electron transport of various parts of the electron transport chain in chloroplasts isolated from triazine-resistant and susceptible biotypes</u> of *Solanum nigrum*.

Reaction	Rate oxygen evolution (a.u).		R/S
	Resistant	Susceptible	
a) $H_2O \rightarrow FeCy$	18.8 ± 1.7	16.1 ± 1.1	1.2
b) $H_2O \rightarrow DQ$	12.6 ± 0.9	26.4 ± 1.5	0.5
c) $H_2O \rightarrow MV$	12.5 ± 1.2	15.0 ± 2.1	0.8

Table 2. <u>Growth and productivity in triazine-susceptible (S) and -resistant (R) biotypes</u> of *Solanum nigrum.*

Week	Biotype	Height (cm)	Leaves (no.)	Leaf area (cm^2)	Seeds (no.)	Root	Shoot	Leaf	100 seeds
							Dry weight (mg)		
3	R	3.7a	6.0a	18.5a	–	18a	9a	36a	–
	S	4.3b	7.0b	26.2b	–	26b	20b	69b	–
5	R	17.0a	23.8a	273.0a	–	437b	330a	774a	–
	S	19.7b	34.2b	318.0b	–	1038b	646b	1180b	–
6	R	33.9a	55.8a	573.0a	–	896a	1134a	1706a	–
	S	44.4b	79.8b	1050.0b	–	1436b	1984b	3298b	–
11	R	–	–	–	5248a	–	–	–	1020a
	S	–	–	–	4800b	–	–	–	996b

Values followed by the same letter do not differ significantly at $\underline{P}$ = 0.05.

REFERENCES

Hirschberg, J., Bleeker, A., Kyle, D.J., McIntosh, L. and Arntzen, C.J. (1984). The molecular basis of triazine-resistances in higher-plant chloroplasts. <u>Zeitscrift für Naturforschung</u> **39c**, 412–419.

Le Baron, H.M. and Gressel, J. (1982). Herbicide Resistance in Plants. Wiley and Sons, New York. pp. 401.

Jansen, M.A.K., Hobe, J.M., Wesselius, J.C. and Rensen, J.J.S. (1986). Comparison of photosynthetic activity and growth performance in triazine-resistant and -susceptible biotypes of *Chenopodium album.* <u>Physiologie Végetale</u> **24**, 475–484.

Stowe, A.E. and Holt, J.S. (1988). Comparison of triazine-resistant and -susceptible biotypes of *Senecio vulgaris* and their F_1 hybrids. <u>Plant Physiologie</u> **87**, 183.

CHARACTERISATION OF TRIAZINE-RESISTANT POPULATIONS OF *AMARANTHUS CRUENTUS*

R. de Prado, E. Romera, C. Dominguez and M. Tena

Departamento de Bioquímica y Biología Molecular E.T.S. Ingenieros Agrónomos, Universidad de Córdoba, Cordoba, Spain

Resistance to s-triazine herbicides has been reported in recent years (LeBaron and Gressel, 1982), and is due to alteration of the herbicide binding site in the chloroplast thylakoid membrane. In this study, we have characterised at both whole plant and chloroplast levels the atrazine resistance response observed in five *Amaranthus cruentus* populations found in different sites in Zaragoza in Northern Spain.

Surveys of cornfields and railways, which have been treated continuously with atrazine in recent years, led to the detection of five *A. cruentus* populations which were no longer controlled by this herbicide. They were provisionally identified as triazine-resistant weed biotypes by field analysis of chlorophyll fluoresence emission, using a commercial portable fluorometer. Resistant plants showed low levels of fluorescence emission during a 24 h period after foliar application of 20 ppm atrazine whereas these plants treated with 20 ppm diuron as well as susceptible (S) control plants treated with 20 ppm atrazine or diuron showed a much higher level of fluorescence emission (Fig. 1). The atrazine-resistant character of the five weed biotypes was further verified in the laboratory by analysis of rapid fluorescence induction curves of intact leaves (Fig. 2).

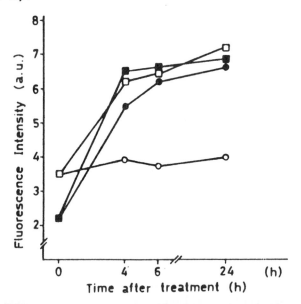

Figure 1. <u>Chlorophyll fluorescence intensity emitted by leaves of resistant (open symbols) and susceptible (closed symbols)</u> *Amaranthus cruentus* <u>biotypes during a 24 h period after treatment with 20 ppm atrazine (o, •) or diuron (□,■) and the fluorescence emission was measured during a 24 h period after treatment.</u>

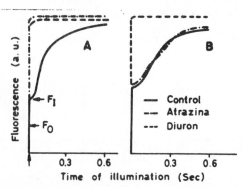

Figure 2. Fluorescence induction curves of susceptible (A) and resistant (B) *Amaranthus cruentus* leaves previously incubated for 4 h in water or 10^{-4}M atrazine or diuron.

Further characterisation of these five *A. cruentus* atrazine-resistant biotypes included determinations of the herbicide concentration required for 50% inhibition of the Hill reaction in isolated chloroplasts (I_{50} values) and for 50% reduction in plant growth and development (ED_{50} values). I_{50} and ED_{50} values for both R and S plants as well as resistant factors (R/S ratios) are shown in Table 1. Chloroplasts of R biotypes showed 500 to 1000 times (values of I_{50} – resistant factors) less sensitivity to inhibition by atrazine than the S control. Similar values have been reported for other s-triazine resistant biotypes(LeBaron and Gressel, 1982). In contrast, ED_{50}–resistant factors values in the 69-87 range, were lower than those described for other s-triazine resistant biotypes. This is apparently a characteristic feature of weed biotypes showing intermediate resistance to s-triazines(Gasquez *et al.*, 1985).

Table 1. I_{50} and ED_{50} values corresponding to five atrazine-resistant biotypes of *Amaranthus cruentus*.

Biotype	$I_{50}(\mu M)$	$I_{50}R/I_{50}S$	ED_{50} (kg ha^{-1})	$ED_{50}R/ED_{50}S$
R_1	205	911	1.290	86
R_2	215	955	1.190	77
R_3	125	556	1.050	70
R_4	180	800	1.030	69
R_5	150	667	1.100	73
S	0.225	–	0.015	–

ACKNOWLEDGEMENTS

The authors thank Dr Zaragoza and Dr Ochoa for the supply of *A. cruentus* seeds.

REFERENCES

Le Baron, H.M. and Gressel, J. (1982). Herbicide Resistance in Plants. Wiley and Sons, New York.

Gasquez, J., Al Mouemar, A. and Darmency, H. (1985). Pesticide Science **16**, 390.

M. Ramsay and T. Hodgkin

Scottish Crop Research Institute, Invergowrie, Dundee, DD2 5DA

The *in vitro* culture of haploid microspores is being examined as a means of producing herbicide-resistant *Brassica napus*, as it permits a large number of haploid single cells to be processed. Currently investigations are therefore being carried out to determine the effect of the herbicide glyphosate on viability of uninucleate/early binucleate microspores, the stage thought to be most suitable for microsporogenesis (Chuong *et al.*, 1988).

Brassica napus microspores were incubated for 15, 30, 60 min and 24 hours in liquid culture medium (Keller, W.A., pers. comm.) containing 0, 5, 50 or 100 mM glyphosate. Viability was assessed by staining with fluorescein diacetate (Heslop-Harrison and Heslop-Harrison, 1970). In addition to uninucleate/binucleate microspores (bud size, 2-3 mm), the effect of these herbicide concentrations on the viability of trinucleate microspores (5-6 mm buds) and mature pollen and leaf protoplasts was also determined. The work carried out on three *Brassica napus* genotypes; the spring oilseed rape variety 'Puma', rapid-cycling *B. napus* (Williams and Hill, 1986) and a synthetically reproduced *B. napus* (Hodgkin, T., in press).

In all genotypes the viability of mature pollen and trinucleate microspores was rapidly reduced by high concentrations of glyphosate (50-100 mM). However, this pattern was not reproduced by uninucleate/binucleate microspores in which all genotypes maintained higher levels of viability at 50 mM glyphosate than were achieved after incubation in either glyphosate-free media or the lower glyphosate concentration of 5 mM. In two genotypes (Puma and rapid-cycling material) these higher levels of viability were maintained up to a concentration of 100 mM. The response of leaf protoplasts to glyphosate was similar to that of mature pollen. Examination of the results by analysis of variance showed that mature pollen and trinucleate microspores exposed to concentrations of glyphosate of 50 mM or above had their viability significantly reduced ($P < 1\%$). However, compared with the glyphosate-free control, viability of uninucleate microspores was significantly enhanced by addition of 50 mM glyphosate ($P < 1\%$) (Fig. 1). Pollen of Puma and microspores showed significantly higher viability than the other genotypes ($P < 1\%$).

These preliminary data suggest the response of *Brassica napus* microspores to glyphosate may alter during development. This could be a reflection of enzymatic differences between developmental stages influencing herbicide action. Unexpected differences in tolerance to glyphosate have recently been reported between callus and differentiated tissue in maize, and were attributed to differences to sensitivity between EPSP synthase isozymes

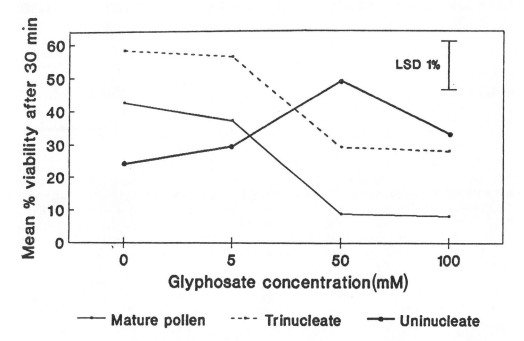

Fig. 1. Viability of _Brassica napus_ microspores after 30 min exposure to glyphosate.

Racchi _et al._, 1989). A comparision of varietal viabilities at the highest glyphosate concentration (100 mM) suggests that genotypic differences are expressed in pollen. The effect of glyphosate on uninucleate/binucleate microspores has implications for the application of microspore culture techniques in the selection of resistance to this herbicide, as it could affect embryo yield and the efficiency of selection for resistance to embryo material, thus necessitating the application of selection techniques subsequent to embryo production (Kenyon _et al._, 1987).

REFERENCES

Chuong, P.V., Deslauriers, C., Kott, L.S. and Beversdorf, W.D. (1988). Effects of donor genotype and bud sampling on microspore cultures of _Brassica napus_. Canadian Journal of Botany **66**, 1653–1657.

Heslop-Harrison, J. and Heslop-Harrison, Y. (1970). Stain Technology **45**:3, 115–120.

Kenyon, P.D., Marshall G. and Morrison, I.N. (1987). Selection for sulfonylurea herbicide tolerance in oilseed rape (_Brassica napus_) using microspore culture. Proceeding 1978 British Crop Protection Conference - Weeds **2**, 871–876.

Racchi, M.I., Forlani. G., Pelanda, R. and Neilsen, E. (1989). XII Eucarpia Congress 1989 26–11 Vortrage fur Pflanzensuchtg 15.

Williams, P.H. and Hill, C.B. (1986). Rapid cycling populations of _Brassica_. Science **282**, 1385–1389.

ISOLATION OF ATRAZINE-RESISTANT CELLS LINES AND REGENERATION OF PLANTLETS IN *CICER ARIETINUM* (L)

S. Rao, K. Basavaraj and M. Madhava Naidu

Department of Postgraduate Studies in Botany, "Jnana Ganga", Gulbarga University, Gulbarga, Karnataka State, India 585106

The s-triazine herbicides such as simazine and atrazine give excellent weed protection in corn and jawar fields, but are highly persistent and extremely toxic and mutagenic to many other plant species (Wuu and Grant, 1966; Sharma and Rao, 1982)   Development of tolerance in food crops against widely used herbicides would facilitate better management of land for diversified food production and would permit chemical weed control in minor crops. Mutants selected for tolerance/resistance to herbicides are valuable for mode of action studies and as a source of genes for genetically-engineered herbicide resistance (Duesing, 1987). The present investigation reports the selection of atrazine-resistant cell lines and regeneration of *Cicer arientinum* L. (chickpea).

Twenty eight-day-old calluses derived from the hypocotyl region were transferred to B_5 medium containing 100-500 ppm atrazine. Calluses generally died within 10 days, but occasionally in some cultures (2%) a portion of callus continued to grow. These surviving pockets of cells were subcultured on medium containing 200 ppm atrazine (4 passages of 25 days each). Resistance was not lost even when the cells were transferred through several passages of the atrazine-free medium and again re-grown on atrazine-containing medium. The resistant cell lines contained higher amounts of proteins than the control cell line. Electrophoretic patterns revealed additional bands of peroxidase isozymes in the resistant cell line (cf. the control cell line). The selected calluses were transferred to shoot-inducing medium and the regenerated plants were transferred to soil by growing them on through sterile vermiculture and high humidity transitional stages. In glasshouse tests, these plants tolerated sprays containing 1000 ppm atrazine - a concentration lethal to non-selected plants.

Selection for herbicide-resistant phenotypes in the field is complicated by difficulties in applying uniform selection pressure. However, there is the possibility of selecting for resistance in tissue culture (Gressel *et al.*, 1978; Brag and Umeil, 1978).   Using this technique, plants resistant to herbicides/pesticides have been reported on several species, for example, *Nicotiana* (Chaleff and Parsons, 1978), *Brassica campestris* (Gressel *et al.*, 1978) and chickpea (Saxena and Beg, 1985; Rao, 1987).

In our investigations using tissue culture techniques, plants resistant to the herbicide, atrazine, have been obtained from callus subjected to selection pressure. The cell line which grew in the presence of toxic doses of herbicide may also arise because of somaclonal variation in the callus (Larkins and Scowcroft, 1981). This phenomenon is being exploited for the varietal improvement of crops. The selected variants generally show altered biochemical pathways Thus, in the present investigations it was noticed that the herbicide-resistant cell line contained increased amount of proteins and additional bands of peroxidase isozymes compared with the wild-type cell lines. Saxena and Beg (1985) and Rao (1987) also reported similar changes in pesticide-resistant cell lines of this species.

ACKNOWLEDGEMENTS

The authors thank Professor V. Rangaswamy, Head of the Department of Botany for his encouragement and facilities provided throughout this investigation.

REFERENCES

Brag, R. and Umeil, N. (1977). Development of tobacco seedlings and callus cultures in the presence of amitrole. Zeitscrift für Pflanzenphysiology **83**, 437–447.

Chaleff, R.S. and Parsons, M.F. (1978). Direct selection *in vitro* for herbicide resistant mutants of *Nicotiana tobacum*. Proceedings of the National Academy of Sciences (USA) **75**, 5104–5107.

Duesing, J.H. (1987). Whole plant and cellular strategies towards herbicide resistance. XIVth International Botanical Congress, Berlin (West), pp. 418.

Gamborg, O.L., Miller, R.A. and Ogima, K. (1968). Nutrient requirements of suspension culture of soyabean root cells. Experimental Cell Research **50**, 151–158.

Gressel, J., Zilkah, S. and Ezra, G. (1978). Herbicide action, resistance and screening in cultures versus plants. In "Frontiers of Plant Tissue Culture" (T. Thorpe, ed.), pp. 427–436. University of Calgary Press, Calgary.

Larkin, P.J. and Scowcroft, W.R. (1981). Somaclonal variations – a novel source of variability from cell cultures for plant improvement. Theoretical and Applied Genetics **60**, 197–214.

Rao, S. (1987). Selection of pesticide resistance cell line in *Cicer arietinum* L. through cell and tissue culture. XIVth International Botanical Congress, Berlin West, 483.

Saxena, R.P. and Beg, M.U. (1985). On the selection of pesticide resistant cell line (Abstract O-43). International Conference on Pesticides, Toxicity, Safety and Risk Assessment, Lucknow.

Sharma, C.B.S.R. and Rao, A.A. (1982). Induced variation in chiasma frequency in barley in response to atrazine and simazine treatments. Indian Journal of Experimental Botany **20**, 97–98.

Wuu, K.D. and Grant, W.F. (1967). Chromosomal aberrations induced in somatic cells of *Vicia faba* by pesticides. The Nucleus **10**, 37–46.

NODULATION AND YIELD OF BEANS (*PHASEOLUS VULGARIS* L.) SHOW DIFFERENTIAL TOLERANCE TO PRE-EMERGENCE HERBICIDES

K.P. Sibuga

Sokoine University of Agriculture, Department of Crop Science, P.O. Box 3005, Morogoro, Tanzania

Field studies on the effect of pendimethalin at 1.0 to 2.0 kg a.i. ha^{-1}, linuron at 0.75 to 1.25 kg a.i. ha^{-1}, alachor at 1.5 to 2.5 kg a.i. ha^{-1}, and Galex, a tank mix of 50 % metobromuron and 50 % metolachlor at 1.5 to 2.5 kg a.i. ha^{-1} on dry matter production and nodulation of beans were conducted under rainfed conditions at Kabete and furrow irrigation at Perkerra. Herbicide treatments reduced shoot dry weights of beans at flowering and occasionally root dry weights, compared to hoe-weeding. Increasing concentrations of all herbicides reduced nodule numbers more than nodule dry weights both under rainfed and irrigated conditions. Compared to hoe-weeding, increasing concentration of linuron reduced nodule numbers the least and that of pendimethalin the most both at Kabete and Perkerra. Nevertheless, bean yields remained high as nodule numbers declined. Herbicide treatments caused very little change in the total nitrogen content of bean tops and the crude protein content of bean grains.

COMPARISON OF THE PHOTOSYNTHETIC CAPACITY AND CHLOROPHYLL FLUORESCENCE FOR TRIAZINE-RESISTANT AND -SUSCEPTIBLE *CHENOPODIUM ALBUM* L.

K. Smis[1], R. Lemeur[1] and R. Bulcke[2]

[1] Laboratory of Plant Ecology, Faculty of Agricultural Sciences, University of Ghent, Coupure Links 653, 9000 Ghent, Belgium
[2] Laboratory of Herbology, Faculty of Agricultural Sciences, University of Ghent, Coupure Links 653, 9000 Ghent, Belgium

Differences in growth and development have been observed between triazine-resistant (R) and triazine-susceptible (S) plants of *Chenopodium album* (Common Lambsquarters) (Radosevich and Holt, 1982; Bulcke *et al.*, 1985; Gressel, 1985; Jansen Mak *et al.*, 1986; Holt, 1988). The resistant biotypes have been shown to have a slower growth rate compared with the susceptible ones. Our own preliminary experiments involving measurements of leaf area and plant dry weight have confirmed this. In addition, the resistant plants flowered later – a phenomenon which explains their stocky form.

As growth is related to leaf photosynthetic capacity it should be possible to relate differences in growth and development between the resistant and susceptible *C. album* to differences in their photosynthetic capacity. Therefore, we compared photosynthesis between the two forms. As photosynthesis measurements require accurate control of environmental factors such as light intensity, air temperature, air humidity and CO_2 concentration, this assessment is complex, difficult and time-consuming.

On the other hand, evaluation of photosynthetic assimilation based on chlorophyll-a fluorescence gives a rapid non-destructive, and more easily performed method for the determination of photosynthetic capacity. Therefore, we compared measurements of chlorophyll-a fluorescence with net photosynthesis rates in order to test the reliability of the fluorescence method for analysing differences in growth potential between biotypes.

The relative differences between the S- and R-biotypes calculated as Δ(S-R)/S, are useful for comparing the functional photosynthetic parameters with the chlorophyll-a fluorescence ratios. These relative values permit the selection of useful parameters for comparing the R- and S-biotypes. As the difference between S and R, relative to S, is largest for the calculated fluorescence ratio (P-I)/I($\pm$ 33.9%), we conclude that this parameter is very sensitive and most adequate for discrimination between the biotypes.

The ratio (P-I)/I might also be useful for comparing the photosynthetic capacity of R and S plants when the light reactions are the primary limiting process for photosynthesis. The relative differences (Δ(S-R)/S) in photosynthetic activity are clear in the case of very low light intensities. The net CO_2 exchange rate for photon flux densities less than 100 μmol m^{-2} s^{-1} leads to the same relative differences as those shown by the most discriminating fluorescence parameter (i.e. 33.9%).

Also, the smaller relative differences in $\Delta(S-R)/S$ for net CO_2 exchange rates at higher photon flux densities confirm that the growth disadvantage of the R-biotype due to a defective electron transfer at photosystem II, diminishes in conditions of less limiting light intensities. This means that competitive differences between biotypes will be stronger in situations of reduced light intensity, as might be the case when R- and S- plants grow within field crops.

Our results clearly show that the fluorescence-induction curve, determined with a fluorometer, allows the detection of differences in photosynthesis rates, especially at low light intensities. Therefore, chlorophyll-a fluorescence is an interesting alternative to CO_2-exchange methods when the complex experimental setup for such photosynthesis measurements is not available.

ACKNOWLEDGEMENTS

This research was supported by the Institute for Encouragement of Scientific Research in Industry and Agriculture (IWONL/IRSIA).

REFERENCES

Bulcke, R., de Vleeschauwer, J., Vercruysse, J. and Stryckers, J. (1985). Comparison between triazine-resistant and -susceptible biotypes of *Chenopodium album* L. and *Solanum nigrum* L. Mededelingen Faculteit Landbouwwetenschappen RUG **50** (2a), 211-220.

Gressel, J. (1985). Herbicide tolerance and resistance : Alteration of site of activity. In "Weed Physiology, Vol. II Herbicide Physiology". (S.O. Duke, ed.), pp. 159-189. CRC Press, Boca Raton, Florida.

Holt, J.S. (1988). Reduced growth, competitiveness and photosynthetic efficiency of triazine-resistant *Senecio vulgaris* from California. Journal of Applied Ecology **25**, 307-318.

Jansen Mak, Hobe, J.H., Wesselius, J.C. and van Rensen, J.J.S. (1986). Comparison of photosynthetic activity and growth performance in triazine-resistant and susceptible biotypes of *Chenopodium album*. Physiologie Vegetale **24**, 475-484.

Radosevich, S.R. and Holt, J.S. (1982). Physiological responses and fitness of susceptible and ressistant weed biotypes to triazine herbicides. In "Herbicide Resistance in Plants" (H.M. LeBaron and J. Gressel, eds). New York; John Wiley and Sons.

Mougou, A., Lemeur, R. and Schalck, J. (1984). Water stress induction using polyethylene glycol in nutrient film technique. Effects on the photosynthetic parameters of three tomato species and one hybrid. Ecol. Plant. **19**, 375-385.

THE GENETICAL ANALYSIS AND EXPLOITATION OF DIFFERENTIAL RESPONSES TO HERBICIDES IN CROP SPECIES

J.W. Snape[1], B.B. Parker[1], D. Leckie[1] and E. Nevo[2]

[1] Institute of Plant Science Research, Cambridge Laboratory, Trumpington, Cambridge, UK
[2] Institute of Evolution, University of Haifa, Haifa, Israel

Crop species show differential responses to successful, widely used herbicides where some varieties are unaffected by application, whilst others show symptoms of damage ranging from a slight reduction in vigour to complete plant death. The elucidation of the genetical control of such responses is important for developing strategies for breeding for herbicide resistance within crop species and also in understanding the modes of action of the herbicides and the evolution of resistance in weed species.

To investigate the phenomenon in cereals, detailed studies of the control of responses of wheat to difenzoquat and to the phenylureas, chlorotoluron, metoxuron and isoproturon have been carried out. These have revealed that the primary differences in response between resistant and susceptible varieties are due to single major genes, although the influence of other "modifier" genes has also been detected. It has also been shown that the responses to chlorotoluron and metoxuron are determined by the same gene.

Studies of related wild grass species have indicated that polymorphisms for response also exist in such species. This suggests that the genes for differential responses have evolved prior to the domestication of the cultivated cereals and not in response to the development and use of the chemicals.

The importance of these results for developing strategies for the incorporation of herbicide resistance in new varieties of crop species by conventional and non-conventional methods of plant breeding will be discussed.

THE INFLUENCE OF NITROGEN FORMS ON THE GROWTH, PHOTOSYNTHESIS AND CHLOROPLAST ACTIVITY OF TRIAZINE-RESISTANT AND -SUSCEPTIBLE PLANTS OF *ERIGERON CANADENSIS* L.

R. Stanek[1] and J. Lipecki[2]

[1] Department of Plant Physiology
[2] Department of Pomology, Agricultural University, Lublin, Poland

It is well established that crop plants can utilise both NO_3^- and NH_4^+ as N sources but the physiological responses to these ions varies considerably according to species (Haynes and Goh, 1978). However, it has been very difficult to find information on the effect of these ions on weeds that show resistance to triazine herbicides.

Table 1. The growth, photosynthesis and chloroplast activity of plants treated with NH_4^+ or NO_3^-

Biotype	Treatments	RGR ($g\ g^{-1}\ d^{-1}$)	Photosynthesis ($mg\ CO_2\ dm^{-2}b\ h^{-1}$)	Chloroplast activity (μM reduced DCPIP mg^{-1} chlorophyll h^{-1})	
				$(-)NH_4Cl$	$(+)NH_4Cl$
Resistant	NO_3^-	0.25b*	36.5b	13.2c	18.6b
	NH_4^+	0.14a	25.2a	7.0a	10.0a
Susceptible	NO_3^-	0.31c	48.8c	15.2d	30.2d
	NH_4^+	0.15a	29.5a	9.0b	16.7c

Chloroplast activity (at illumination 300 $\mu E\ m^{-2}\ s^{-1}$) with (+) or without (–) addition of NH_4Cl to reaction mixture.
* The values within the range of column followed by the same letters do not differ significantly on the 5% level of probability.

The results of our preliminary pot experiments (unpublished) showed that feeding with a mixture of NH_4^+:NO_3^- ions increased both dry weight and CO_2 assimilation of the resistant and susceptible biotypes of *Erigeron canadensis* L., whereas feeding with the nitrate or ammonium ions alone decreased the growth and photosynthesis. To explain which of the two ions is preferred by plants, we carried out an experiment in a growth chamber. Vernalised plants with eight fully developed leaves were grown in water culture (pH 6) with NH_4^+ or NO_3^- ions and other essential elements for 14 days under illumination of 165 $\mu Em^{-2}\ s^{-1}$ (photoperiod 16 h), day/night temperature 25°/20°C and 70% relative air humidity. The nitrate-treated plants had a greater relative growth rate (RGR) and higher photosynthesis (measurements with an IRGA-Beckman 865 in the open system) than ammonium-treated plants (Table 1). Ammonium ions resulted in a strong inhibition of the photosynthesis and growth for the susceptible biotype compared with the resistant one. The Hill activity (Sane *et al.*, 1970) of chloroplasts from NO_3^--treated plants was higher than with the NH_4^+-fed ones. Addition of 5 mM NH_4Cl to samples containing chloroplasts (40 μg chlorophyll and 50 mM Tricine pH 7) stimulated electron flow from water to 2,6-dichlorophenolindophenol (DCPIP) due to uncoupling of phosphorylation of electron transport. This stimulation was greater in the susceptible chloroplasts than in the resistant ones. Other authors (Rashid and van Rensen, 1987) reported a similar effect in *Chenopodium album* L. plants. Our results suggest that the resistant plants of *Erigeron canadensis* are a little more tolerant to ammonium ions.

REFERENCES

Haynes, R.J. and Goh, K.M. (1978). Ammonium and nitrate nutrition of plants. <u>Biological Reviews</u> **53**, 465–510.

Rashid, A. and van Rensen, J.J.S. (1987). Uncoupling and photoinhibition in chloroplasts from triazine-resistant and a susceptible *Chenopodium album* biotype. <u>Pesticide Biochemistry and Physiology</u> **28**, 325–332

Sane, P.W., Goodchild, D.J. and Park, R.B. (1970). Characterization of chloroplast Photosystems 1 and 2 separated by a non-detergent method. <u>Biochimica et Biophysica Acta</u> **216**, 162–178.

INVESTIGATION OF THE SELECTIVITY MECHANISM OF PLANTS TO THE HERBICIDE, CLOMAZONE

M.R. Weimer, D.H. Buhler and N.E. Balke

Department of Agronomy, University of Wisconsin, Madison 53706, USA

Herbicide selectivity and/or resistance can result from a number of factors including differences in rates of absorption and subsequent translocation, tissue and subcellular localisation of the herbicide, metabolism to products of modified phytotoxicity, and differences in target site sensitivity. Like most herbicides, clomazone causes different degrees of toxicity among various plant species.

Several annual broadleaf and grass weeds are quite sensitive to clomazone whereas soybean is very tolerant. *Abutilon theophrasti* (Velvetleaf) is very sensitive to clomazone. One proposed mechanism of action of clomazone is inhibition of terpenoid synthesis by inhibition of prenyl pyrophosphate synthesis and conversion, the early steps in the terpenoid pathway (Sandmann and Boger, 1987). This mechanism is consistent with the symptomology of clomazone, namely pale green or white foliage (Duke *et al.*, 1985).

The aim of our research has been to determine the selectivity mechanism of clomazone by conducting experiments in which velvetleaf (sensitive) and soybean (tolerant) were compared. The initial objective was to determine if differences in clomazone uptake, translocation, and/or metabolism could explain differential toxicity of clomazone to soybean and velvetleaf.

Roots of seedlings grown hydroponically in Hoagland's nutrient solution were exposed to 1 μM ^{14}C-clomazone (13.75 μCi μmol^{-1}). Two separate ^{14}C labels of clomazone (methylene and carbonyl) were applied to individual seedlings. Plants were harvested 24, 48 or 72 h after clomazone treatment. Roots, stems and leaves were separated, fresh weights determined, and tissue homogenised in 80% methanol. Tissue debris was removed by vacuum filtration. Concentration of ^{14}C was determined in sub-samples of the filtrates by liquid scintillation spectrometry. Remaining filtrate was concentrated by evaporation under nitrogen gas, and clomazone and clomazone metabolites were separated by high performance liquid chromatography.

Uptake of clomazone from the nutrient solution and translocation out of the roots were similar for both plants. Clomazone did not accumulate in roots but rather was rapidly translocated to shoots of both species. Concentration of unmetabolised clomazone was higher or similar in both roots and shoots of soybean compared to velvetleaf, indicating that velvetleaf metabolised clomazone more rapidly than did soybean. Thus, it appears that detoxication of clomazone by soybean is not responsible for the selectivity observed between velvetleaf and soybean. These results are consistent with previous work on other plant species (Weston and Barrett, 1989).

Other mechanisms of selectivity, such as bioactivation and/or differences in target site sensitivity may explain differential toxicity of clomazone. Treatment of clomazone metabolites with β-glucosidase demonstrated that metabolism of clomazone proceeds by molecular cleavage and conjugation in both species. One of the aglycones produced by β-glucosidase treatment co-chromatographed with a 2-chlorobenzylalcohol standard.

At this point of our research, we conclude that velvetleaf does not produce a unique clomazone metabolite that results in bioactivation of clomazone by this species. However, because structure elucidation of the metabolites has not yet been accomplished, we cannot unequivocally eliminate bioactivation as a possible mechanism of selectivity.

Because of the failure of the above results to explain the mechanism of clomazone selectivity, we have recently initiated studies to determine possible differences in target site sensitivity of velvetleaf and soybean to clomazone and/or a clomazone metabolite.

REFERENCES

Duke, S.O., Kenyon, W.H. and Paul, R.N. (1985). FMC 57020 effects on chloroplast development in pitted morning glory (*Ipomoea lacunosa*). Weed Science **33**, 786-794.

Sandmann, G. and Boger, P. (1987). Interconversion of prenyl pyrophosphates and subsequent reactions in the presence of FMC 57020. Zeitschrift für Naturforschung **42c**. 803-807.

Weston, L.A. and Barrett, M. (1989). Tolerance of tomato (*Lycopersicon esculentum*) and bell pepper (*Capsicum annum*) to clomazone. Weed Science **37**, 285-289.

POPULATION SHIFTS FOLLOWING CHEMICAL HERBICIDE APPLICATIONS TO WHEAT FIELDS

F. Wen-xu and T. Hong-Yuan

Plant Protection Institute, Shanghai Academy of Agricultural Sciences,
Shanghai 201106, Peoples Republic of China

This poster deals with the population shifts between dominant and subordinate weed species in wheat fields following the application of herbicides. The results showed that the removal of one or several weed species from weed communities may exert a strong influence on the associated weed populations. The continuous application of a single type of herbicide may lead to the subordinate weed species becoming dominant as the former dominant species is removed and thereby cause even greater damage to the crop. This response of weed populations to the continuous application of a single type of herbicide emphasises the need to consider all components of the weed population in relation to effective applications, if weed population shifts, resistance to specific herbicides and other side-effects are to be avoided.

Three types of weed community in Shanghai were studied over two years. Selective herbicides were used to remove dominant, subordinate and all weed species respectively. Measurements were made of individual plants, biomass, plant height, seed output per unit area and per single plant, and of the theoretical crop yield, seed weight and spikelets per unit area. De Wit's replacement series competition model was also used to assess the relationships between the plants.

The removal of former dominant graminaceous weed species led to a marked increase of the above parameters for broadleaved (BL) weeds. In certain situations, the number of plants was 2-fold, biomass 6-7-fold and seed ouput per unit area 7-fold greater compared with the control plots. Even the removal of subordinate graminaceous weeds led to increases (60%) of dominant BL weeds compared with the controls. Conversely, the increase of subordinate graminaceous weeds was relatively less after the removal of dominant BL weeds, while the removal of subordinate BL weeds had no influence on the dominant graminaceous weed population. It seems that BL weeds may be severely suppressed by the existence of cohabiting graminaceous weeds and may recover rapidly after their removal. Competition experiments using de Wit's model also showed that graminaceous weeds are the stronger competitor against BL weeds.

CHARACTERISATION OF CYCLOHEXANEDIONE AND ARYLOXYPHENOXYPROPIONATE-TOLERANT MAIZE MUTANTS SELECTED FROM TISSUE CULTURE

D.L. Wyse, W.B. Parker, L.C. Marshall, D.A. Somers, J.W. Gronwald and B.G. Gengenbach

Department of Agronomy and Plant Genetics, University of Minnesota and USDA-ARS, St. Paul, Minnesota 55108, USA

Acetyl-CoA carboxylase (ACCase; E.C.6.4.1.2) from most grasses including maize is inhibited by cyclohexanedione and aryloxyphenoxypropionate herbicides, whereas the enzyme from dicotyledons is insensitive to herbicide inhibition (Burton *et al.*, 1987; Secor and Cseke, 1988; Rendina and Felts, 1988; Walker *et al.*, 1988). This differential inhibition of ACCases by these herbicides accounts for their selective control of grass weeds in dicotyledonous crops. Herbicide-mediated inhibition of ACCase is thought to deplete cellular malonyl-CoA which is consistent with the observed inhibition of acetate incorporation of ACCase into fatty acids in herbicide-treated tissue of susceptible grass species. The involvement of ACCase in herbicide tolerance has also been demonstrated in comparisons of tolerant and susceptible grass species. Red fescue is tolerant to both herbicide families and exhibits ACCase activity that is substantially less sensitive to inhibition by sethoxydim, a cyclohexanedione, and haloxyfop, an aryloxyphenoxypropionate, compared with susceptible tall fescue (Stoltenberg *et al.*, 1989).

Maize genotypes with naturally-occurring tolerance to rates of sethoxydim or haloxyfop that control annual grasses have not been identified. To investigate the role of ACCase in determining the susceptibility of grasses to cyclohexanedione and aryloxyphenoxypropionate herbicides, we have selected and characterised herbicide-tolerant plants from maize tissue cultures. Tissue culture *per se* spontaneously induces genetic variation and the number of cells subjected to selection is thereby maximised.

In a preliminary study, we selected sethoxydim-tolerant cell cultures which demonstrated that variation for herbicide-tolerance at the cell level could be obtained in maize. Sethoxydim-tolerant cell cultures were also cross-tolerant to haloxyfop. ACCase activity in the variant cell lines was elevated more than two-fold compared to wild-type and exhibited wild-type herbicide inhibition kinetics. Increased ACCase activity in the tolerant cell lines was due to overproduction of the ACCase enzyme. Overproduction of ACCase had not previously been reported as a mechanism of tolerance to these herbicides.

Friable, embryonic maize callus obtained from the cross (A188 x B73) were used to select for sethoxydim and haloxyfop tolerance. Two sethoxydim-tolerant (S1 and S2) and one haloxyfop-tolerant (H1) maize tissue cultures were selected. S1 and S2 were tolerant to >100-fold sethoxydim and to 7- and 44-fold the sethoxydim and haloxyfop concentrations, respectively, that were lethal to wild-type tissue cultures. H1 was 24-fold more tolerant to haloxyfop but was not sethoxydim-tolerant compared with the control. ACCase activities of the unselected control, S1, S2 and H1 were similar in the absence of herbicide. ACCase activity from S1 and S2 was inhibited 50% (I_{50}) by sethoxydim concentrations 4- and 40-fold higher and by haloxyfop concentrations 3.5- and 9-fold higher, respectively, than I_{50}'s for ACCase activity from the unselected control or H1. The ACCase I_{50} value from H1 treated with haloxyfop was 10-fold greater than the unselected control. Control regenerated plants were killed by 0.05 kg ha^{-1} sethoxydim and 0.02 kg ha^{-1} haloxyfop. S1 and S2 regenerated plants were injured by 0.11 kg ha^{-1} and higher sethoxydim treatments, respectively, but survived 0.44 kg ha^{-1} sethoxydim. Regenerated plants from S1, S2 and H1 were injured by 0.01 kg ha^{-1} and killed by 0.10 kg ha^{-1} haloxyfop. Sethoxydim, cycloxydim and quizalofop were potent inhibitors of wild-type ACCase from homozygous maize plants. The mutant ACCase I_{50} from homozygous maize plants was 122 to 175 time higher than the I_{50} for wild-type ACCase. ACCase from homozygous mutant (S2) corn plants was cross-tolerant to other cyclohexanedione and aryloxyphenoxypropionic acid herbicides.

These studies have shown that sethoxydim- and haloxyfop-tolerance in the regenerable tissue cultures was associated with an altered form of ACCase that was less sensitive to herbicide inhibition. Tolerance was also expressed in regenerated plants and their progeny demonstrating that a heritable mutation confers herbicide tolerance. S2 and H1 regenerates were heterozygous for a dominant allele(s) at a single gene(s) which conferred sethoxydim tolerance. The tolerance of the S2 line to sethoxydim, cycloxydim and quizalofop is due to an insensitive form(s) of ACCase. The homozygous S2 plants have shown complete tolerance to sethoxydim at 0.8 kg ha^{-1} in 1989 field trials.

REFERENCES

Burton, J.D., Gronwald, J.W., Somers, D.A., Connelly, J.A., Gengenbach, B.G. and Wyse, D.L. (1987). Inhibition of plant acetyl-coenzyme A carboxylase by the herbicides sethoxydim and haloxyfop. Biochemistry and Biophysics Research Communications 148, 1039-1044.

Rendina, A.R. and Felts, J.M. (1988). Cyclohexanedione herbicides are selective and potent inhibitors of acetyl-CoA carboxylase from grasses. Plant Physiology 86, 983-986.

Secor, J. and Cseke, C. (1988). Inhibition of acetyl-CoA carboxylase activity by haloxyfop and tralkoxydim. Plant Physiology 86, 10-12.

Stoltenberg, D.E., Gronwald, J.W., Wyse, D.L., Burton, J.D., Somers, D.A. and Gengenbach, B.G. (1989). Effect of sethoxydim and haloxyfop on acetyl-CoA carboxylase activity in Festuca species. Weed Science 37, 512-516.

Walker, K.A., Ridley, S.M., Lewis, T. and Harwood, J.L. (1988). Fluazifop, a grass-selective herbicide which inhibits acetyl-CoA carboxylase in sensitive plant species. Biochemical Journal 254, 307-310.

XENOBIOTIC METABOLISM IN HIGHER PLANTS: ARYL HYDROXYLATION OF DICLOFOP BY A CYTOCHROME P_{450} ENZYME FROM WHEAT MICROSOMES

A. Zimmerlin and F. Durst

Laboratoire d'Enzymologie Cellulaire et Moléculaire, Université Louis Pasteur, Institut de Botanique, CNRS UA 1182, 28 rue Goethe, F-67083, Strasbourg Cedex, France

There is ample evidence that many instances of herbicide resistance are based on metabolism. In very few cases, however, have the enzymes involved been identified and characterised.

The herbicide diclofop-methyl is used for post-emergent control of wild oat and other grasses in cereal crops such as wheat. Diclofop-methyl is rapidly hydrolysed to diclofop in both resistant and susceptible plants. Aryl hydroxylation by NIH-shift of the free acid producing essentially (2,5-dichloro-4-hydroxyphenyl)diclofop is the basis of this selectivity (Shimabukuro *et al.*. 1979; Tanaka *et al.*, 1988).

A microsomal fraction from etiolated wheat shoots catalysed the aryl hydroxylation of diclofop. The reaction required O_2 and NADPH as co-factors. It was inhibited by inhibitors of NADPH-cytochrome $\underline{c}$ (P_{450}) reductase, by anti-reductase antibodies and by carbon monoxide; the latter was reversed by light. Apparent K_m values of 9.3 ± 0.8 µM for diclopfop and 16.3 ± 2.3 µM for NADPH were determined. The activity, which was very low in etiolated wheat shoots, was strongly enhanced upon aging of seedlings on a phenobarbital solution (Table 1). This enzyme may constitute a factor in the resistance of wheat towards diclofop.

Table 1. Effect of phenobarbital treatment of etiolated wheat shoots on *t*-cinnamic acid 4-hydroxylase (CA4H), lauric acid in chain-hydroxylase (IC-LAH), diclofop aryl hydroxylase (DIAH) activities and cytochrome P_{450} content.

Treatment	Enzyme activities*			P_{450} content
	CA4H	IC-LAH	DIAH	
Water	36.00	0.43	0.25	244
Phenobarbital	35.50	7.85	3.90	297
Stimulation factor	x1	x18.2	x15.6	x1.2

* CA4H, IC-LAH and DIAH are expressed as pkat mg^{-1} and P_{450} content as pmoles $protein^{-1}$.

Chlorotoluron, diuron, chlorsulfuron, 2,4-D and bentazon which are detoxified presumably by cytochrome P_{450}-dependent N-demethylation and/or aryl hydroxylation (Frear *et al.*, 1969; Fonne, 1985) were at least very weak inhibitors of diclofop aryl hydroxylase *in vitro*. Moreover, microsomes which actively metabolise diclofop did not oxidize chlorsulfuron, 2,4-D, bentazon or chlorotoluron suggesting the existence of other hydroxylases.

REFERENCES

Fonne, R. (1985). "Intervention du Cytochrome P-450 des Végétaux supérieurs dans l'Oxydation de Compasés Exogènes: l'Aminopyrine et le Chlortoluron". D. Phil. Thesis. Université Louis Pasteur.

Frear, D.S., Swanson, H.R. and Tanaka, F.S. (1969). *N*-Demethylation of substituted 3-(phenyl)-1-methylureas: isolation and characterisation of a microsomal mixed function oxidase from cotton. Phytochemistry **8**, 2157-2169.

Shimabukuro, R.H., Walsh, W.C. and Hoerauf, R.A. (1979). Metabolism and selectivity of diclofop-methyl in wild oat and wheat. Journal of Agriculture and Food Chemistry **27**, 615-623.

Tanaka, F.S., Wein, R.G., Hoffer, B.L. and Shimabukuro, R.H. (1988). Application of NMR spectrometry in the identification of the isomeric hydroxylated metabolites of methyl-(4-[2,4-dichlorophenoxy]propanoate (diclofop-methyl) in wheat seedlings. Abstract 19th National ACS Meetings Toronto, Canada.

ARSENOVIC, Dr. M., Faculty of Agriculture, Institute for Plant Protection, V. Vlhovica 2, 21000 Novi Sad, YUGOSLAVIA.

BALKE, Professor N.E., 259 Moore Hall, University of Wisconsin, Madison, WI 53706, U.S.A.

BAUGHAN Dr. P.J. Agrichem Limited, Fenland Industrial Estate, Station Road, Whitttlesey, Cambs. PE7 2EY

BENYAMINI, Mr. Y., Dept. of Field & Vegetable Crops, Faculty of Agriculture, The Hebrew University of Jerusalem, Rehovot 76100, ISRAEL.

BERZSENYI, Dr. Z.K., Agricultural Research Institute, Hungarian Academy of Sciences, HUNGARY.

BHALLA, Dr. P.L., Research Group, ICI Australia, Newsom Street, Ascot Vale, Victoria 3032, AUSTRALIA.

BINGHAM, Dr. I., School of Agriculture, Department of Crop Science, 581 King Street, Aberdeen, AB9 1UD, U.K.

BOCION, Mr. P.F., Dr. R. Maag Ltd., Agrochemicals, CH-8157 Dielsdorf, SWITZERLAND.

BOTTERMAN, Dr. J., Plant Genetic Systems NV, Plateanstraat 22, 9000 Gent, BELGIUM.

BOWYER, Professor J., Dept. of Biochemistry, Royal Holloway & Bedford New College, Egham Hill, Surrey, TW20 0EX, U.K.

BOZORGIPOUR, R., AFRC IPSR, Cambridge Laboratory, Maris Lane, Trumpington, Cambridge, CB2 2LQ, U.K.

BRENT, Dr. K.J., Long Ashton Research Station, University of Bristol, Long Ashton, Bristol, BS18 9AF, U.K.

BRIGHT, Dr. S.W.J., ICI Seeds, Jealott's Hill Research Station, Bracknell, Berks. RG12 6EY, U.K.

BROWN, Prof. A.E., Dept. of Botany & Microbiology, Funchess Hall, Aubury University, Auburn, Alabama, U.S.A. 36849-4201.

BRYAN, Dr. I.B., Weed Science, ICI Agrochemicals, Jealott's Hill, Bracknell, Berkshire, RG12 6EY, U.K.

BULCKE, Dr. R.A.J., Faculty of Agricultural Sciences, State University Gent, Coupure Links, 653-B-9000 Gent, BELGIUM.

BUTCHART, Mr. J.E., Ciba-Geigy Agrochemicals, Whittlesford, Cambridge, CB2 4QT, U.K.

BUTLER, Mr. N.A.J., Chipman Ltd., The Goodsyard, Horsham, West Sussex, RN12 2NR, U.K.

CABANNE, Dr. F., Laboratoire des Herbicides (INRA), BV 1540, 21034 Dijon Cedex, FRANCE.

CARSON, Mr. C.M., Dow Chemical USA, P.O. Box 1706, Midland, MI 48640, USA.

CASELEY, Dr. J.C., Long Ashton Research Station, University of Bristol, Long Ashton, Bristol, BS18 9AF, U.K.

CHAUVEL, Mr. B., Ministere de L'Agriculture, I.N.R.A., Laboratoire de Malhherbologie, B.V. 1540, 21034 Dijon, Cedex, FRANCE.

CLAY, Mr. D.V., Long Ashton Research Station, University of Bristol, Long Ashton, Bristol, BS18 9AF, U.K.

COBB, Professor A.H., Dept. of Life Sciences, Nottingham Polytechnic, Clifton Lane, Nottingham, NG11 8NS, U.K.

CODD, Dr. T., CAB International, Wallingford, Oxon, OX10 8DE, U.K.

COPPING, Dr. L.G., Dow Agricultural Products Research & Development, Letcombe Laboratory, Letcombe Regis, Wantage, Oxon, OX12 9JT, U.K.

COUPLAND, Dr. D., Long Ashton Research Station, University of Bristol, Long Ashton, Bristol, BS18 9AF, U.K.

COURDURIES, Mr. P., Plant Sciences Dept., West of Scotland College, Auchincruive, Ayr, KA6 5HW, U.K.

CUBITT, Mr. I.R., Nickerson International Seed Co., Cambridge Science Park, Milton Road, Cambridge. CB4 4GZ

CURE, Mr. B., ITCF, Station Expérimentale Boigneville, 91720 Maisse, FRANCE.

CUSSANS, Mr. G.W., Long Ashton Research Station, University of Bristol, Long Ashton, Bristol, BS18 9AF, U.K.

DARMENCY, Mr. H., Ministere de L'Agriculture, I.N.R.A., Laboratoire de Malhherbologie, B.V. 1540, 21034 Dijon, Cedex, FRANCE.

DE PRADO, Dr. R., Departmento de Bioquimica y Biologia Molecular, Escuel Tecnica Superior de Ingenieros Agronos, Aptd 3048, E-14808-Cordoba, SPAIN.

DIETMAR, Dr. K., Agrolinz Agrarchemikalien GMBH, Welserstr 42, A-4020 Linz, AUSTRIA.

DIETZ, Dr. A., Biologische Bundesanstalt, Institut fur Biochemie, Messeweg 11, D-3300 Braunschweig, FEDERAL REPUBLIC OF GERMANY.

DODGE, Dr. A.D., School of Biological Sciences, University of Bath, Bath, BA2 7AY, U.K.

DOMINGUEZ, Mr. C., Departmento de Bioquimica y Biologia Molecular, Escuel Tecnica Superior de Ingenieros Agronos, Aptd 3048, E-14808-Cordoba, SPAIN.

DUNN, Mr. S.M., 46 Culverland Road, Exeter, EX4 6JJ, U.K.

EVANS, Dr. I., ICI Seeds, Plant Biotechnology Section, Jealott's Hill Research Station, Bracknell, Berks. RG12 6EY, U.K.

FIGEYS, Professor H.P., Dept. of Organic Chemistry, Faculty of Sciences, Free University of Brussels, 50 Av. F.D. Roosevelt, 1050 Brussels, BELGIUM.

FLEMONS, Mr. G.P., Rhone Poulenc Agriculture, Fyfield Road, Ongar, Essex, CM5 0HW, U.K.

FLÜH, Dr. M., c/o Ciba-Geigy GmbH, Liebigstrasse 51/53, D-6000 Frankfurt/M.1, FEDERAL REPUBLIC OF GERMANY.

FOX, Ms. M., Butterworth Scientific Limited, P.O. Box 63, Westbury House, Bury Street, Guildford, Surrey, GU2 5BH, U.K.

FREAR, Dr. D.S., Biosciences Research Laboratory, P.O. Box 5674, State University Station, Fargo, North Dakota 58105-5674, U.S.A.

FROUD-WILLIAMS, Dr. R., Dept. of Agricultural Botany, University of Reading, Reading, U.K.

GARSTANG, Mr. J.R., Agronomy Dept., Block A, ADAS Government Buildings, Coley Park, Reading, Berks. RG1 6DT, U.K.

GASQUEZ, Dr. J., INRA, Laboratoire de Malherbologie, BV 1540, 21034 Dijon, Cedex, FRANCE.

GRAHAM, Mr. K., Du Pont (UK) Ltd., Agricultural Products Dept., Wedgerwood Way, Stevenage, Herts. SG1 4QN

GRAY, Mr. N.R., Dow Chemical Company Ltd., Letcombe Regis, Nr. Wantage, Oxon, OX12 9JT, U.K.

GREENFIELD, Mr. A., ADAS, Government Buildings, Marston Road, New Marston, Oxford, OX3 0TP, U.K.

GRESSEL, Professor J., Dept. of Plant Genetics, The Weizmann Inst. of Science, Rehovot, ISRAEL.

GRIFFITHS, Mr. G., Agrichem Limited, Fenland Industrial Estate, Station Road, Whittlesey, Cambs. PE7 2EY, U.K.

GROFFBARD, Dr. E., 29 Perrott Close, Northleigh, Oxfordshire. OX8 6RU

GRUENHOLZ, Mr. P., ICI-Zeltia, S.A. Agroquimicos, Costa Brava, 13, 28034
    - Madrid, SPAIN.
GUNNARSEN, Ms. M.O., Dept. of Crop Science, KVL Thorvaldsensvej 40,
    DK-1871, Frederiksberg C. DENMARK.
HACK, Miss C., Plant Sciences Dept., West of Scotland College,
    Auchincruive, Ayr, KA6 5HW, U.K.
HADLEIGH, Mr. R., Long Ashton Research Station, University of Bristol,
    Long Ashton, Bristol, BS18 9AF, U.K.
HALLAHAN, Dr. D.L., Biochemistry & Physiology Dept., AFRC IACR,
    Rothamsted Experimental Station, Harpenden, Herts. AL5 2JQ, U.K.
HARTLEY, Dr. J., Hawkes Bay Agricultural Research Centre, Box 85,
    Hastings, NEW ZEALAND.
HARTNETT, Ms. M.E., Agricultural Products Dept., E.I. DuPont de Nemours
    & Co., Experimental Station, 402 Wilmington, DE 19880-0402, U.S.A.
HARVEY, Mrs. K., Schering Agrochemicals, Chesterford Park Research
    Station, Saffron Walden, Essex, U.K.
HATZIOS, Professor K.K., 203 PMB Building, Dept. of Plant Pathology,
    Physiology and Weed Science, VPI & State University, Blacksburg,
    Virginia 24061-0330, U.S.A.
HAWKES, Dr. T.R., Biochemistry Dept., ICI Agrochemicals, Jealott's Hill
    Research Station, Bracknell, Berkshire, RG12 6EY, U.K.
HEAP, Dr. I., c/o Waite Agricultural Research Institute, Agronomy Dept.,
    Box 1, Glen Osmond 5064, Adelaide, SOUTH AUSTRALIA.
HEINISCH, Dr. H., SKW Trostberg AG, CFZ-Biochemie, Postfach 1262, 8223
    Trostberg, FEDERAL REPUBLIC OF GERMANY.
HILL, Mr. A., Dept. of Agricultural Botany, Queens University, Newforge
    Lane, Belfast, BT9 5PX, U.K.
HOAD, Dr. G.V., Long Ashton Rearch, University of Bristol, Long Ashton,
    Brisotl, BS18 9AF, U.K.
HODSON, Mr. M., Farm Protection Ltd., Glaston Park, Glaston, Oakham,
    Leics. LE15 9BX, U.K.
HORWOOD, Mrs. P., Butterworth Scientific Limited, P.O. Box 63, Westbury
    House, Bury Street, Guildford, Surrey, GU2 5BH, U.K.
HOWATT, Mr. S., Dept. of Biology, Nova Scotia Agricultural College, P.O.
    Box 550, Truro, Nova Scotia. B2N 5E3, CANADA
HUMPHREYS, Dr. M.O., AFRC-IGAP, Welsh Plant Breeding Station, Plas
    Gogerddan, Aberystwyth, Dyfed, SY23 3EB, U.K.
IVANY, Mr. J.A., Weed Control Research Station, Box 1210, Charlottetown,
    PEI, CANADA, CIA 7M8
IWANZIK, Dr. W., c/o CIBA-GEIGY AG, CH-4332 Stein, SWITZERLAND
JEPSON, Dr. I., ICI Seeds, Jealott's Hill Research Station, Bracknell,
    Berks. RG12 6EY, U.K.
dESSOP, Mr. C.D., Uniroyal Chemical Ltd., Brooklands Farm, Cheltenham
    Road, Evesham, WR11 6LW, U.K.
JOHANN, Dr. G., Schering Aktiengesellschaft, PF-FO Biologie, Dept.
    Herbologie u- Pflanzenphysiologie, D - 1000 Berlin 28,
    Gollanczstr. 57-101, FEDERAL REPUBLIC OF GERMANY.
JOHNSTON, Mr. D.T., Northern Ireland Horticultural and Plant Breeding
    Station, Loughall, Armagh, BT61 8JB, U.K.
JOLLANDS, Mr. P., Hallsannery, Bideford, Devon, EX39 5HE, U.K.
JONES, Dr. O.T.G., Biochemistry Dept., University of Bristol, Bristol,
    BS1 1TD, U.K.
KEMP, Dr. M.S., Long Ashton Research Station, University of Bristol,
    Long Ashton, Bristol, BS18 9AF, U.K.
KLEVORN, Dr. T.B., Monsanto Technical Center Europe, Rue Laid Burniat,
    Parc Scientifique, B-1348 Louvain la Neuve, BELGIUM.

KOHLE, Dr. H., BASF AG, APE-FP Landwirtschaftliche Versuchsstation, D-6703, Limburgerhof, HOLLAND.

KOLLMAN, Dr. G., Rohm and Haas Company, Chesterfield House, Bloomsbury Way, LONDON, WC1A 2TP, U.K.

KOVACS, Mrs. H., Research Institute for Heavy Chemical Industires, NEVIKI, H-8201 Veszprém, P.O.B. 160, HUNGARY.

KREUZ, Dr. K., Ciba-Geigy Ltd., R-1040 P 21, CH-4002 Basel, SWITZERLAND.

LAINSBURY, Mr. M.A., BASF United Kingdom Limited, Lady Lane, Hadleigh, Ipswich, Suffolk, IP7 6BQ, U.K.

LALOVA, Professor M., Plant Protection Institute, Kostinbrod 2230, Sofia POB 238, BULGARIA.

LAMOUREAUX, Dr. G.L., Biosciences Research Laboratory, P.O. Box 5674, State University Station, Fargo, North Dakota, U.S.A. 58105.

Le BARON, Dr. H.M., Ciba-Geigy Corporation, North Carolina, USA.

LEHOCZKI, Dr. E., Dept. of Biophysics, József Attila University, H-6722, Szeged, HUNGARY.

LIPECKI, Dr. J., Agricultural University, 20-069 Lublin, Leszczynskiego 7, POLAND.

LÓRIK, Mr. E., North Hungarian Chemical Works, Sajóbábony, H-3792, HUNGARY.

LÜRSSEN, Dr. K., Bayer AG, Pflanzenschutz Anwendungstechnik, Biologische Forschung, D 5090 Leverkusen - Bayerwerk, FEDERAL REPUBLIC OF GERMANY.

LUKE, Mr. A.J., Rhone-Poulenc Ag. Company, P.O. Box 12014, Research Triangle Park, Princeton, New Jersey 27709, U.S.A.

LUTMAN, Dr. P.J.W., Long Ashton Research Station, University of Bristol, Long Ashton, Bristol, BS18 9AF, U.K.

MACDONALD, Miss M., ICI Seeds, Plant Biotechnology Section, Jealott's Hill Research Station, Bracknell, Berks. RG12 6EY, U.K.

MAKEPEACE, Mr. R.J., Oxford Agricultural Consultants Ltd., 1 Murdock Road, Bicester, Oxon, OX6 7PP, U.K.

MALLORY-SMITH, Mrs. C., Plant Soil & Entomological Sciences, University of Idaho, Moscow, ID 83843, U.S.A.

MARSHALL, Dr. G., Dept. of Plant Sciences, West of Scotland College, Auchincruive, Ayr, KA6 5HW, U.K.

MATSUNAKA, Dr. S., Faculty of Agriculture, Kobe University, Nada-ku, Kobe 657, JAPAN.

MAURER, Mr. W., Ciba-Geigy Limited, Agricultural Division, Weed Control Sector, P.O. Box, CH-4002 Basel, SWITZERLAND.

MAXWELL, Mr. B., Forest Science Dept., Oregon State University, Corvallis, OR 97331, U.S.A.

MAYALL, Mr. J., CAB International, Wallingford, Oxon, OX10 8DE, U.K.

McELWEE, Dr. M., University of Strathclyde, Biology Division, Todd Centre, Taylor Street, Glasgow, G4 0NR, U.K.

MICHIEKA, Professor R.W., University of Nairobi, Dept. of Crop Science, Box 29053, Nairobi, KENYA.

MONTEIRO, Ms. A., Departamento de Botânica, Instituto Superior de Agronomia, Tapada da Ajuda, 1399 Lisboa Codex, PORTUGAL.

MORRISON, Professor I.N., Dept. of Plant Science, University of Manitoba, Winnipeg, Manitoba, R3T 2N2, CANADA,

MORTIMER, Dr. A.M., Dept. of Environmental & Evolutionry Biology, University of Liverpool, P.O. Box 147, Liverpool. L69 3BX

MOSS, Mr. G.I., 'Netherleigh', Bridgewater Street, Whitchurch, Shropshire, SY13 1QJ, U.K.

MOSS, Mr. S.R., Long Ashton Research Station, University of Bristol, Long Ashton, Bristol, BS18 9AF, U.K.

NAGY, Dr. J., Északmagyarországi Vegyimüvek, H-3792, Sajóbábony, HUNGARY.

NEWHOUSE, Dr. K., American Cyanamid, P.O. Box 400, Princeton, New Jersey, 08540, U.S.A.

NIKOLOVA, Dr. G.K., Plant Protection Institute, Kostinbrid 2230, Sofia, 1618, P.O. Box 68, BULGARIA.

NOYÉ, Mr. G., Institute of Weed Control, Flakkebjerg, DK 4200, Slagelse, DENMARK.

ORLANDO, Mme D., ITCF, Station Expérimentale Boigneville, 91720 Maisse, FRANCE.

ORSON, Mr. J.H., MAFF National Cereals Specialst, MAFF/ADAS, Brooklands Avenue, Cambridge, U.K.

OSEI-BONSU, Dr. K., Cocoa Research Institute, P.O. Box 8, Tafo-Akim, GHANA.

OWEN, Dr. W.J., Dow Chemical Co. Ltd., Letcombe Laboratory, Letcombe Regis, Wantage, Oxon, OX12 9JT, U.K.

PADGETTE, Mr. S.R., Monsanto, 700 Chesterfield Village Pkwy, St. Louis, MO, U.S.A 63198.

PALLETT, Dr. K.E., Rhone Poulenc Agriculture, Ongar Research Centre, Fyfield Road, Ongar, Essex, CM5 OHW, U.K.

PARHAM, Dr. M.R., Weed Science Section, ICI Agrochemicals, Jealott's Hill Research Station, Bracknell, Berkshire, RG12 6EY, U.K.

PEARSON, Dr. D.P.J., ICI Agrochemicals, Jealott's Hill Research Station, Bracknell, Berkshire, RG12 6EY, U.K.

POLGE, Mr. N., School of Biological Sciences, University of Bath, Claverton Down, Bath, BA2 7AY, U.K.

POPAY, Dr. I., Maftech, P.O. Box 1654, Palmerston North, NEW ZEALAND.

POWLES, Dr. S.B., Dept. of Agronomy, Waite Agricultural Research Institute, University of Adelaide, S.A. 5064, AUSTRALIA.

POZSGAI, Dr. J., Research Institute for Sugar Produktion, Sopronhorpács, H-9463, HUNGARY.

PUTWAIN, Dr. P., Dept. of Environmental & Evolutionary Biology, University of Liverpool, P.O. Box 147, Liverpool, L69 3BX, U.K.

QUADRANTI, Dr M., Ciba-Geigy Limited, Agricultural Division, Weed Control Section, P.O. Box, CH-4002 Nasel, SWITZERLAND.

QUINN, Dr. J.P., Dept. Biology, Queen's University, Belfast, BT7 1NN, U.K.

RADOSEVITCH, Dr. S., Forest Science Dept., Oregon State University, Corvallis, OR 97331, U.S.A.

RAMSAY, Mrs. M., Potato & Brassica Genetics, Scottish Crop Research Institute, Invergowrie, Dundee, DD2 5DA, U.K.

RAO, Dr. S., Dept. of Botany, Gulbarga University, Gulbarga - 585 106, Karnataka, INDIA.

READ, Mr. M., Hoechst UK Ltd., Agricultural Division, East Winch Hall, East Winch, Kings Lynn, Norfolk, P.E. 32 1HN, U.K.

REED, Mr. W.T., Agricultural Products Department, WM3-224 - Barley Mill Plaza, Wilmington, DE, U.S.A. 19898

ROMERA, Mr. E., Departmento de Bioquimica y Biologia Molecular, Escuel Tecnica Superior de Ingenieros Agronos, Aptd 3048, E-14808-Cordoba, SPAIN.

ROSE, Mr. P.W., Bayer UK Limited, Agrochem Business Groups, Eastern Way, Bury St. Edmunds, Suffolk, IP32 7AH, U.K.

ROTTEVEEL, Dr. T., Plantenziektenkundige Dienst, P.O. Box 9102, 6700 HC, Wageningen, THE NETHERLANDS.

RUBIN, Dr. B., Faculty of Agriculture, p.O. Box 12, Rehovot, ISRAEL.

RYAN, Dr. P.J., Ciba-Geigy Agrochemicals, Whittlesford, Cambridge, CB2 4QT, U.K.

SAARI, Dr. L., Agricultural Products Dept., Experimental Station E402/5332, P.O. Box 80402, E.I. DuPont Co. Inc., Wilmington, DE19803-0402, U.S.A.

SAMPSON, Mr. G., Dept. of Biology, Nova Scotia Agricultural College, P.O. Box 550, Truro, Nova Scotia. B2N 5E3, CANADA.

SAWICKI, Dr. R.M., Rothamsted Experimental Station, AFRC-IACR, Harpenden, Herts. AL5 2JQ, U.K.

SCALLA, Dr. R., Laboratoire des Herbicides, INRA, B.V. 1540, 21034 Dijon, FRANCE.

SEILER, Dr. A., c/o Ciba-Geigy AG, CH-4332 Stein, SWITZERLAND.

SHANER, Dr. D.L., American Cyanamid, P.O. Box 400, Princeton, New Jersey 08540, U.S.A.

SHEWRY, Professor P.R., University of Bristol, Long Ashton Research Station, Long Ashton, Bristol, BS18 9AF, U.K.

SIBUGA, Ms. K.P., Sokoine University of Agriculture, Dept. of Crop Science, P.O. Box 3005, Morogoro, Tanzania.

SMIS, Mr. K., Faculty of Agricultural Sciences (Laboratory of Plant Ecology), Coupure Links, 653, 900 Gent, BELGIUM.

SMIT, Mr Z.K., Faculty of Agriculture, Institute for Plant Protection, V. Vlhovica 2, 21000 Novi Sad, YUGOSLAVIA.

SMITH, Mr. J., Purcombe Bungalow, Whitchurch, Canonicorum, Bridport, Dorset, DT6 6RL, U.K.

SNAPE, Dr. J., AFRC-IPRS, Cambridge Laboratory, Maris Lane, Trumpington, Cambridge, CB2 2JB, U.K.

STEPHENS, Mr. R.J., School of Biological Sciences, University of Bath, Claverton Down, Bath, BA2 7AY, U.K.

STRIDE, Mr. C.D., 1 Beale Close, Bussage, Stroud, Gloucester, GL6 8DF, U.K.

SUBRAMANIAN, Dr. M., Dow Chemical U.S.A., 2800 Mitchell Drive, Walnut Creek, CA 94598, U.S.A.

SUTTON, Mr. P.B., ICI Agrochemicals, Woolmead House West, 3 Bear Lane, Farnham, Surrey, GU9 8UB, U.K.

SZIGETI, Dr. Z., Dept. of Plant Physiology, Eötvös Loránd University, Budapest P.O.B. 330., H-1445, HUNGARY.

THILL, Professor D.C., Dept. of Plant, Soil & Entomological Sciences, University of Idaho, Moscow, ID 83843, U.S.A.

THOMAS, Dr. W.D., ICI Agrochemicals, Jealott's Hill Research Station, Bracknell, Berks. RG12 6EY, U.K.

TISSUT, Mr. M., 30 Av. M. Berthelot, 38100 Grenoble, FRANCE.

TREBST, Professor A., Rhur-Universität der Pflanzen, Universitätssrasse 150, 4630 Bochum 1, Gebäude ND 3/130, FEDERAL REPUBLIC OF GERMANY.

TRINKS, Dr. K., Biological Research C., Hoechst AG, DH-872 N, 6230 Frankfurt M. 80, FEDERAL REPUBLIC OF GERMANY.

van DOORNE, Miss L.E., Plant Sciences Dept., West of Scotland College, Auchincruive, Ayr, KA6 5HW, U.K.

van HOOGSTRATEN, Dr. S.D., c/o Cyanamid International, Rue du Bosquet 15, B-1348 Mont-St-Guibert, BELGIUM.

van OORSCHOT, Mr. J.L.P., Centrum voor Agrobiologisch Onderzoek (CABO), Bornsesteeg 65 - P.O. Box 14, 6700 AA, Wageningen, THE NETHERLANDS.

VARADI, Mr. G., Research Institute for Viticulture and Enology, H-6000 Kecskemet, P.O.B. 25, HUNGARY.

VAUGHN, Dr. M., Rochester Institute of Technology, New York, U.S.A.

VERITY, Dr. J., Schering Ag., Chesterford Park Research Station, Saffron Walden, Essex, CB10 1XL. U.K.

WERCK-REICHHART, Dr. Daniele, Laboratoire d'enzymologie cellulaire et moleculaire, 28 rue GOETHE, Institut de Botanique, 67000 Strasbourg, FRANCE.

WHITE, Dr. B.G., ICI Agrochemicals, Jealott's Hill Research Station, Bracknell, Berkshire, RG12 6EY, U.K.

WHITEHEAD, Mr. R., Schering Agriculture, Nottingham Road, Stapleford, Nottingham,
NG9 8AJ, U.K.

WHITEHOUSE, Dr. P., Shell Research Ltd., Sittingbourne Research Centre, Sittingbourne, Kent, U.K.

WHITLEY, Miss E., Cyanamid, Dernford Farm, London Road, Stapleford, Cambridge, CB2 3EA, U.K.

WILLIAMSON, Professor M., Dept. of Biology, University of York, York, YO1 5DD, U.K.

WINFIELD, Mr. J., MAFF, PSD 'G', Harpenden Laboratory, Hatching Green, Harpenden, Herts. AL5 2BD, U.K.

WOOLACOTT, Dr. B., MAFF, PSD, Harpenden Laboratory, Hatching Green, Harpenden, Herts. AL5 2BD, U.K.

WUERZER, Dr. B., BASF Aktiengesellschaft, Landwirtschaftliche Versuchsstation, Postfach 2 20, Abt. APE/FH, 6703 Limburgerhof, FEDERAL REPUBLIC OF GERMANY.

WYSE, Professor D., University of Minnesota, 1991 Batford Circle, 138 Borlaug Hall, St. Paul, Minnesota 55108, U.S.A.

ZOHNER, Dr. A., Agrolinz Agrarchemikalien GMBH, Welserstr 42, A-4020 Linz, AUSTRIA.

501

505